GROUP THEORY AND QUANTUM MECHANICS

International Series in Pure and Applied Physics

Leonard I. Schiff, *Consulting Editor*

Allis and Herlin   *Thermodynamics and Statistical Mechanics*
Becker   *Introduction to Theoretical Mechanics*
Clark   *Applied X-rays*
Collin   *Field Theory of Guided Waves*
Evans   *The Atomic Nucleus*
Finkelnburg   *Atomic Physics*
Ginzton   *Microwave Measurements*
Green   *Nuclear Physics*
Gurney   *Introduction to Statistical Mechanics*
Hall   *Introduction to Electron Microscopy*
Hardy and Perrin   *The Principles of Optics*
Harnwell   *Electricity and Electromagnetism*
Harnwell and Livingood   *Experimental Atomic Physics*
Harnwell and Stephens   *Atomic Physics*
Henley and Thirring   *Elementary Quantum Field Theory*
Houston   *Principles of Mathematical Physics*
Hund   *High-frequency Measurements*
Kennard   *Kinetic Theory of Gases*
Lane   *Superfluid Physics*
Leighton   *Principles of Modern Physics*
Lindsay   *Mechanical Radiation*
Livingston and Blewett   *Particle Accelerators*
Middleton   *An Introduction to Statistical Communication Theory*
Morse   *Vibration and Sound*
Morse and Feshbach   *Methods of Theoretical Physics*
Muskat   *Physical Principles of Oil Production*
Present   *Kinetic Theory of Gases*
Read   *Dislocations in Crystals*
Richtmyer, Kennard, and Lauritsen   *Introduction to Modern Physics*
Schiff   *Quantum Mechanics*
Seitz   *The Modern Theory of Solids*
Slater   *Introduction to Chemical Physics*
Slater   *Quantum Theory of Matter*
Slater   *Quantum Theory of Atomic Structure, Vol. I*
Slater   *Quantum Theory of Atomic Structure, Vol. II*
Slater   *Quantum Theory of Molecules and Solids, Vol. I*
Slater and Frank   *Electromagnetism*
Slater and Frank   *Introduction to Theoretical Physics*
Slater and Frank   *Mechanics*
Smythe   *Static and Dynamic Electricity*
Stratton   *Electromagnetic Theory*
Thorndike   *Mesons: A Summary of Experimental Facts*
Tinkham   *Group Theory and Quantum Mechanics*
Townes and Schawlow   *Microwave Spectroscopy*
White   *Introduction to Atomic Spectra*

The late F. K. Richtmyer was Consulting Editor of the series from its inception in 1929 to his death in 1939. Lee A. DuBridge was Consulting Editor from 1939 to 1946; and G. P. Harnwell from 1947 to 1954.

GROUP THEORY AND QUANTUM MECHANICS

Michael Tinkham

Professor of Physics
University of California
Berkeley, California

McGraw-Hill Book Company

New York San Francisco Toronto London

Group Theory and Quantum Mechanics

Library of Congress Catalog Card Number 63-19773

64895

PREFACE

This book has developed from a set of mimeographed notes, written to accompany a series of 44 lectures which comprise a semester course on group theory and the quantum mechanics of atoms, molecules, and solids. The course has been given by the author to graduate students at the University of California since 1957. Since this course is followed by another dealing entirely with solid-state theory, the applications to solid-state problems in the book are largely limited to areas in which the apparatus of group theory is particularly essential. Because of the resulting emphasis on atomic and molecular problems, many chemists take the course, typically about a third of the class. Since students are assumed to have had a year of quantum theory as a prerequisite to the course, such standard elementary examples as the hydrogen atom and the harmonic oscillator have been dispensed with in the interest of brevity. However, a real attempt has been made throughout to keep the treatment as simple as possible and understandable to students with a variety of backgrounds. In particular, the group theory is developed on the assumption that the student has no previous knowledge of it, and elegant formal elaborations have been avoided in favor of presenting only a useful core of ideas as simply as possible. The treatment throughout is definitely intended to be useful to experimentalists who desire a better understanding of the current theoretical picture of the structure of atoms, molecules, and solids.

The book is intended for use as an introductory text rather than as a definitive treatise. Relatively little completely original work is presented. Rather, an attempt has been made to collect in a compact and unified form some of the scattered material on standard methods which is needed for an understanding of current research work.

The book divides logically into two parts. The first five chapters are primarily devoted to the introduction of methods, illustrated by physical examples, whereas the last three chapters present a systematic treatment of the quantum theory of atoms, molecules, and solids, additional methods being introduced as needed to handle specific problems as they arise.

After a brief introductory chapter, the formal theory of finite groups and their representation is developed in Chapters 2 and 3 in a manner which draws heavily on Wigner's classic "Group Theory." Chapter 4 is devoted to the illustration of the group theory. This is done by developing the crystallographic point groups which are basic for both solid-state and molecular theory. The classic work of Bethe on the splitting of atomic terms in crystalline fields of the various symmetries is then presented, including the introduction of the crystal double groups. This beautiful and illuminating work has not been readily accessible before, though basic for the theory of magnetic materials. As a final illustration of the point groups, we consider the directed-valence theory of hybrid orbitals.

Chapter 5 is devoted to the theory of systems with full rotational symmetry. First we develop the relation between angular momentum and transformation properties under rotation. Then, closely following Wigner, the homomorphism between the full rotation group and the unimodular unitary group is used to develop all the representations of the rotation group. From these, the Wigner or Clebsch-Gordan coefficients are derived in general form and applied through the Wigner-Eckart theorem to matrix elements of tensor operators. Finally, the Racah coefficients are introduced to handle more general problems arising in the application of tensor operators. This latter treatment is based on the monograph of Rose, "Elementary Theory of Angular Momentum." The chapter concludes with a discussion of space- and time-inversion symmetries and their consequences in parity conservation and Kramers' degeneracy.

The systematic presentation of atomic structure in Chapter 6 draws heavily on the monumental treatise by Condon and Shortley, "The Theory of Atomic Structure." In particular, the determinantal method developed by Slater is used for calculation of L-S-term energies and multiplet structure. However, the group-theoretical viewpoint is emphasized where it provides helpful insight, rather than being minimized. The Hartree-Fock method is treated, including Slater's modified form, and some reference is made to recent work on the unrestricted Hartree-Fock method. The rigorous group-theoretical derivation of the Landé g-factor formula is given, as is a discussion of the Paschen-Back transition in the Zeeman effect. The discussion of magnetic hyperfine structure follows that of Ramsey ("Nuclear Moments") but again makes more use of group-theoretical arguments. The electric hyperfine structure is treated as an application of the Racah coefficients, although the connection with the usual elementary approach is indicated.

Chapter 7 on molecular structure makes considerable use of material from "Quantum Chemistry" by Eyring, Walter, and Kimball. It opens with a discussion of the Born-Oppenheimer approximation. The rest of the chapter provides a systematic following through of the approximation. First the electronic energy levels for the problem of "clamped nuclei" are attacked, the method of molecular orbitals being emphasized. The group theory is employed in classifying and constructing these. With the electronic energy used as an effective potential, vibrational motion is discussed, together with the symmetry properties of combination and overtone levels in polyatomic molecules. The rotational energy of a rigid molecule is then considered, and a discussion of centrifugal distortion effects is given. At the end of the chapter a variety of cross terms which involve a higher-order Born-Oppenheimer approximation are indicated.

Chapter 8, on solid-state physics, treats two important topics of solid-state theory in which group-theoretical symmetry considerations are particularly useful, namely, electronic energy band theory and magnetic crystal symmetry. The general periodic-potential problem is set up, and some standard approximate approaches are discussed briefly, though the problems of actual quantitative calculations are not dealt with. The principal emphasis is on the group-theoretical discussion of the symmetry classification of wave functions in crystals, as initiated by Bouckaert, Smoluchowski, and Wigner and extended by many others. The introduction of spin-orbit coupling and the implications of time-reversal symmetry in band theory are indicated. This leads to a consideration of magnetic materials, in which time-reversal symmetry in the usual sense is not effective. The Shubnikov groups are developed, following Tavger and Zaitsev, and illustrated with examples. Finally, the Landau theory of second-order phase transitions is presented and applied to treat the change in symmetry of a magnetic material upon ordering.

As this summary indicates, a wide range of material, which introduces many techniques, is covered. An emphasis on the application of group theory to give insight into physical problems serves as a unifying principle. Of necessity, the treatment is rather brief, since with some few omissions the material is intended for a one-semester course. Accordingly, little material has been included merely for completeness. Rather, an attempt has been made to pick out some of the main features of the picture, things which should be part of the working understanding of nature for physicists and chemists. It is hoped that this approach will help the reader to acquire the new concepts and advanced methods and give him the sort of perspective which seems to be so easy to miss. He should then be able to continue with confidence to further independent study of particular topics required in his research work.

For the benefit of those who are familiar with the mimeographed version of this book, which has been widely circulated since 1957, we might note that the text has been completely rewritten in an effort to improve its clarity and completeness. New material added includes the entire final chapter on band theory and magnetic-crystal symmetry groups, a major expansion of the section on time-reversal symmetry, a major expansion of the treatment of molecular vibrations, a short discussion of the 3-j coefficients, a treatment of the Zeeman effect in atomic spectra, and problems at the end of each chapter. The new material, together with the general expansion of the old upon rewriting, accounts for the increase by about one-half in the length of the present volume compared with the notes.

The author wishes to acknowledge the contributions of Professors C. Kittel and R. Karplus, the notes from whose previous lecture courses provided the starting point for this book. Thanks are also due to many persons, particularly Professors L. I. Schiff, J. H. Van Vleck, and L. C. Allen, for their encouragement when the work was progressing slowly. The support of a John Simon Guggenheim Memorial Foundation grant and the hospitality of the Massachusetts Institute of Technology provided the author the opportunity to go over the proofs of the book with the care necessary to eliminate errors.

Finally, it is a pleasure to thank Dr. V. Heine for his very thorough and helpful scrutiny of the entire manuscript, which has eliminated many obscurities and some errors. The author is grateful for this help, and that of other reviewers of the manuscript, in improving the presentation, but he, of course, accepts all responsibility for the final result.

M. Tinkham

CONTENTS

Preface *V*

I Introduction *I*

1-1   The Nature of the Problem *1*

1-2   The Role of Symmetry *3*

2 Abstract Group Theory **6**

2-1   Definitions and Nomenclature *6*

2-2   Illustrative Examples *7*

2-3   Rearrangement Theorem *8*

2-4   Cyclic Groups *9*

2-5   Subgroups and Cosets *9*

2-6   Example Groups of Finite Order *10*

2-7   Conjugate Elements and Class Structure *12*

2-8   Normal Divisors and Factor Groups *13*

2-9   Class Multiplication *15*

Exercises *16*

References *17*

3 Theory of Group Representations **18**

3-1   Definitions *18*

3-2   Proof of the Orthogonality Theorem *20*

3-3   The Character of a Representation *25*

3-4   Construction of Character Tables   *28*
3-5   Decomposition of Reducible Representations   *29*
3-6   Application of Representation Theory in Quantum Mechanics   *31*
3-7   Illustrative Representations of Abelian Groups   *37*
3-8   Basis Functions for Irreducible Representations   *39*
3-9   Direct-product Groups   *43*
3-10 Direct-product Representations within a Group   *46*
      Exercises   *47*
      References   *48*

4   Physical Applications of Group Theory   50
4-1   Crystal-symmetry Operators   *51*
4-2   The Crystallographic Point Groups   *54*
4-3   Irreducible Representations of the Point Groups   *62*
4-4   Elementary Representations of the Three-dimensional Rotation Group   *65*
4-5   Crystal-field Splitting of Atomic Energy Levels   *67*
4-6   Intermediate Crystal-field-splitting Case   *69*
4-7   Weak-crystal-field Case and Crystal Double Groups   *75*
4-8   Introduction of Spin Effects in the Medium-field Case   *78*
4-9   Group-theoretical Matrix-element Theorems   *80*
4-10 Selection Rules and Parity   *82*
4-11 Directed Valence   *87*
4-12 Application of Group Theory to Directed Valence   *89*
      Exercises   *92*
      References   *93*

5   Full Rotation Group and Angular Momentum   94
5-1   Rotational Transformation Properties and Angular Momentum   *94*
5-2   Continuous Groups   *98*
5-3   Representation of Rotations through Eulerian Angles   *101*
5-4   Homomorphism with the Unitary Group   *103*
5-5   Representations of the Unitary Group   *106*
5-6   Representation of the Rotation Group by Representations
      of the Unitary Group   *109*
5-7   Application of the Rotation-representation Matrices   *111*
5-8   Vector Model for Addition of Angular Momenta   *115*
5-9   The Wigner or Clebsch-Gordan Coefficients   *117*
5-10 Notation, Tabulations, and Symmetry Properties
      of the Wigner Coefficients   *121*
5-11 Tensor Operators   *124*

5-12   The Wigner-Eckart Theorem   *131*
5-13   The Racah Coefficients   *133*
5-14   Application of Racah Coefficients   *137*
5-15   The Rotation-Inversion Group   *139*
5-16   Time-reversal Symmetry   *141*
5-17   More General Invariances   *147*
       Exercises   *151*
       References   *153*

6   Quantum Mechanics of Atoms   *154*
6-1   Review of Elementary Atomic Structure and Nomenclature   *155*
6-2   The Hamiltonian   *157*
6-3   Approximate Eigenfunctions   *157*
6-4   Calculation of Matrix Elements between Determinantal
       Wavefunctions   *162*
6-5   Hartree-Fock Method   *167*
6-6   Calculation of L-S-term Energies   *170*
6-7   Evaluation of Matrix Elements of the Energy   *173*
6-8   Eigenfunctions and Angular-momentum Operations   *178*
6-9   Calculation of Fine Structure   *181*
6-10   Zeeman Effect   *188*
6-11   Magnetic Hyperfine Structure   *193*
6-12   Electric Hyperfine Structure   *201*
       Exercises   *206*
       References   *208*

7   Molecular Quantum Mechanics   *210*
7-1   Born-Oppenheimer Approximation   *210*
7-2   Simple Electronic Eigenfunctions   *213*
7-3   Irreducible Representations for Linear Molecules   *216*
7-4   The Hydrogen Molecule   *219*
7-5   Molecular Orbitals   *220*
7-6   Heitler-London Method   *223*
7-7   Orthogonal Atomic Orbitals   *226*
7-8   Group Theory and Molecular Orbitals   *227*
7-9   Selection Rules for Electronic Transitions   *233*
7-10   Vibration of Diatomic Molecules   *234*
7-11   Normal Modes in Polyatomic Molecules   *238*
7-12   Group Theory and Normal Modes   *242*
7-13   Selection Rules for Vibrational Transitions   *248*

7-14   Molecular Rotation   *250*

7-15   Effect of Nuclear Statistics on Molecular Rotation   *252*

7-16   Asymmetric Rotor   *255*

7-17   Vibration-Rotation Interaction   *257*

7-18   Rotation-Electronic Coupling   *260*

Exercises   *264*

References   *266*

8   Solid-state Theory   267

8-1   Symmetry Properties in Solids   *267*

8-2   The Reciprocal Lattice and Brillouin Zones   *270*

8-3   Form of Energy-band Wavefunctions   *275*

8-4   Crystal Symmetry and the Group of the k Vector   *279*

8-5   Pictorial Consideration of Eigenfunctions   *281*

8-6   Formal Consideration of Degeneracy and Compatibility   *284*

8-7   Group Theory and the Plane-wave Approximation   *290*

8-8   Connection between Tight- and Loose-binding Approximations   *293*

8-9   Spin-orbit Coupling in Band Theory   *295*

8-10   Time Reversal in Band Theory   *297*

8-11   Magnetic Crystal Groups   *299*

8-12   Symmetries of Magnetic Structures   *303*

8-13   The Landau Theory of Second-order Phase Transitions   *309*

8-14   Irreducible Representations of Magnetic Groups   *311*

Exercises   *313*

References   *315*

Appendix

A   Review of Vectors, Vector Spaces, and Matrices   *317*

B   Character Tables for Point-symmetry Groups   *323*

C   Tables of c^k and a^k Coefficients for s, p, and d Electrons   *331*

Index   335

1 : INTRODUCTION

1-1 The Nature of the Problem

The basic task of quantum theory in the study of atoms, molecules, and solids consists in solving the time-independent Schrödinger equation

$$H\psi_n = E_n\psi_n$$

to determine the energy eigenvalues E_n and the corresponding eigenfunctions ψ_n. The reason for the preeminent position of energy eigenfunctions in the theory is, of course, that they form the stationary states of isolated systems in which energy must be conserved. The science of spectroscopy is devoted to discovering the eigenvalues E_n experimentally, so that one may work back to infer the internal structure of the system which produces them. Moreover, because of their simple time dependence $(e^{-iE_n t/\hbar})$,

1

they can be used to build up a description of the time evolution of non-stationary systems as well.   Thus a major share of the theoretical work in extranuclear physics is devoted to finding eigenstates and eigenvalues of the Hamiltonian operator.

Apart from spin and other relativistic effects, the Hamiltonian operator H is obtained from the classical expression for the energy $H(p_i, q_i)$ by making the usual operator replacement $\mathbf{p}_i \rightarrow (\hbar/i)\mathbf{\nabla}_i$.   Since we shall be dealing with low-energy phenomena, we may introduce the spin effects through the standard two-component Pauli matrix treatment, without resorting to the four-component Dirac equation.   In other words, there is really no longer any serious difficulty in principle in writing down a basic Hamiltonian which is accurate enough for all practical purposes in extranuclear physics. (Within the nucleus, of course, the situation is quite different.)   The difficulty is simply that the eigenvalue problems to be solved are very hard and complicated.

This is illustrated by writing down our Hamiltonian.

$$H = \sum_{\text{el}} \frac{\mathbf{p}_j^2}{2m} + \sum_{\text{nuc}} \frac{\mathbf{p}_K^2}{2M_K} - \sum_{\text{nuc,el}} \frac{Z_K e^2}{r_{Kj}} + \sum_{\text{nuc}} \frac{Z_K Z_L e^2}{r_{KL}} + \sum_{\text{el}} \frac{e^2}{r_{ij}}$$

$$+ H_{so} + H_{ss} + H_{hfs} + H_{\text{ext}}$$

where H_{so}, H_{ss}, H_{hfs}, and H_{ext} refer to spin-orbit, spin-spin, hyperfine-structure, and external-field couplings.   The eigenfunctions of this operator will be functions of the space and spin coordinates of all the electrons and nuclei.   Even for as simple a system as the oxygen molecule, O_2^{16}, this will involve a 54-dimensional spatial function together with 16 spin functions. It is clear that exact "textbook" solutions cannot be expected except for such simple cases as harmonic oscillators, single-particle central-field problems, and noninteracting particles in a square-walled box.

One might propose a direct brute-force numerical solution by use of high-speed digital computers.   The impossibility of this approach may be seen by considering the difficulty of simply stating the wavefunction obtained, which is a $3N$-dimensional function (if there are N electrons and the nuclei are assumed fixed).   For reasonable accuracy, we would need to divide each axis into, say, 100 units.   Thus we would need to tabulate $(100)^{3N} = 10^{6N}$ entries.   For something as simple as a four-electron system, this already exceeds the number of molecules in a mole, and our libraries could hardly contain the millions of volumes required to record the result.   Needless to say, no human mind could comprehend a result hidden in such a mass of numbers.   By making the sweeping, but reasonable, physical approximation of treating each particle as moving independently in an average field from the others, the problem can be reduced to large, but more manageable, dimensions. The detailed interactions could then be reintroduced as a correction, if a more refined calculation were required.

In any case, it is clear that the only practical approach to real physical problems, of any except the very simplest sort, will be an approximate one. The correct answer is approached by successively improved approximations, obtained by including successively smaller correction terms in the Hamiltonian and by increasing the accuracy with which any given approximation to the Schrödinger equation is satisfied. In many cases, this procedure can be carried far enough to yield a description adequate to explain all observed physical properties of the system under consideration, while remaining simple enough to be qualitatively helpful.

1-2 The Role of Symmetry

Being faced by the task of efficiently simplifying a problem so that it may be solved, it is clearly advantageous first to seek out any simplifications which can be made rigorously on the basis of symmetry. Only after these have been fully exploited should one resort to approximations which reduce the generality and accuracy of the final answer. To assist us in the search for the full symmetry-based simplification of the problem, we draw upon the resources of group theory. This provides us with a systematic calculus for exploiting symmetry properties to the fullest extent. Since we require only elementary group theory for our purposes, we shall be able to first develop this machinery from the very beginning. This involves a certain amount of "capital investment" in the form of learning some new formal methods. However, once this investment has been made, we are in a position to adduce very general results in many problems in an economical way, and to gain new insight. It seems inefficient to proceed in many quantum-mechanical problems without this tool at our disposal.

The symmetry which we aim to exploit via group theory is, in most cases, the symmetry of the Hamiltonian operator. To discuss this symmetry, it is convenient to introduce the concept of *transformation operators* which induce some particular coordinate transformation in whatever follows them. For example, the inversion operator i reverses the signs of all coordinates, taking $\mathbf{r}$ into $-\mathbf{r}$. Other common operators are those inducing reflections, rotations, translations, or permutations of coordinates of particles. Such an operator will be a *symmetry* operator appropriate to a given Hamiltonian if the Hamiltonian looks the same after the coordinate transformation as it did before; in other words, if the Hamiltonian is invariant under the transformation. For example, if the potential-energy expression is an even function of coordinates about the origin, it will be invariant under inversion. Since the kinetic-energy operator involves only second derivatives, it is always even.   Hence the entire Hamiltonian is invariant under inversion if the potential is.   This has useful consequences, as we shall soon see.

To continue this discussion, it is useful to note that if an operator R leaves the Hamiltonian invariant, $RH\psi = HR\psi$, for any ψ.   That is to say, R will commute with H if it leaves H invariant.   But if R commutes with H, a general theorem tells us that there are no matrix elements of H between eigenstates of R having different eigenvalues for the operator R.   The proof is simple.   Namely, in the equation

$$RH = HR$$

we may expand the products in a matrix representation based on eigen-functions of R.   Thus

$$\sum_j R_{ij}H_{jk} = \sum_j H_{ij}R_{jk}$$

But since the representation is based on eigenfunctions of R, R has a diagonal matrix and each sum reduces to a single term, namely,

$$R_{ii}H_{ik} = H_{ik}R_{kk}$$

or $$(R_{ii} - R_{kk})H_{ik} = 0$$

Thus $H_{ik} = 0$ if i and k refer to different eigenvalues of R, as stated.   The significance of this result to us is that, in searching for eigenfunctions that diagonalize the Hamiltonian operator, the search can be made separately within the classes of functions having different eigenvalues of a commuting symmetry operator since no off-diagonal matrix elements of H will connect functions of different symmetry.   For example, considering inversion symmetry, we can find all possible energy eigenfunctions by considering only functions which are either even or odd under inversion.   Roughly speaking, this cuts the detailed work in half, besides leading to some qualitative information about the solutions with practically no work at all.

If there are several mutually commuting symmetry operators, all of which commute with H, we can then choose basis functions which are simultaneous eigenfunctions of all these symmetry operators.   It then follows that there are no matrix elements of H between states which differ in their classification according to any of the symmetry operators.   Consequently, we must be able to find a complete set of eigenfunctions of H which are also eigenfunctions of our complete set of mutually commuting symmetry operators.   Thus we may restrict our search for eigenfunctions of H to functions having a definite symmetry under this set of operators.

It is often the case that, after finding the largest group of mutually commuting symmetry operators, there may be additional symmetry operators which commute with the Hamiltonian but which do not commute with all the previous symmetry operators.   Although this larger group of operators is not completely mutually commuting, some additional information can be obtained by using these symmetry operators. We shall show later that, although it is no longer possible to work with functions which are eigenfunctions of all the (larger group of) operators, finite sets of functions can be found such that the effect of a symmetry operation on one function of the set produces only a linear combination of functions within the set. These linear combinations are described by matrices which replace the simple eigenvalues described above as the entity describing the symmetry properties of the functions. The dimensionality of these matrices will turn out to give the degeneracy of the quantum-mechanical eigenfunctions, and the labels characterizing the various matrices and rows within the matrices will form the "good quantum numbers" for the system. These are the generalizations of the parity, or even versus oddness, "quantum number" in the simple example of inversion symmetry treated above. These matrices are the subject of the theory of group representations, which we shall consider in Chap. 3.

The results of group theory which we shall obtain are of considerable practical value, since the method of approximate solution almost invariably consists in expanding the solution in terms of a "suitable" set of approximate functions. Our group theoretical study of the symmetry of the system will enable us to choose zero-order functions of the correct symmetry so as to eliminate most off-diagonal matrix elements of H, leaving a largely simplified problem. Since we shall also see that symmetry determines selection rules, the selection rules governing transitions between these eigenfunctions can also be determined by group-theoretical arguments, without explicit integrations to compute matrix elements. Of course, there is also a quantitative calculation left to find the actual eigenfunctions, the eigenvalues, and the transition probabilities. However, the group theory will usually provide very considerable aid in reducing the scale of this residual calculation as well as giving some rigorous qualitative results with practically no effort at all.

We now proceed to develop the group-theoretical machinery before actually considering in detail the physical problems which are our primary concern.

2 : ABSTRACT GROUP THEORY

2-1 Definitions and Nomenclature

By a group we mean a set of *elements* $A, B, C, \ldots$ such that a form of *group multiplication* may be defined which associates a third element with any ordered pair. This multiplication must satisfy the requirements:

1. The product of any two elements is in the set; i.e., the set is *closed* under group multiplication.

2. The *associative law* holds; for example, $A(BC) = (AB)C$.

3. There is a *unit element* E such that $EA = AE = A$.

4. There is in the group an inverse A^{-1} to each element A such that $AA^{-1} = A^{-1}A = E$.

For the present we shall restrict our attention primarily to *finite groups*. These contain a finite number h of group elements, where h is said to be the *order* of the group. If group multiplication is commutative, so that $AB = BA$ for all A and B, the group is said to be *Abelian*.

2-2 Illustrative Examples

An example of an Abelian group of infinite order is the set of all positive and negative integers including zero. In this case, ordinary addition serves as the group-multiplication operation, zero serves as the unit element, and $-n$ is the inverse of n. Clearly the set is closed, and the associative law is obeyed.

An example of a non-Abelian group of infinite order is the set of all $n \times n$ matrices with nonvanishing determinants. Here the group-multiplication operation is matrix multiplication, and the unit element is the $n \times n$ unit matrix. The inverse matrix of each matrix may be constructed by the usual methods,[1] since the matrices are required to have nonvanishing determinants.

A physically important example of a finite group is the set of covering operations of a symmetrical object. By a covering operation, we mean a rotation, reflection, or inversion which would bring the object into a form indistinguishable from the original one. For example, all rotations about the center are covering operations of a sphere. In such a group the product AB means the operation obtained by first performing B, then A. The unit operation is no operation at all, or perhaps a rotation through 2π. The inverse of each operation is physically apparent. For example, the inverse of a rotation is a rotation through the same angle in the reverse sense about the same axis.

As a complete example, which we shall often use for illustrative purposes, consider the non-Abelian group of order 6 specified by the following *group-multiplication table*:

	E	A	B	C	D	F
E	E	A	B	C	D	F
A	A	E	D	F	B	C
B	B	F	E	D	C	A
C	C	D	F	E	A	B
D	D	C	A	B	F	E
F	F	B	C	A	E	D

The meaning of this table is that each entry is the product of the element labeling the row times the element labeling the column. For example, $AB = D \neq BA$. This table results, for example, if we take our elements to be the following six matrices, and if ordinary matrix multiplication is

[1] See Appendix A and references cited there.

used as the group-multiplication operation:

$$E = \begin{pmatrix} 1 & 0 \\ 0 & 1 \end{pmatrix} \qquad A = \begin{pmatrix} 1 & 0 \\ 0 & -1 \end{pmatrix} \qquad B = \begin{pmatrix} -\dfrac{1}{2} & \dfrac{\sqrt{3}}{2} \\ \dfrac{\sqrt{3}}{2} & \dfrac{1}{2} \end{pmatrix}$$

$$C = \begin{pmatrix} -\dfrac{1}{2} & -\dfrac{\sqrt{3}}{2} \\ -\dfrac{\sqrt{3}}{2} & \dfrac{1}{2} \end{pmatrix} \qquad D = \begin{pmatrix} -\dfrac{1}{2} & \dfrac{\sqrt{3}}{2} \\ -\dfrac{\sqrt{3}}{2} & -\dfrac{1}{2} \end{pmatrix} \qquad F = \begin{pmatrix} -\dfrac{1}{2} & -\dfrac{\sqrt{3}}{2} \\ \dfrac{\sqrt{3}}{2} & -\dfrac{1}{2} \end{pmatrix}$$

Verification of the table is left as a simple exercise.

The very same multiplication table could be obtained by considering the group elements $A, \ldots, F$ to represent the proper covering operations of

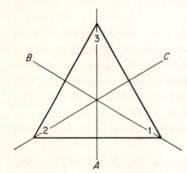

Fig. 2-1. *Symmetry axes of equilateral triangle.*

an equilateral triangle as indicated in Fig. 2-1. The elements A, B, and C are rotations by π about the axes shown. Element D is a clockwise rotation by $2\pi/3$ in the plane of the triangle, and F is a counterclockwise rotation through the same angle. The numbering of the corners destroys the symmetry so that the position of the triangle can be followed through successive operations. If we make the convention that we consider the rotation axes to be kept fixed in space (not rotated with the object), it is easy to verify that the multiplication table given above describes this group as well.

Two groups obeying the same multiplication table are said to be *isomorphic*.

2-3 Rearrangement Theorem

In the multiplication table in the example above, each column or row contains each element once and only once. This rule is true in general and is called the *rearrangement theorem*. Stated more formally, in the sequence

$$EA_k, A_2A_k, A_3A_k, \ldots, A_hA_k,$$

each group element A_i appears exactly once (in the form A_rA_k). The elements are merely rearranged by multiplying each by A_k.

PROOF: For any A_i and A_k, there exists an element $A_r = A_iA_k^{-1}$ in the group since the group contains inverses and is closed. Since $A_rA_k = A_i$ for this particular A_r, A_i must appear in the sequence at least once. But there are h elements in the group and h terms in the sequence. Hence there is no opportunity for any element to make more than a single appearance.

2-4 Cyclic Groups

For any group element X, one can form the sequence

$$X, X^2, X^3, \ldots, X^{n-1}, X^n = E$$

This is called the *period* of X, since the sequence would simply repeat this period over and over if it were extended. (Eventually we must find repetition, since the group is assumed to be finite.) The integer n is called the *order* of X, and this period clearly forms a group as it stands, although it need not exhaust all the elements of the group with which we started. Hence it may be said to form a *cyclic group* of order n. If it is indeed only part of a larger group, it is referred to as a cyclic *subgroup*.[1] We note that all cyclic groups must be Abelian.

In our standard example of the triangle, the period of D is D, $D^2 = F$, $D^3 = DF = E$. Thus D is of order 3, and D, F, E form a cyclic subgroup of our entire group of order 6.

2-5 Subgroups and Cosets

Let $\mathscr{S} = E, S_2, S_3, \ldots, S_g$ be a *subgroup* of order g of a larger group $\mathscr{G}$ of order h. We then call the set of g elements $EX, S_2X, S_3X, \ldots, S_gX$ a *right coset* $\mathscr{S}X$ if X is not in $\mathscr{S}$. (If X were in $\mathscr{S}$, $\mathscr{S}X$ would simply be the subgroup $\mathscr{S}$ itself, by the rearrangement theorem.) Similarly, we define the set $X\mathscr{S}$ as being a *left coset*. These cosets cannot be subgroups, since they cannot include the identity element. In fact, a coset $\mathscr{S}X$ contains *no* elements in common with the subgroup $\mathscr{S}$.

The proof of this statement is easily given by assuming, on the contrary, that for some element S_k we have $S_kX = S_l$, a member of $\mathscr{S}$. Then $X = S_k^{-1}S_l$, which is in the subgroup, and $\mathscr{S}X$ is not a coset at all, but just $\mathscr{S}$ itself.

---

[1] Although the concept is introduced here in connection with cyclic groups, subgroups need not be cyclic. Any subset of elements within a group which in itself forms a group is called a subgroup of the larger group.

Next we note that two right (or left) cosets of subgroup $\mathscr{S}$ in $\mathscr{G}$ either are identical or have no elements in common.

PROOF: Consider two cosets $\mathscr{S}X$ and $\mathscr{S}Y$.   Assume that there exists a common element $S_k X = S_l Y$.   Then $XY^{-1} = S_k^{-1}S_l$, which is in $\mathscr{S}$.   Therefore $\mathscr{S}XY^{-1} = \mathscr{S}$, by the rearrangement theorem.   Postmultiplying both sides by Y leads to $\mathscr{S}X = \mathscr{S}Y$.   Thus the two cosets are completely identical if a single common element exists.

If we combine the results of the preceding paragraphs, we can prove the following theorem: *The order g of a subgroup must be an integral divisor of the order h of the entire group.*   That is, $h/g = l$, where the integer l is called the *index* of the subgroup $\mathscr{S}$ in $\mathscr{G}$.

PROOF: Each of the h elements of $\mathscr{G}$ must appear either in $\mathscr{S}$ or in a coset $\mathscr{S}X$, for some X.   Thus each element must appear in one of the sets $\mathscr{S}, \mathscr{S}X_2, \mathscr{S}X_3, \ldots, \mathscr{S}X_l$, where we have listed all the *distinct* cosets of $\mathscr{S}$ together with $\mathscr{S}$ itself.   But we have shown that there are no elements common to any of these collections of g elements.   Hence it must be possible to divide the total number of elements h into an integral number of sets of g each, and consequently $h = l \times g$.

As an example, consider the subgroup $\mathscr{S} = A, E$ of our illustrative group of order 6.   The right cosets with B and D are identical, namely, $\mathscr{S}B = \mathscr{S}D = B, D$.   Also $\mathscr{S}C = \mathscr{S}F = C, F$.   We note that, as proved in general, these cosets contain no common elements unless entirely identical and they contain no elements in common with $\mathscr{S}$.   Also, the order (2) of the subgroup *is* an integral divisor of the order (6) of the group.   To generalize, the order of *any* cyclic subgroup formed by the period of some group element must be a divisor of the order of the group.

2-6   Example Groups of Finite Order

1. *Groups of order* 1.   The only example is the group consisting solely of the identity element E.

2. *Groups of order* 2.   Again there is only one possibility, the group $(A, A^2 = E)$.   This is an Abelian group, and in physical applications A might represent reflection, inversion, or an interchange of two identical particles.

3. *Groups of order* 3.   In this case, if we start with two elements A and E, it must be that $A^2 = B \neq E$.   Otherwise, if A^2 were to equal E, then (A, E) would form a subgroup of order 2 in a group of order 3, which would violate our theorem.   Thus the only possibility is the cyclic group $(A, A^2 = B, A^3 = E)$.

4. *Groups of order* 4.   With order 4 we begin to have more than one possible distinct group-multiplication table of given order.   The two possibilities here are (1) the cyclic group $(A, A^2, A^3, A^4 = E)$ and (2) the so-called *Vierergruppe* (A, B, C, E) whose multiplication table is:

	E	A	B	C
E	E	A	B	C
A	A	E	C	B
B	B	C	E	A
C	C	B	A	E

Both these groups are Abelian, and in both cases we can pick out subgroups of order 2, as allowed by our theorem. A physical example of the cyclic group of order 4 is provided by the four fold rotations about an axis. On the other hand, the *Vierergruppe* is the rotational-symmetry group of a rectangular solid, if A, B, C are taken to be the rotations by π about the three orthogonal symmetry axes.

5. *Groups of prime order.* These must all be cyclic Abelian groups. Otherwise the period of some element would have to appear as a subgroup whose order was a divisor of a prime number. This general result allows us to note at once that there can be only single groups of order 1, 2, 3, 5, 7, 11, 13, etc.

6. *Permutation groups (of factorial order).* One group of order $n!$ can always be set up based on all the permutations of n distinguishable things. (Of course, others, such as a cyclic group, can also be found.) A permutation can be specified by a symbol such as

$$\begin{pmatrix} 1 & 2 & 3 & \cdots & n \\ \alpha_1 & \alpha_2 & \alpha_3 & \cdots & \alpha_n \end{pmatrix}$$

where $\alpha_1, \alpha_2, \ldots, \alpha_n = 1, 2, \ldots, n$, except for order. The permutation described by this symbol is one in which the item in position i is shifted to the position indicated in the lower line. Successive permutations form the group-multiplication operation. As an example, our standard example group of order 6 can be viewed as the permutation group of the three numbered corners of the triangle. The permutations may be expressed in the above notation as

$$E = \begin{pmatrix} 1 & 2 & 3 \\ 1 & 2 & 3 \end{pmatrix} \qquad A = \begin{pmatrix} 1 & 2 & 3 \\ 2 & 1 & 3 \end{pmatrix} \qquad B = \begin{pmatrix} 1 & 2 & 3 \\ 1 & 3 & 2 \end{pmatrix}$$

$$C = \begin{pmatrix} 1 & 2 & 3 \\ 3 & 2 & 1 \end{pmatrix} \qquad D = \begin{pmatrix} 1 & 2 & 3 \\ 3 & 1 & 2 \end{pmatrix} \qquad F = \begin{pmatrix} 1 & 2 & 3 \\ 2 & 3 & 1 \end{pmatrix}$$

For example, operator A interchanges corners 1 and 2, whereas D replaces 1 by 3, 2 by 1, and 3 by 2, corresponding to a clockwise rotation by $2\pi/3$.

Applying B followed by A leads to

$$AB = \begin{pmatrix} 1 & 2 & 3 \\ 2 & 1 & 3 \end{pmatrix} \begin{pmatrix} 1 & 2 & 3 \\ 1 & 3 & 2 \end{pmatrix} = \begin{pmatrix} 1 & 2 & 3 \\ 3 & 1 & 2 \end{pmatrix} = D$$

which is consistent with the group-multiplication table worked out previously.

Because of the identity of like particles, permutation of them leaves the Hamiltonian invariant. Accordingly, the permutation group plays an important role in quantum theory.

2-7 Conjugate Elements and Class Structure

An element B is said to be *conjugate* to A if

$$B = XAX^{-1} \qquad \text{or} \qquad A = X^{-1}BX$$

where X is some member of the group. Clearly this is a reciprocal property of the pair of elements. Further, if B and C are both conjugate to A, they are conjugate to each other.

PROOF: Assume that

$$B = XAX^{-1} \qquad \text{and} \qquad C = YAY^{-1}$$

Then $$A = Y^{-1}CY$$

and $$B = XY^{-1}CYX^{-1} = (XY^{-1})C(XY^{-1})^{-1}$$

$$= ZCZ^{-1}$$

[In this proof we have used the fact that the inverse of the product of two group elements is the product of the inverses of the elements in inverse order. This is clearly true, since $(RS)(S^{-1}R^{-1}) = R(SS^{-1})R^{-1} = RR^{-1} = E.$]

The properties of conjugate elements given above allow us to collect all mutually conjugate elements into what is called a *class* of elements. The class including A_i is found by forming all products of the form

$$EA_iE^{-1} = A_i, A_2A_iA_2^{-1}, \ldots, A_hA_iA_h^{-1}$$

Of course, some elements may be found several times by this procedure. By proceeding in this way, we can divide all the elements of the group among the various distinct classes. Luckily, we may usually avoid this rather tedious method by using physical-symmetry considerations, as shown below. For example, in the group of covering operations of an equilateral triangle, the two rotations by $2\pi/3$ form a class, the three rotations by π form a class, and, as always, *the identity element is in a class by itself.* The latter follows, since $AEA^{-1} = AA^{-1} = E$ for all A. Note that E is the *only class* which is also a *subgroup*, since all other classes must lack the identity element.

In Abelian groups, each element is in a class by itself, since $XAX^{-1} = AXX^{-1} = AE = A$.

If the group elements are represented by matrices, the traces of all elements in a class must be the same. This follows, since in this case the operation of conjugation becomes that of making a similarity transformation, which leaves the trace invariant.[1]

Physical interpretation of class structure. In physical applications the group elements can often be considered to be symmetry operations which are the covering operations of a symmetrical object. In this case, the operation $B = X^{-1}AX$ is the net operation obtained by first rotating the object to some equivalent position by X, next carrying out the operation A, and then undoing the initial rotation by X^{-1}. Thus B must be an operation of the same physical sort as A, such as a rotation through the same angle, but performed about some different (but physically equivalent) axis which is related to the axis of A by the group operation X^{-1}. This is the significance of operators being in the same class.

As a concrete example, consider the covering operations of the equilateral triangle indicated in Fig. 2-1. If we consider the conjugation of A with D, we have $D^{-1}AD = C$. To follow this through in detail, D rotates the triangle clockwise by $2\pi/3$ so that vertex 2 instead of 3 lies on axis A; next the rotation by π about the A axis interchanges 1 and 3; finally $D^{-1} = F$ rotates the triangle back $2\pi/3$ counterclockwise. This sequence leaves precisely the result of a single rotation by π about axis C, which is an axis equivalent to A but rotated $2\pi/3$ counterclockwise by the symmetry operator D^{-1}.

2-8　Normal Divisors and Factor Groups

If a subgroup $\mathscr{S}$ of a larger group $\mathscr{G}$ consists entirely of complete classes, it is called an *invariant subgroup*, or *normal divisor*. By consisting of complete classes, we mean that, if an element A is in $\mathscr{S}$, then all elements $X^{-1}AX$ are in $\mathscr{S}$, even when X runs over elements of $\mathscr{G}$ which are not in $\mathscr{S}$. Such a subgroup is called invariant because by the rearrangement theorem it is unchanged (except for order) by conjugation with any element of $\mathscr{G}$.

To allow a compact discussion, we introduce the notion of a *complex* such as $\mathscr{K} = (K_1, K_2, \ldots, K_n)$, which is a collection of group elements disregarding order. Such a complex can be multiplied by a single element or by another complex. For example,

$$\mathscr{K}X = (K_1X, K_2X, \ldots, K_nX)$$

and
$$\mathscr{K}\mathscr{R} = (K_1R_1, K_1R_2, \ldots, K_1R_m, \ldots, K_nR_m)$$

[1] See Appendix A.

Elements are considered to be included only once, regardless of how often they are generated.

We can now state our argument concisely by treating sets of elements as complexes. First, a subgroup is defined by the property of closure, that is, $\mathscr{S}\mathscr{S} = \mathscr{S}$. Second, if $\mathscr{S}$ is an *invariant* subgroup, then $X^{-1}\mathscr{S}X = \mathscr{S}$, for all X in the group $\mathscr{G}$. From this it follows that $\mathscr{S}X = X\mathscr{S}$, or, in words, the left and right cosets of an invariant subgroup are identical.

In Sec. 2-5 we have shown that there are a finite number $(l-1)$ of distinct cosets for any subgroup $\mathscr{S}$. We may denote each of these as a complex, and if $\mathscr{S}$ is an invariant subgroup, we have, for example, $\mathscr{K}_i = \mathscr{S}K_i = K_i\mathscr{S}$. Note that $\mathscr{S}K_i = \mathscr{S}K_j$ if K_i and K_j are group elements in the same coset, since we are not concerned with the order in which the elements of the complex appear. Together with the subgroup $\mathscr{S}$, this set of $(l-1)$ distinct complexes can themselves be regarded as the elements of a smaller group (of order $l = h/g$) on a higher level of abstraction. This new group is called the *factor group* of $\mathscr{G}$ with respect to the *normal divisor* (or invariant subgroup) $\mathscr{S}$. In this factor group, $\mathscr{S}$ forms the unit element. We can see this by considering

$$\mathscr{S}\mathscr{K}_i = \mathscr{S}(\mathscr{S}K_i) = (\mathscr{S}\mathscr{S})K_i = \mathscr{S}K_i = \mathscr{K}_i$$

Group multiplication works out as shown in the following example,

$$\mathscr{K}_i\mathscr{K}_j = (\mathscr{S}K_i)(\mathscr{S}K_j) = K_i\mathscr{S}\mathscr{S}K_j = K_i\mathscr{S}K_j = \mathscr{S}(K_iK_j) = (\mathscr{K}_i\mathscr{K}_j)$$

where the last expression refers to the complex which is the coset associated with the product K_iK_j. The concept of factor groups and normal divisors will prove useful in analyzing the structure of groups.

Isomorphy and homomorphy. We have already introduced the concept of isomorphy by noting that two groups having the same multiplication table are called isomorphic. This means that there is a one-to-one correspondence between the elements $A, B, \ldots$ of one group and those $A', B', \ldots$ of the other, such that $AB = C$ implies $A'B' = C'$, and vice versa.

Two groups are said to be *homomorphic* if there exists a correspondence between the elements of the two groups of the sort $A \leftrightarrow A_1', A_2', \ldots$. By this we mean that, if $AB = C$, then the product of any A_i' with any B_j' will be a member of the set C_k'. In general, a homomorphism is a many-to-one correspondence, as indicated here. It specializes to an isomorphism if the correspondence is one-to-one. For example, the group containing the single element E is homomorphic to any other group, since, in view of the fact that each group element is represented by E, group multiplication reduces simply to $EE = E$. A much less trivial example is provided by the homomorphic relation between any group and one of its factor groups (if it has one). The invariant subgroup $\mathscr{S}$ corresponds to all the members of $\mathscr{S}$, and the cosets $\mathscr{K}_i = \mathscr{S}K_i$ correspond to all members of the coset (including K_i and all other group elements having the same coset, which are just the members of $\mathscr{K}_i$). Thus, if $\mathscr{S}$ is of order g, there is a g-to-one correspondence between the original group elements and the elements of the factor group.

2-9 Class Multiplication

In this section we consider a different form of multiplication of collections of group elements in which we do keep track of the number of times an element appears. That is, $\mathscr{R} = \mathscr{K}$ implies that each element appears as often in $\mathscr{R}$ as in $\mathscr{K}$. In this notation

$$X^{-1}\mathscr{C}X = \mathscr{C} \tag{2-1}$$

where $\mathscr{C}$ is any complete *class* of the group and X is any element of the group. (PROOF: Each element produced on the left must appear on the right because they are all conjugate to elements in $\mathscr{C}$ and hence are in $\mathscr{C}$ by the definition of a class. But each element on the left is different, because of the uniqueness of group multiplication, as is each on the right. These two statements are consistent only if the two sides of the equation are equal.)

The converse of this theorem is also true: any collection $\mathscr{C}$ obeying (2-1) for all X in the group is comprised wholly of complete classes. (PROOF: First subtract all complete classes from both sides and denote any remainder by $\mathscr{R}$. Now consider any element R_i of $\mathscr{R}$ on the left in $X^{-1}\mathscr{R}X = \mathscr{R}$. Since this is assumed true for all X, $\mathscr{R}$ must by definition include the complete class of R. Thus $\mathscr{C}$ must be composed of complete classes.)

If we now apply the theorem (2-1) to the product of two classes, we have

$$\mathscr{C}_i\mathscr{C}_j = X^{-1}\mathscr{C}_iXX^{-1}\mathscr{C}_jX$$

$$= X^{-1}(\mathscr{C}_i\mathscr{C}_j)X$$

for all X. Then, upon applying the converse theorem, it follows that $\mathscr{C}_i\mathscr{C}_j$ consists of complete classes. This may be expressed formally by writing

$$\mathscr{C}_i\mathscr{C}_j = \sum_k c_{ijk}\mathscr{C}_k \tag{2-2}$$

where c_{ijk} is the integer telling how often the complete class $\mathscr{C}_k$ appears in the product $\mathscr{C}_i\mathscr{C}_j$.

An an example, in the symmetry group of the triangle whose class structure we noted earlier, let $\mathscr{C}_i = E$; $\mathscr{C}_2 = A$, B, C; and $\mathscr{C}_3 = D$, F. Then $\mathscr{C}_1\mathscr{C}_2 = \mathscr{C}_2$; $\mathscr{C}_1\mathscr{C}_3 = \mathscr{C}_3$; $\mathscr{C}_2\mathscr{C}_2 = 3\mathscr{C}_1 + 3\mathscr{C}_3$; $\mathscr{C}_2\mathscr{C}_3 = 2\mathscr{C}_2$.

EXERCISES

2-1 Consider the symmetry group of the proper covering operations of a square (D_4). This consists of eight elements:

E = the identity

A, B, C, D = 180° rotations about the corresponding labeled axes in Fig. 2-2 which are considered fixed in *space*, not on the body

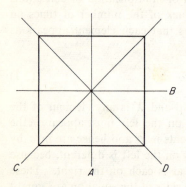

Fig. 2-2. *Symmetry axes of square.*

F, G, H = clockwise rotations in plane of the paper by $\pi/2$, π, and $3\pi/2$, respectively

(*a*) From the geometry, work out the multiplication table of the group; take advantage of the rearrangement theorem to check your result.

(*b*) From the nature of the operations, divide the group elements into classes. If in doubt, check by using the multiplication table [from (*a*)] and the definition of conjugate elements.

(*c*) Write down all the subgroups of the complete group. Note that the orders of the subgroups must be divisors of 8. Which of these subgroups are invariant subgroups (normal divisors)?

(*d*) Work out the cosets of the normal divisors.

(*e*) Work out the group-multiplication tables of the factor groups corresponding to the nontrivial normal divisors of the group.

(*f*) Determine the coefficients c_{ijk} appearing in all class multiplication products.

2-2 List the symmetries of a general rectangle. Work out the multiplication table, and divide the elements into classes.

2-3 Use the multiplication table for the symmetry group of the triangle to verify in several cases the rule for the inverse of a product.

2-4 Consider the group of order $(p - 1)$ obtained by taking as group elements the integers $1, 2, \ldots, (p - 1)$ and as group multiplication ordinary multiplication modulo p, where p is a prime number. (Modulo p means that $m + np$ is considered to be equal to m, where m and n are any integers.)

(*a*) Show that this is a group, and work out the multiplication table when $p = 7$.

(*b*) Prove in general that $A^{p-1} = E$, for all elements A of the group. In this way you have proved Fermat's number-theoretical theorem that $n^p = n(\mathrm{mod}\ p)$, where n is an integer and p is a prime.

(c) Check the theorem for $p = 7$ and $n = 2, 3, 5$.

2-5 Prove that all elements in the same class have the same order when used to generate a cyclic group.

2-6 Show that there is a homomorphism between the cyclic groups of order 4 and 2.

2-7 Prove that a group is Abelian if, and only if, the correspondence of each element to its inverse forms an isomorphism.

2-8 Prove that $c_{ijk} = c_{jik}$ in Eq. (2-2). In other words, prove that $\mathscr{C}_i\mathscr{C}_j = \mathscr{C}_j\mathscr{C}_i$, even if the group is not Abelian.

REFERENCES

BIRKHOFF, G., and S. MACLANE: "A Survey of Modern Algebra," chap. 6, The Macmillan Company, New York, 1949.

LEDERMANN, W.: "Introduction to the Theory of Finite Groups," 2d ed., Oliver & Boyd Ltd., Edinburgh and London, 1953.

SPEISER, A.: "Die Theorie der Gruppen von endlicher Ordnung," 3d ed., Springer-Verlag OHG, Berlin, 1937; reprinted by Dover Publications, Inc., New York, 1945.

WIGNER, E. P.: "Group Theory and Its Application to the Quantum Mechanics of Atomic Spectra," chaps. 7 and 8, Academic Press Inc., New York, 1959.

3 | THEORY OF GROUP REPRESENTATIONS

3-1 Definitions

By a *representation* of an abstract group we mean in general any group composed of concrete mathematical entities which is homomorphic to the original group. However, we shall restrict our attention to representation by square matrices, with matrix multiplication as the group multiplication operation. That is, we associate a matrix $\boldsymbol{\Gamma}(A)$ with each group element A in such a way that

$$\boldsymbol{\Gamma}(A)\boldsymbol{\Gamma}(B) = \boldsymbol{\Gamma}(AB)$$

These matrices then satisfy the group-multiplication table and in every way "represent" the abstract group elements. Clearly this is possible only if $\boldsymbol{\Gamma}(E) = \mathbf{E}$, the unit matrix. The number of rows and columns in the matrix is called the *dimensionality* of the representation.

If each matrix is different, then the two groups are isomorphic rather than merely homomorphic and the representation is said to be *true*, or *faithful*. On the other hand, if several elements correspond to a single matrix, one can easily see that all elements corresponding to the unit matrix form an invariant subgroup of the full group. Similarly, the elements corresponding to each of the other matrices of the representation form the distinct cosets of the invariant subgroup, and the matrices form a true representation of the factor group of this invariant subgroup.

Obviously the 2×2 matrices introduced in Sec. 2-2 form a true matrix representation of our standard example group of order 6. Another representation of the same group can be obtained by taking the determinant of each matrix, because of the fact that

$$|\mathbf{\Gamma}(A)| \cdot |\mathbf{\Gamma}(B)| = |\mathbf{\Gamma}(A)\mathbf{\Gamma}(B)| = |\mathbf{\Gamma}(AB)|$$

This operation reduces the matrices to ordinary numbers, namely, ± 1, giving a one-dimensional representation. This representation is no longer true, since there are only two distinct "matrices," whereas there are six group elements. A still simpler one-dimensional representation is obtained by representing each element by $+1$. This corresponds to the rather trivial homomorphism which can be set up by associating the unit element with all members of the original group, and it is often called the *identical representation*. We shall soon be able to prove that these three form the only possible, essentially different representations of the example group. After we establish the close connection between these matrix representations and quantum-mechanical eigenfunctions, it will become clear that such unconditional statements of group theory will have great practical value.

In discussing the various possible matrix representations of a group, it is important to note that a similarity transformation leaves matrix equations unchanged. That is, if we define $\mathbf{\Gamma}'(A) = \mathbf{S}^{-1}\mathbf{\Gamma}(A)\mathbf{S}$, then

$$\mathbf{\Gamma}'(A)\mathbf{\Gamma}'(B) = [\mathbf{S}^{-1}\mathbf{\Gamma}(A)\mathbf{S}][\mathbf{S}^{-1}\mathbf{\Gamma}(B)\mathbf{S}] = \mathbf{S}^{-1}\mathbf{\Gamma}(A)\mathbf{\Gamma}(B)\mathbf{S}$$
$$= \mathbf{S}^{-1}\mathbf{\Gamma}(AB)\mathbf{S} = \mathbf{\Gamma}'(AB)$$

and the transformed matrices $\mathbf{\Gamma}'$ form a representation if the $\mathbf{\Gamma}$ matrices do. However, the infinity of representations related to each other in this way for various matrices $\mathbf{S}$ differ only in that they are stated with respect to different coordinate axes of some sort, and hence all are considered to be *equivalent*.

Reducible and irreducible representations.   Clearly one can take two (or more) representations and construct from them a new representation by combining the matrices into larger matrices. For a typical element, we could form

$$\mathbf{\Gamma}(A) = \begin{pmatrix} \mathbf{\Gamma}^{(1)}(A) & \mathbf{0} \\ \mathbf{0} & \mathbf{\Gamma}^{(2)}(A) \end{pmatrix}$$

where $\mathbf{\Gamma}^{(1)}(A)$ and $\mathbf{\Gamma}^{(2)}(A)$ are the matrices representing element A in the original representations and $\mathbf{\Gamma}(A)$ represents A in the new larger representation. However, such an artificially enlarged matrix representation is said to be *reducible*. This reducibility might be concealed by applying a similarity transformation which scrambles rows and columns, leaving an equivalent representation which is *not* in block form. Thus our real criterion for reducibility is that it be possible to reduce the matrices representing *all* the elements of the group to block form (with the same block structure) by the *same* similarity transformation. If this *cannot* be done, a representation is said to be *irreducible*, meaning that it cannot be expressed in terms of representations of lower dimensionality. It is customary to indicate the structure of reducible representations by giving the irreducible representations which form the blocks after reduction to block form. In our example, we would write $\Gamma = \Gamma^{(1)} + \Gamma^{(2)}$; more generally, $\Gamma = \Sigma\, a_i \Gamma^{(i)}$, where the a_i are integers telling how often $\Gamma^{(i)}$ appears in Γ. (Observe that this notation does *not* refer to matrix addition.) In many quantum-mechanical applications each irreducible representation will display the transformation properties of a set of degenerate eigenfunctions. Hence, as we shall see, the number of irreducible representations may give the number of distinct energy levels, a very useful piece of information.

3-2 Proof of the Orthogonality Theorem

We commence this section by proving several lemmas leading up to the orthogonality theorem which is central to the development of the theory of group representations. The method of proof given here follows closely that in Wigner's classic book.[1] A review of the various properties of matrices which are used here may be found in Appendix A.

LEMMA: Any representation by matrices with nonvanishing determinants is equivalent through a similarity transformation to a representation by *unitary* matrices.

PROOF: For simplicity in notation let the matrix representing the element A_i be written $\mathbf{A}_i$. We can then construct a Hermitian matrix $\mathbf{H}$ by

$$\mathbf{H} = \sum_{i=1}^{h} \mathbf{A}_i \mathbf{A}_i{}^\dagger \tag{3-1}$$

since each term is already Hermitian. (In our notation a matrix is Hermitian if $\mathbf{H}^\dagger = \mathbf{H}$, or $H_{ji}^* = H_{ij}$.) But it is well known[2] that any Hermitian matrix can be diagonalized by the unitary transformation made up from

---

[1] E. P. Wigner, "Gruppentheorie und ihre Anwendung auf die Quantenmechanik der Atomspektren," Friedr. Vieweg & Sohn, Brunswick, Germany, 1931; revised and translated edition, Academic Press Inc., New York, 1959.

[2] See Appendix A.

the orthonormal eigenvectors found by solving the associated secular equation. Thus we can write

$$\mathbf{d} = \mathbf{U}^{-1}\mathbf{H}\mathbf{U} = \sum_i \mathbf{U}^{-1}\mathbf{A}_i\mathbf{A}_i^{\dagger}\mathbf{U}$$

$$= \sum_i \mathbf{U}^{-1}\mathbf{A}_i\mathbf{U}\mathbf{U}^{-1}\mathbf{A}_i^{\dagger}\mathbf{U} = \sum_i \mathbf{A}_i'\mathbf{A}_i'^{\dagger} \qquad (3\text{-}2)$$

where the primed matrices differ from the original ones by the unitary transformation $\mathbf{U}$. Consideration of a typical element shows that not only is $\mathbf{d}$ diagonal but it has only real positive diagonal elements. This enables us to form other real positive diagonal matrices for $\mathbf{d}^{\frac{1}{2}}$ and $\mathbf{d}^{-\frac{1}{2}}$ by simply taking the appropriate power of the elements of $\mathbf{d}$. Taking advantage of the commutation of diagonal matrices, we can then write (3-2) as

$$\mathbf{E} = \mathbf{d}^{-\frac{1}{2}} \sum_i \mathbf{A}_i'\mathbf{A}_i'^{\dagger}\, \mathbf{d}^{-\frac{1}{2}}$$

where $\mathbf{E}$ is the unit matrix. We can also define a new set of doubly transformed matrices by $\mathbf{A}_j'' = \mathbf{d}^{-\frac{1}{2}}\mathbf{A}_j'\mathbf{d}^{\frac{1}{2}}$. We conclude the proof by showing that these matrices $\mathbf{A}_j''$ are unitary.

To do this, consider

$$\mathbf{A}_j''\mathbf{A}_j''^{\dagger} = \mathbf{d}^{-\frac{1}{2}}\mathbf{A}_j'\mathbf{d}^{\frac{1}{2}}[\mathbf{d}^{-\frac{1}{2}}\sum_i \mathbf{A}_i'\mathbf{A}_i'^{\dagger}\mathbf{d}^{-\frac{1}{2}}]\mathbf{d}^{\frac{1}{2}}\mathbf{A}_j'^{\dagger}\mathbf{d}^{-\frac{1}{2}}$$

$$= \mathbf{d}^{-\frac{1}{2}}\sum_i \mathbf{A}_j'\mathbf{A}_i'\mathbf{A}_i'^{\dagger}\mathbf{A}_j'^{\dagger}\mathbf{d}^{-\frac{1}{2}}$$

$$= \mathbf{d}^{-\frac{1}{2}}\sum_i \mathbf{A}_j'\mathbf{A}_i'(\mathbf{A}_j'\mathbf{A}_i')^{\dagger}\mathbf{d}^{-\frac{1}{2}}$$

$$= \mathbf{d}^{-\frac{1}{2}}\sum_k \mathbf{A}_k'\mathbf{A}_k'^{\dagger}\mathbf{d}^{-\frac{1}{2}} = \mathbf{E}$$

In making the change to the sum on k we have used the rearrangement theorem. Since we have shown that $\mathbf{A}_j''\mathbf{A}_j''^{\dagger} = \mathbf{E}$, we have shown that $\mathbf{A}_j''$ is unitary. Hence we can always construct a unitary representation from any given one by forming

$$\mathbf{A}_j'' = \mathbf{d}^{-\frac{1}{2}}\mathbf{U}^{-1}\mathbf{A}_j\mathbf{U}\mathbf{d}^{\frac{1}{2}} \qquad (3\text{-}3)$$

where $\mathbf{U}$ and $\mathbf{d}$ have been defined above in (3-2).

SCHUR'S LEMMA: Any matrix which commutes with all matrices of an irreducible representation must be a constant matrix (i.e., a multiple of $\mathbf{E}$). Thus, if a nonconstant commuting matrix exists, the representation is reducible, whereas if none exists, the representation is irreducible.

PROOF: On the basis of our first lemma, we can restrict our attention to unitary representations. Let $\mathbf{M}$ be a matrix which commutes with all matrices of the representation. Then

$$\mathbf{A}_i\mathbf{M} = \mathbf{M}\mathbf{A}_i \qquad i = 1, 2, \ldots, h$$

and taking the adjoint of both sides,

$$\mathbf{M}^{\dagger}\mathbf{A}_i^{\dagger} = \mathbf{A}_i^{\dagger}\mathbf{M}^{\dagger}$$

Pre- and postmultiplying the second of these by A_i leads to

$$A_i M^\dagger = M^\dagger A_i$$

Thus, if M commutes, $M^\dagger$ also commutes. From this, it follows that the two Hermitian matrices $H_1 = M + M^\dagger$ and $H_2 = i(M - M^\dagger)$ also commute with all A_i. If we can now show that a commuting Hermitian matrix is a constant, then it follows that $M = H_1 - iH_2$ is also a constant.

Confining our attention to Hermitian commuting matrices, we can always reduce them to diagonal form by a unitary transformation: $d = U^{-1}MU$. If we define a transformed $A_i' = U^{-1}A_iU$, then

$$A_i' d = d A_i'$$

by the invariance of matrix equations under unitary transformations. We now must show that d is not only diagonal but also constant. To do this, consider the $\mu\nu$ element of the matrix, namely,

$$(A_i')_{\mu\nu} d_{\nu\nu} = d_{\mu\mu} (A_i')_{\mu\nu}$$

or $\qquad\qquad (A_i')_{\mu\nu}(d_{\nu\nu} - d_{\mu\mu}) = 0 \qquad i = 1, 2, \ldots, h$

Now, if $d_{\nu\nu} \neq d_{\mu\mu}$, so that the matrix is not constant, then $(A_i')_{\mu\nu}$ must be zero for all A_i' and our transformation U has brought A_i to block form, showing that the representation was in fact reducible. On the other hand, if we assume the representation was irreducible, then it follows that $d_{\nu\nu} = d_{\mu\mu}$ and any commuting matrix must be a constant.

LEMMA: If we are given two irreducible representations of the same group $\boldsymbol{\Gamma}^{(1)}(A_i)$ and $\boldsymbol{\Gamma}^{(2)}(A_i)$ of dimensionality l_1 and l_2, and if a rectangular matrix M exists such that

$$M\boldsymbol{\Gamma}^{(1)}(A_i) = \boldsymbol{\Gamma}^{(2)}(A_i)M \qquad i = 1, 2, \ldots, h \qquad (3\text{-}4)$$

then (1) if $l_1 \neq l_2$, $M = 0$, or (2) if $l_1 = l_2$, $M = 0$, or else $|M| \neq 0$. In the latter case, M has an inverse, $M\boldsymbol{\Gamma}^{(1)}(A_i)M^{-1} = \boldsymbol{\Gamma}^{(2)}(A_i)$, and the two representations are equivalent.

PROOF: As shown in the first lemma, we may confine our attention to unitary representations. Also, we may assume $l_1 \leq l_2$ without loss of generality. Then, taking the adjoint of (3-4), we have

$$\boldsymbol{\Gamma}^{(1)}(A_i)^\dagger M^\dagger = M^\dagger \boldsymbol{\Gamma}^{(2)}(A_i)^\dagger$$

or $\qquad\qquad \boldsymbol{\Gamma}^{(1)}(A_i^{-1})M^\dagger = M^\dagger \boldsymbol{\Gamma}^{(2)}(A_i^{-1}) \qquad (3\text{-}5)$

by the unitary property which implies $\boldsymbol{\Gamma}^{(1)}(A_i)^\dagger = \boldsymbol{\Gamma}^{(1)}(A_i)^{-1} = \boldsymbol{\Gamma}^{(1)}(A_i^{-1})$. If we now premultiply both sides of (3-5) by M and use the fact that (3-4) holds for A_i^{-1} as well as A_i, since it holds for all group elements, we find

$$\boldsymbol{\Gamma}^{(2)}(A_i^{-1})MM^\dagger = MM^\dagger \boldsymbol{\Gamma}^{(2)}(A_i^{-1}) \qquad (3\text{-}6)$$

Thus the matrix $\mathbf{MM}^\dagger$ commutes with all the matrices of the representation and hence must be a multiple of the unit matrix, by Schur's lemma. That is,

$$\mathbf{MM}^\dagger = c\mathbf{E} \tag{3-7}$$

Consider first the case when $l_1 = l_2$, so that $\mathbf{M}$ is a square matrix. Taking the determinant of (3-7) then yields $|\mathbf{M}|^2 = c^{l_1}$. Now, if $c \neq 0$, $|\mathbf{M}| \neq 0$, $\mathbf{M}$ has an inverse and the two representations are equivalent. On the other hand, if $c = 0$, then $\mathbf{MM}^\dagger = \mathbf{0}$. In terms of components, this means that $\sum_\lambda M_{\mu\lambda} M_{\lambda\nu}{}^\dagger = \sum_\lambda M_{\mu\lambda} M_{\nu\lambda}{}^* = 0$, for all μ and ν. Taking in particular $\mu = \nu$, we have $\sum_\lambda |M_{\mu\lambda}|^2 = 0$, which is possible only if all $M_{\mu\lambda} = 0$, or in other words, if $\mathbf{M} = \mathbf{0}$.

Now consider the case when $l_1 < l_2$, so that $\mathbf{M}$ has l_1 columns and l_2 rows. We can fill $\mathbf{M}$ out to a square $l_2 \times l_2$ matrix $\mathbf{N}$ by inserting $(l_2 - l_1)$ columns of zeros. Inspection then shows that $\mathbf{NN}^\dagger \equiv \mathbf{MM}^\dagger$. Since $\mathbf{N}$ clearly has zero determinant, so does $\mathbf{NN}^\dagger$ and hence $\mathbf{MM}^\dagger$. But by (3-7) $\mathbf{MM}^\dagger$ is a constant matrix, which we now see has $c = 0$, since the determinant vanishes. From this it follows that $\mathbf{M} = \mathbf{0}$. This completes the full proof of our lemma.

The great orthogonality theorem. If we consider all the *inequivalent*, *irreducible*, *unitary* representations of a group, then

$$\sum_R \Gamma^{(i)}(R)^*_{\mu\nu} \Gamma^{(j)}(R)_{\alpha\beta} = \frac{h}{l_i} \delta_{ij} \delta_{\mu\alpha} \delta_{\nu\beta} \quad \blacktriangleright \tag{3-8}$$

where in the summation R runs over all group elements $E, A_2, \ldots, A_h$ and l_i is the dimensionality of $\Gamma^{(i)}$.

PROOF: We first consider the case of two inequivalent representations $\Gamma^{(1)}$ and $\Gamma^{(2)}$. Then we may construct a matrix $\mathbf{M}$ satisfying our third lemma by forming

$$\mathbf{M} = \sum_R \Gamma^{(2)}(R)\mathbf{X}\Gamma^{(1)}(R^{-1})$$

where $\mathbf{X}$ is a completely arbitrary matrix having l_1 columns and l_2 rows. To show that this $\mathbf{M}$ satisfies (3-4), take

$$\Gamma^{(2)}(S)\mathbf{M} = \sum_R \Gamma^{(2)}(S)\Gamma^{(2)}(R)\mathbf{X}\Gamma^{(1)}(R^{-1})$$

$$= \sum_R \Gamma^{(2)}(S)\Gamma^{(2)}(R)\mathbf{X}\Gamma^{(1)}(R^{-1})\Gamma^{(1)}(S^{-1})\Gamma^{(1)}(S)$$

$$= \sum_R \Gamma^{(2)}(SR)\mathbf{X}\Gamma^{(1)}(R^{-1}S^{-1})\Gamma^{(1)}(S)$$

$$= [\sum_R \Gamma^{(2)}(SR)\mathbf{X}\Gamma^{(1)}(SR)^{-1}]\Gamma^{(1)}(S)$$

$$= [\sum_R \Gamma^{(2)}(R)\mathbf{X}\Gamma^{(1)}(R^{-1})]\Gamma^{(1)}(S)$$

by the rearrangement theorem; so finally

$$\mathbf{\Gamma}^{(2)}(S)\mathbf{M} = \mathbf{M}\mathbf{\Gamma}^{(1)}(S)$$

Therefore, our lemma shows that $\mathbf{M} = \mathbf{0}$.   But then

$$M_{\alpha\mu} = 0 = \sum_R \sum_{\kappa\lambda} \Gamma^{(2)}(R)_{\alpha\kappa} X_{\kappa\lambda} \Gamma^{(1)}(R^{-1})_{\lambda\mu}$$

Since $\mathbf{X}$ is arbitrary, we may set all $X_{\kappa\lambda} = 0$ except $X_{\beta\nu} = 1$.   Then

$$\sum_R \Gamma^{(2)}(R)_{\alpha\beta} \Gamma^{(1)}(R^{-1})_{\nu\mu} = 0 \tag{3-9}$$

Using the unitary property of $\mathbf{\Gamma}^{(1)}$, we have the equivalent form

$$\sum_R \Gamma^{(1)}(R)_{\mu\nu}^* \Gamma^{(2)}(R)_{\alpha\beta} = 0 \tag{3-9a}$$

which completes the proof of the δ_{ij} factor in (3-8).

Next we consider the case when $i = j = 1$, say.   In analogy with the previous paragraph, we may construct a matrix $\mathbf{M}$ which commutes with all matrices of the representation by forming

$$\mathbf{M} = \sum_R \mathbf{\Gamma}^{(1)}(R)\mathbf{X}\mathbf{\Gamma}^{(1)}(R^{-1})$$

Then by Schur's lemma $\mathbf{M} = c\mathbf{E}$.   Thus, taking the $\mu\mu'$ element,

$$\sum_{\kappa,\lambda} \sum_R \Gamma^{(1)}(R)_{\mu\kappa} X_{\kappa\lambda} \Gamma^{(1)}(R^{-1})_{\lambda\mu'} = c\delta_{\mu\mu'}$$

Choosing $X_{\kappa\lambda} = 0$ except $X_{\nu\nu'} = 1$ reduces this to

$$\sum_R \Gamma^{(1)}(R)_{\mu\nu} \Gamma^{(1)}(R^{-1})_{\nu'\mu'} = c_{\nu\nu'}\delta_{\mu\mu'} \tag{3-10}$$

We have put subscripts on the constant c to indicate the particular choice of $\mathbf{X}$.   Now, choose $\mu' = \mu$, and sum on μ.   This yields, on interchanging the order of the factors,

$$\sum_R \sum_\mu \Gamma^{(1)}(R^{-1})_{\nu'\mu} \Gamma^{(1)}(R)_{\mu\nu} = c_{\nu\nu'} \sum_\mu \delta_{\mu\mu}$$

or

$$\sum_R \Gamma^{(1)}(R^{-1}R)_{\nu'\nu} = l_1 c_{\nu\nu'}$$

since μ runs over the l_1 rows of the $\Gamma^{(1)}$ representation.   We may reduce the left member further by noting that

$$\sum_R \Gamma^{(1)}(R^{-1}R)_{\nu'\nu} = \sum_R \Gamma^{(1)}(E)_{\nu'\nu} = h\Gamma^{(1)}(E)_{\nu'\nu} = h\delta_{\nu'\nu}$$

Therefore $c_{\nu\nu'} = h\delta_{\nu'\nu}/l_1$.   Substituting back into (3-10), we have

$$\sum_R \Gamma^{(1)}(R)_{\mu\nu} \Gamma^{(1)}(R^{-1})_{\nu'\mu'} = \frac{h}{l_1} \delta_{\mu'\mu}\delta_{\nu'\nu} \tag{3-11}$$

By using the unitary property of $\mathbf{\Gamma}^{(1)}$, this becomes

$$\sum_R \Gamma^{(1)}(R)^*_{\mu'\nu'}\Gamma^{(1)}(R)_{\mu\nu} = \frac{h}{l_1}\,\delta_{\mu'\mu}\delta_{\nu'\nu} \qquad (3\text{-}11a)$$

Combining (3-9a) and (3-11a), we have proved the general theorem (3-8). Moreover, (3-9) and (3-11) provide a generalization of (3-8) for nonunitary representations.

Geometric interpretation. To appreciate the significance of this result, it is helpful to interpret it as stating the orthogonality of a set of vectors in "group-element space." This is an h-dimensional vector space in which the axes or components are labeled by the various group elements $R = E, A_2, \ldots, A_h$. The vectors themselves are labeled by three indices—the representation index i and the subscripts $\mu\nu$, indicating row and column within the representation matrix. The theorem then states that all these vectors are mutually orthogonal in this h-dimensional space.

From this result we may readily draw an extremely important conclusion. If we count up the number of these orthogonal vectors, we find $\sum_i l_i^2$, where i runs over all the distinct irreducible representations, since there are l_i^2 entries in a matrix of dimensionality l_i. But clearly the maximum number of orthogonal vectors in an h-dimensional vector space is just h. Thus it follows that $\sum_i l_i^2 \leq h$. In fact we shall soon prove that the equality always holds. This gives us the dimensionality theorem

$$\sum_i l_i^2 = h \qquad \blacktriangleright \quad (3\text{-}12)$$

which is essential for working out the irreducible representations of any group. For example, in our example group of order 6, we have $h = 6$, and we have already found three irreducible representations—one of dimensionality 2 and two of dimensionality 1. Since $2^2 + 1^2 + 1^2 = 6$, this simple theorem tells us that it is impossible for any other distinct irreducible representations to exist.

3-3 The Character of a Representation

Because all matrix representations related to each other through unitary transformations are equivalent, it is clear that there is a large degree of arbitrariness in the actual forms of the matrices. This makes it worthwhile to seek a way of characterizing any given representation which is invariant under such transformations. This immediately suggests using the traces of the matrices, since these are invariant. Accordingly, we define the *character* of the jth representation as being the set of h numbers $\chi^{(j)}(E)$, $\chi^{(j)}(A_2), \ldots, \chi^{(j)}(A_h)$, where

$$\chi^{(j)}(R) = \mathrm{Tr}\,\mathbf{\Gamma}^{(j)}(R) = \sum_{\mu=1}^{l_j}\Gamma^{(j)}(R)_{\mu\mu} \qquad (3\text{-}13)$$

Since the matrix representations of all elements in the same class are related by similarity transformations (by definition of conjugate group elements), the invariance of traces shows that all elements in the same class have the same character. This enables us to specify the character of any given representation by simply giving the trace of one matrix from each class of group elements. We denote this by $\chi^{(j)}(\mathscr{C}_k)$ for the kth class.

We can profitably apply the orthogonality theorem to the character in the following way. If we set $\nu = \mu$ and $\beta = \alpha$ in Eq. (3-8), we have

$$\sum_R \Gamma^{(i)}(R)^*_{\mu\mu}\Gamma^{(j)}(R)_{\alpha\alpha} = \frac{h}{l_i}\delta_{ij}\delta_{\mu\alpha}$$

Now summing over μ and α and use of (3-13) yields

$$\sum_R \chi^{(i)}(R)^*\chi^{(j)}(R) = \frac{h}{l_i}\delta_{ij}\sum_{\mu\alpha}\delta_{\mu\alpha}$$

$$= h\delta_{ij} \tag{3-14}$$

Thus, the characters form a set of orthogonal vectors in group-element space. Collecting the group elements according to classes, within which the $\chi^{(j)}(R)$ are the same, we can rewrite (3-14) as

$$\sum_k \chi^{(i)}(\mathscr{C}_k)^*\chi^{(j)}(\mathscr{C}_k)N_k = h\delta_{ij} \qquad \blacktriangleright \quad (3-15)$$

where N_k is the number of elements in the class $\mathscr{C}_k$ and the sum now runs over *classes*.

Written in the form (3-15), our result shows that the characters of the various irreducible representations also form an orthogonal vector system in the vector space where axes are labeled by classes $\mathscr{C}_k$ rather than group elements R. Since the number of mutually orthogonal vectors in a space cannot exceed its dimensionality, it follows that the number of irreducible representations cannot exceed the number of classes. In fact it can be shown that they are always equal. Thus,

Number of irreducible representations = number of classes $\qquad \blacktriangleright \quad (3-16)$

This rule, together with (3-12), enables us to work out the number and dimensionality of the irreducible representations from the numbers of group elements and classes. Thus these numerological results are of great importance in applications. Applying them to our example group of order 6, we recall that there are three classes: $\mathscr{C}_1 = E$; $\mathscr{C}_2 = A, B, C$; and $\mathscr{C}_3 = D, F$. By (3-16), this implies that there are just three irreducible representations. Then, since $2^2 + 1^2 + 1^2 = 6$ is the only solution of (3-12) for the case in which the sum of three squares must equal 6, we conclude that there must be one two-dimensional and two one-dimensional irreducible representations. Of course we already had noted these results for this case, but the method is very effective in more difficult examples where the representations are by no means obvious.

Character tables. It is convenient to display the characters of the various representations in a *character table* for any given group. The columns are labeled by the various classes, preceded by the number N_k of elements in the class. The rows are labeled by the irreducible representations, and the entries in the table are the $\chi^{(j)}(\mathscr{C}_k)$. Thus in our example group we have the table

	$\mathscr{C}_1$	$3\mathscr{C}_2$	$2\mathscr{C}_3$
$\Gamma^{(1)}$	1	1	1
$\Gamma^{(2)}$	1	-1	1
$\Gamma^{(3)}$	2	0	-1

as may be verified by using the explicit representations noted in Sec. 3-1. Note that the rows of the table are orthogonal if we use the N_k weighting factors as prescribed by (3-15). We also note that the columns form orthogonal vectors. This is not an accident, and we now proceed to prove it to be true in general.

Second orthogonality relation for characters. Since the number of classes equals the number of irreducible representations, we may form a square matrix $\mathbf{Q}$ which has the same form as the character table, namely,

$$\mathbf{Q} = \begin{pmatrix} \chi^{(1)}(\mathscr{C}_1) & \chi^{(1)}(\mathscr{C}_2) & \cdots \\ \chi^{(2)}(\mathscr{C}_1) & \chi^{(2)}(\mathscr{C}_2) & \cdots \\ \cdots\cdots\cdots\cdots\cdots\cdots \end{pmatrix}$$

Now consider the related matrix $\mathbf{Q}'$ defined by

$$\mathbf{Q}' = \begin{pmatrix} \dfrac{\chi^{(1)}(\mathscr{C}_1)^* N_1}{h} & \dfrac{\chi^{(2)}(\mathscr{C}_1)^* N_1}{h} & \cdots \\ \dfrac{\chi^{(1)}(\mathscr{C}_2)^* N_2}{h} & \dfrac{\chi^{(2)}(\mathscr{C}_2)^* N_2}{h} & \cdots \\ \cdots\cdots\cdots\cdots\cdots\cdots\cdots\cdots \end{pmatrix}$$

Then a typical element of the product is

$$(\mathbf{QQ}')_{ij} = \sum_k \frac{\chi^{(i)}(\mathscr{C}_k)\chi^{(j)}(\mathscr{C}_k)^* N_k}{h} = \delta_{ij}$$

by the first orthogonality relation for characters, (3-15). Thus $\mathbf{Q}' = \mathbf{Q}^{-1}$. But any matrix commutes with its inverse. Therefore, we also must have

$(Q'Q)_{ij} = \delta_{ij}$. Carrying out the multiplication in the reverse order leads directly to our result,

$$\sum_i \chi^{(i)}(\mathscr{C}_k)^* \chi^{(i)}(\mathscr{C}_l) = \frac{h}{N_k} \delta_{kl} \qquad \blacktriangleright \text{(3-17)}$$

Since (3-17) is actually a direct consequence of (3-15), it contains no fundamentally new information. However, as we shall see directly, it is often a convenient aid in setting up character tables by inspection.

3-4 Construction of Character Tables

Although the character table gives much less information about a group than would a complete set of matrices for the irreducible representations, it does give enough information for many purposes. Thus it is highly desirable to be able to obtain the character table of a group directly, without first working out explicit representation matrices. In fact, in the simple cases of most interest it is possible to work out the table by inspection with the aid of a few simple rules based on the results of the previous sections. These rules are collected here for convenience.

1. The *number* of irreducible representations equals the number of classes of group elements. The latter is easily found by considering the nature of the operations or, more mechanically, by computing conjugate elements with the aid of the group-multiplication table.

2. The *dimensionalities* l_i of the irreducible representations are then determined by the fact that $\sum_i l_i^2 = h$. In most cases this has a unique solution. Since the identity element must be represented by a unit matrix, the trace of a matrix of the identity class is simply l_i. This determines the first column of the table, $\chi^{(i)}(E) = l_i$. Also, since we always have the one-dimensional representation (referred to as *totally symmetric, identical,* or *invariant*) in which each group element is represented by unity, we can always fill in the first row by $\chi^{(1)}(\mathscr{C}_k) = 1$.

3. The rows of the table must be orthogonal and normalized to h, with weighting factor N_k, the number of elements in $\mathscr{C}_k$. That is,

$$\sum_k \chi^{(i)}(\mathscr{C}_k)^* \chi^{(j)}(\mathscr{C}_k) N_k = h \delta_{ij} \qquad \text{(3-15)}$$

4. The columns of the table must be orthogonal vectors normalized to h/N_k. That is,

$$\sum_i \chi^{(i)}(C_k)^* \chi^{(i)}(\mathscr{C}_l) = \frac{h}{N_k} \delta_{kl} \qquad \text{(3-17)}$$

5. Elements within the ith row are related by

$$N_j \chi^{(i)}(\mathscr{C}_j) N_k \chi^{(i)}(\mathscr{C}_k) = l_i \sum_l c_{jkl} N_l \chi^{(i)}(\mathscr{C}_l) \qquad \text{(3-18)}$$

where the c_{jkl} are the constants defined by the expression governing class multiplication $\mathscr{C}_j\mathscr{C}_k = \sum_l c_{jkl}\mathscr{C}_l$. These constants may be determined given the group-multiplication table, as explained in Sec. 2-9.

PROOF: Recall that any class satisfies $X^{-1}\mathscr{C}_kX = \mathscr{C}_k$, or, equivalently, $\mathscr{C}_kX = X\mathscr{C}_k$, for all X. Now consider this equation in terms of a matrix representation. If we introduce the matrix $\mathscr{C}_k$, which is the sum of the matrices of all the elements in the class $\mathscr{C}_k$, then it follows that $\mathscr{C}_k\mathbf{X} = \mathbf{X}\mathscr{C}_k$, because matrix multiplication by $\mathbf{X}$ is linear and matrix addition is commutative. But if this is so, Schur's lemma states that $\mathscr{C}_k$ must be a constant matrix, say, $\eta_k\mathbf{E}$. In this case, one readily sees that the matrix form of the equation defining the c_{jkl} reduces to $\eta_j\eta_k = \sum_l c_{jkl}\eta_l$. We now evaluate η_k by taking the trace of $\mathscr{C}_k$ in two ways and comparing, namely,

$$\text{Tr } \mathscr{C}_k = \text{Tr } \eta_k\mathbf{E} = \eta_k l_i$$

and
$$\text{Tr } \mathscr{C}_k = \text{Tr } \sum_{\mu=1}^{N_k} \mathbf{A}_\mu = N_k\chi^{(i)}(\mathscr{C}_k)$$

Thus
$$\eta_k = \frac{N_k\chi^{(i)}(\mathscr{C}_k)}{l_i}$$

and rule 5 follows directly.

The first three rules usually suffice to work out a character table, but the last two often facilitate the process of inspection. For example, by use of these rules the complete character table for the group of the equilateral triangle, as given above, could be quickly found without any knowledge of the explicit matrices at all. The first row and column are fixed by rule 2, and the four integers to complete the table so as to satisfy rules 3 and 4 are readily picked out. In case the characters should be nonintegral, as may occur in various instances, the procedure is less simple. However, if a set of integers satisfying the rules can be found, one may normally take it to be the proper character table.

3-5 Decomposition of Reducible Representations

Clearly the character of a reducible representation Γ is the sum of the characters of the component irreducible representations $\Gamma^{(j)}$. This can be seen by supposing that the representation has been brought into block form by a suitable similarity transformation. In that case, the trace of the large matrix is simply the sum of the traces of the submatrices in the blocks along the diagonal. Thus we can write

$$\chi(R) = \sum_j a_j\chi^{(j)}(R) \tag{3-19}$$

where χ is the character of Γ and a_j is the number of times $\Gamma^{(j)}$ appears in Γ. Since the $\chi^{(j)}(R)$ form an orthogonal vector system, the expansion coefficients a_i can be determined as usual by taking the scalar product with $\chi^{(i)}(R)$. Thus, using (3-14),

$$\sum_R \chi(R)\chi^{(i)}(R)^* = \sum_R \sum_j a_j \chi^{(j)}(R)\chi^{(i)}(R)^* = h a_i$$

and therefore

$$a_j = h^{-1} \sum_R \chi^{(j)}(R)^* \chi(R) = h^{-1} \sum_k N_k \chi^{(j)}(\mathscr{C}_k)^* \chi(\mathscr{C}_k) \qquad \blacktriangleright \quad (3\text{-}20)$$

We conclude that the number of times the various irreducible representations appear in a given reducible representation is uniquely determined by the character of the reducible representation, assuming that the character table of the group is known. We shall find this result central to the enumeration of residual degeneracies when a representation is rendered reducible by decreasing the size of the symmetry group.

The regular representation. Given the multiplication table of a group, we can always form a reducible representation called the *regular representation* as follows: Write down the multiplication table, rearranging rows so that they correspond to the *inverses* of the elements labeling the columns. In this way one naturally obtains only the identity element E along the principal diagonal. The matrix of the regular representation for the group element R is then obtained by replacing R by unity and all other elements by zero in the resulting table. For our example group, the rearranged multiplication table and a typical matrix of $\Gamma^{(\text{reg})}$ are shown below:

	E	A	B	C	D	F
E^{-1}	E	A	B	C	D	F
A^{-1}	A	E	D	F	B	C
B^{-1}	B	F	E	D	C	A
C^{-1}	C	D	F	E	A	B
D^{-1}	F	B	C	A	E	D
F^{-1}	D	C	A	B	F	E

$$\Gamma^{(\text{reg})}(A) = \begin{pmatrix} 0 & 1 & 0 & 0 & 0 & 0 \\ 1 & 0 & 0 & 0 & 0 & 0 \\ 0 & 0 & 0 & 0 & 0 & 1 \\ 0 & 0 & 0 & 0 & 1 & 0 \\ 0 & 0 & 0 & 1 & 0 & 0 \\ 0 & 0 & 1 & 0 & 0 & 0 \end{pmatrix}$$

Evidently, in general $\chi^{(\text{reg})}(E) = h$, and $\chi^{(\text{reg})}(R) = 0$ for $R \neq E$, since by construction only $\Gamma^{(\text{reg})}(E)$ has nonzero elements on the diagonal, and it has unity h times.

Before proceeding, we should confirm that the matrices defined above do in fact form a representation. That is, we must show that

$$\Gamma^{(\text{reg})}(BC) = \Gamma^{(\text{reg})}(B)\Gamma^{(\text{reg})}(C)$$

or in component form

$$\Gamma^{(\text{reg})}(BC)_{A_k^{-1}A_i} = \sum_{A_j} \Gamma^{(\text{reg})}(B)_{A_k^{-1}A_j} \Gamma^{(\text{reg})}(C)_{A_j^{-1}A_i}$$

where the subscripts label the rows and columns in the rearranged multiplication table.   By construction, we know that

$$\Gamma^{(\text{reg})}(B)_{A_k^{-1}A_j} = \begin{cases} 1 & \text{if } A_k^{-1}A_j = B \\ 0 & \text{otherwise} \end{cases}$$

and a similar relation holds for $\boldsymbol{\Gamma}^{(\text{reg})}(C)$.   Thus the indicated sum over A_j will vanish unless for some A_j both $\Gamma^{(\text{reg})}(B)_{A_k^{-1}A_j}$ and $\Gamma^{(\text{reg})}(C)_{A_j^{-1}A_i}$ are simultaneously nonzero.   This will occur if, and only if,

$$BC = (A_k^{-1}A_j)(A_j^{-1}A_i) = A_k^{-1}A_i,$$

in which case the sum equals unity.   But this coincides with the definition of the left member $\Gamma^{(\text{reg})}(BC)_{A_k^{-1}A_i}$.   Hence the matrices defined above do form a representation.

CELEBRATED THEOREM: This theorem states that the regular representation contains each irreducible representation a number of times equal to the dimensionality of the irreducible representation.

PROOF: We apply our formula (3-20) for the decomposition of representations.   Thus a_j, the number of times $\Gamma^{(j)}$ appears in $\Gamma^{(\text{reg})}$, is given by

$$a_j = h^{-1} \sum_R \chi^{(j)}(R)^* \chi^{(\text{reg})}(R)$$
$$= h^{-1}\chi^{(j)}(E)h$$
$$= l_j$$

We now use this theorem to prove that the equality sign in (3-12) must in fact hold.   By construction, the dimensionality of the regular representation is equal to h, the order of the group.   But it also must equal the sum of the dimensionalities of all the irreducible representations of which it is composed.   By the celebrated theorem, the latter is $\sum_j l_j \times l_j = \sum_j l_j^2$.   Therefore,

$$\sum_j l_j^2 = h$$

This result removes the possible inequality sign left in our earlier argument based simply on the number of possible orthogonal vectors in a space of given dimension.

3-6   Application of Representation Theory in Quantum Mechanics

Transformation operators.   Let us now depart from our formal development of the theory of group representations to examine the relation of this theory to quantum mechanics, which provides our reason for presenting the theory in the first place.   In the applications which we shall consider, the group of interest is the group of symmetry operators which leave the

Hamiltonian of the problem invariant. Each such operator can be specified by giving a real orthogonal transformation of coordinates $\mathbf{R}$ such that the new coordinates $\mathbf{x}'$ are related to the original ones $\mathbf{x}$ by

$$\mathbf{x}' = \mathbf{R}\mathbf{x} \qquad (3\text{-}21)$$

or in terms of components

$$x_i' = \sum_j R_{ij} x_j$$

Depending on the particular form of $\mathbf{R}$, it may represent a rotation of coordinates, a reflection, an inversion, or any combination of these. In all these cases, $\mathbf{R}$ is a real orthogonal matrix, and hence $\mathbf{R}^{-1} = \mathbf{R}^\dagger = \tilde{\mathbf{R}}$, where $\tilde{\mathbf{R}}$ is the transpose of $\mathbf{R}$. Thus we can write the inverse transformation as

$$x_i = \sum_j R^{-1}{}_{ij} x_j' = \sum_j R_{ji} x_j'$$

As discussed earlier, such a set of matrices $\mathbf{R}$ form a group under matrix multiplication. Although the present discussion will be given in the language of such orthogonal transformations, the results are readily carried over when the symmetry operations include, e.g., permutation of the coordinates of identical particles and translation of coordinates.

For our purposes, we now introduce a new group isomorphic to this group of coordinate transformations, in which the group elements are transformation operators which operate on *functions* rather than *coordinates*. We denote the operator which corresponds to $\mathbf{R}$ by P_R and follow Wigner's convention in defining P_R by requiring that the following be satisfied identically in $\mathbf{x}$:

$$P_R f(\mathbf{R}\mathbf{x}) = f(\mathbf{x})$$

Equivalently,

$$P_R f(\mathbf{x}) = f(\mathbf{R}^{-1}\mathbf{x}) \qquad (3\text{-}22)$$

That is, P_R changes the functional form of $f(\mathbf{x})$ in such a way as to compensate for the change of variable $\mathbf{R}$.

As an example, let $\mathbf{R}$ be the transformation to X', Y', Z' axes rotated by 90° about the X axis. Then $x' = x$, $y' = z$, $z' = -y$, and the matrices concerned are

$$\mathbf{R} = \begin{pmatrix} 1 & 0 & 0 \\ 0 & 0 & 1 \\ 0 & -1 & 0 \end{pmatrix}, \quad \mathbf{R}^{-1} = \tilde{\mathbf{R}} = \begin{pmatrix} 1 & 0 & 0 \\ 0 & 0 & -1 \\ 0 & 1 & 0 \end{pmatrix}$$

Thus

$$\mathbf{R}^{-1}\mathbf{x} = \begin{pmatrix} 1 & 0 & 0 \\ 0 & 0 & -1 \\ 0 & 1 & 0 \end{pmatrix} \begin{pmatrix} x \\ y \\ z \end{pmatrix} = \begin{pmatrix} x \\ -z \\ y \end{pmatrix}$$

and from our definition above

$$P_R f(x, y, z) = P_R f(\mathbf{x}) = f(\mathbf{R}^{-1}\mathbf{x}) = f(x, -z, y)$$

For example, if we let this P_R operate on the three orthogonal p-like functions for an atom with nucleus at the origin, we obtain

$$P_R\{x\varphi(r)\} = x\varphi(r)$$
$$P_R\{y\varphi(r)\} = -z\varphi(r)$$
$$P_R\{z\varphi(r)\} = y\varphi(r)$$

since $r = (x^2 + y^2 + z^2)^{\frac{1}{2}}$ is invariant under such a rotation. We note that the contours of these functions are rotated clockwise with respect to the axes, whereas the axes are rotated counterclockwise. These two rotations compensate, as required by the definition of P_R, and one can consider the operation from either point of view. However, we shall emphasize the viewpoint in which the transformation operator P_R rotates the contours of the function, so that we minimize the chance of confusion with several sets of coordinate axes.

Before proceeding, let us verify that the group of operators P_R is in fact isomorphic to the group of coordinate transformations **R**. To see this, we need to show that

$$P_S P_R = P_{SR}$$

where S and R are two transformations. We consider the successive operations in detail. First,

$$P_R f(\mathbf{x}) = f(\mathbf{R}^{-1}\mathbf{x}) = g(\mathbf{x})$$

where $g(\mathbf{x})$ is the new function which incorporates the effect of $\mathbf{R}^{-1}$ into the functional form. [In our example above, if $f(\mathbf{x}) = y\varphi(r)$, then $g(\mathbf{x}) = -z\varphi(r)$.] Now we apply the second operator P_S, obtaining

$$P_S[P_R f(\mathbf{x})] = P_S g(\mathbf{x}) = g(\mathbf{S}^{-1}\mathbf{x}) = f[\mathbf{R}^{-1}(\mathbf{S}^{-1}\mathbf{x})]$$
$$= f[(\mathbf{SR})^{-1}\mathbf{x}] = P_{SR} f(\mathbf{x})$$

Thus the transformation P_{SR} arising from the product SR is the product of the transformation P_S and P_R applied in the proper order.

The group of the Schrödinger equation. Now let us consider that special group of operators P_R which commute with the Hamiltonian operator H for any given problem. These will be the operators arising from transformations which leave the Hamiltonian invariant. For example, if the potential energy in an atom depends only on the distance from the nucleus and is independent of the angular coordinates, then any rotation or reflection leaves the potential unchanged. To illustrate precisely what we mean by this, consider a simple Coulomb potential $V = -e^2/r = -e^2/(x^2 + y^2 + z^2)^{\frac{1}{2}}$. Now, under the example transformation treated above, this becomes $V = -e^2/[x^2 + (-z)^2 + y^2]^{\frac{1}{2}}$, which is in fact the same as the expression with which we began. A similar argument could be applied to the kinetic energy operator $-(\hbar^2/2m)(\partial^2/\partial x^2 + \partial^2/\partial y^2 + \partial^2/\partial z^2)$.   Clearly, if an operator leaves H invariant, it is immaterial whether it appears to the left or right of it, that is, $P_R H\psi = HP_R\psi$, and the two operators commute.   The set of all such operators which commute with H are said to form *the group of the Schrödinger equation*.   They certainly do form a group because inverse coordinate transformations exist and because the product of two operators which leave H invariant will also leave it invariant since the product simply indicates that the two operate in succession.   In the alternate language, the product of two operators which commute with H will also commute with H.

If we apply one of these commuting transformation operators to the Schrödinger equation, we have

$$P_R H\psi_n = P_R E_n\psi_n$$

or

$$HP_R\psi_n = E_n P_R\psi_n$$

since P_R commutes with H and of course with the eigenvalue E_n.   From this result we conclude that any function $P_R\psi_n$ obtained by operating on an eigenfunction ψ_n by a symmetry operator from the group of the Schrödinger equation will also be an eigenfunction having the same energy as the original one.   Thus, given any eigenfunction, we can generate other eigenfunctions degenerate with it by application of all the symmetry operators which commute with H.   If this procedure yields *all* the degenerate functions, the degeneracy is said to be *normal*.   For example, given one atomic p function, we can generate the other two degenerate with it by making rotations of coordinates, which commute with the atomic Hamiltonian because of its spherical symmetry.   (Depending on the particular choice of p functions, it may be necessary to use linear combinations of rotated functions, but this is a trivial extension.)   Any degenerate functions which cannot be obtained in this way are said to comprise an *accidental degeneracy*, meaning one with no obvious origin in symmetry.   A classical example is the degeneracy in the hydrogen atom of states of different angular momentum l but the same principal quantum number n, for example, of $2s$ and $2p$ functions.   Deeper study usually shows either that the degeneracy is not exact or else that a hidden symmetry in the Hamiltonian can be found which "explains" the degeneracy.   In the example of hydrogen, Fock[1] has shown that the degeneracy can be considered to arise from a four-dimensional rotational symmetry of the Hamiltonian in momentum space.

Representations.   Let us assume that the eigenvalue E_n is l_n-fold degenerate (excluding any accidental degeneracies).   Then we may choose a set of l_n orthonormal eigenfunctions belonging to E_n.   By our result above, operation with any commuting P_R on any one of the l_n functions produces

[1] V. Fock, *Z. Physik*, **98**, 145 (1935).

another function having the same energy, which accordingly must be expressible as a linear combination of this complete set of degenerate functions. In the language of algebra, the l_n degenerate functions form basis vectors in an l_n-dimensional vector space. This space is a subspace of the entire Hilbert space of eigenfunctions of H, a subspace invariant under all the operations of the group of the Schrödinger equation. Thus, the effect of each of these transformation operators on any function in this subspace can be represented by a matrix which can be worked out by considering the effect of the operator on each of the basis functions in turn. These matrices Γ are defined formally by

$$P_R \psi_v^{(n)} = \sum_{\kappa=1}^{l_n} \psi_\kappa^{(n)} \Gamma^{(n)}(R)_{\kappa v} \qquad (3\text{-}23)$$

where the sum runs over the l_n degenerate eigenfunctions $\psi_\kappa^{(n)}$ having the same energy E_n as $\psi_v^{(n)}$. These l_n-dimensional matrices then form an l_n-dimensional representation of the group of the Schrödinger equation. Such a representation can be based on each set of degenerate eigenfunctions. These representations are irreducible since (excluding accidental degeneracies) there is always an operator in the group which transforms each function into any other degenerate with it. Thus no smaller matrices could express the most general transformation. To prove that the matrices defined in (3-23) actually form a representation of the group, we consider two successive operations. For simplicity we suppress the index n, which denotes the particular representation.

$$P_{SR}\psi_v = P_S P_R \psi_v = P_S \sum_\kappa \psi_\kappa \Gamma(R)_{\kappa v}$$

$$= \sum_\kappa (P_S \psi_\kappa) \Gamma(R)_{\kappa v} = \sum_{\kappa,\lambda} \psi_\lambda \Gamma(S)_{\lambda\kappa} \Gamma(R)_{\kappa v}$$

$$= \sum_\lambda \psi_\lambda [\Gamma(S)\Gamma(R)]_{\lambda v}$$

But by definition of $\Gamma(SR)$,

$$P_{SR}\psi_v = \sum_\lambda \psi_\lambda \Gamma(SR)_{\lambda v}$$

Therefore $$\Gamma(SR) = \Gamma(S)\Gamma(R)$$

and the matrices do in fact form a representation of the group. Thus we conclude that *the set of l_n degenerate eigenfunctions $\psi_v^{(n)}$ of energy E_n form basis functions for an l_n-dimensional irreducible representation $\Gamma^{(n)}$ of the group of the Schrödinger equation.* One can readily show that the representation is *unitary* if one chooses an orthonormal set of basis functions.

   PROOF: If we denote the Hermitian scalar product of ψ and φ by $(\psi, \varphi) = \int \psi^* \varphi \, d\tau$, and if we assume the basis functions ψ_κ to be orthonormal, then

$$\delta_{\kappa v} = (\psi_\kappa, \psi_v) = (P_R \psi_\kappa, P_R \psi_v)$$

since in the second form the integral may be considered simply to be carried out in a rotated coordinate system. Using the definition of the representation matrices, we find

$$\delta_{\kappa\nu} = \left(\sum_{\lambda} \psi_{\lambda} \Gamma(R)_{\lambda\kappa}, \sum_{\mu} \psi_{\mu} \Gamma(R)_{\mu\nu} \right)$$

$$= \sum_{\lambda,\mu} (\psi_{\lambda}, \psi_{\mu}) \Gamma(R)_{\lambda\kappa}^{*} \Gamma(R)_{\mu\nu}$$

$$= \sum_{\lambda} \Gamma(R)_{\lambda\kappa}^{*} \Gamma(R)_{\lambda\nu} = \sum_{\lambda} \Gamma(R)_{\kappa\lambda}^{\dagger} \Gamma(R)_{\lambda\nu}$$

$$= [\mathbf{\Gamma}(R)^{\dagger} \mathbf{\Gamma}(R)]_{\kappa\nu}$$

Thus $\mathbf{\Gamma}(R)^{\dagger} \mathbf{\Gamma}(R) = \mathbf{E}$, and the matrix representation is unitary.

Group theory and quantum numbers. Finally, let us consider the effect of choosing a different set of linearly independent basis functions ψ_{μ}' which are linear combinations of the first set. That is,

$$\psi_{\mu}' = \sum_{\nu=1}^{l} \psi_{\nu} \alpha_{\nu\mu} \quad \text{or} \quad \psi_{\kappa} = \sum_{\lambda=1}^{l} \psi_{\lambda}' \alpha^{-1}{}_{\lambda\kappa}$$

Then
$$P_R \psi_{\mu}' = P_R \sum_{\nu} \psi_{\nu} \alpha_{\nu\mu} = \sum_{\kappa,\nu} \psi_{\kappa} \Gamma(R)_{\kappa\nu} \alpha_{\nu\mu}$$

$$= \sum_{\lambda,\kappa,\nu} \psi_{\lambda}' \alpha^{-1}{}_{\lambda\kappa} \Gamma(R)_{\kappa\nu} \alpha_{\nu\mu} = \sum_{\lambda} \psi_{\lambda}' [\alpha^{-1} \mathbf{\Gamma}(R) \alpha]_{\lambda\mu}$$

$$= \sum_{\lambda} \psi_{\lambda}' \Gamma'(R)_{\lambda\mu}$$

where $\mathbf{\Gamma}'$ is the new representation matrix. Thus we see that

$$\mathbf{\Gamma}'(R) = \alpha^{-1} \mathbf{\Gamma}(R) \alpha \tag{3-24}$$

and the different choice of basis functions merely produces a representation *equivalent* to the old one. Thus, within a similarity transformation, *there is a unique representation of the group of the Schrödinger equation corresponding to each eigenvalue of the Hamiltonian.* A set of eigenfunctions can always be classified uniquely according to the irreducible representation to which it belongs, i.e., the one for which the eigenfunctions form a set of basis vectors.

If a particular choice of matrices (within the range allowed by the similarity transformation) is made, then a function may be characterized even more precisely by giving its row index within the representation. In this way group theory provides "good quantum numbers" for any problem in the form of the labels of the representations and the rows within each one. The associated degeneracy is simply the dimensionality of the representation. Thus, by finding the dimensionalities of all the irreducible representations of the group of the Schrödinger equation (as described in Sec. 3-4), we are able to determine unequivocally the degrees of (nonaccidental) degeneracy possible in any problem. From this observation it follows that a perturbation can lift degeneracies if, and only if, its inclusion in the Hamiltonian reduces the symmetry group and hence changes the possible irreducible representations. Moreover, if representation matrices are worked out, they contain the transformation properties of all eigenfunctions under all the symmetry operators of the group.

   An example. As an example using our standard group of order 6, imagine that we are interested in finding the eigenstates of an electron moving in the potential field of three protons located at the corners of an equilateral triangle. In this case the group of the Schrödinger equation contains all six rotational operations considered in this example group. (It also contains six more operations involving a reflection in the plane, but as we shall see later these extra operations do not affect the degeneracies.) This group has three irreducible representations of dimensionality 1, 1, and 2. Thus only nondegenerate and doubly degenerate states are possible. All higher degeneracy is excluded by group theory alone. Those eigenfunctions belonging to the identical representation $\Gamma^{(1)}$ are invariant under all the group operations. Similarly, basis functions of $\Gamma^{(2)}$ are invariant under E, D, and F, but they change sign under the 180° rotations A, B, and C. This can be seen from the character table, since for a one-dimensional representation $\chi(R) = \Gamma(R)$. The doubly degenerate eigenfunctions of $\Gamma^{(3)}$ symmetry can be chosen so as to transform between themselves in accordance with the 2×2 matrix representation. For example, an eigenfunction ψ_1 belonging to the first row of $\Gamma^{(3)}$ will transform into $-\frac{1}{2}\psi_1 + (\sqrt{3}/2)\psi_2$ under P_B. Of course, a different choice of the degenerate eigenfunctions would transform according to a set of matrices related to those given above by a similarity transformation, which will also be unitary if the new linear combinations are chosen to preserve orthonormality.

3-7 Illustrative Representations of Abelian Groups

In an Abelian group, each element forms a class by itself. Therefore the number of classes, and hence the number of irreducible representations, equals the number of group elements h. But in this case, (3-12) requires that

$$\sum_{i=1}^{h} l_i^2 = h$$

This is possible only by choosing $l_1 = l_2 = \cdots = l_h = 1$. Thus *an Abelian group of order h has h one-dimensional representations and no others.* Each of these representations is simply a set of complex numbers, one number being associated with each group element. Note that the absence of any larger irreducible representations implies that there are no degeneracies if the symmetry group of the Hamiltonian is Abelian.

Cyclic groups. Cyclic groups are Abelian groups with elements $A_1 = A$, $A_2 = A^2, \ldots, A^h = E$. Let the number representing A itself in some representation of this group be denoted $\Gamma(A) = r$. Then $\Gamma(A_n) = \Gamma(A)^n = r^n$, and in particular $\Gamma(E) = 1$ requires that

$$\Gamma(A)^h = r^h = 1$$

or
$$r = e^{2\pi i p/h} \qquad p = 1, 2, 3, \ldots, h$$

since these h values of r form the hth roots of unity. In this way we have found all h of the irreducible representations. They can be written in the form

$$\Gamma^{(p)}(A) = e^{2\pi i p/h} \qquad p = 1, 2, 3, \ldots, h \qquad \blacktriangleright \quad (3\text{-}25)$$

Bloch's theorem. The cyclic group of order h is the symmetry group of the Hamiltonian for a periodic potential with h periods in a ring or in a linear arrangement with periodic boundary conditions applied at walls h periods apart. In this application the group element A represents a displacement through one period. For definiteness, let A represent a displacement by a in coordinates (or by $-a$ in the contours of the function) so that $P_A \psi(x) = \psi(x + a)$. By our general theorem, all eigenfunctions of a Hamiltonian having this symmetry must transform according to some representation of the group. For example, all solutions from the pth representation must have the property

$$\psi_p(x + a) = P_A\psi_p(x) = \Gamma^{(p)}(A)\psi_p(x) = e^{2\pi i p/h}\psi_p(x)$$

Upon introducing the total length $L = ah$, this can be written as

$$\psi_p(x + a) = e^{2\pi i p a/L}\psi_p(x) = e^{ika}\psi_p(x)$$

where k is related to p by $k = 2\pi p/L$. Relabeling the function with the equivalent index k, we have

$$\psi_k(x + a) = e^{ika}\psi_k(x) \qquad (3\text{-}26)$$

This equation gives the transformation property imposed by the translational-symmetry group. As is easily seen, any function $\psi_k(x)$ satisfying (3-26) can be written in the form

$$\psi_k(x) = u_k(x)e^{ikx} \qquad (3\text{-}27)$$

where $u_k(x)$ is periodic with the period a. This result is the celebrated Bloch theorem of solid-state physics. Clearly it is based purely on symmetry through the machinery of group theory. Therefore it is a rigorous result, free of special approximations, and k defined in this way is a "good quantum number" for characterizing eigenfunctions for a periodic potential.

Two-dimensional rotation group. Any group composed of rotations about a fixed axis is clearly an Abelian group. If all angles of rotation φ are allowed, the group is of infinite order. However, it is so simple that we can handle it anyway. Being Abelian, its representations are just numbers. The group-multiplication property for successive rotations implies that the representations satisfy

$$\Gamma(\varphi_1)\Gamma(\varphi_2) = \Gamma(\varphi_1 + \varphi_2)$$

This can be satisfied only by an exponential relation of the sort

$$\Gamma^{(m)}(\pm\varphi) = e^{im\varphi} \tag{3-28}$$

the plus sign referring to a rotation of axes and the negative sign to a rotation of the contours of the function. We choose the latter convention in discussing rotations. The representation index m is restricted to the values $m = 0, \pm 1, \pm 2, \ldots$ by the requirement that $\Gamma^{(m)}(2\pi) = \Gamma^{(m)}(E) = 1$. It is readily verified that any function satisfying

$$\psi_m(r, \theta, \varphi - \varphi_0) = P_{\varphi_0}\psi_m(r, \theta, \varphi)$$
$$= \Gamma^{(m)}(\varphi_0)\psi_m(r, \theta, \varphi) = e^{-im\varphi_0}\psi_m(r, \theta, \varphi)$$

depends on φ only through a factor $e^{im\varphi}$. Hence any eigenfunction for a Hamiltonian whose symmetry group is the group of all rotations about an axis must have the form

$$\psi_m(r, \theta, \varphi) = f(r, \theta)e^{im\varphi} \qquad m = 0, \pm 1, \pm 2, \ldots$$

This result is a familiar consequence of the fact that angular momentum along an axis of symmetry is conserved and hence has a good quantum number m associated with it.

3-8 Basis Functions for Irreducible Representations

The foregoing simple examples have given a small indication of the way in which irreducible-representation labels serve as good quantum numbers in physical applications. We now wish to develop methods for dealing with the representations of dimensionality greater than 1, which arise when the symmetry group contains noncommuting elements leading to the possibility of degeneracy. In this case we need *two* labels for a basis function, one for the irreducible representation and one for the row (or column) within the representation. Naturally the second label retains a definite significance only as long as we confine ourselves to a particular choice of representation matrices from among all the equivalent sets related by a similarity transformation.

Let a basis function belonging to the κth row of the jth irreducible representation be denoted $\varphi_\kappa^{(j)}$. The other functions $\varphi_\lambda^{(j)}$ required to complete the basis for the representation are called the *partners* of the given function. Then by definition the result of operating with any element of the group on $\varphi_\kappa^{(j)}$ can be expressed as a linear combination of $\varphi_\kappa^{(j)}$ and its partners as follows,

$$P_R \varphi_\kappa^{(j)} = \sum_{\lambda=1}^{l_j} \varphi_\lambda^{(j)} \Gamma^{(j)}(R)_{\lambda\kappa} \qquad (3\text{-}29)$$

where l_j is the dimensionality of the representation. Now, if we multiply through by $\Gamma^{(i)}(R)_{\lambda'\kappa'}^*$, sum over R, and use the great orthogonality theorem (3-8), we obtain

$$\sum_R \Gamma^{(i)}(R)_{\lambda'\kappa'}^* P_R \varphi_\kappa^{(j)} = \frac{h}{l_j} \delta_{ij} \delta_{\kappa\kappa'} \varphi_{\lambda'}^{(j)} \qquad (3\text{-}30)$$

From this equation we conclude that application of the operator

$$\mathscr{P}_{\lambda\kappa}^{(j)} = \frac{l_j}{h} \sum_R \Gamma^{(j)}(R)_{\lambda\kappa}^* P_R \qquad (3\text{-}31)$$

to a basis function has the property of yielding zero unless the function being operated on belongs to the κth row of $\Gamma^{(j)}$. Moreover, we see that, if this condition *is* satisfied, then the result of the operation is $\varphi_\lambda^{(j)}$. This gives us a prescription for generating all the partners of any given basis function. Also, if we set $\lambda = \kappa$, we obtain

$$\mathscr{P}_{\kappa\kappa}^{(j)} \varphi_\kappa^{(j)} = \varphi_\kappa^{(j)} \qquad (3\text{-}32)$$

In other words, $\varphi_\kappa^{(j)}$ is an eigenfunction of $\mathscr{P}_{\kappa\kappa}^{(j)}$ with eigenvalue unity. This property serves to identify uniquely the labels of any basis function. Note that, since $\mathscr{P}_{\kappa\kappa}^{(j)}$ is a linear operator, any linear combination of functions belonging to the κth row of $\Gamma^{(j)}$ (but coming from different choices of basis functions) such as $a\varphi_\kappa^{(j)} + b\psi_\kappa^{(j)}$ will also belong to that row and representation.

THEOREM: If $\Gamma^{(1)}, \Gamma^{(2)}, \ldots, \Gamma^{(c)}$ are all of the distinct irreducible representations of a group of operators P_R, then any function F in the space operated on by P_R can be decomposed into a sum of the form

$$F = \sum_{j=1}^{c} \sum_{\kappa=1}^{l_j} f_\kappa^{(j)} \qquad (3\text{-}33)$$

where $f_\kappa^{(j)}$ belongs to the κth row of the jth irreducible representation.

PROOF: Consider the set of all functions $F, F_2', \ldots, F_h'$ formed by operation with the operators $P_E, P_{A_2}, \ldots, P_{A_h}$ on F. Discard all functions which are not linearly independent of the others, and orthogonalize the remainder (e.g., by the Schmidt procedure). Denote the resulting set of n functions by $F, F_2, \ldots, F_n$. These functions form the basis for a unitary representation of the group, since the result of successive operations must always be expressible as a linear combination of the set. This follows since the equation $P_S P_R F = P_{SR} F$ is in the space spanned by the set, by construction.

Call this n-dimensional representation Γ, so that

$$P_R F_k = \sum_{i=1}^{n} F_i \Gamma(R)_{ik}$$

There are now two possibilities: either Γ is one of the irreducible representations, or it is not. If it is, then F has been shown to belong to a particular row of that representation and the theorem is proved. If it is not, then it can be brought to block form in terms of the chosen irreducible representation matrices by a similarity transformation, so that

$$\alpha^{-1}\Gamma(R)\alpha = \begin{pmatrix} \mathbf{\Gamma}^{(i)}(R) & 0 & 0 \\ 0 & \mathbf{\Gamma}^{(j)}(R) & 0 \\ 0 & 0 & \cdots \end{pmatrix} \tag{3-34}$$

The matrix α now defines a set of functions F_k'' which transform according to (3-34) and hence are functions of the type $f_\kappa^{(j)}$ for various values of j and κ. Using the inverse α^{-1}, we may express the F_k, and in particular F, as linear combinations of the F_k'' or $f_\kappa^{(j)}$. This completes the proof.

Having established the validity of (3-33), we may now use the operators $\mathscr{P}_{\kappa\kappa}^{(j)}$ defined in (3-31) to determine the individual terms in the sum in (3-33). We have noted that

$$\mathscr{P}_{\kappa\kappa}^{(j)} f_{\kappa'}^{(j')} = \delta_{\kappa\kappa'} \delta_{jj'} f_\kappa^{(j)} \tag{3-35}$$

Therefore

$$\mathscr{P}_{\kappa\kappa}^{(j)} F = f_\kappa^{(j)} \tag{3-36}$$

and $\mathscr{P}_{\kappa\kappa}^{(j)}$ is a *projection operator* which projects out the part of any function which belongs to the κth row of the jth representation. Such a projection operator is called *idempotent* because of the property that

$$\mathscr{P}_{\kappa\kappa}^{(j)} \mathscr{P}_{\kappa\kappa}^{(j)} = \mathscr{P}_{\kappa\kappa}^{(j)}$$

i.e., all powers of the operator are equal.

We now are able to form a set of basis functions for any representation at will. Starting with an arbitrary function F, we can project out one $f_\kappa^{(j)}$, which after normalization is a suitable basis function $\varphi_\kappa^{(j)}$. Then use of the transfer operators $\mathscr{P}_{\lambda\kappa}^{(j)}$ yields all its partners, since $\mathscr{P}_{\lambda\kappa}^{(j)} \varphi_\kappa^{(j)} = \varphi_\lambda^{(j)}$. Van Vleck has appropriately styled this procedure as the *basis-function generating machine*.

THEOREM: Two functions which belong to different irreducible representations or to different rows of the same unitary representation are orthogonal.

PROOF: Consider the scalar product

$$(\varphi_\kappa^{(j)}, \psi_{\kappa'}^{(j')}) = (P_R \varphi_\kappa^{(j)}, P_R \psi_{\kappa'}^{(j')})$$

$$= \sum_{\lambda, \lambda'} \Gamma^{(j)}(R)_{\lambda\kappa}^* \Gamma^{(j')}(R)_{\lambda'\kappa'} (\varphi_\lambda^{(j)}, \psi_{\lambda'}^{(j')})$$

We may sum the right member over all R and divide by h, since it must be independent of R. Applying the great orthogonality theorem, we obtain

$$(\varphi_\kappa^{(j)}, \psi_{\kappa'}^{(j')}) = \delta_{jj'}\delta_{\kappa\kappa'} \sum_{\lambda=1}^{l_j} (\varphi_\lambda^{(j)}, \psi_\lambda^{(j)})l_j^{-1}$$

Thus, not only have we proved the theorem, but we have also shown that the scalar product $(\varphi_\kappa^{(j)}, \psi_\kappa^{(j)})$ is independent of κ.

The results derived above have required knowledge of the complete representation matrices $\mathbf{\Gamma}^{(j)}(R)$ or at least all the diagonal values $\Gamma^{(j)}(R)_{\kappa\kappa}$. We can get similar but less detailed results with knowledge of only the characters of the representations. To do this, we set $\lambda = \kappa$ in (3-31) and sum over κ. This defines a new projection operator

$$\mathscr{P}^{(j)} = \sum_\kappa \mathscr{P}_{\kappa\kappa}^{(j)} = \frac{l_j}{h} \sum_R \chi^{(j)}(R)^* P_R \tag{3-37}$$

Following arguments similar to those above, we see that any function $f^{(j)}$ expressible as a sum of functions belonging only to rows within the jth representation will satisfy $\mathscr{P}^{(j)}f^{(j)} = f^{(j)}$ and that

$$\mathscr{P}^{(j)}F = f^{(j)} \tag{3-38}$$

That is, from an arbitrary function $\mathscr{P}^{(j)}$ projects out the part belonging to the jth representation. These results are of course unaffected by similarity transformations which scramble the basis functions, and hence the rows, in any given representation.

As a rather trivial example of these results, consider the group consisting of the identity and the reflection operator σ which takes x into $-x$. This group has two classes, hence two one-dimensional irreducible representations. The character table is

	E	σ
$\Gamma^{(1)}$	1	1
$\Gamma^{(2)}$	1	-1

Hence, the projection operator for $\Gamma^{(1)}$ is $\mathscr{P}^{(1)} = (\frac{1}{2})(P_E + P_\sigma)$, and that for $\Gamma^{(2)}$ is $\mathscr{P}^{(2)} = (\frac{1}{2})(P_E - P_\sigma)$. Operating on an arbitrary function $F(x)$, these yield $\mathscr{P}^{(1)}F(x) = (\frac{1}{2})[F(x) + F(-x)]$ and $\mathscr{P}^{(2)}F(x) = (\frac{1}{2})[F(x) - F(-x)]$. Clearly these projected functions are, respectively, even and odd under reflection, as required for them to belong to $\Gamma^{(1)}$ and $\Gamma^{(2)}$, respectively. Our other theorems are illustrated by the facts that any function can be expressed as the sum of odd and even parts constructed as above and that any odd function is orthogonal to any even one.

The techniques developed in this section are useful in setting up proper zero-order symmetry orbitals which belong to a given row of a given representation. We shall later prove that, since the Hamiltonian commutes with the entire group of symmetry operators, then it can have matrix elements only between functions of the same representational classification. This leads to a maximum reduction in the size of the secular equations that must be solved.

3-9   Direct-product Groups

It often occurs that the complete symmetry of the system under consideration can be broken up into two or more types such that the operators of one type commute with those of any other. An important example is that in which the two types of operators operate on entirely different coordinates. For example, in H_2O we can permute the two protons, permute the 10 electrons, and rotate the molecule as a whole. One of each of these three types of operations could be carried out in any order with the same final result. A second example is the separation of orbital and spin operators. A third example is the inversion operator and the group of proper rotations.

Although such cases could be treated with no special attention, we can simplify our work considerably by taking advantage of these properties by introducing the concept of *direct-product groups*. If we have two groups of operators, $\mathscr{G}_a = E, A_2, \ldots, A_{h_a}$ and $\mathscr{G}_b = E, B_2, \ldots, B_{h_b}$ such that all of $\mathscr{G}_a$ commute with all of $\mathscr{G}_b$, then

$$\mathscr{G}_a \times \mathscr{G}_b = E, A_2, \ldots, A_{h_a}, B_2, A_2B_2, \ldots, A_{h_a}B_2, \ldots, A_{h_a}B_{h_b}$$

forms a group of $h_a h_b$ elements, assuming that the only common element in the groups is the identity. To check that this is so, we consider the multiplication of two elements

$$A_k B_l A_{k'} B_{l'} = (A_k A_{k'})(B_l B_{l'})$$

since B_l and $A_{k'}$ commute. But the right member is found in the direct-product group defined above since $\mathscr{G}_a$ and $\mathscr{G}_b$ are separately groups closed under multiplication. Therefore $\mathscr{G}_a \times \mathscr{G}_b$ is also closed, as required.

Representations. It is natural to suppose that the direct-product matrices of the irreducible representations of the component groups might form irreducible representations of the direct-product group. This in fact is the case. As described in Appendix A, the direct product of two matrices $\mathbf{A}$ and $\mathbf{B}$ is a matrix $\mathbf{A} \times \mathbf{B}$ whose elements are all the products of an element of $\mathbf{A}$ and one of $\mathbf{B}$. Each element bears a double set of subscripts. For example, the element obtained from $A_{ij}B_{kl}$ is labeled $(A \times B)_{ik,jl}$, the two initial indices being given before the comma. The fact that these direct-product matrices in fact have the requisite group-multiplication property follows from the commutivity of ordinary matrix multiplication and the direct-product operation. Thus

$$\mathbf{\Gamma}^{(a \times b)}(A_k B_l)\mathbf{\Gamma}^{(a \times b)}(A_{k'} B_{l'}) = [\mathbf{\Gamma}^{(a)}(A_k) \times \mathbf{\Gamma}^{(b)}(B_l)][\mathbf{\Gamma}^{(a)}(A_{k'}) \times \mathbf{\Gamma}^{(b)}(B_{l'})]$$
$$= [\mathbf{\Gamma}^{(a)}(A_k)\mathbf{\Gamma}^{(a)}(A_{k'})] \times [\mathbf{\Gamma}^{(b)}(B_l)\mathbf{\Gamma}^{(b)}(B_{l'})]$$
$$= \mathbf{\Gamma}^{(a)}(A_k A_{k'}) \times \mathbf{\Gamma}^{(b)}(B_l B_{l'})$$
$$= \mathbf{\Gamma}^{(a \times b)}(A_k A_{k'} B_l B_{l'})$$

By an application of Schur's lemma it can be shown[1] that the direct product of two *irreducible* representations forms an *irreducible* representation of the direct product group. We now show that we get *all* the irreducible representations of the direct-product group in this way. Let $l_1{}^a, l_2{}^a, \ldots$ and $l_1{}^b, l_2{}^b, \ldots$ be the dimensionalities of the irreducible representations of $\mathscr{G}_a$ and $\mathscr{G}_b$. By our dimensionality theorem (3-12),

$$\sum_i (l_i{}^a)^2 = h_a, \quad \text{and} \quad \sum_j (l_j{}^b)^2 = h_b.$$

By the definition of the direct-product matrices, their dimensionalities will be $l_{ij} = l_i{}^a l_j{}^b$. Then applying (3-12) to the product group

$$\sum_{i,j} l_{ij}{}^2 = \sum_{i,j} (l_i{}^a l_j{}^b)^2 = \sum_i (l_i{}^a)^2 \sum_j (l_j{}^b)^2$$
$$= h_a h_b = h$$

where h is the order of the direct-product group. Since $\sum_{i,j} l_{ij}{}^2 = h$, there can be no other irreducible representations than those expressible as direct products. Note that these direct-product representations carry a double set of representation and row labels. This illustrates how extra quantum numbers arise when there are extra independent degrees of freedom characterized by extra commuting symmetry operators.

Class structure and characters. The class structure of the product group is easily obtained from knowledge of the class structure of the component groups since elements from $\mathscr{G}_a$ commute with those of $\mathscr{G}_b$. Therefore, the number of classes is simply the product of the numbers of classes in $\mathscr{G}_a$ and in $\mathscr{G}_b$, in agreement with the number of irreducible representations in the product group.

An important observation is that the character of any direct-product representation is the product of the characters of the component representations. This is proved by simple inspection of the character,

$$\chi^{(a \times b)}(A_k B_l) = \sum_{i,j} \Gamma^{(a \times b)}(A_k B_l)_{ij,ij}$$
$$= \sum_{i,j} \Gamma^{(a)}(A_k)_{ii} \Gamma^{(b)}(B_l)_{jj} = [\sum_i \Gamma^{(a)}(A_k)_{ii}][\sum_j \Gamma^{(b)}(B_l)_{jj}]$$
$$= \chi^{(a)}(A_k)\chi^{(b)}(B_l)$$

[1] Wigner, *op. cit.*, chap. 16.

This allows us to write out the character table of any group which can be expressed as a direct product with knowledge of only the character tables of the smaller groups from which it is composed.   Naturally this provides a great practical simplification.

Example.   Let us consider the direct-product group composed of the rotation group of the equilateral triangle (called D_3 in standard notation) and the group $\mathscr{S} = (E, \sigma_h)$.   In the latter group, σ_h is the operation of reflection in the plane of the triangle.   In standard notation, the direct-product group is called D_{3h}.   It is the symmetry group for an equilateral triangle of finite thickness, so that we have an additional six inequivalent positions in which the "numbers on the triangle" have been reflected through onto the other surface.   There are now 12 group elements, the original 6 from D_3, each multiplied both by the identity and by σ_h.   This is what we mean by the notation $D_{3h} = D_3 \times \mathscr{S}$.   The elements of D_3 and $\mathscr{S}$ commute because it makes no difference whether the triangle is first rotated and then subjected to reflection in the plane, or vice versa.

The multiplication table of $\mathscr{S}$ is

	E	σ_h
E	E	σ_h
σ_h	σ_h	E

The group is Abelian; so there are two classes and two one-dimensional irreducible representations.   Upon denoting the two representations Γ^+ and Γ^- for even and odd, the character table is simply

$\mathscr{S}$	E	σ_h
Γ^+	1	1
Γ^-	1	-1

Multiplying this table by that for D_3 given in Sec. 3-3, we obtain the character table for the product group D_{3h}.

D_{3h}	E	(A, B, C)	(D, F)	σ_h	$\sigma_h(A, B, C)$	$\sigma_h(D, F)$
$\Gamma^{(1+)}$	1	1	1	1	1	1
$\Gamma^{(2+)}$	1	-1	1	1	-1	1
$\Gamma^{(3+)}$	2	0	-1	2	0	-1
$\Gamma^{(1-)}$	1	1	1	-1	-1	-1
$\Gamma^{(2-)}$	1	-1	1	-1	1	-1
$\Gamma^{(3-)}$	2	0	-1	-2	0	1

For clarity, we have labeled the columns by giving the actual group elements in the class rather than using arbitrary class labels $\mathscr{C}_1, \ldots, \mathscr{C}_6$. Now that the table is worked out, we could dispense completely with the observation that the group D_{3h} may be viewed as a direct product and simply treat it as an ordinary group of order 12. However, the distinction is very worthwhile in reducing the effort required to work out the table. We note in passing that going from D_3 to D_{3h} has not changed the degeneracies (one or two) which are possible in view of the dimensionalities of the irreducible representations. This verifies the assertion made in the example treated at the end of Sec. 3-6.

3-10 Direct-product Representations within a Group

A distinct, although related, example of the use of the direct product is in the formation of a new representation Γ of a given group from two representations $\Gamma^{(1)}$ and $\Gamma^{(2)}$ of the same group. This is done by using as basis functions all possible products of the basis functions from the two initial representations. Let the two bases be

$$\varphi_1, \ldots, \varphi_n \quad \text{and} \quad \psi_1, \ldots, \psi_m$$

where n and m are the dimensionalities of $\Gamma^{(1)}$ and $\Gamma^{(2)}$, respectively. Then the functions $\varphi_\kappa \psi_\lambda$ form a basis for an nm-dimensional representation Γ. This may be verified by considering the effect of a transformation operator,

$$P_R(\varphi_\kappa \psi_\lambda) = \sum_{\kappa'} \varphi_{\kappa'} \Gamma^{(1)}(R)_{\kappa'\kappa} \sum_{\lambda'} \psi_{\lambda'} \Gamma^{(2)}(R)_{\lambda'\lambda}$$

$$= \sum_{\kappa',\lambda'} \varphi_{\kappa'} \psi_{\lambda'} [\Gamma^{(1)}(R)_{\kappa'\kappa} \Gamma^{(2)}(R)_{\lambda'\lambda}]$$

$$= \sum_{\kappa',\lambda'} \varphi_{\kappa'} \psi_{\lambda'} \Gamma(R)_{\kappa'\lambda',\kappa\lambda}$$

where $\mathbf{\Gamma}(R) = \mathbf{\Gamma}^{(1)}(R) \times \mathbf{\Gamma}^{(2)}(R)$. Thus the direct-product matrices $\mathbf{\Gamma}(R)$ form a representation of the group based on the functions $\varphi_\kappa \psi_\lambda$.

As in the previous section, the character is obtained by taking the product of the characters of the component representations. That is,

$$\chi(R) = \chi^{(1)}(R)\chi^{(2)}(R)$$

where χ is the character for the representation Γ.

An important difference between these direct-product representations and those treated in the previous section is that in the present case the representations are in general reducible. This is obviously true, in view of the limited total number of distinct *irreducible* representations for any given group. If one then forms a new nm-dimensional representation, it must in general be reducible to a sum of irreducible representations. We express this schematically by writing

$$\Gamma^{(i)} \times \Gamma^{(j)} = \sum_k a_{ijk} \Gamma^{(k)} \tag{3-39}$$

As before, this notation means that, if $\Gamma^{(i)}(R) \times \Gamma^{(j)}(R)$ is brought to block form, $\Gamma^{(k)}(R)$ appears a_{ijk} times along the diagonal. We may find the coefficients a_{ijk} by using our standard formula (3-20) for the decomposition of a reducible representation. Since

$$\chi(R) = \chi^{(i)}(R)\chi^{(j)}(R) = \sum_k a_{ijk}\chi^{(k)}(R)$$

$$a_{ijk} = h^{-1} \sum_R \chi^{(i)}(R)\chi^{(j)}(R)\chi^{(k)}(R)^*$$

$$= h^{-1} \sum_l N_l \chi^{(i)}(\mathscr{C}_l)\chi^{(j)}(\mathscr{C}_l)\chi^{(k)}(\mathscr{C}_l)^* \qquad \blacktriangleright \tag{3-40}$$

Note that if, as is usual, the $\chi^{(k)}(\mathscr{C}_l)$ are real, then a_{ijk} is independent of the order of the indices.

This technique of decomposing a direct-product representation will prove of great usefulness in determining selection rules and in other applications.

EXERCISES

3-1 (a) How many inequivalent irreducible representations of D_4, the symmetry group of the square considered in Exercise 2-1, exist? What are their dimensionalities?

(b) Work out the character table of this group by inspection, using the rules given in Sec. 3-4. Verify the formula based on the c_{ijk} in several instances.

3-2 Show in general that $\sum_R \chi^{(j)}(R) = 0$ for all representations except the identical representation $\Gamma^{(1)}$, in which all elements are represented by 1. What is the value of the sum for $\Gamma^{(1)}$?

3-3 Write out the matrices of the regular representation of the group D_4 for the elements E, F, G, H. Using these matrices, verify by direct matrix multiplication that $FG = H$.

3-4 Using the regular representation, prove that

$$\sum_j l_j \chi^{(j)}(R) = h$$

if $R = E$ and is zero otherwise. In this l_j is the dimensionality of the jth representation. This result is an additional aid in working out the character tables.

3-5 Work out the orthogonal transformation matrices $R_x(\theta_x)$, $R_y(\theta_y)$, and $R_z(\theta_z)$ for rotations of axes by angles θ_x, θ_y, θ_z about the x, y, z axes, respectively. Take the sense of rotation so that the y axis is rotated toward x in R_z, etc. Using them, compute the matrices $R_x(\theta_x)R_y(\theta_y)$ and $R_y(\theta_y)R_x(\theta_x)$. Note that these rotations do not commute. What is the matrix for $R_xR_y - R_yR_x$? Show that this matrix vanishes as θ^2 if $\theta_x = \theta_y = \theta$ is much less than 1 and corresponds to the *change* $(R_z - E)$ produced by a rotation through θ^2 about the z axis.

3-6 (a) Find the transformed functions $P_R f$, $P_S f$, $P_S(P_R f)$, $P_{SR} f$, and $P_{RS} f$, where $f = x$, $R = R_y(\theta_y)$, and $S = R_z(\theta_z)$. Note that $\mathbf{R}^{-1} = \mathbf{R}^\dagger = \tilde{\mathbf{R}}$. Generalizing from your results, you may observe that the three p-like functions x, y, z form a basis for a three-dimensional representation of the complete rotation group since

$$P_R p_i = \sum_j p_j \Gamma(R)_{ji}$$

where $i, j = x, y, z$ and R represents any arbitrary rotation. This property is true of the $(2l + 1)$ polynomials equivalent to the spherical harmonics corresponding to any l and of the spherical harmonics themselves.

(b) Repeat (a) for $f = xy$. Find a convenient set of partners to xy in a basis for representing an arbitrary rotation.

3-7 (a) Form a representation of D_3, the rotational symmetry group of the equilateral triangle, which has $F = x^2 z g(r)$ as part of its basis. The z axis is the three-fold axis, and $g(r)$ is such as to assure radial convergence in the normalization integral. What are the other basis functions?

(b) If this representation is reducible, into what irreducible representations does it reduce?

(c) If your basis functions are not orthonormal, choose a new set of linear combinations of them that are. (You need orthonormalize only the angular part.) By applying the corresponding similarity transformation to the matrices of your representation, transform it into a unitary representation. If you have chosen your orthonormal functions wisely, the matrices will now be in block form. If not, reduce them to block form by a suitable unitary transformation to a new orthonormal basis.

(d) Express $F = x^2 z g(r)$ explicitly as a sum of parts $\sum_{j,\kappa} f_\kappa^{(j)}$, each of which transforms according to a particular row of one of the irreducible representations.

3-8 Write out the character table of D_{4h}, taking advantage of the fact that $D_{4h} = D_4 \times i$, where i is the group containing only the inversion and the identity.

3-9 (a) Find unitary matrices for all the irreducible representations of D_4. (HINT: Consider the transformation properties of the coordinates x and y in the plane of the square.)

(b) How could you now obtain matrices for all the irreducible representations of D_{4h}?

3-10 Work out all direct products $\Gamma^{(i)} \times \Gamma^{(j)} = \sum_k a_{ijk} \Gamma^{(k)}$ of the irreducible representations of the group D_3. Note that a_{ijk} is independent of the subscript order since all the characters involved are real.

3-11 Repeat Exercise 3-10 for the groups D_4 and O, O being the octahedral group discussed in Sec. 4-6, where a character table is given.

REFERENCES

EYRING, H., J. WALTER, and G. E. KIMBALL: "Quantum Chemistry," chap. 10, John Wiley & Sons, Inc., New York, 1944.

HAMERMESH, M.: "Group Theory," Addison-Wesley Publishing Company, Inc., Reading, Mass., 1962.

HEINE, V.: "Group Theory in Quantum Mechanics," Pergamon Press, New York, 1960.

MURNAGHAN, F. D.: "The Theory of Group Representations," chaps. 1–3, The Johns Hopkins Press, Baltimore, 1938.

SPEISER, A.: "Die Theorie der Gruppen von Endlicher Ordnung," chaps. 11–13, 3d ed., Springer-Verlag OHG, Berlin, 1937; reprinted by Dover Publications, Inc., New York, 1945.

VAN DER WAERDEN, B. L.: "Die Gruppentheoretische Methode in der Quanten-mechanik," Springer-Verlag OHG, Berlin, 1932.

WEYL, H.: "Gruppentheorie und Quantenmechanik," S. Hirzel Verlag, Leipzig, 1928; translated version, "Theory of Groups and Quantum Mechanics," Dover Publications, Inc., New York, 1950.

WIGNER, E.: "Gruppentheorie und ihre Anwendung auf die Quantenmechanik der Atomspektren," chaps. 9, 11, 12, and 16, Friedr. Vieweg & Sohn, Brunswick, Germany, 1931; reprinted by J. W. Edwards, Publisher, Incorporated, Ann Arbor, Mich., 1944; revised and translated edition, "Group Theory," Academic Press Inc., New York, 1959.

4 : PHYSICAL APPLICATIONS OF GROUP THEORY

In this chapter we shall give a number of illustrative applications of group theory to physical problems. We start by discussing the crystallographic point groups, which play a central role in the theory of molecules and crystalline solids. Then, after a brief introduction to the full rotation group, which is the symmetry group of an isolated atom, we consider the lifting of the orientational degeneracy of the free atom when it is placed into a crystalline environment. This topic, first treated by Bethe in a classic paper, serves as an excellent illustration of the use of the theory of group representations and especially of group characters in physical problems. It also serves as a foundation for the theory of energy bands in solids, which we shall treat in a later chapter. With this crystal-field-splitting example completed, we continue by proving some group-theoretical theorems about matrix elements which determine selection rules. We conclude the chapter with an illustration of the use of group theory in the theory of directed-valence orbitals of specified symmetry.

4-1 Crystal-symmetry Operators

A crystalline solid is a regular array of identical unit cells such that (with edge effects neglected) the crystal is invariant under lattice translations by $\mathbf{T} = n_1\mathbf{a}_1 + n_2\mathbf{a}_2 + n_3\mathbf{a}_3$, where n_1, n_2, and n_3 are integers and $\mathbf{a}_1$, $\mathbf{a}_2$, and $\mathbf{a}_3$ are the primitive translations which define a unit cell. These translations are, however, in general not the only covering operations of the crystal. There are also covering operations, such as rotations, which can be carried out by holding a point (called the origin) fixed. The complete set of covering operations is called the *space group* of the crystal. The group of operations which is obtained by setting all translations in the space-group elements equal to zero is called the *point group* of the crystal. It has been shown[1] that there are only 32 point groups consistent with translational symmetry and that there is a total of 230 possible space groups in all. In a molecule there is no requirement of translational symmetry, and so there is no limit on the number of possible point groups. In practice, though, the most frequently met molecular point groups are among the 32 crystallographic ones. Hence, if we become familiar with them, we are well prepared for the symmetry of molecular as well as crystalline structures.

The fundamental covering operations of which we build up the point groups are the following:

1. Rotations about axes through the origin
2. Reflections in planes containing the origin
3. The inversion, which takes $\mathbf{r}$ into $-\mathbf{r}$

The inversion operator is actually equivalent to a rotation by π followed by reflection in a plane perpendicular to the axis of rotation. However, it enters so frequently that we list it as one of our fundamental operations. In fact, many workers take the inversion as a fundamental operation and the reflections as derived.

Successive operations. For working out the group multiplication properties of these operations it is worth noting the results of various successive operations which frequently occur.

1. The product of two rotations is again a rotation.
2. The product of two reflections is a rotation by an angle $2\varphi_{AB}$ around the line of intersection of the reflection planes A and B, where φ_{AB} is the angle between the planes. For example, in Fig. 4-1a, the angle φ_{13} between the initial and final positions is $2\alpha + 2\beta$, which is twice the angle $\varphi_{AB} = (\alpha + \beta)$.

[1] F. Seitz, *Z. Krist.*, **88**, 433 (1934); **90**, 289 (1935); **91**, 336 (1935); **94**, 100 (1936).

3. The product of a rotation and a reflection in a plane A containing the axis of rotation O is a reflection in another plane B passing through the axis. The angle between the planes is half the angle of rotation, as may be seen in Fig. 4-1b.

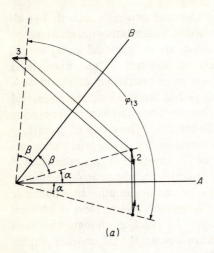

(a)

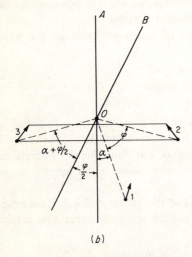

(b)

Fig. 4-1. (a) *Diagram illustrating how two reflections produce a rotation;* (b) *diagram illustrating how rotation and reflection produce a reflection.*

4. The product of two rotations by π about intersecting axes u and v is a rotation about an axis perpendicular to u and v, the angle of rotation being twice the angle between u and v. This can be illustrated by a diagram quite similar to Fig. 4-1a.

Commuting operations. Because of the special group-theoretical and quantum-mechanical significance of commuting operations, it is worth noting that the following pairs of operations commute:

1. Two rotations about the same axis.

2. Two reflections in perpendicular planes.

3. Two rotations by π about perpendicular axes.

4. A rotation and a reflection in a plane perpendicular to the axis of rotation. The resulting combination is called an *improper rotation*, or a *rotary reflection*. An improper rotation by π is identical with the inversion operation.

5. The inversion and any rotation or reflection.

Notation. We now list the standard Schoenflies notation for the various types of symmetry operations which can occur in the crystallographic point groups.

E = identity.

C_n = rotation through $2\pi/n$. We shall show that in solids n can take on only the values 1, 2, 3, 4, and 6.

σ = reflection in a plane.

σ_h = reflection in the "horizontal" plane, i.e., the plane through the origin perpendicular to the axis of highest rotation symmetry.

σ_v = reflection in a "vertical" place, i.e., one passing through the axis of highest symmetry.

σ_d = reflection in a "diagonal" plane, i.e., one containing the symmetry axis and bisecting the angle between the twofold axes perpendicular to the symmetry axis. This is just a special kind of σ_v.

S_n = improper rotation through $2\pi/n$.

$i = S_2$ = inversion.

Interrelations of symmetry elements. As a final list of general observations concerning these symmetry elements, we now give some necessary interrelations of symmetry elements. These observations help to reduce the amount of information which is needed to determine the complete symmetry of the problem. The verification of the rules may be carried out by elementary geometrical considerations. We list the following rules:

1a. The intersection of two reflection planes must be a symmetry axis. If the angle φ between the planes is π/n, the axis is n-fold.

1b. If a reflection plane contains an n-fold axis, there must be $n - 1$ other reflection planes at angles of π/n.

2a. Two twofold axes separated by an angle of π/n require a perpendicular n-fold axis.

2b. A twofold axis and an n-fold axis perpendicular to it require $n - 1$ additional twofold axes separated by angles of π/n.

3. An even-fold axis, a reflection plane perpendicular to it, and an inversion center are interdependent. Any two of these elements implies the existence of the third.

4-2   The Crystallographic Point Groups

Having collected a large number of the general properties of the sort of symmetry operations which comprise the point groups, we now proceed to enumerate and describe the 32 point groups which exhaust the possible symmetries of a crystal.   These groups fall into two general categories—the simple rotation groups and the groups of higher symmetry.   In the former, there is one symmetry axis (called Z) of higher symmetry than any other. In the latter, there is no unique axis of highest symmetry, but more than one n-fold axis, where $n > 2$.   While we are at it, we also include a third type of point group, namely, groups containing C_∞.   These groups form the point groups of linear molecules but are inconsistent with the translational symmetry required for crystalline solids.

We begin our enumeration with the simple rotation groups.   These groups are easily visualized by means of *stereographic projections*.   To form these, we project onto the XY plane the position of a mark at a general point on a unit sphere which is subjected to the various symmetry operations of the group.   If the mark is above the plane, we use a $+$; if below, a $\bigcirc$. The rotation axes are indicated by characteristic symbols.   The projections of all 27 crystallographic point groups of the simple rotation type are collected in Fig. 4-2.   We now proceed to consider the various possibilities. We follow the Schoenflies system of notation, in which the groups are labeled in a way which suggests the independent group elements required to determine all the elements of the group.

C_n ■   These are the point groups in which the only symmetry consists of a single n-fold axis of symmetry.   These groups are cyclic Abelian groups of order n.   The group C_6, for example, contains C_6, $(C_6)^2 = C_3$, $(C_6)^3 = C_2$, $(C_6)^4 = (C_3)^{-1}$, $(C_6)^5 = (C_6)^{-1}$, $(C_6)^6 = E$.

We now prove that the only cases consistent with the translational symmetry of a crystalline solid are C_1, C_2, C_3, C_4, and C_6.   The proof is based on the existence of a translation of minimum length which is perpendicular to the rotation axis.   Such perpendicular translational symmetry elements must exist in the space group of any crystal, because one may construct them from *any* translational symmetry element

$$\mathbf{T} = n_1\mathbf{a}_1 + n_2\mathbf{a}_2 + n_3\mathbf{a}_3.$$

This is done by rotating $\mathbf{T}$ about the rotational axis to a new position $\mathbf{T}'$ and then taking $\mathbf{T} - \mathbf{T}'$, which must also be a translational-symmetry element. We then may select the shortest of these perpendicular translations and call it $\mathbf{R}$.   Now applying C_n to $\mathbf{R}$ yields another allowed translation $\mathbf{R}'$ of the same magnitude as $\mathbf{R}$, but rotated through $2\pi/n$.   But $\mathbf{R} - \mathbf{R}'$ must also be an allowed translation, since the translation group is closed under

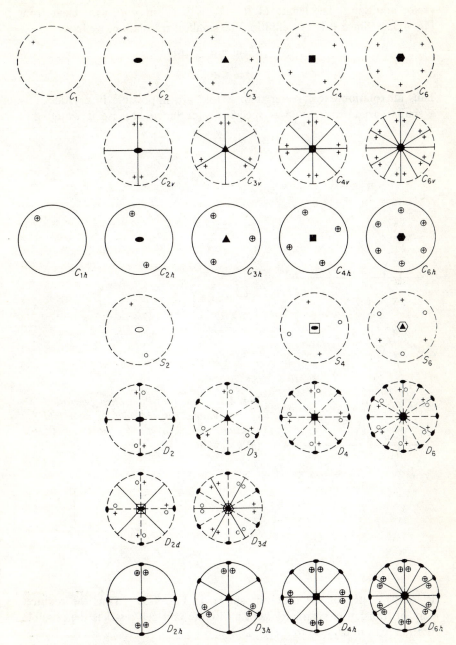

Fig. 4-2. *Stereographic projections of simple point groups.*

vector addition.   The length of $\mathbf{R} - \mathbf{R}'$ is $2\,|R|\sin(\pi/n)$, as is clear from Fig. 4-3a.   Since $\mathbf{R}$ is by definition the smallest translation, we have

$$2\,|R|\sin\pi/n \geqq |R|$$

or
$$n \leqq 6$$

Thus all rotation axes greater than sixfold are impossible in crystals.   By a similar argument, $n = 5$ is specifically excluded because it would give

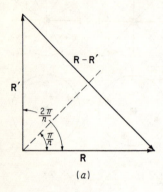

(a)

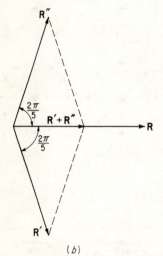

(b)

Fig. 4-3.   *Diagrams illustrating generation of translations perpendicular to an n-fold rotation axis.   (a) $n = 4$, $|\mathbf{R}\text{-}\mathbf{R}'| > |\mathbf{R}|$; (b) $n = 5$, $|\mathbf{R}' + \mathbf{R}''| < |R|$.*

$|\mathbf{R}' + \mathbf{R}''| < |\mathbf{R}|$ for the rotations shown in Fig. 4-3b.   Thus the requirement of translation symmetry with a discrete unit cell restricts the possible rotation axes in crystals to $n = 1, 2, 3, 4, 6$.

C_{nv} ■ These groups contain a σ_v reflection plane in addition to the C_n axis.   By rule 1b at the end of the previous section, this implies the existence of n reflection planes, separated by an angle of π/n around the C_n

axis. Solid radial lines on the stereographic projections indicate the vertical reflection planes.

C_{nh} ■ These groups contain a σ_h reflection as well as the C_n axis. The reflection plane is indicated on the stereographic projection by making the unit circle solid rather than broken. Reflection in this plane takes $+$ into $\bigcirc$. Note that all groups of the sort $C_{2n,h}$ automatically include the inversion element i.

S_n ■ These groups contain an n-fold axis for improper rotations. If n is odd, these groups are identical with C_{nh} and hence they are not considered. If n is even, they form distinct groups, each of which includes $C_{n/2}$ as a subgroup. The cases occurring in crystals are thus S_2, which is equivalent to simple inversion symmetry, S_4, and S_6, which may be considered as the direct-product group $C_3 \times i$. The axes of improper rotations are shown on stereographic projections as hollow figures.

D_n ■ These groups have n twofold axes perpendicular to the principal C_n axis. In D_2, therefore, there are three mutually perpendicular twofold axes. The group D_2 is sometimes called V for the German *Vierergruppe*, mentioned previously in Sec. 2-6.

D_{nd} ■ These groups contain the elements of D_n together with diagonal reflection planes σ_d bisecting the angles between the twofold axes perpendicular to the principal rotation axis.

D_{nh} ■ These groups contain the elements of D_n plus the horizontal reflection plane σ_h. Hence, D_{nh} has twice as many elements as D_n.

In working out character tables and in other applications, it is useful to note that some of the above-mentioned groups are conveniently expressed as direct products of simpler groups with the group of the inversion. These are

$$
\begin{array}{ll}
C_{2h} = C_2 \times i & D_{2h} = D_2 \times i \\
C_{4h} = C_4 \times i & D_{4h} = D_4 \times i \\
C_{6h} = C_6 \times i & D_{6h} = D_6 \times i \\
S_6 = C_3 \times i & D_{3d} = D_3 \times i
\end{array}
\tag{4-1}
$$

We now proceed to consider the groups of higher symmetry. The five groups in this category—T, T_d, T_h, O, O_h—have no unique axis of highest symmetry but rather have more than one axis which is at least of threefold symmetry. In order to make all these rotation axes consistent with translational crystal symmetry, the crystal must belong to the *cubic* crystal system, in which the fundamental translational vectors are mutually perpendicular and of equal length. Accordingly, it will be convenient to consider all these groups in conjunction with a unit cube.

T ■ This is the smallest of the groups of higher symmetry. It consists of the 12 proper rotational operations which take a regular tetrahedron into itself. These operations can be readily visualized if we consider the tetrahedron *abcd* inscribed in a cube as shown in Fig. 4-4. The X, Y, and Z axes are normal to cube faces, and the origin is at the center of the cube. The covering operations of the tetrahedron are then seen to be E; three C_2's about the X, Y, and Z axes, respectively; and eight C_3's about the body diagonals of the cube. The latter actually comprise two distinct classes of four elements each. These classes may be considered to be, respectively, the clockwise and counterclockwise rotations by $2\pi/3$ about the four three-fold axes leaving the tetrahedron normally at the center of the faces. Despite the fact that all these are threefold rotations about face-center axes, these two types of operations form distinct classes because these three-fold axes are "directed." That is, there is no operation in the group (such as a perpendicular twofold axis) which will reverse the line of the axis.

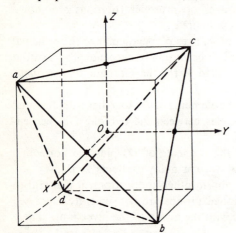

Fig. 4-4. *Tetrahedron inscribed in cube.*

T_d ■ The full tetrahedral group T_d contains all the covering operations of a regular tetrahedron, including reflections. It is the symmetry group of CH_4, for example. Upon applying the theorem proved in Sec. 2-5, the fact that T is a subgroup of T_d implies that the order of T_d must be an integral multiple of 12, the order of T. In fact T_d has 24 elements. Six of the extra elements are diagonal reflection planes normal to a cube face and passing through a tetrahedral edge such as ac or ab. The existence of these six σ_d's implies the existence of six S_4's about the (positive and negative) X, Y, and Z directions. This completes the group. Note that the reflection planes in T_d allow one to interchange clockwise and counterclockwise rotations. Hence, unlike the situation in T, all eight C_3's are in the same class in the group T_d.

T_h ■ We can form another group of order 24 based on T by taking the direct product of T with the group of the inversion operator. The nature of all its elements follows from this description. Note that a regular tetrahedron does *not* have the symmetry T_h since it has no inversion symmetry.

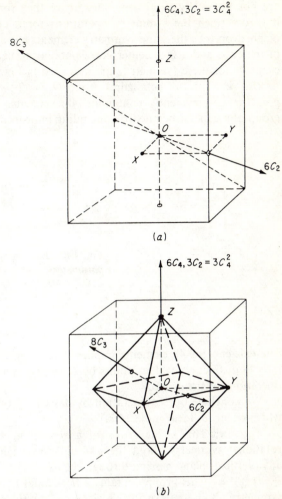

Fig. 4-5. (a) *Rotational symmetries of a cube;* (b) *rotational symmetries of a regular octahedron.*

O ■ One of the most important of all the point groups is the octahedral group O, the group of proper rotations which take a cube or an octahedron into itself. This group contains the following 24 elements: E; eight C_3's about cube body diagonals; three C_2's about the X, Y, Z axes; six C_4's about the X, Y, Z axes; and six C_2's about axes through the origin parallel to face diagonals. Note that all the C_3's are in the same class since a counterclockwise rotation about a given corner is clockwise about the diametrically opposite corner, which may be exchanged with it by one of the twofold rotations from the group. Clearly the two types of C_2 operations mentioned are not physically equivalent, and hence they form separate classes. We have described the symmetry operations for a cube. The corresponding octahedron with the same symmetry elements has a vertex at the face centers of the cube and a face center corresponding to each cube corner. Typical elements for these symmetries are indicated in Fig. 4-5.

O_h ■ The full octahedral group $O_h = O \times i$ is the largest of all the point groups, evidently containing 48 elements. It is the full symmetry group of a cube or octahedron, including improper rotations and reflections.

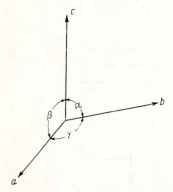

Fig. 4-6. *Labeling of unit-cell parameters.*

The existence of 48 elements implies that a general point may be mapped into 48 distinct equivalent points by successive application of the symmetry operations of the group.

We conclude our enumeration by describing two groups containing C_∞ which occur in linear molecules.

$C_{\infty v}$ ■ This is the group of a general linear molecule. It has full rotational symmetry about the molecular axis and reflection symmetry in any (vertical) plane passing through the axis.

$D_{\infty h}$ ■ This group contains a horizontal reflection plane and twofold symmetry about any axis in this plane passing through the molecular center. These symmetries imply in addition the existence of inversion symmetry and vertical reflection planes. $D_{\infty h}$ is the symmetry group of a homonuclear diatomic molecule or a symmetric linear molecule of the *ABA* type, such as CO_2.

Crystal systems. For applications it is convenient to tabulate the 32 crystallographic point groups according to the crystal system to which they belong. That is, we give the properties of the translational unit cell which are required for compatibility with the point-group symmetry. The unit cell is specified by the three translations a, b, c and three angles α, β, γ as shown in Fig. 4-6. In Table 4-1, we also list the number of elements in each point group.

Table 4-1

Classification of the point groups according to crystal systems

System	Unit cell	Groups	Number of symmetry elements
Triclinic	$a \neq b \neq c$ $\alpha \neq \beta \neq \gamma$	C_1 $S_2(C_i)$	1 2
Monoclinic	$a \neq b \neq c$ $\alpha = \gamma = \pi/2 \neq \beta$	C_{1h} C_2 C_{2h}	2 2 4
Orthorhombic	$a \neq b \neq c$ $\alpha = \beta = \gamma = \pi/2$	C_{2v} $D_2(V)$ $D_{2h}(V_h)$	4 4 8
Tetragonal	$a = b \neq c$ $\alpha = \beta = \gamma = \pi/2$	C_4 S_4 C_{4h} $D_{2d}(V_d)$ C_{4v} D_4 D_{4h}	4 4 8 8 8 8 16
Rhombohedral (trigonal)	$a = b = c$ $\alpha = \beta = \gamma < 2\pi/3 \neq \pi/2$	C_3 $S_6(C_{3i})$ C_{3v} D_3 D_{3d}	3 6 6 6 12
Hexagonal	$a = b \neq c$ $\alpha = \beta = \pi/2, \gamma = 2\pi/3$	C_{3h} C_6 C_{6h} D_{3h} C_{6v} D_6 D_{6h}	6 6 12 12 12 12 24
Cubic	$a = b = c$ $\alpha = \beta = \gamma = \pi/2$	T T_h T_d O O_h	12 24 24 24 48

4-3 Irreducible Representations of the Point Groups

Given the number of elements in each point group and its class structure, the number and dimensionality of its irreducible representations are fixed by the rules set down in Sec. 3-4. Moreover, we can also work out the complete character tables by the methods cited there. This has been done, and the results are tabulated in various references such as the books of Margenau and Murphy and of Eyring, Walter, and Kimball. For convenience we reproduce these results in Appendix B. Not all 32 groups are tabulated separately since some are direct products of simpler groups with the group of the inversion. It will also be noted that a number of sets of isomorphous groups have been included individually. This has been done to reduce the confusion of multiple labeling in the tables.

As an example, consider the table for our standard example group D_3. As given in the Appendix, the character table is:

D_3			E	$2C_3$	$3C_2'$
$x^2 + y^2, z^2$		A_1	1	1	1
	R_z, z	A_2	1	1	-1
(xz, yz) $(x^2 - y^2, xy)$	(x, y) (R_x, R_y)	E	2	-1	0

The columns are labeled according to the number and type of operations forming each class. In particular, C_2' refers to a twofold axis perpendicular to the principal threefold axis. Had there been other inequivalent twofold axes, they would have been labeled C_2''. The labels of the irreducible representations are A or B for one-dimensional representations. Those having a character $+1$ under the principal rotation C_n are labeled A; those with -1 are labeled B. Two-dimensional representations are labeled E, and three-dimensional ones are labeled T. When inversion symmetry is present, it is customary to use g and u as subscripts for even and odd (*gerade* and *ungerade*), respectively.

The other columns list the coordinates, quadratic forms of coordinates, and rotations R_x, R_y, R_z about the coordinate axes to indicate the representation according to which they transform. These assignments can be checked by use of the projection operators $\mathscr{P}^{(j)}$ introduced in Sec. 3-8 or by more elementary inspection procedures.

Classification of functions by transformation properties. Let us consider some explicit examples of this important concept of classifying functions or operators as belonging to some particular irreducible representation. The simplest case arises in a group having only one-dimensional representations, since then the characters actually are the complete representation matrices.

Consider, for example, the group C_{2v}. Its character table is:

C_{2v}			E	C_2	σ_v	σ_v'
x^2, y^2, z^2	z	A_1	1	1	1	1
xy	R_z	A_2	1	1	-1	-1
xz	R_y, x	B_1	1	-1	1	-1
yz	R_x, y	B_2	1	-1	-1	1

Our notation here is such that σ_v is reflection in the xz plane and σ_v' is reflection in the yz plane. Then we can write out symbolically the effect of each of these operations (or its inverse, which is the same in this example) on the three coordinate functions as follows:

$$P_E \begin{pmatrix} x \\ y \\ z \end{pmatrix} = \begin{pmatrix} x \\ y \\ z \end{pmatrix} \quad P_{C_2} \begin{pmatrix} x \\ y \\ z \end{pmatrix} = \begin{pmatrix} -x \\ -y \\ z \end{pmatrix} \quad P_{\sigma_v} \begin{pmatrix} x \\ y \\ z \end{pmatrix} = \begin{pmatrix} x \\ -y \\ z \end{pmatrix} \quad P_{\sigma_v'} \begin{pmatrix} x \\ y \\ z \end{pmatrix} = \begin{pmatrix} -x \\ y \\ z \end{pmatrix}$$

Comparing the results with the character table, and using the fact that $\chi^{(i)}(R) = \Gamma^{(i)}(R)$ for a one-dimensional representation, we have, for example,

$$P_{C_2}x = \Gamma^{(B_1)}(C_2)x$$

or more generally

$$P_R x = \Gamma^{(B_1)}(R)x$$

Similarly,

$$P_R y = \Gamma^{(B_2)}(R)y$$

and

$$P_R z = \Gamma^{(A_1)}(R)z$$

We summarize these results by saying that, under the group C_{2v}, x transforms according to B_1, y according to B_2, and z according to A_1. From these results it follows that xy transforms according to $B_1 \times B_2 = A_2$, etc.

The confirmation of assignments by the projection operator technique may be illustrated by letting $\mathscr{P}^{(A_2)}$ operate on xy. This gives

$$\mathscr{P}^{(A_2)}xy = \frac{l_{A_2}}{h} \sum_R \chi^{(A_2)}(R)P_R xy$$

$$= \tfrac{1}{4}\{xy + (-x)(-y) - x(-y) - (-x)y\}$$

$$= xy$$

Thus xy does indeed belong to A_2.

To illustrate the situation which arises with representations of higher order, we now consider our standard example D_3. We take the z axis as the threefold axis, the x axis as the A rotation axis (see Fig. 2-1), and the y axis as the axis perpendicular to both of these. Now, if we apply operators to the simple functions x, y, z, we can work out a matrix representation.

For example,

$$P_E\begin{pmatrix} x \\ y \\ z \end{pmatrix} = \begin{pmatrix} x \\ y \\ z \end{pmatrix} \qquad P_D\begin{pmatrix} x \\ y \\ z \end{pmatrix} = \begin{pmatrix} -\dfrac{x}{2} + \dfrac{\sqrt{3}y}{2} \\ -\dfrac{\sqrt{3}x}{2} - \dfrac{y}{2} \\ z \end{pmatrix} \qquad P_A\begin{pmatrix} x \\ y \\ z \end{pmatrix} = \begin{pmatrix} x \\ -y \\ -z \end{pmatrix}$$

Note that in working out the effect of P_D we must be careful of the sense of the 120° rotation because $D \neq D^{-1}$. Thus $P_D f(\mathbf{x}) = f(\mathbf{D}^{-1}\mathbf{x})$; or, in words, we rotate the coordinates counterclockwise to represent a clockwise rotation of the object.

If we write out the result of all six operations of the group in this way, we note that z always transforms only into a multiple of itself. Hence z alone forms a basis for a one-dimensional representation of the group. Since the character under A is seen to be -1, $\chi(C_2') = -1$ and the representation is identified as A_2 by reference to the character table. With x and y the situation is more complicated since cross terms occur under the operation D. Hence x and y together form a basis for a two-dimensional representation which is readily seen to be irreducible and hence must be E. The matrices can be obtained by inspecting the results above, and they are found to coincide with those introduced in defining the group originally in Sec. 2-2. Again care is required to keep the indices straight. For example, labeling the rows by x, y,

$$P_D x = x\Gamma(D)_{xx} + y\Gamma(D)_{yx}$$

Thus, $\Gamma(D)_{xx} = -\frac{1}{2}$ and $\Gamma(D)_{yx} = -\sqrt{3}/2$, and $\mathbf{\Gamma}(D)$ coincides with the matrix D introduced earlier. This agreement is by no means accidental. Because only one two-dimensional representation of D_3 exists, the matrices found here could at most differ from those given before by a similarity transformation corresponding to the choice of two independent linear combinations of x and y as basis functions. This example illustrates a common and powerful method of finding the matrices of the irreducible representations of some given group.

Although the matrices may be derived by consideration of simple functions like x and y, this in no way means that their applicability is so restricted. There are in general infinitely many pairs of functions which will transform in the same way. For example, since $x^2 + y^2$ and z^2 belong to A_1, the totally symmetric representation, any function $\varphi(x^2 + y^2, z^2)$ is invariant under all operations of D_3. Hence, if we consider $x\varphi$ and $y\varphi$, for any such φ, we have two functions which transform in the same way as x and y. In this way a relatively small number of matrices can describe the transformation properties of *all* possible eigenfunctions for a given problem.

4-4 Elementary Representations of the Three-dimensional Rotation Group

In the next section we shall begin to investigate the sorts of eigenfunctions and degeneracies which are consistent with the symmetries experienced by electrons in a crystal. In many cases the perturbing influence of the crystalline environment is small enough to make it worthwhile to consider the free atom or ion as a starting point. In this section we prepare for this application by giving a brief survey of the elementary representations of the group of all rotations in three dimensions, which is the rotational-symmetry group for an atom in free space. These are the odd-dimensional representations corresponding to integral angular momentum, for which the spherical harmonics form convenient sets of basis functions. We shall give a much more detailed treatment of this group in the next chapter.

The set of all rotations in three-dimensional space obviously forms an infinite group—the covering operations of a sphere. There are also an infinite number of classes, all rotations through the same angle (about whatever axis) being in the same class because another rotation of the group connects the two rotation axes. Therefore, there must also be an infinite number of irreducible representations. We can find the representations of odd dimensionality by taking the spherical harmonics

$$Y_l^m(\theta, \varphi) \sim P_l^m(\theta)e^{im\varphi}$$

as basis functions. It is a well-known property of the spherical harmonics that if we shift the polar axis we can express the resulting $Y_l^m(\theta', \varphi')$ in terms of a linear combination of all the $Y_l^{m'}(\theta, \varphi)$ of *the same l*. That is, we can write

$$P_R Y_l^m = \sum_{m'} \Gamma^{(l)}(R)_{m'm} Y_l^{m'} \tag{4-2}$$

[If $m = 0$, the $\Gamma^{(l)}(R)_{m'0}$ can be determined by using the addition theorem for spherical harmonics to be simply $\sim Y_l^{-m'}(\theta'', \varphi'')$, where θ'' and φ'' are the angles giving the direction of the new polar axis.] This means that the $2l + 1$ functions Y_l^m form a basis for a representation of the full rotation group. One might imagine that the representation is irreducible, since all $2l + 1$ functions enter in (4-2), and that is in fact the case. We shall see shortly that these spherical harmonic representations exhaust the possible odd-dimensional representations.

Characters. If we rotate the contours of the function by α about the z axis (or make a rotation of axes by $-\alpha$), the effect upon the spherical harmonic is simple, namely,

$$P_\alpha Y_l^m(\theta, \varphi) = Y_l^m(\theta, \varphi - \alpha) = e^{-im\alpha} Y_l^m(\theta, \varphi) \tag{4-3}$$

Thus the representation of such a rotation is the diagonal matrix

$$\mathbf{\Gamma}^{(l)}(\alpha) = \begin{pmatrix} e^{-il\alpha} & 0 & \cdots & 0 \\ 0 & e^{-i(l-1)\alpha} & \cdots & 0 \\ \multicolumn{4}{c}{\dotfill} \\ 0 & 0 & \cdots & e^{il\alpha} \end{pmatrix} \qquad (4\text{-}4)$$

Hence the character is

$$\chi^{(l)}(\alpha) = \text{Tr } \mathbf{\Gamma}^{(l)}(\alpha) = e^{-il\alpha} + \cdots + e^{il\alpha} \qquad (4\text{-}5)$$

This sum can be carried out explicitly as follows by factoring out a geometric series:

$$\chi^{(l)}(\alpha) = e^{-il\alpha} \sum_{k=0}^{2l} (e^{i\alpha})^k$$

$$= e^{-il\alpha} \frac{e^{i(2l+1)\alpha} - 1}{e^{i\alpha} - 1}$$

$$= \frac{e^{i(l+\frac{1}{2})\alpha} - e^{-i(l+\frac{1}{2})\alpha}}{e^{i\alpha/2} - e^{-i\alpha/2}}$$

$$= \frac{\sin (l + \frac{1}{2})\alpha}{\sin (\alpha/2)}. \qquad \blacktriangleright (4\text{-}6)$$

We have previously noted that all rotations by α are in the same class, regardless of axis. Hence, they must all have the character calculated here for a simple special case.

Finally, we show that there can be no other inequivalent irreducible representations of odd order. If there were such, they would have to have a character orthogonal to the characters of all those found above. But if they are orthogonal to all of them, they must be also orthogonal to the difference of two successive ones such as $[\chi^{(l)}(\alpha) - \chi^{(l-1)}(\alpha)]$. However, this difference is just $2 \cos l\alpha$. Thus any additional representation would need to have a character $\chi(\alpha)$ orthogonal to a Fourier series of cosines. Since rotations by $\pm\alpha$ are in the same class, χ must be an even function of α. Hence, there is no possibility of any new $\chi(\alpha)$ being orthogonal to a cosine series, and *the spherical harmonics of order l form the basis for the only distinct irreducible representation of dimensionality $(2l + 1)$ of the full rotation group.* Its character is given by (4-5) or (4-6) above.

Although this proof may seem so general as to exclude *any* other representations, we shall in fact be led to consider even-dimensional representations, corresponding to half-integral angular momenta. It turns out that the formulas derived above still hold for the characters. Hence the new characters are orthogonal to those of the odd-dimensional representations *if* the domain of α is extended from $\pm\pi$ to $\pm 2\pi$. This curious extension is required because of the double-valued nature of the basis functions of half-integral angular momentum. As we shall show, rotation by 2π produces a sign change, and rotation by 4π serves as the identity. Since all observable quantities depend on bilinear combinations of these functions, no physical anomalies arise from the resulting sign ambiguity.

4-5 Crystal-field Splitting of Atomic Energy Levels

When an atom or ion is located, not in free space, but in a crystal, it is subjected to various inhomogeneous electric fields which destroy the isotropy of free space. Thus the symmetry group is reduced from that of the full three-dimensional rotation group plus the inversion to some finite group of rotations through finite angles and perhaps also reflections. This reduced group allows the originally irreducible representations of the full rotation group based on spherical harmonics (angular-momentum eigenfunctions) to be reduced with respect to this subgroup. This reduction in the dimensionality of the irreducible representations causes the degeneracy associated with complete rotational symmetry to be lifted. Hence the free-ion energy levels are split by the crystalline electric field. As first pointed out by Bethe,[1] the degree of residual degeneracy is determined from the symmetry by group theory with ultimate accuracy since no perturbation-theory approximation is involved. However, to calculate the magnitudes of the splittings or the order of the levels, it is of course necessary to employ more specific knowledge of the crystalline field. Given such knowledge, semiquantitative predictions of the splittings can be made by perturbation methods. In addition to these zero-field splittings, the additional splitting produced by an external magnetic field can be elegantly handled by the use of "spin Hamiltonians." These may be obtained by perturbation theory or simply guessed on the basis of symmetry. The spin-Hamiltonian method is presented in review papers by Bleaney and Stevens and by Bowers and Owen,[2] and it has been extremely successful in describing the results of paramagnetic-resonance experiments. We shall not go into it here, however, as it generally entails a more restricted and specific approach than required by purely group-theoretical considerations.[3]

As mentioned above, the splittings are most accurately studied by microwave paramagnetic resonance. However, they were first studied by means of magnetic-susceptibility and heat-capacity experiments and in optical absorption in solids. The magnitude of the splittings can range from

---

[1] H. Bethe, *Ann. Physik*, **3**, 133 (1929).

[2] B. Bleaney and K. W. H. Stevens, *Rept. Progr. Phys.*, **16**, 108 (1953); K. D. Bowers and J. Owen, *ibid.*, **18**, 304 (1955).

[3] G. F. Koster and H. Statz, *Phys. Rev.*, **113**, 445 (1959).

~ 0.01 cm^{-1} up to $\sim 50,000$ cm^{-1}, depending on the ion involved and the nature of the crystalline field. This great range makes the use of different methods valuable for handling different cases. We distinguish three cases depending on whether the crystal-field energies are small compared with the spin-orbit coupling (rare-earth ions), large compared with spin-orbit coupling but small compared with term separations (iron group), or large compared even with the term separations (covalent complexes). After some general remarks about the origin of the crystal field, we shall return to treat some of these cases.

Form of the crystal field. In any given crystalline environment, one can estimate the potential near the ion in question by replacing each neighboring ion by a point charge. Taking our origin of coordinates at the center of the magnetic ion, we can then make a Taylor's-series expansion of the resulting potential. An ion of charge e at $x = -a, y = z = 0$ will contribute

$$V(x, y, z) = e[(a + x)^2 + y^2 + z^2]^{-\frac{1}{2}}$$

$$= \frac{e}{a}\left(1 - \frac{x}{a} - \frac{r^2}{2a^2} + \frac{3}{2}\frac{x^2}{a^2} + \cdots\right)$$

Putting another charge at $x = +a$ gives a situation with inversion symmetry, causing all odd powers of x to drop out. If we add four more charges at $\pm a$ on the y and z axes to form a regular octahedron of charges, we get further simplifications. The first term with less than spherical symmetry is

$$V_c = \frac{35}{4}\frac{e}{a^5}(x^4 + y^4 + z^4 - \tfrac{3}{5}r^4) \tag{4-7}$$

which is of course invariant under all the operations of the full octahedral group O_h. This is often referred to as the cubic-field term, as it is the leading nonspherical term in the potential of cubic symmetry. If the symmetry were less complete than this, e.g., if the charges were at $\pm c \neq \pm a$ along the z axis, then nonspherical terms would remain in the second order, as well as higher-order terms.

Note that all these potentials must obey Laplace's equation, since they are being produced by charges outside the region of interest. This restricts their forms to linear combinations of expressions of the sort $r^l Y_l^m(\theta, \varphi)$. For example, if there is axial symmetry about the z axis, the leading term is

$$V_a \sim r^2 Y_2^0(\theta) \sim 3z^2 - r^2 \tag{4-8}$$

If there is only orthorhombic symmetry, the leading term is

$$V_r \sim Ax^2 + By^2 - (A + B)z^2 \tag{4-9}$$

This V_r is a mixture of Y_2^0 and $(Y_2^2 + Y_2^{-2}) \sim (x^2 - y^2)$, both of which are invariant under D_2.

The method of estimation outlined above gives a reasonable estimate of the magnitude of the crystalline field, but it is far from rigorous. Errors arise from overlap, nonspherical charge distributions, and exchange effects.[1] These effects do *not* affect the symmetry properties, however. If x, y, and z are equivalent, they will be equivalent even in the presence of these subtleties and the leading term in the effective potential will still act like V_c. However, in the absence of extremely elaborate calculations, the coefficient will have to be determined experimentally from the magnitude of some splitting which it produces. If we are content with those results which can actually be determined theoretically from rigorous symmetry properties, we need not try to compute potentials and consider perturbation theory at all. We can obtain these results directly by the use of group theory.

4-6 Intermediate Crystal-field-splitting Case

Since more work has been done on iron-group ions than on those of any other sort, we first consider the iron-group case in which the crystal-field splitting is intermediate between the spin-orbit energy and the separation of L-S terms in the free atom. This means that L and S will remain approximately good quantum numbers, but J will not. Since the crystal field acts only on the orbital motion (position) of the electrons, it will split the $(2L + 1)$-fold orbital degeneracy but will affect the spin degeneracy only via the (weak) spin-orbit coupling. Thus we initially ignore spin completely and work out the orbital splittings. Later on we shall show how the spin is to be handled.

We proceed by example. Consider the case of an ion in a crystal at a site where it is surrounded by a regular octahedron of negative ions. This is a reasonable approximation of the real situation in a large number of instances. As noted above, $V_c \sim (x^4 + y^4 + z^4 - 3r^4/5)$ would be the leading term in the potential, but we shall use only the symmetry of the situation. This is described by the group O_h, but for the present we shall consider only the 24 proper rotations comprising the group O. Later (Sec. 4-10) we shall show that inclusion of the inversion to form O_h introduces nothing new. For convenience, the character table of O is reproduced here, including a variety of labels for the various irreducible representations which have been introduced in the literature and are still in use. Those given in the first two columns are used by Bouckaert, Smoluchowski, and Wigner and by Von der Lage and Bethe.[2] Both are systems of notation in which a prime changes the parity or inversion symmetry, and the labels given

---

[1] See, for example, J. C. Phillips, *J. Phys. Chem. Solids*, **11**, 226 (1959); A. J. Freeman and R. E. Watson, *Phys. Rev.*, **120**, 1254 (1960).

[2] L. P. Bouckaert, R. Smoluchowski, and E. P. Wigner, *Phys. Rev.*, **50**, 58 (1936); F. C. Von der Lage and H. Bethe, *Phys. Rev.*, **71**, 612 (1947).

all refer to the even parity case. The third column is the notation used in Bethe's original paper, and the fourth is the conventional one used in most character tables for molecular applications. Upon recalling the discussion

O				E	$8C_3$	$3C_2$	$6C_2$	$6C_4$
Γ_1	α	Γ_1	A_1	1	1	1	1	1
Γ_2	β'	Γ_2	A_2	1	1	1	-1	-1
Γ_{12}	γ	Γ_3	E	2	-1	2	0	0
Γ'_{15}	δ'	Γ_4	T_1	3	0	-1	-1	1
Γ'_{25}	ϵ	Γ_5	T_2	3	0	-1	1	-1

of Sec. 4-2, the $8C_3$'s are rotations about face centers of the octahedron (body diagonals of the complementary cube); the $3C_2$'s and $6C_4$'s are rotations about the corners of the octahedron (face centers of the cube); and the $6C_2$'s are about the remaining symmetry axes through midpoints of edges. These axes are illustrated in Fig. 4-5.

Now consider an atomic term with orbital angular momentum L. The $(2L + 1)$ spherical harmonics Y_L^M, which are degenerate in isotropic space, form the basis for a representation which we have called $\Gamma^{(L)}$. We now change notation to the conventional one, $D^{(L)}$ or D_L, introduced by Wigner (D for *Darstellung*) for representations of the rotation group. Using Eq. (4-6), we can compute the characters of all the classes of group elements. The results are simply

$$\chi(C_2) = \chi(\pi) = (-1)^L$$

$$\chi(C_3) = \chi\left(\frac{2\pi}{3}\right) = \begin{cases} 1 & L = 0, 3, \ldots \\ 0 & L = 1, 4, \ldots \\ -1 & L = 2, 5, \ldots \end{cases}$$

$$\chi(C_4) = \chi\left(\frac{\pi}{2}\right) = \begin{cases} 1 & L = 0, 1, 4, 5, \ldots \\ -1 & L = 2, 3, 6, 7, \ldots \end{cases}$$

(4-10)

Since the spherical harmonics of order L form a basis for representing the group of *all* rotations, they certainly do so for the finite group of rotations O. In fact, we expect that we may be able to reduce the representation into smaller irreducible ones for this smaller group. To do this formally, we merely need the character of D_L together with the decomposition formula (3-20). For convenience, we tabulate the characters of the first few of these (perhaps) reducible representations of O. Then, using (3-20),

$$D_L = \sum_i a_i \Gamma_i \qquad \text{where} \qquad a_i = (24)^{-1} \sum_k N_k \chi_i(\mathscr{C}_k) \chi_L(\mathscr{C}_k)$$

For the simpler cases, we can decompose the representation by inspection by noting what rows from the irreducible representation table need to be

O	E	$8C_3$	$3C_2$	$6C_2$	$6C_4$
D_0	1	1	1	1	1
D_1	3	0	-1	-1	1
D_2	5	-1	1	1	-1
D_3	7	1	-1	-1	-1
D_4	9	0	1	1	1

added together to produce the desired row in the D_L table. Let us list the first few cases:

$D_0 = A_1$ cannot split since it is only one-dimensional.

$D_1 = T_1$, by comparison of the characters. Since D_1 remains a single irreducible representation, a P state ($L = 1$) is not split by a cubic field.

$D_2 = E + T_2$. D_2 *must* split, since there are no five-dimensional *irreducible* representations of O. The actual decomposition shows that a D state ($L = 2$) is split into a twofold and a threefold degenerate level in a cubic field.

$D_3 = A_2 + T_1 + T_2$. Thus an F state is split into a nondegenerate and two threefold degenerate states.

$D_4 = A_1 + E + T_1 + T_2$. Thus a G state splits into a nondegenerate state, a doubly degenerate state, and two triply degenerate states.

These examples show that the degrees of residual degeneracy of an atom in a field of given symmetry may be worked out almost trivially by using group theory. Moreover, the various irreducible representations completely determine the symmetry or transformation properties of the various sets of degenerate eigenfunctions.

Additional splitting in field of lower symmetry. In most actual crystals there are at least small departures from cubic symmetry at the lattice site of a magnetic ion. We could handle such cases by a method exactly analogous to that employed above for the group O, but using the appropriate smaller group. However, we get much more insight into the final level structure by considering the problem in steps, first working out the splitting of the free-ion terms into the irreducible representations of the cubic (octahedral) group, then working out the additional smaller splitting of these representations under the field of lower symmetry.

As an example, we have shown that an $L = 2$ state splits into $E + T_2$ under the group O. Now let us assume that the octahedron of ions producing the crystal field is distorted by an elongation along one of the threefold axes. This reduces the rotational symmetry to simply the group D_3, which is a subgroup of O. To find the additional splitting of the E and T_2 levels, we need only decompose these representations, which may be reducible with respect to the subgroup D_3, though irreducible under O. To facilitate this reduction, we write down the character table of D_3 and add to it the characters of the various irreducible representations of O, treated as representations of D_3. This can be done if we note by geometrical inspection that the C_2's of D_3 in this case are the same as the $6C_2$ type of twofold rotations of O. In this way we obtain the following table:

D_3		E	$2C_3$	$3C_2$
Irreducible	A_1	1	1	1
representations	A_2	1	1	-1
of D_3	E	2	-1	0
	A_1	1	1	1
Irreducible	A_2	1	1	-1
representations	E	2	-1	0
of O	T_1	3	0	-1
	T_2	3	0	1

Comparing the two halves of the table, we see that the one- and two-dimensional representations A_1, A_2, and E are also irreducible under D_3 (even retaining the same labels); so the corresponding energy levels do not split any further under the deviation from cubic symmetry. Of course, the one-dimensional representations could not split in any case, but the result for E is less trivial. Next we note that T_1 and T_2 must split, because there are no three-dimensional representations of D_3. Inspection of the tables or use of the decomposition formula (3-20) shows that

$$T_1 \rightarrow E + A_2$$
$$T_2 \rightarrow E + A_1$$

Thus each triply degenerate level in a cubic field splits into a doubly degenerate and a nondegenerate level under a trigonal distortion along one of the threefold axes of O. (If the trigonal axis were *not* one of these axes, then the residual symmetry in the presence of both fields would be smaller than either group and might contain only the identity. In this case, all the orbital degeneracy would be lifted.)

The results which we have obtained are shown schematically in Fig. 4-7. Naturally it is impossible to know the order of the levels without knowledge of at least the sign of the perturbing fields. Such information can be obtained, as noted above, from a rough estimate of the crystalline potential. By perturbation methods, one can find the *ratios* of splittings produced by a given potential (such as the ratio $[E(T_2) - E(T_1)]/[E(T_1) - E(A_2)]$ for the $L = 3$ case) even if the magnitude of the crystal field is known only roughly. We shall not go into this here, but it has been treated by Bethe and by more recent authors such as Bleaney and Stevens.

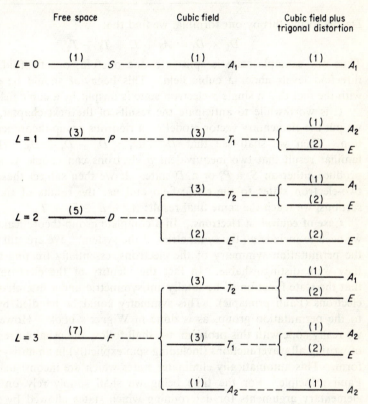

Fig. 4-7. *Scheme of crystal-field splittings for various orbital angular-momentum states. The parenthetical numbers give the degeneracies of the various levels.*

Case of two inequivalent electrons. If we are considering an atom with two electrons outside filled shells, we can form a direct product representation of the symmetry group with basis functions $\psi(1)\varphi(2)$. If the electrons are "inequivalent," i.e., if they belong to different nl subshells, the exclusion principle is satisfied for all possible product functions and no complications arise. (More precisely, a nonvanishing antisymmetrized product can be formed from any such pair. See below.)

Consider the example of a $2p3p$ configuration, in which we have two inequivalent p electrons, situated in a cubic field. The character of the nine-dimensional representation $\mathscr{D} = D_1 \times D_1$ is simply found by taking the product of the character of D_1 with itself.   The result is:

O	E	$8C_3$	$3C_2$	$6C_2$	$6C_4$
$D_1 \times D_1$	9	0	1	1	1

Decomposing this by our formula, we find that

$$D_1 \times D_1 = A_1 + E + T_1 + T_2$$

Thus our example state is split into one onefold, one twofold, and two threefold levels under a cubic field.   This behavior should be contrasted with the fact that a single p-electron state is unsplit by a cubic field.

It is worthwhile to anticipate the results of the next chapter, in which we put the elementary vector model on a rigorous group-theoretical foundation.   Then we shall see that $D_1 \times D_1 = D_0 + D_1 + D_2$.   This is the familiar result that two inequivalent p electrons can couple vectorially to produce either an S, a P, or a D state.   If we then subject these coupled, two-electron states to the cubic field and use the results of the previous sections, we reach the same final result: $A_1 + E + T_1 + T_2$.

Case of equivalent electrons.   In a complete group-theoretical approach, we need to include *all* the symmetries of the system.   We are still neglecting the permutation symmetry of the electrons, essentially treating them as if they were distinguishable.   In fact the identity of the electrons requires that the state function Ψ be totally antisymmetric under interchange of two electrons (Pauli principle).   This symmetry could be handled by recourse to the permutation group, as is done in Wigner's book.   However, when we really cope with this problem, we shall follow the usual current practice of writing all wavefunctions (including spin explicitly) in an antisymmetrized form.   This automatically eliminates states which are inconsistent with the Pauli principle.   For the time being we shall simply rely on the usual elementary arguments for determining which states allowed by the vector model are also consistent with the Pauli principle.   For example, with two equivalent p electrons (an np^2 configuration), the only allowed triplet state is a ^3P state.   Since the definition of a P state is basically that its three degenerate substates transform under rotations like a single p electron, we can simply apply all the results of our previous sections.   Thus $^3P \rightarrow {}^3T_1$, and the triplet state is not split in a cubic field. ^1S and ^1D states can also be formed from an np^2 configuration, but according to Hund's rules, they will have higher energies than the triplet state.   In the presence of a cubic field, the ^1S is unsplit, but the ^1D splits into $^1E + {}^1T_2$, as noted above.

Physical picture for splittings.   Consider a cubic field produced by six ions located at the vertices of a regular octahedron surrounding the ion under consideration.   The coordinate axes are chosen to pass through these ions. Under these circumstances it is obvious that the three p functions (p_x, p_y, p_z) are subjected to completely equivalent fields, and no splitting can occur.   Now, if we lower the symmetry by pulling the ions on the z axis further out, we introduce a term of the type $(3z^2 - r^2)$ in the potential and make the p_z orbital inequivalent to the p_x, p_y pair.   This leaves us with two separate representations: a one-dimensional one based on p_z and a two-dimensional one based on (p_x, p_y).   These smaller sets of functions now form bases for representations because we have reduced the size of the symmetry group by removing all symmetry operators which interchanged the z direction with x or y.

If we consider d functions, there *is* a splitting by a cubic field.   A suitable choice of d functions for cubic symmetry is the set of three which transform like (xy, yz, zx) and form a basis for T_2, together with the set of two transforming like $(x^2 - y^2)$, $(3z^2 - r^2)$ which form a basis for E.   (This set of five functions differs from the set of $Y_2{}^m$'s only by a unitary transformation.)   Considering contours of these functions, we see that the T_2 functions have lobes which avoid the surrounding octahedron of ions, whereas the E functions point straight at them.   Therefore, if the ions are repulsive centers, the T_2 functions lie lower in energy than those of E.   If one considers the various operations of the cubic group O, one finds that they merely scramble the T_2 and E functions (separately) among themselves.   No mixing occurs between T_2 and E functions, as it would under an *arbitrary* three-dimensional rotation.   This is the group-theoretical basis for the possibility of decomposing D_2 into $E + T_2$ under the subgroup O of the full rotation group.   A pictorial representation of these functions is given in Fig. 8-7.

4-7   Weak-crystal-field Case and Crystal Double Groups

In this case the crystal field is weak compared with the spin-orbit coupling. Hence the *total* angular momentum (including spin) J stays nearly a good quantum number.   This allows us to restrict our consideration to a single level of a multiplet, such as $^4F_{\frac{3}{2}}$.   If there is an even number of electrons, $J = |\mathbf{L} + \mathbf{S}|$ must be integral, $(2J + 1)$ is odd, and we may handle the splittings by exactly the same procedure as used when we considered only the orbital angular momentum L.   If there is an odd number of electrons, however, J is half integral, $(2J + 1)$ is even, and we have to deal with representations of the full rotation group which are not offered by the spherical harmonics.   In fact we must introduce the concept of *spinors* to handle these cases.   These introduce the two-valued representations mentioned at the end of Sec. 4-4.   These will be dealt with in detail in the next chapter.   For the time being we shall follow a procedure introduced by Bethe which allows us to obtain all our results by a simple extension of our standard analysis by introducing the concept of *crystal double groups*.

Crystal double groups.   In this approach we anticipate the fact (which we prove in the next chapter) that our formula (4-6) for the character of a rotation is valid whether the angular momentum J is integral or half integral. Thus, for a rotation through an angle α, we have

$$\chi_J(\alpha) = \frac{\sin(J + \frac{1}{2})\alpha}{\sin(\alpha/2)} \qquad (4\text{-}11)$$

If we now consider a rotation by $(\alpha + 2\pi)$, we find

$$\chi_J(\alpha + 2\pi) = \frac{\sin(J + \frac{1}{2})(\alpha + 2\pi)}{\sin(\alpha/2 + \pi)} = (-1)^{2J}\chi_J(\alpha)$$

Therefore, for half integral J we have a *double-valued* character, since a rotation by 2π should leave everything unchanged, whereas for these cases $\chi_J(\alpha + 2\pi) = -\chi_J(\alpha)$.   Note, however, that even for half integral J

$$\chi_J(\alpha \pm 4\pi) = \chi_J(\alpha)$$

so that the representations act as if a rotation through 4π should be considered the identity.   Also note that

$$\chi_J(\pi + 2\pi) = \chi_J(\pi) = 0$$

showing that the character for a twofold rotation is *single*-valued.

We cope with this strange situation by introducing the fiction that the crystal is taken into itself, not under a rotation by 2π, but only after a rotation of 4π.   We introduce a new group element R which represents a rotation by 2π and has the property $R \neq E$ but $R^2 = E$.   The augmented group contains twice as many elements as the original one and is called the crystal double group.   The double group has more classes than the original, but not necessarily twice as many, since $\chi_J(RC_2) = \chi_J(C_2) = 0$, making it possible that RC_2 is in the same class as C_2.   Opechowski[1] has shown by a more thorough investigation that RC_2 is in the same class as C_2 if, and only if, there is another twofold axis perpendicular to the one in question.   This method of reducing the double-valued representations (i.e., two distinct matrix representations of each symmetry operator, differing in sign) to an apparently single-valued situation bears some analogy to the construction of several-sheeted Riemann surfaces in the discussion of multiple-valued functions.   Actually, by using the homomorphism of the three-dimensional rotation group to the unitary group discussed in the next chapter, one can provide quite a straightforward interpretation of the double group.   But we shall not go into this here.

Within the framework of the double-group method, we can handle the half integral J's by the same procedure as we developed for the ordinary

[1] W. Opechowski, *Physica*, **7**, 552 (1940).   See also V. Heine, "Group Theory in Quantum Mechanics," p. 137, Pergamon Press, New York, 1960.

case. As an example, we treat the octahedral double group O'. This consists of 48 elements: each of the 24 in O together with the 24 new elements obtained by combining each of them with R. In addition to the five classes of O, we have the classes R, $R \cdot C_3$, $R \cdot C_4$. (The RC_2's remain in the same class with the C_2's because there are perpendicular C_2 axes.) Thus we need to add three irreducible representations and 24 elements. This can be done consistently with the dimensionality theorem (3-12) if we keep the five ordinary representations of O ($\Gamma_1 \cdots \Gamma_5$ in Bethe's notation) and add three new ones Γ_6, Γ_7, Γ_8 of dimensionality 2, 2, and 4.

In setting up the character table of O' we take advantage of the close relation between O' and O. First of all, we see that we can get the characters of all the "normal" single-valued representations $\Gamma_1 \cdots \Gamma_5$ by simply taking $\chi(RA) = \chi(A)$, where A is any group element. This procedure assures satisfaction of the requirements on orthonormality between rows. In the double-valued representations, we take $\chi(RA) = -\chi(A)$. This assures that the R column is normalized properly and orthogonal to the E column. We also take $\chi(RC_2) = \chi(C_2) = 0$ since $\chi(RC_2) = \chi(C_2)$, because they are in the same class, while $\chi(RA) = -\chi(A)$ for all A in the double-valued representations. The value zero is also required in order to give the correct column normalization. In the exceptional case when there is no perpendicular C_2 axis, leaving RC_2 and C_2 in different classes, both these arguments fail, and $\chi(RC_2) = -\chi(C_2) \neq 0$, in general. With these various aids and rules, one quite readily can work out the character tables, such as that of O' given below.

O'	E	R	$8C_3$	$8RC_3$	$3C_2 + 3RC_2$	$6C_2 + 6RC_2$	$6C_4$	$6RC_4$
Γ_1	1	1	1	1	1	1	1	1
Γ_2	1	1	1	1	1	-1	-1	-1
Γ_3	2	2	-1	-1	2	0	0	0
Γ_4	3	3	0	0	-1	-1	1	1
Γ_5	3	3	0	0	-1	1	-1	-1
Γ_6	2	-2	1	-1	0	0	$\sqrt{2}$	$-\sqrt{2}$
Γ_7	2	-2	1	-1	0	0	$-\sqrt{2}$	$\sqrt{2}$
Γ_8	4	-4	-1	1	0	0	0	0

Application of double groups to term splittings. For integral J, $\chi_J(RC) = \chi_J(C)$, and the character is orthogonal to the characters of all the new double-valued irreducible representations. Accordingly, the decomposition is entirely into the normal single-valued representations, and it works out exactly as it would if the group had not been "doubled."

For half integral J we have seen that $\chi_J(RC) = -\chi_J(C)$ for arbitrary rotations. Thus the character will be orthogonal to those of the single-valued representations, and the decomposition is wholly into the new double-valued representations (Γ_6, Γ_7, and Γ_8 for O').

Evaluating characters for half integral J with Eq. (4-11), we find:

O'	E	R	$8C_3$	$8RC_3$		$3C_2 +$ $3RC_2$	$6C_2 +$ $6RC_2$	$6C_4$	$6RC_4$	
D_J	$(2J+1)$	$-(2J+1)$	1	-1	$(J=\frac{1}{2},\frac{7}{2},\cdots)$	0	0	$\sqrt{2}$	$-\sqrt{2}$	$(J=\frac{1}{2},\frac{9}{2},\cdots)$
			-1	1	$(J=\frac{3}{2},\frac{9}{2},\cdots)$			0	0	$(J=\frac{3}{2},\frac{7}{2},\frac{11}{2},\cdots)$
			0	0	$(J=\frac{5}{2},\frac{11}{2}\cdots)$			$-\sqrt{2}$	$\sqrt{2}$	$(J=\frac{5}{2},\frac{13}{2},\cdots)$

If we use these results in the standard decomposition formula, we find the results tabulated below:

$$D_{\frac{1}{2}} \to \Gamma_6$$

$$D_{\frac{3}{2}} \to \Gamma_8$$

$$D_{\frac{5}{2}} \to \Gamma_7 + \Gamma_8$$

$$D_{\frac{7}{2}} \to \Gamma_6 + \Gamma_7 + \Gamma_8$$

$$D_{\frac{9}{2}} \to \Gamma_6 + 2\Gamma_8$$

. . . .

Thus levels with $J = \frac{1}{2}$ or $\frac{3}{2}$ are unsplit in a cubic field; a $J = \frac{5}{2}$ level splits into a doublet and a quartet, and so on. We note that all the extra double-valued representations are of even dimension, and hence of dimension greater than 1. Therefore all levels are at least twofold degenerate. This result can be proved (see Sec. 5-16) in general for half-integral angular momenta (odd number of electrons) for an electric field of any symmetry since it is a consequence of time reversal symmetry. This is known as *Kramers' theorem*.[1] It plays an important role in the theoretical interpretation of paramagnetic-resonance experiments.

4-8   Introduction of Spin Effects in the Medium-field Case

In our previous treatment of this case (Sec. 4-6), we neglected spin effects completely and merely found the orbital-level structure. In addition to the residual degeneracies found there, we still had the $(2S + 1)$-fold spin degeneracy. If the spin-orbit and spin-spin couplings are now introduced so that the spins can "feel" the symmetry of the lattice through the orbital

[1] H. Kramers, *Proc. Acad. Sci. Amsterdam*, **33**, 959 (1936).

motion, second-order effects arise which may split the spin degeneracy. We can find out about this by introducing direct-product representations based on products of orbital and spin functions.

For example, consider the splitting of a ^4F term in a cubic crystal field. As noted earlier, the orbital F state splits according to $D_3 = A_2 + T_1 + T_2 \equiv \Gamma_2 + \Gamma_4 + \Gamma_5$. Each of these representations is now to be combined

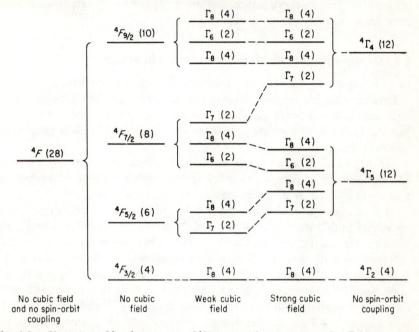

Fig. 4-8. *Variation of level structure of ^4F term with increasing cubic-field strength compared with spin-orbit coupling strength. The parenthetical numbers give the degeneracies of the various levels.*

with the representation $D_{\frac{3}{2}} = \Gamma_8$ of the spin. Using the decomposition formula to decompose the direct-product representations, we find

$$\Gamma_2 \times \Gamma_8 = \Gamma_8$$
$$\Gamma_4 \times \Gamma_8 = \Gamma_6 + \Gamma_7 + 2\Gamma_8$$
$$\Gamma_5 \times \Gamma_8 = \Gamma_6 + \Gamma_7 + 2\Gamma_8$$

This shows that the Γ_2 level keeps its fourfold spin degeneracy, whereas the Γ_4 level, for example, splits into four distinct levels of degeneracy 2, 2, 4, and 4.

It is instructive to combine these results with the results of the previous two sections in a schematic diagram (Fig. 4-8) showing the level structure of a ^4F term under a cubic field of various strengths compared with the spin-orbit coupling. Note that the irreducible representations of the double group serve as rigorous good quantum numbers under all conditions shown. However, if either the crystal field or the spin-orbit coupling is set to zero, a number of these representations coalesce into larger representations, as shown. In any case, of course, the total number of states remains invariant under all types of splitting. We note that a remarkable amount of information about the level structure has been obtained by group-theoretical considerations of symmetry alone.

4-9 Group-theoretical Matrix-element Theorems

In Sec. 3-8 we proved that two functions $\varphi_\kappa^{(j)}$ and $\psi_{\kappa'}^{(j')}$ which belong to different irreducible representations or to different rows of the same unitary representation are orthogonal. Moreover, we showed that the scalar product $(\varphi_\kappa^{(j)}, \psi_\kappa^{(j)})$ was independent of κ. This result is easily generalized to the following theorem:

Matrix elements of an operator H which is invariant under all operations of a group vanish between functions belonging to different irreducible representations or to different rows of the same unitary representation.

This theorem follows from the simple orthogonality result quoted first, by simply noting that the invariance of H means that H belongs to Γ_1, the identical representation. In this case, $H\varphi_\kappa^{(j)}$ belongs to the direct-product representation, $\Gamma_1 \times \Gamma^{(j)} = \Gamma^{(j)}$, as can be shown formally by noting that $\chi_1(R) = 1$ for all operators R. Physically speaking, it is apparent that application of an operator which is so symmetrical as to be invariant under all group operations will leave the symmetry of $\varphi_\kappa^{(j)}$ completely unchanged with respect to all group operations. Hence $H\varphi_\kappa^{(j)}$ also belongs to the κth row of the jth irreducible representation, and the matrix-element theorem for invariant operators follows from the orthogonality proof. Moreover, the equality for all κ of the scalar products $(\varphi_\kappa^{(j)}, \psi_\kappa^{(j)})$ carries over to the equality of the diagonal-type matrix elements $(\varphi_\kappa^{(j)}, H\psi_\kappa^{(j)})$. Hence, if we choose the set $\psi_\kappa^{(j)}$ to be actually identical with the $\varphi_\kappa^{(j)}$ and take H to be the Hamiltonian operator, we have the result that all basis-function partners within a given irreducible representation have the same energy. This is consistent with the way we originally connected the representation theory with quantum mechanics. Of equal importance is the fact that the absence of *any* matrix elements of H connecting states of different symmetry (representation indices) guarantees the rigor of these group-theoretical quantum numbers, regardless of the order to which one might carry perturbation theory. Accordingly, our symmetry-based results on splittings and selection rules will retain their ultimate accuracy.

Generalized Unsöld theorem. In connection with the preceding theorem about invariant operators, it is of some interest to consider the following theorem, which prescribes a method of generating invariant functions. The theorem states that the sum $\sum_{\kappa=1}^{l_j} |\varphi_\kappa^{(j)}|^2$ is invariant under all operations of the group for which the $\varphi_\kappa^{(j)}$ form the basis of an irreducible representation. This is a generalization of the theorem originally given by Unsöld, which states that $\sum_{m=-l}^{l} |Y_l^m(\theta, \varphi)|^2$ is invariant under all rotations; i.e., closed shells of atoms have spherical symmetry.

We prove the theorem by evaluating the effect on the sum of applying an arbitrary transformation operator from the group. Thus, we consider

$$
\begin{aligned}
P_R \sum_\kappa |\varphi_\kappa^{(j)}|^2 &= P_R \sum_\kappa \varphi_\kappa^{(j)*} \varphi_\kappa^{(j)} \\
&= \sum_{\mu\lambda\kappa} \varphi_\mu^{(j)*} \Gamma^{(j)}(R)_{\mu\kappa}^* \varphi_\lambda^{(j)} \Gamma^{(j)}(R)_{\lambda\kappa} \\
&= \sum_{\mu\lambda} \varphi_\mu^{(j)*} \varphi_\lambda^{(j)} \sum_\kappa \Gamma^{(j)}(R)_{\lambda\kappa} \Gamma^{(j)}(R^{-1})_{\kappa\mu}
\end{aligned}
$$

since the representation is unitary. Noting that the sum on κ is just $\Gamma^{(j)}(E)_{\lambda\mu} = \delta_{\lambda\mu}$, we find that

$$
P_R \sum_\kappa |\varphi_\kappa^{(j)}|^2 = \sum_\mu |\varphi_\mu^{(j)}|^2 = \sum_\kappa |\varphi_\kappa^{(j)}|^2
$$

and the theorem is proved.

As a simple example of the theorem, recall that the three functions x, y, and z form a basis for the representation T_1 of the group O. The sum of the squares of these is $x^2 + y^2 + z^2 = r^2$, which is invariant under *all* rotations and therefore certainly under all the rotations comprising O.

A less trivial example is provided by the functions xy, yz, and zx which form a basis for T_2 of O. In this case, the sum of squares is $x^2y^2 + y^2z^2 + z^2x^2$. This function is invariant under all operations of O and hence belongs to $A_1 (\equiv \Gamma_1)$. Note that

$$
\tfrac{2}{5}(x^2 + y^2 + z^2)^2 - 2(x^2y^2 + y^2z^2 + z^2x^2) = (x^4 + y^4 + z^4 - \tfrac{3}{5}r^4)
$$

Thus, apart from an additive term of spherical symmetry, the invariant function generated here is equivalent to the form of the cubic field potential V_c given in Eq. (4-7).

Matrix-element selection rules. In the first part of this section, we proved an important theorem on matrix elements of invariant operators. In many problems one is also interested in matrix elements of perturbations which are not invariant under all group operations. These may be static perturbations which lower the symmetry below that of the unperturbed problem, or they may be time-dependent perturbations by radiation which induce transitions. The symmetry considerations are the same in either case.

What we seek is a selection rule which restricts the variety of matrix elements between different irreducible representations according to the symmetry of the perturbing operator H'. To obtain this selection rule, we first determine the irreducible representation $\Gamma_{H'}$ according to which H' transforms. [In a general case, H' may contain parts of different symmetries. We may always decompose it, however, according to an analog of Eq. (3-33).] Having found $\Gamma_{H'}$, we note that $H'\varphi_\kappa^{(j)}$ will belong to representations included in $\Gamma_{H'} \times \Gamma^{(j)}$. Then, by the basis-function orthogonality theorem, *the matrix element* $(\psi_{\kappa'}^{(j')}, H'\varphi_\kappa^{(j)})$ *must be zero unless* $\Gamma^{(j')}$ *is found in* $\Gamma_{H'} \times \Gamma^{(j)}$. An equivalent statement is that the matrix element must vanish unless $\Gamma^{(j')*} \times \Gamma_{H'} \times \Gamma^{(j)}$ includes Γ_1, the identical representation. The latter form follows from the observation that a matrix element is invariant under any unitary transformation P_R of the integrand $\psi_{\kappa'}^{(j')*}H'\varphi_\kappa^{(j)}$. If we sum over all P_R and use the orthogonality of all other $\Gamma^{(j)}(R)_{\kappa\lambda}$ to $\Gamma_1(R) = 1$, we have the result.

Note that the theorem proved earlier in the section is a special case of the general matrix-element theorem just proved. We should also emphasize that this theorem does *not* guarantee the *existence* of any matrix elements. Rather, it *excludes* the possibility of some kinds. To make a more complete prediction based on symmetry alone, one would need to take advantage of the row index κ as well as the representation label j. For example, if $H' \sim x$ and $\varphi_\kappa^{(j)} \sim y$, we have $\Gamma_{H'} = \Gamma^{(j)} = T_1$ of the group O. Taking $T_1 \times T_1$, we obtain $T_1 + T_2 + E + A_1$. Yet $H'\varphi_\kappa^{(j)} \sim xy$ belongs to T_2 alone. Thus, only if $\Gamma^{(j')} = T_2$ can the matrix element $(\psi_{\kappa'}^{(j')}, H'\varphi_\kappa^{(j)})$ differ from zero, and then only if κ' is the xy row index of T_2. In the absence of such a detailed examination, one can say only that, if one considered functions with all the possible symmetries allowed by the representation labels alone, then matrix elements of all the types allowed by the theorem would occur (barring any fortuitous cancellations for specially chosen values of parameters).

4-10   Selection Rules and Parity

The transition probability per unit time in the presence of a perturbation is given by the famous "Golden Rule formula"[1]

$$w = \frac{2\pi}{\hbar} \, \rho(k) \, |H'_{km}|^2 \tag{4-12}$$

where $\rho(k)$ is the density of final states in energy and H'_{km} is the matrix element of the perturbative Hamiltonian H' which connects the initial state

---

[1] L. I. Schiff, "Quantum Mechanics," 2d ed., p. 199, McGraw-Hill Book Company, Inc., New York, 1955. Note that in using this formula a consistent normalization volume must be used for $\rho(k)$ and H'_{km}.

m with the final state k.   For electric-dipole transitions, $H' = e\mathbf{r} \cdot \mathbf{E} \sim e\mathbf{p} \cdot \mathbf{A}/c$. Since the field $\mathbf{E}$ is determined externally by the radiation inducing the transition, the transition probability under given conditions is proportional to $|\mathbf{r}_{km}|^2$ or $|\mathbf{p}_{km}|^2$.   Similarly, for magnetic-dipole transitions, $w \sim |\mathbf{\mu}_{km}|^2 = \beta^2 |(2\mathbf{S} + \mathbf{L})_{km}|^2$, where $\mathbf{\mu}$ is the magnetic-moment operator and β is the Bohr magneton.   For electric-quadrupole transitions, we have to find matrix elements of quadratic forms of coordinates which couple to gradients in the exciting field.   For most purposes, however, we may confine our attention to the dipole coupling terms.   We now consider how our matrix-element selection rule governs the allowed dipole transitions.

Electric-dipole transitions.   The electric-dipole-moment operator $e\mathbf{r}$ is a *polar vector* (V).   This means that its components transform like the co-ordinates.   For example, in a rotation through φ about the z axis

$$\Gamma^V(\varphi) = \begin{pmatrix} \cos \varphi & -\sin \varphi & 0 \\ \sin \varphi & \cos \varphi & 0 \\ 0 & 0 & 1 \end{pmatrix} \tag{4-13a}$$

gives the transformation properties of the x, y, z components.   From this we see that the character is

$$\chi^V(\varphi) = 1 + 2 \cos \varphi \tag{4-14a}$$

This result is true for rotations by φ about any axis, since they are all in the same class.

If we now combine an inversion with the rotation by φ, then the matrix changes sign to

$$\Gamma^V(\varphi, i) = \begin{pmatrix} -\cos \varphi & \sin \varphi & 0 \\ -\sin \varphi & -\cos \varphi & 0 \\ 0 & 0 & -1 \end{pmatrix} \tag{4-13b}$$

so that
$$\chi^V(\varphi, i) = -(1 + 2 \cos \varphi) \tag{4-14b}$$

Knowing the character under an arbitrary proper or improper rotation, we can find what representation a polar vector belongs to for any given group.   Of course, the three-dimensional representation Γ^V may decompose into more than one irreducible representation.   In fact it *must* decompose if the group under consideration has no three-dimensional irreducible representations.

Magnetic-dipole transitions.   The magnetic-dipole-moment operator $\mathbf{\mu}$ is an *axial vector* (A), or *pseudovector*.   This means that it transforms like a (polar) vector under rotations but is *invariant under inversion*.   This is true because, for example, $\mu \sim L \sim r \times p \sim mr \times dr/dt$. The two sign changes as $r \to -r$ cancel, leaving the product invariant. An axial vector is really a skew-symmetric tensor of rank 2. From these properties,

$$\Gamma^A(\varphi, i) = \Gamma^A(\varphi) = \begin{pmatrix} \cos\varphi & -\sin\varphi & 0 \\ \sin\varphi & \cos\varphi & 0 \\ 0 & 0 & 1 \end{pmatrix} \tag{4-15}$$

and

$$\chi^A(\varphi, i) = \chi^A(\varphi) = 1 + 2\cos\varphi \tag{4-16}$$

Examples. In the group $O_h = O \times i$, we know that (x, y, z) and hence any polar vector V transforms as T_{1u}. Similarly, an axial vector A transforms as T_{1g} since it has the same rotational properties as V but is even under inversion. These results can be checked by using the characters above together with the character table of O_h.

In our example group D_3, we do not have inversion symmetry, and parity is not a good quantum number. V and A act the same under all operations of the group. Using (4-16), $\chi(E) = 3$, $\chi(C_2) = 1 + 2\cos\pi = -1$, $\chi(C_3) = 1 + 2\cos(2\pi/3) = 0$. Decomposing this character, we find

$$\Gamma^V = \Gamma^A = A_2 + E$$

Actually, we have noted earlier that z belongs to A_2 and (x, y) belongs to E. Thus we can say that A_2 determines the selection rules for radiation polarized along z, while E determines selection rules for radiation polarized in the xy plane.

Inversion and parity. Until now, we have skirted the question of the inversion operator i. We have concentrated our attention on the point groups composed only of proper rotations. In many groups which actually interest us, we do in fact have inversion symmetry. In handling these, we take advantage of the fact that i commutes with all other operators in our group. This allows us to consider the full group as a direct product and to write the character table directly. This point was illustrated rather fully in the example of Sec. 3-9, except that in that example we considered the group (σ_h, E) rather than (i, E). Thus, in the present instance, we combine the two tables

$\mathcal{G}$	$\mathcal{C}_1 \cdots \mathcal{C}_r$
Γ_1	
$\cdots$	χ
Γ_r	

and

i	E	i
g	1	1
u	1	-1

to obtain the table

$\mathscr{G} \times i$	$\mathscr{C}_1 \cdots \mathscr{C}_r$	$i\mathscr{C}_1 \cdots i\mathscr{C}_r$
Γ_{1g} $\cdots$ Γ_{rg}	χ	χ
Γ_{1u} $\cdots$ Γ_{ru}	χ	$-\chi$

In the free atom we certainly have inversion symmetry in the Hamiltonian (for all the terms concerning the electrons at least!). Therefore parity is a good quantum number, and all atomic states will be either pure g or pure u. Actually the conventional notation for parity in atoms is $+$ and $-$ rather than g and u. The parity of a one-electron wavefunction is $(-)^l$, where l is the orbital angular-momentum quantum number. In a many-electron atom, therefore, it is $\prod_k (-)^{l_k}$, where k runs over all the electrons. In other words, $\chi(i) = \prod_k (-1)^{l_k}$.

Now, if we place such an atomic state in a crystal field, we may calculate the character of its representation with respect to the operators of the full symmetry group, including the inversion (if present). When this character is decomposed into those of the irreducible representations of the crystal group, it is obvious that only g representations will be found if the atomic state was g (or $+$) and similarly for u (or $-$). This follows from the "evenness" and "oddness" of the upper and lower halves of the character table, respectively. *Thus we conclude that g terms split only into g terms and u terms split only into u terms.* Moreover, since the left and right halves of the character table are identical, except for a possible uniform sign change, by using both halves we merely repeat each term in the sum which occurs in the orthonormality relations and the decomposition formulas. *Thus we can work out all splittings successfully by simply considering the group of proper rotations.* This validates our previous work on the cubic group, for example, where we restricted our attention to O, not O_h.

Parity, or inversion symmetry, does play a vital role in the determination of selection rules, however. The electric-dipole operator belongs to a u representation. Therefore our selection-rule theorem tells us that transitions are forbidden between states of the same parity (since $\Gamma_u \times \Gamma$ has the opposite parity to Γ and therefore cannot contain any representations of the same parity as Γ). But the argument of the preceding paragraph shows that the parity is invariant under crystal field splittings.   Therefore, *if the crystal field has inversion symmetry, all electric-dipole transitions within the level structure arising from a single free-atom term are forbidden.*

On the other hand, since the magnetic-moment operator is an *axial* vector, it is a Γ_g.   Therefore, *magnetic-dipole transitions within a term are not forbidden by parity.*   Many of these transitions will be forbidden by other symmetry properties, however.   We can find those which are *not* forbidden by considering direct products, as shown in the previous section.

For example, in D_3 we found that μ_z belonged to A_2, and (μ_x, μ_y) to E. Then it is not forbidden that z-polarized radiation will induce transitions between any Γ_i and $\Gamma_i \times A_2$.   Taking all possible products,

$$A_1 \times A_2 = A_2 \qquad A_2 \times A_2 = A_1 \qquad E \times A_2 = E$$

Hence the allowed transitions are

$$A_2 \leftrightarrow A_1 \qquad E \leftrightarrow E$$

For radiation polarized in the xy plane,

$$A_1 \times E = E \qquad A_2 \times E = E \qquad E \times E = A_1 + A_2 + E$$

and transitions are allowed between all states and those belonging to E.

Many of the cases of importance in crystal-field splittings have been treated along these lines by Kittel and Luttinger.[1]   Actually more recent theoretical research[2] indicates that the optical transitions between crystal-field split levels which cause crystals of paramagnetic salts to have characteristic colors are not predominantly magnetic-dipole transitions but instead are weak electric-dipole transitions.   These are possible because vibrations of the ions in the physical crystal destroy the perfect inversion symmetry of the ideal crystal, leaving parity not quite a good quantum number.   Since electric-dipole transitions are normally of order $(1/\alpha)^2 \sim (137)^2$ times as intense as magnetic-dipole transitions, even a small breakdown in the parity selection rule allows the electric-dipole transitions to dominate.

Atomic selection rules.   The above argument showing that electric-dipole transitions require a parity change is valid *a fortiori* in atoms where the inversion symmetry is perfect (with interatomic interactions and collisions neglected).   In a one-electron atom, where $\chi(i) = (-1)^l$, this requires a change of an odd number of units in l.   We shall see in the next chapter that rotational symmetry requires that $|\Delta l| \leq 1$ for a dipole transition.

---

[1] C. Kittel and J. M. Luttinger, *Phys. Rev.*, **73**, 162 (1948).

[2] A. D. Liehr and C. J. Ballhausen, *Phys. Rev.*, **106**, 1161 (1957); R. A. Satten, *J. Chem. Phys.*, **27**, 286 (1957), **29**, 658 (1958); C. J. Ballhausen, "Introduction to Ligand Field Theory," McGraw-Hill Book Company, Inc., New York, 1962.

Combining these two requirements, we obtain the selection rule $\Delta l = \pm 1$ for one-electron atoms, which is well obeyed in practice.

4-11   Directed Valence

In this section we want to explain the symmetries found in molecular binding in terms of those of the available atomic orbitals. For example, why does carbon like tetrahedral and trigonal bonding? Alternatively, given a set of bond angles determining a configuration, we wish to find out what types of orbitals play a role in the binding. This type of question—that of chemical valence—can be approached from two different approximate methods: *molecular orbitals* (Mulliken, Lennard-Jones, and others), in which one-electron eigenfunctions extending over the entire molecule are taken as basis functions; and the Heitler-London-Pauling method of *valence orbitals*, localized in the region between the cores of the atoms being bonded. When carried to a high enough order of approximation, both lead to the same results. Both have their own strong points. For the present discussion, which is intended primarily as another example of the application of group theory, we shall use the valence orbital method. We shall give a more thorough discussion of molecular binding in Chap. 7.

In this method the binding energy may be viewed as arising from the fact that interference terms in the electronic wavefunction, arising from the exchange equivalence of all electrons, cause charge to pile up in the region of low potential energy between the nuclei. The extent to which this occurs is roughly proportional to the amount of overlap between the functions based on the two centers. This leads to Pauling's *overlap criterion* of bond strength. Thus the problem to which we shall apply our group-theoretical machinery is the working out of all possible sets of atomic orbitals from which a set of these overlapping, directed valence orbitals can be formed.

Tetrahedral carbon. As an elementary example of the things we are looking for, let us look into the familiar tetrahedral bonding of carbon, as found in CH_4, for example. The ground state of atomic C is $1s^2 2s^2 2p^2$. We expect the $1s$ shell to be undisturbed by binding; so we have to deal only with $2s$ and $2p$ functions. Neglecting the different radial dependences, and normalizing to 4π over a sphere, these have the forms

$$s = 1 \qquad p_x = \sqrt{3} \sin \theta \cos \varphi$$

$$p_y = \sqrt{3} \sin \theta \sin \varphi$$

$$p_z = \sqrt{3} \cos \theta$$

where θ and φ are the usual polar coordinates. We wish to form four linear combinations of these four functions which are equivalent in all ways except that each is directed along a different one of the four tetrahedral directions. A little consideration of symmetry properties leads to

$$\psi_{111} = \tfrac{1}{2}(s + p_x + p_y + p_z)$$

$$\psi_{1,-1,-1} = \tfrac{1}{2}(s + p_x - p_y - p_z)$$

$$\psi_{-1,1,-1} = \tfrac{1}{2}(s - p_x + p_y - p_z)$$

$$\psi_{-1,-1,1} = \tfrac{1}{2}(s - p_x - p_y + p_z)$$

The indices on ψ indicate the direction numbers of the major lobe of the function with respect to the axes of the cube in which the tetrahedron is inscribed. These four orthogonal functions are called the tetrahedral sp^3 *hybrid orbitals*. Under the operation of the group T_d these functions transform only among themselves. Thus they form a basis for a (reducible) representation of T_d. We shall use this property in our later discussion.

A simple measure of overlap and hence of bond strength is the numerical value of the angular function along the bond axis where it is at a maximum. Along the (111) direction, $\cos \theta = 1/\sqrt{3}$, $\sin \theta = \sqrt{2/3}$, and $\sin \varphi = \cos \varphi = 1/\sqrt{2}$. Therefore $p_x = p_y = p_z = \sqrt{3}/\sqrt{3} = 1$ along the (111) direction. Combining this with $s = 1$, we have $\psi_{111}(111 \text{ axis}) = \tfrac{1}{2}[1 + 1 + 1 + 1] = 2$. It is easily verified that all four of the hybrid orbitals have this value along their own axes. This must be true if they have the required degree of symmetry. Further analysis shows that these orbitals give the highest bond strength of any which can be formed from s and p functions. This explains why the tetrahedral bonds of C are so common in nature despite the fact that it is necessary to promote a $2s$ electron from the $2s^2 2p^2$ ground configuration to a $2p$ electron in order to form these functions. The extra binding more than compensates the promotion energy.

Trigonal carbon. Another observed bonding scheme for carbon is the trigonal bond. In this we form three equivalent coplanar bonds separated by angles of 120°. These can be formed as sp^2 hybrids from s, p_x, p_y.

$$\psi_1 = \sqrt{\tfrac{1}{3}}s + \sqrt{\tfrac{2}{3}}p_x \qquad \psi_{2,3} = \sqrt{\tfrac{1}{3}}s - \sqrt{\tfrac{1}{6}}p_x \pm \sqrt{\tfrac{1}{2}}p_y$$

These are orthonormal and transform into each other under the symmetry operations of D_{3h}. The bond strength is $\psi_1(\theta = \pi/2, \varphi = 0) = \sqrt{\tfrac{1}{3}} + \sqrt{2} = 1.99$. Since this is only slightly below the strength of the tetrahedral bonds, these bonds in the plane are strong. However, perpendicular to the plane we have only p_z, with a bond strength of $\sqrt{3} = 1.73$. This difference accounts qualitatively for the fact that graphite consists of layers with strong internal bonds but weak interlayer bonding.

Another aspect of the bonding problem arises here, namely, the π bond. This is contrasted with the σ bonds which we have been considering, in which the charge is axially symmetric about the bond axis. For example, in ethylene (C_2H_4), we have what chemists call a *double bond* between the two carbons. The first bond is an ordinary σ bond formed by the overlap of two trigonal orbitals, one from each C. The second is a π bond formed by the overlap of the edges of the p_z charge clouds, which are actually pninting normal to the plane of the molecule. These have a relatively weak overlap and so contribute less to the binding energy. [C—C bond energy is ~59 kcal/mole; C=C is ~100 (= 59 + 41) kcal/mole.] The extra bonding is adequate to prevent relative rotation of the two CH_2 groups, which would reduce the overlap of the p_z functions. This ensures that the molecule stays planar.

4-12 Application of Group Theory to Directed Valence

Now that we have built up some feeling for what we are trying to find, let us examine the problem from a group-theoretical viewpoint. The key to the problem of finding suitable directed orbitals is that they must transform into each other under the group operations. Therefore they form a basis for a (generally reducible) representation of the group. By simple geometrical considerations we can work out the character of the bond representation. Knowing the characters of all the irreducible representations based on free-atom functions, we can easily see from which atomic functions suitable hybrids can be formed.

We illustrate with the trigonal-bond case where the full symmetry group is D_{3h}. The character table (from Sec. 3-9, after relabeling and rearranging to the conventional order) is:

	D_{3h}	E	σ_h	$2C_3$	$2S_3$	$3C_2'$	$3\sigma_v$
z^2	A_1'	1	1	1	1	1	1
	A_2'	1	1	1	1	-1	-1
	A_1''	1	-1	1	-1	1	-1
z	A_2''	1	-1	1	-1	-1	1
$(x, y), (xy, x^2 - y^2)$	E'	2	2	-1	-1	0	0
(xz, yz)	E''	2	-2	-1	1	0	0
	σ	3	3	0	0	1	1
	π	6	0	0	0	-2	0

In this we have listed how the p-like functions (x, y, z) and the d-like functions $(z^2, x^2 - y^2, xy, yz, zx)$ transform. Of course s-like functions are invariant under all group operations and so belong to A_1'. We have also included the characters of the σ-bond and π-bond representations. These characters are simply the number of orbitals going into themselves minus the number going into their negatives under a group operation. This fact follows directly from the definition of the representation matrices. We now consider these characters in more detail.

The three σ bonds can be represented by arrows along the bond axes, as shown in Fig. 4-9a. Under E, they all go into themselves, so $\chi^{(\sigma)}(E) = 3$;

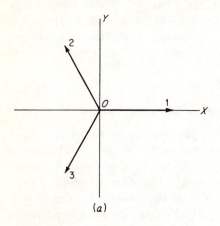

(a)

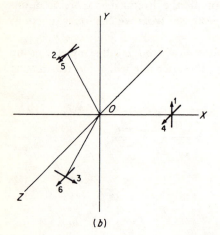

(b)

Fig. 4-9. *Schematic representations of* (a) *the three σ bonds and* (b) *the six π bonds which arise in a case of trigonal symmetry. In both cases, bonds* 1, 2, 3 *lie in the xy plane. The arrows representing bonds* 4, 5, 6 *are perpendicular to the xy plane.*

similarly for σ_h, since the arrows (bonds) are axially symmetric. Under C_3 and S_3, the threefold rotations change each arrow into another, leaving no diagonal elements in the matrix representation. Thus $\chi^{(\sigma)}(C_3) = \chi^{(\sigma)}(S_3) = 0$. Under either C_2' or σ_v, two arrows are interchanged, while the third goes into itself. Therefore $\chi^{(\sigma)}(C_2') = \chi^{(\sigma)}(\sigma_v) = 1$. From this character system we see that

$$\sigma = A_1' + E'$$

Considering the representations based on the various atomic functions, we see that the set of three σ-bonding orbitals can be made up in four different ways from s, p, and d functions:

$$sp^2 \qquad s + (p_x, p_y)$$
$$sd^2 \qquad s + (d_{xy}, d_{x^2-y^2})$$
$$dp^2 \qquad d_{z^2} + (p_x, p_y)$$
$$d^3 \qquad d_{z^2} + (d_{xy}, d_{x^2-y^2})$$

The group theory assures us that these are the only possible suitable combinations of these functions for the trigonal symmetry.

The π-bonds can be represented (Fig. 4-9b) by arrows perpendicular to the bond axis, the arrow indicating the direction of the positive lobe of the transverse function. This scheme suggests the symmetry of functions on the ligand atoms required to overlap with the corresponding central-atom function. In our D_{3h} example, we have six such bonds: 1, 2, 3 are in the xy plane; 4, 5, 6 are parallel to z and point out of the page. By considering the effect of the group operations on these arrows, we work out the character as before. For example, $\chi^{(\pi)}(E) = 6$ because all the arrows go into themselves; $\chi^{(\pi)}(\sigma_h) = 0$ since 1, 2, 3 go into themselves, while 4, 5, 6 go into their negatives. Using the complete character listed as π in the table, we find

$$\pi = A_2' + A_2'' + E' + E''$$

A little inspection shows that the bonds 1, 2, 3 already form a complete basis for $A_2' + E'$, while 4, 5, 6 form a basis for $A_2'' + E''$. This is natural, since 1, 2, 3 are even under σ_h, while 4, 5, 6 are odd, and the representations with single primes are even, while those with a double prime are odd. Since no A_2' functions can be formed from s, p, or d functions, we can properly form only the 4, 5, 6 type of π bond, which is odd on reflection in the xy plane. These require $A_2'' + E''$ and hence may be formed only as pd^2 hybrids from the functions p_z, d_{xz}, d_{yz}. The suitable combinations are readily seen to be analogous to those for the trigonal σ bonds, namely,

$$\phi_1 = \sqrt{\tfrac{1}{3}}p_z + \sqrt{\tfrac{2}{3}}d_{xz} \qquad \phi_{2,3} = \sqrt{\tfrac{1}{3}}p_z - \sqrt{\tfrac{1}{6}}d_{xz} \pm \sqrt{\tfrac{1}{2}}d_{yz}$$

These functions are odd on reflection in the xy plane and are concentrated in lobes pointing along the three trigonal directions. They are called *strong* π-bonding orbitals because they require none of the atomic functions already used to form the σ bonds. It is also possible to form *weak* π bonds of the 1, 2, 3 type (even on reflection in the xy plane) from the E' functions (p_x, p_y) or $(d_{xy}, d_{x^2-y^2})$. These are weak because the E' functions are likely to be already partially used up in forming the σ bonds.

Working in the way shown in these examples, one can prepare an exhaustive table of the electronic configurations which are consistent with various bond symmetries. Such a table is given in the book of Eyring, Walter, and Kimball.[1] Group theory is incapable of deciding which of the various *possible* configurations will have the lowest energy and will therefore be found in nature. This must be decided by rules like Pauling's bond-strength criterion or by detailed calculations considering actual matrix elements of the Hamiltonian. However, the group theory *does* give considerable guidance in setting up functions of suitable symmetry to serve as a point of departure for detailed calculations.

EXERCISES

4-1 According to what irreducible representation of O does $xyzf(r)$ transform? Explain.

4-2 (*a*) Calculate the crystal-field potential up through terms of fourth degree resulting from point charges $-e$ located at $x = \pm a$, $y = \pm b$, and $z = \pm c$.

(*b*) Show that this may be written as a sum of six independent terms: a constant term, two homogeneous second-degree polynomials, and three homogeneous fourth-degree polynomials, each satisfying Laplace's equation $\nabla^2\varphi = 0$, and each invariant under the symmetry operations of the group D_{2h}. What are the coefficients in terms of a, b, and c? [HINT: Note that $r^l Y_l{}^m(\theta, \varphi)$ satisfies Laplace's equation; and verify that $(Y_l{}^m + Y_l{}^{-m})$, with l and m even integers, has the symmetry of D_{2h}.]

(*c*) Specialize your results for the cases $a = b \neq c$ (tetragonal field) and $a = b = c$ (cubic field).

4-3 The splitting of terms of $L = 0, 1, 2,$ and 3 under a crystal-field perturbation of octahedral symmetry is given in the text. Determine the further splitting of each term under a small added perturbation having the symmetry group D_4 (tetragonal group). You may assume that the D_4 axis coincides with one of the fourfold axes of the group O. The case $a = b \neq c$ above is an example of such a field, but make the calculation purely group theoretically so that it is independent of the approximations of this model.

4-4 (*a*) Work out the character table of the tetragonal double group.

(*b*) Use it to determine the additional splitting of the levels of Exercise 4-3 if the atom has an electronic spin of $\frac{3}{2}$ and the spin-orbit coupling is taken into account.

4-5 Using the results of Exercise 3-11, work out the selection rules for magnetic-dipole transitions between the various irreducible representations of O and D_4 found in Exercise 4-3.

4-6 (*a*) According to what irreducible representation of O does $H' = Ax^2 + By^2 - (A + B)z^2$ transform?

(*b*) If one considered this rhombic field as a perturbation applied to a situation of cubic symmetry, what sorts of matrix elements are allowed by symmetry?

---

[1] H. Eyring, J. Walter, and G. E. Kimball, "Quantum Chemistry," p. 231, J. Wiley & Sons, Inc., New York, 1944; see also G. E. Kimball, *J. Chem. Phys.*, **8**, 188 (1940); and J. H. Van Vleck and A. Sherman, *Revs. Mod. Phys.*, **7**, 174 (1935).

4-7 According to what irreducible representation of O_h does $z(x^2 - y^2)$ transform? What are its partners?

4-8 (*a*) What representations of D_{4h} are required to form four equivalent σ bonds directed from a central atom toward four others at the corners of a square?

(*b*) What electronic configurations using only s, p, and d orbitals will form such hybrids?

(*c*) Find explicitly a set of four equivalent sp^2d hybrid σ bonds.

REFERENCES

BALLHAUSEN, C. J.: "Introduction to Ligand Field Theory," McGraw-Hill Book Company, Inc., New York, 1962.

BELL, DOROTHY G.: Group Theory and Crystal Lattices, *Revs. Modern Phys.*, **26**, 106 (1954).

BETHE, H. A.: Termaufspaltung in Kristallen, *Ann. Physik*, **3**, 133 (1929).

BUERGER, M. J.: "Elementary Crystallography," John Wiley & Sons, Inc., New York, 1956.

EYRING, H., J. WALTER, and G. E. KIMBALL: "Quantum Chemistry," app. VII, John Wiley & Sons, Inc., New York, 1944.

FICK, E.: Die Termaufspaltung in elektrostatischen Kristallfeldern, *Z. Physik*, **147**, 307 (1957).

KOSTER, G. F.: Space Groups and Their Representations, "Solid State Physics," vol. 5, p. 174, Academic Press Inc., New York, 1957.

LANDAU, L. D., and E. M. LIFSHITZ: "Quantum Mechanics, Non-relativistic Theory," chap. XII, Addison-Wesley Publishing Company, Inc., Reading, Mass., 1958.

OPECHOWSKI, W.: Sur les groupes cristallographiques "doubles," *Physica*, **7**, 552 (1940).

PHILLIPS, F. C.: "An Introduction to Crystallography," Longmans Green & Co., Inc., New York, 1946.

ROSENTHAL, JENNY E., and G. M. MURPHY: Group Theory and the Vibrations of Polyatomic Molecules, *Revs. Modern Phys.*, **8**, 317 (1936).

5 FULL ROTATION GROUP AND ANGULAR MOMENTUM

5-1 Rotational Transformation Properties and Angular Momentum

Before developing the representations of the full three-dimensional rotation group, let us review the underlying relation between the rotational transformation properties of a wavefunction and its angular momentum.

The *linear* momentum $\mathbf{p}$ of a system may be defined most basically as that quantity whose conservation is a consequence of the translational isotropy of space. Now, the isotropy of space means that the spatial gradient of the Hamiltonian H is zero and hence that ∇ commutes with H. But commutation of an operator with H is the requirement for conservation of that operator in quantum mechanics. Hence we associate ∇ with $\mathbf{p}$ and are led to the usual Schrödinger operator replacement $\mathbf{p} = (\hbar/i)\nabla$.

Considering for simplicity only the x component, we have

$$p_x \psi = \frac{\hbar}{i} \frac{\partial \psi}{\partial x} = \lim_{\delta x \to 0} \frac{\hbar}{i} \frac{\psi(x) - \psi(x - \delta x)}{\delta x}$$

$$= \lim_{\delta x \to 0} i\hbar \frac{P_{\delta x} - 1}{\delta x} \psi(x)$$

where $P_{\delta x}$ is the transformation operator which shifts the contours of $\psi(x)$ by δx with respect to fixed axes (or shifts the axes by $-\delta x$ if the function is considered to be stationary). An alternative way of writing this equation is

$$P_{\delta x} = 1 - \frac{i(\delta x)p_x}{\hbar}$$

where δx is treated as infinitesimal. These equations state the relationship between the linear-momentum operator and the transformation operator $P_{\delta x}$ for translational displacement.

In a similar manner, we may *define* the x component of the *angular momentum* as the operator whose conservation follows from the invariance of the Hamiltonian under rotations about the x axis, i.e., from $\partial H / \partial \theta_x = 0$. In complete analogy to the linear momentum, we then obtain a definition of angular momentum in terms of infinitesimal rotations,

$$J_x \psi = \lim_{\delta \theta_x \to 0} i\hbar \frac{P_{\delta \theta_x} - 1}{\delta \theta_x} \psi \tag{5-1}$$

or

$$P_{\delta \theta_x} = 1 - \frac{i(\delta \theta_x)}{\hbar} J_x \tag{5-2}$$

where $P_{\delta \theta_x}$ is the transformation operator which rotates the contours of the function by $\delta \theta_x$ about the x axis in the sense which makes a right-hand-screw advance along x. From this definition we can readily find the transformation operator P_{θ_x} for a finite rotation as follows: Let $\theta_x = n(\delta \theta_x)$, and let $n \to \infty$ while $(\delta \theta_x) \to 0$ in such a way as to hold θ_x constant. Then we can write P_{θ_x} as the result of $n = \theta_x / \delta \theta_x$ iterations of $P_{\delta \theta_x}$,

$$P_{\theta_x} \psi = \lim_{\delta \theta_x \to 0} (P_{\delta \theta_x})^{\theta_x / \delta \theta_x} \psi$$

$$= \lim_{\delta \theta_x \to 0} \left[1 - \frac{i(\delta \theta_x)}{\hbar} J_x \right]^{\theta_x / \delta \theta_x} \psi$$

$$= e^{-i\theta_x J_x / \hbar} \psi \tag{5-3a}$$

or

$$P_{\theta_x} \psi = \sum_{k=1}^{\infty} \frac{(-i\theta_x J_x / \hbar)^k}{k!} \psi \tag{5-3b}$$

where we have used the definition of the exponential e^y as the limit of $(1 + x)^{y/x}$ as $x \to 0$. The second form (5-3b) indicates how one could go about calculation with an exponential operator by Taylor's-series expansion. Clearly a result analogous to (5-3) is true for rotations about any axis, since we have made no special assumptions about the x axis. We have already made use of this result for the z component of the angular momentum in Sec. 4-4 while working out the character of the representations based on the spherical harmonics. Since the z axis was chosen for the axis of quantization, J_z had a definite value $m\hbar$ and we simply found

$$P_{\theta_z}\psi_m{}^{(j)} = e^{-im\theta_z}\psi_m{}^{(j)} \tag{5-4}$$

Note that this result follows directly from the relation between J_z and infinitesimal rotations. Hence it is valid whenever J_z is quantized. In particular, it holds also for half-integral j values where the spherical harmonics no longer serve as basis functions. This observation justifies our use of the formula (4-11) for the character $\chi_j(\alpha)$ for half-integral as well as integral values of j. The double-valued characters noted there are consequences of the fact that, for *half-integral j* values, $P_{2\pi}\psi_m{}^{(j)} = -\psi_m{}^{(j)}$, so that a rotation by 2π leaves the wavefunctions changed in phase.

We can use (5-3) to define the transformation operator for an arbitrary rotation about an arbitrary axis. In many applications, however, it is convenient to express the arbitrary rotation as a sequence of rotations about, say, the x, y, and z axes. In this case we simply operate successively with several operators of the type (5-3). For example, if the desired rotation θ can be carried out by a rotation by θ_y about the y axis followed by a rotation by θ_x about the x axis, we have

$$\begin{aligned} P_\theta\psi &= P_{\theta_x} P_{\theta_y}\psi \\ &= e^{-i\theta_x J_x/\hbar}e^{-i\theta_y J_y/\hbar}\psi \end{aligned}$$

Now, finite rotations do not in general commute. In fact, geometric inspection (Fig. 5-1) shows that, if one made the above rotations in the opposite order, one would need to perform a further rotation of magnitude $\theta_x\theta_y$ about the z axis to reach the same final position. (This statement is valid only to order θ^2, on the assumption that θ_x and θ_y are small.) In other words, to second order,

$$P_{\theta_x} P_{\theta_y} = P_{(\theta_x\theta_y)_z} P_{\theta_y} P_{\theta_x}$$

or $\qquad e^{-i\theta_x J_x/\hbar}e^{-i\theta_y J_y/\hbar} = e^{-i\theta_x\theta_y J_z/\hbar}e^{-i\theta_y J_y/\hbar}e^{-i\theta_x J_x/\hbar}$

If we now expand the exponentials to second order in the θ's, we find identical equality for the constant and linear terms. However, preserving the order and canceling identical terms, we obtain the requirement

$$\frac{-\theta_x\theta_y J_x J_y}{\hbar^2} = \frac{-i\theta_x\theta_y J_z}{\hbar} - \frac{\theta_y\theta_x J_y J_x}{\hbar^2}$$

from the second-order terms.   Upon canceling common factors, this becomes

$$J_x J_y - J_y J_x = i\hbar J_z \tag{5-5}$$

Thus the familiar commutation relations of the angular-momentum operators are seen to form the condition which must be satisfied if the transformation operator associated with a rotation about both x and y is to be

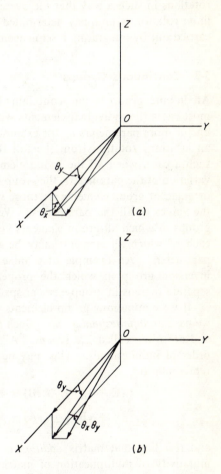

Fig. 5-1. *Illustration of noncommutativity of finite rotations. (a) shows $P_{\theta_x} P_{\theta_y}$ acting on a vector lying along the x axis; (b) shows that $P_{(\theta_x \theta_y)_z} P_{\theta_y} P_{\theta_x}$ yields the same result to second order in the angles. [It is characteristic of this simple illustrative example that P_{θ_x} has no effect in (b). The resulting equivalence is generally valid, however.]*

uniquely defined regardless of the path by which the total rotation is accomplished.

  We may generalize from this example to obtain the other two angular-momentum commutation relations by cyclic interchange of x, y, and z. The three commutation relations obtained in this way completely determine the properties of the angular-momentum operators, and these operators may in fact be defined directly in this way.   Purely algebraic development from the commutation relations has been used almost exclusively by Condon and Shortley,[1] and more recently by Racah and others, to deduce many properties of quantum-mechanical systems.   However, in our group-theoretical development the emphasis is entirely on the transformation properties. In that framework, the commutation relations serve as the *integrability conditions* relating the transformation operators associated with the various rotations in such a way that the overall transformation associated with any finite rotation is uniquely determined regardless of the path by which it is carried out by integrating a sequence of infinitesimal rotations.

5-2 Continuous Groups

An infinite group is one containing an infinite number of elements.   In most cases the individual elements will be specified by giving the values of one or more parameters p_i, for example, the three Eulerian angles specifying an arbitrary rotation from the full three-dimensional rotation group.   In such a case, when all the group elements may be obtained by continuous variation of the parameters, the group is said to be *continuous*.   Neighboring or adjacent group elements are those which differ by only small amounts in the values of all the parameters.   We can also consider *mixed continuous* groups in which all group elements lie in a finite number of regions, within each of which all elements may be obtained by continuous variation of parameters.   An example of a mixed continuous group is the rotation-inversion group in which the proper and improper rotations form two separate regions in group-element space.

If we confine our group elements to those which neighbor the identity, we have an *infinitesimal group*.   Such groups were studied extensively by Lie and are often called *Lie groups*.   All such groups are Abelian to the first order of infinitesimals.   This may be readily seen by considering a matrix representation,

$$(E + \epsilon A)(E + \epsilon B) = E + \epsilon(A + B) + \epsilon^2 AB$$

$$(E + \epsilon B)(E + \epsilon A) = E + \epsilon(B + A) + \epsilon^2 BA$$

and recalling that matrix *addition* is always commutative.   The noncommutativity of multiplication of matrices other than the unit matrix enters only in the second order of infinitesimals.   Obviously the infinitesimal rotations form a Lie group.

The Hurwitz invariant integral.   In developing the theory of finite groups we made heavy use of the rearrangement theorem.   This theorem states

---

[1] E. U. Condon and G. H. Shortley, "Theory of Atomic Spectra," Cambridge University Press, New York, 1935.

that, as R runs over all the elements of the group, SR also runs over all elements, including each just once.   From this we concluded that

$$\sum_R J(R) = \sum_R J(SR) \tag{5-6}$$

where $J(R)$ is any quantity depending only upon the element R.   This statement was central to the proof of the orthogonality theorems on which most of our formal structure rests.   When we consider continuous groups, we naturally convert such a sum to an integral over the parameters which specify the group elements.   The simple proof given for the finite groups fails, however, because one cannot set up a simple correspondence of one element to another.   Rather, one can only deal with a density of group elements in a region defined by certain values of the parameters.   Then one obtains a so-called "Hurwitz integral" of the sort

$$\sum_R J(R) \rightarrow \int J(R)\, dR = \int J(R) g(R)\, dp_1 \cdots dp_n$$

$$= \int J(R\{p_1, \ldots, p_n\}) g(R\{p_1, \ldots, p_n\})\, dp_1 \cdots dp_n \tag{5-7}$$

where $R\{p_1, \ldots, p_n\}$ is the group element fixed by the parameter values $p_1, \ldots, p_n$ and $g(R)$ is the density of group elements in the parameter space in the neighborhood of R.   The desired result is now that

$$\int J(SR)\, dR = \int J(R)\, dR \tag{5-8}$$

This will hold provided that the density of group elements R_i in group space is such that the density of points SR_i will always be the same as that of the points R_i.   The problem is to find a suitable $g(R)$ to convert the integral into one which can actually be carried out explicitly in parameter space. Since there is always an arbitrary normalization factor in the desired invariant density in parameter space $g(R)$, it is convenient to normalize to g_0, the density in the neighborhood of the identity.   We may then obtain the desired $g(R)$ by noting that R takes each group element from a volume element V_0 near E to a new element in a volume V near R.   This volume is determined by simply operating with R on all elements in V_0, as illustrated in Fig. 5-2.   Since we have the same number of points in each volume, we have

$$g(R) = \frac{g_0 V_0}{V} \tag{5-9}$$

and $g(R)$ is determined by the contraction of the volume element V_0 as it is moved from E to R. This contraction is given by the Jacobian of the transformation associated with R, namely,

$$g(R) = g_0 \left\{ \frac{\partial[p_1(RE\{e_1, \ldots, e_n\}), \ldots, p_n(RE\{e_1, \ldots, e_n\})]}{\partial[e_1, \ldots, e_n]} \right\}^{-1} \quad (5\text{-}10)$$

where the derivatives in the Jacobian are to be evaluated at the values of the parameters $e_i = p_i(E)$ corresponding to the identity.

The actual calculation of the density function from (5-10) is usually laborious. However, as soon as it is seen that such an invariant density

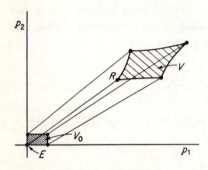

Fig. 5-2. *Illustration of transformation of group elements in V_0 at E into those in V at R in a two-dimensional parameter space.*

exists, it follows that all the orthogonality theorems carry over to the case of continuous groups. The proper weighting functions $g(R)$ for a given parametrization can then be found by elementary means. In any case, we have the orthogonality theorem

$$\int \Gamma^{(j)}(R)^*_{\kappa\lambda}\Gamma^{(j')}(R)_{\kappa'\lambda'}\, dR = \frac{\delta_{jj'}\delta_{\kappa\kappa'}\delta_{\lambda\lambda'}}{l_j}\int dR \quad (5\text{-}11)$$

where l_j is the dimensionality of the representation $\Gamma^{(j)}$ and $\int dR$ depends upon the normalization choice for $g(R)$. The corresponding result for the orthogonality of the characters is

$$\int \chi^{(j)}(R)^*\chi^{(j')}(R)\, dR = \delta_{jj'}\int dR \quad (5\text{-}12)$$

As an example of how (5-12) can be carried out in practice, we consider the character for rotations which we worked out earlier, namely,

$$\chi^{(j)}(\alpha) = \frac{\sin\left(j + \tfrac{1}{2}\right)\alpha}{\sin\left(\alpha/2\right)}$$

This is the character for all elements in the class of rotations by α, regardless of the axis of rotation. Thus we need a new kind of density function $g'(\alpha)$

which includes the effect of integrating $g(R)$ over all possible Eulerian angles, for example, which are consistent with an overall rotation by α. This would be hard to compute using (5-10). However, if we know that (5-12) must be valid, we may proceed to infer $g'(\alpha)$. Thus, we require

$$
\int_0^{2\pi} \frac{\sin\left[(j+\tfrac{1}{2})\alpha\right] \sin\left[(j'+\tfrac{1}{2})\alpha\right] g'(\alpha)\, d\alpha}{\sin^2(\alpha/2)}
$$

$$
= \int_0^{2\pi} \left[\cos(j-j')\alpha - \cos(j+j'+1)\alpha\right] \left[\frac{g'(\alpha)}{2\sin^2(\alpha/2)}\right] d\alpha
$$

$$
= \delta_{jj'} \int_0^{2\pi} g'(\alpha)\, d\alpha
$$

Since the cosine terms will integrate to $\delta_{jj'}$ if multiplied only by a constant, we are led to take $g'(\alpha) \sim \sin^2(\alpha/2)$. This is a reasonable form since the solid angle into which rotations by angles between α and $\alpha + d\alpha$ can be made goes to zero as $\alpha \to 0$ and as $\alpha \to 2\pi$. Such arguments are obviously not rigorous, but at least the result is plausible.

5-3 Representation of Rotations through Eulerian Angles

The most convenient manner of expressing an arbitrary rotation of axes is by the three Eulerian angles. We follow Rose's convention[1] and define the angles as follows. Starting with an x, y, z coordinate system, rotate by α about the z axis, and denote the new axes x', y', z'. Then rotate by β about the y' axis, and denote the resulting axes x'', y'', z''. Finally rotate by γ about the z'' axis, and label the final axes x''', y''', z'''. These rotations are illustrated in Fig. 5-3. Note that the successive rotations here are made about *rotated* axes, not the set x, y, z of space-fixed axes. Since it is often convenient to consider all rotations with respect to a permanent set of space-fixed axes, we now note that the same final rotated axes x''', y''', z''' could have been obtained by carrying out the rotations in reverse order about fixed axes. That is, we could have rotated by γ about z, then by β about y, and finally by α about z again. This equivalence is readily verified by considering some examples of successive finite rotations. It can also be proved rigorously by considerations of the sort used in discussing the elements conjugate to each other in a class. Consider first a simple case where $\gamma = 0$. Then we have only two rotations to consider, and we wish to establish the equality

$$
P_z(\alpha)P_y(\beta) = P_{y'}(\beta)P_z(\alpha)
$$

[1] M. E. Rose, "Elementary Theory of Angular Momentum," chap. IV, J. Wiley & Sons, Inc., New York, 1957.

where $P_i(\theta)$ represents a rotational transformation about the ith axis by θ. But this is the same as requiring that

$$P_y(\beta) = P_z^{-1}(\alpha)P_{y'}(\beta)P_z(\alpha)$$

In this form, the statement is that $P_y(\beta)$ is the same rotation as could be accomplished by first using $P_z(\alpha)$ to move the y axis into the y' position,

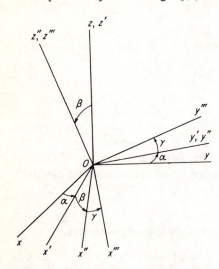

Fig. 5-3.   *The Eulerian angles.*

performing a rotation $P_{y'}(\beta)$ about it there, and then using $P_z^{-1}(\alpha)$ to return the y axis to its normal position.   In a similar way, one can see that

$$P_z(\gamma) = P_z^{-1}(\alpha)P_{y'}^{-1}(\beta)P_{z''}(\gamma)P_{y'}(\beta)P_z(\alpha)$$

Combining this with the above results shows that

$$P_z(\alpha)P_y(\beta)P_z(\gamma) = P_{z''}(\gamma)P_{y'}(\beta)P_z(\alpha) \tag{5-13}$$

and we can accomplish the desired rotation always using fixed axes.   The actual operator is then simply

$$P(\alpha, \beta, \gamma) \equiv P_z(\alpha)P_y(\beta)P_z(\gamma) = e^{-i\alpha J_z/\hbar}e^{-i\beta J_y/\hbar}e^{-i\gamma J_z/\hbar} \tag{5-14}$$

What we now seek are the matrices $D^{(j)}(\alpha, \beta, \gamma)_{m'm}$ which describe how the angular-momentum eigenfunctions of angular momentum j and z component m (in units of $\hbar$) transform under the general rotation operator (5-14).   Let us denote one of these eigenfunctions by $\varphi_m^{(j)}$.   (If j is integral, these transform simply as the spherical harmonics.)   Then by definition of the representation matrices

$$P(\alpha, \beta, \gamma)\varphi_m^{(j)} = \sum_{m'} \varphi_{m'}^{(j)} D^{(j)}(\alpha, \beta, \gamma)_{m'm} \tag{5-15}$$

Because of our choice of z-axis quantization, J_z is a diagonal operator, and the effect of the rotations by α and γ is easily written by using the exponential form of the rotation operator. However, the rotation $P_y(\beta)$ will have a nondiagonal representation, which we denote $\mathbf{d}^{(j)}(\beta)$. Combining the three, we find

$$P(\alpha, \beta, \gamma)\varphi_m^{(j)} = \sum_{m'} \varphi_{m'}^{(j)} e^{-im'\alpha} d^{(j)}(\beta)_{m'm} e^{-im\gamma} \tag{5-16}$$

or

$$D^{(j)}(\alpha, \beta, \gamma)_{m'm} = e^{-im'\alpha} d^{(j)}(\beta)_{m'm} e^{-im\gamma} \tag{5-17}$$

Thus the problem of finding the representation matrices of the full rotation group has been reduced to that of finding $\mathbf{d}^{(j)}(\beta)$. These matrices will now be determined for arbitrary integral or half-integral angular momentum j by an indirect procedure due to H. Weyl, which utilizes the homomorphism between the group of three-dimensional rotations and the group of 2×2 unitary matrices with determinant 1.

5-4   Homomorphism with the Unitary Group

It is apparent that the set of all 2×2 unitary matrices with determinant $+1$ forms a group, since it is closed under multiplication, it has a unit element, and it has an inverse for each element. The most general matrix in this group may be written as

$$\mathbf{u} = \begin{pmatrix} a & b \\ -b^* & a^* \end{pmatrix} \tag{5-18}$$

with the subsidiary condition $aa^* + bb^* = |a|^2 + |b|^2 = 1$ following from the requirement that the determinant be $+1$. Since a and b are in general complex, this leaves three independent parameters, which we shall presently relate to the three Eulerian angles.

To do this, we introduce the three Pauli matrices, which in the phase convention of Condon and Shortley take the form

$$\sigma_x = \begin{pmatrix} 0 & 1 \\ 1 & 0 \end{pmatrix} \qquad \sigma_y = \begin{pmatrix} 0 & -i \\ i & 0 \end{pmatrix} \qquad \sigma_z = \begin{pmatrix} 1 & 0 \\ 0 & -1 \end{pmatrix} \tag{5-19}$$

Evidently, we can express *any* 2×2 matrix with vanishing trace as a linear combination of these three. Denoting the coefficient of σ_x by x, etc., we have, for the general matrix $\mathbf{h}$,

$$\mathbf{h} \equiv \begin{pmatrix} h_{11} & h_{12} \\ h_{21} & h_{22} \end{pmatrix} = x\sigma_x + y\sigma_y + z\sigma_z \equiv \mathbf{r} \cdot \boldsymbol{\sigma} = \begin{pmatrix} z & x - iy \\ x + iy & -z \end{pmatrix} \tag{5-20}$$

Solving for the coefficients x, y, z, we find

$$x = \frac{h_{21} + h_{12}}{2} \qquad y = \frac{h_{21} - h_{12}}{2i} \qquad z = h_{11} = -h_{22} \qquad (5\text{-}21)$$

We also note that, if x, y, and z are real, then $\mathbf{h}$ is Hermitian.

Now let us transform $\mathbf{h}$ into a new matrix $\mathbf{h}'$ by a unitary transformation using a matrix $\mathbf{u}$ from the 2×2 unitary group. Then

$$\mathbf{h}' = \mathbf{u}\mathbf{h}\mathbf{u}^{-1} \equiv \mathbf{r}' \cdot \boldsymbol{\sigma} \qquad (5\text{-}22)$$

defines a new set of coordinates x', y', z' related to the first set x, y, z. If we carry out explicitly the matrix multiplication indicated in (5-22), using the definitions (5-18) and (5-20), we find

$$x' = \frac{h'_{21} + h'_{12}}{2}$$
$$= \frac{1}{2}(a^2 + a^{*2} - b^2 - b^{*2})x$$
$$+ \frac{i}{2}(-a^2 + a^{*2} - b^2 + b^{*2})y - (ab + a^*b^*)z$$

$$y' = \frac{h'_{21} - h'_{12}}{2i} \qquad\qquad\qquad\qquad (5\text{-}23)$$
$$= \frac{i}{2}(a^2 - a^{*2} + b^{*2} - b^2)x$$
$$+ \frac{1}{2}(a^2 + a^{*2} + b^2 + b^{*2})y + i(a^*b^* - ab)z$$

$$z' = h'_{11} = -h'_{22} = (a^*b + ab^*)x$$
$$+ i(a^*b - ab^*)y + (aa^* - bb^*)z$$

Thus, *associated with each 2×2 matrix $\mathbf{u}$ there is a 3×3 matrix $\mathbf{R}(\mathbf{u})$ which transforms x, y, z into x', y', z'.* The elements of $\mathbf{R}(\mathbf{u})$ may be read out from (5-23), which can be expressed more compactly as

$$\mathbf{r}' = \mathbf{R}(\mathbf{u})\mathbf{r} \qquad (5\text{-}24)$$

or, in terms of components, $r'_i = \sum_j [\mathbf{R}(\mathbf{u})]_{ij} r_j$.

We can see that $\mathbf{R}(\mathbf{u})$ actually represents a simple rotation of coordinates by making several observations. First, $\mathbf{R}(\mathbf{u})$ is real, as is evident from inspection of the elements displayed in (5-23). Second, we note that lengths are preserved since

$$|h'| = |u| \cdot |h| \cdot |u^{-1}| = |h|$$

and $\qquad |h| = -z^2 - (x + iy)(x - iy) = -(x^2 + y^2 + z^2)$

Thus $\qquad\qquad x^2 + y^2 + z^2 = x'^2 + y'^2 + z'^2$

But if lengths are preserved, all angles must be also and the transformation must be either a proper or an improper rotation. The inversion is excluded, however, since all elements may be obtained by continuous variation of the parameters (a, b) in $\mathbf{u}$ from the values $(1, 0)$ which yield the identity transformation. Hence the determinant of $\mathbf{R(u)}$ must be $+1$ for all values of the parameters, since there is no way for it to jump discontinuously to -1. Therefore we may conclude that $\mathbf{R(u)}$ *is a real orthogonal matrix representing a proper rotation of coordinates.*

Explicit relation between u and R(u). If we choose $\mathbf{u}$ to be diagonal, its most general form is

$$\mathbf{u}_1(\alpha) = \begin{pmatrix} e^{i\alpha/2} & 0 \\ 0 & e^{-i\alpha/2} \end{pmatrix} \tag{5-25a}$$

By using the general formula (5-23), this leads to the matrix

$$R_1(\alpha) = \begin{pmatrix} \cos \alpha & \sin \alpha & 0 \\ -\sin \alpha & \cos \alpha & 0 \\ 0 & 0 & 1 \end{pmatrix} \tag{5-25b}$$

which is precisely the rotation matrix for a rotation of coordinates by an angle α about z.

If instead we choose $\mathbf{u}$ to be real, the most general form is

$$\mathbf{u}_2(\beta) = \begin{pmatrix} \cos \dfrac{\beta}{2} & \sin \dfrac{\beta}{2} \\ -\sin \dfrac{\beta}{2} & \cos \dfrac{\beta}{2} \end{pmatrix} \tag{5-26a}$$

This produces the rotation matrix

$$R_2(\beta) = \begin{pmatrix} \cos \beta & 0 & -\sin \beta \\ 0 & 1 & 0 \\ \sin \beta & 0 & \cos \beta \end{pmatrix} \tag{5-26b}$$

which describes a rotation through β about the y axis.

With these two components, we can express the rotation of coordinates through an arbitrary set of Eulerian angles by the product matrix

$$\mathbf{u}(\alpha, \beta, \gamma) = \mathbf{u}_1(\gamma)\mathbf{u}_2(\beta)\mathbf{u}_1(\alpha)$$

If we carry out the indicated multiplications, we obtain

$$\mathbf{u}(\alpha, \beta, \gamma) = \begin{pmatrix} e^{i(\alpha+\gamma)/2} \cos \dfrac{\beta}{2} & e^{i(\gamma-\alpha)/2} \sin \dfrac{\beta}{2} \\ -e^{-i(\gamma-\alpha)/2} \sin \dfrac{\beta}{2} & e^{-i(\alpha+\gamma)/2} \cos \dfrac{\beta}{2} \end{pmatrix} \tag{5-27}$$

Thus we have proved that *an arbitrary rotation can be induced by a suitable matrix from the unitary group*. However, we shall now show that the matrix $-\mathbf{u}$ represents the same rotation, so that the representation is not unique, but "double-valued."

This ambiguity of sign arises from the defining relation $\mathbf{h}' = \mathbf{u}\mathbf{h}\mathbf{u}^{-1}$. Since this is quadratic in $\mathbf{u}$, the resulting $\mathbf{h}'$ is unchanged by a uniform sign change in $\mathbf{u}$ (which changes the sign of $\mathbf{u}^{-1}$ also). Another evidence of the ambiguity is the appearance of half angles in (5-27). Since the Eulerian angles specifying a rotation are determined only within multiples of 2π, the half angles are determined only within π. But all trigonometric functions and complex exponentials change sign when the argument is changed by π. Therefore the sign of the matrix $\mathbf{u}$ suffers an ambiguity.

We conclude: *there is a two-fold homomorphism between the group of 2×2 unitary matrices with $|\mathbf{u}| = 1$ and the group of three-dimensional pure rotations. The two matrices $\mathbf{u}$, $-\mathbf{u}$ are associated with a rotation $\mathbf{R}(\mathbf{u})$, and we have formulas establishing the relation in either direction*. Having established this homomorphism, we now work out all the representations of the unitary group by a simple method. Because of the homomorphism, these same matrices will serve as the representation matrices for the three-dimensional rotation group.

5-5   Representations of the Unitary Group

Clearly the matrices $\mathbf{u}$ themselves form a two-dimensional representation of the group. We may denote the two basis functions by ξ_1 and ξ_2, so that by definition

$$\xi_i' = P(\mathbf{u})\xi_i = \sum_{j=1}^{2} \xi_j u_{ji} \tag{5-28a}$$

or, written out fully,

$$(\xi_1', \xi_2') = (a\xi_1 - b^*\xi_2,\ b\xi_1 + a^*\xi_2) \tag{5-28b}$$

Starting with (5-28), we can generate representations of higher dimensionality by taking the basis functions to be the $(n + 1)$ monomials of degree n, namely, $f_p'^{(n)} = \xi_1{}^p \xi_2{}^{n-p}$. In that case, we have

$$P(\mathbf{u})f_p'^{(n)}(\xi_1, \xi_2) = f_p'^{(n)}(\xi_1', \xi_2') = (a\xi_1 - b^*\xi_2)^p(b\xi_1 + a^*\xi_2)^{n-p}$$

$$= \sum_{p'} f_{p'}'^{(n)} U'^{(n)}(\mathbf{u})_{p'p}$$

and the last equality can be written because the polynomial in the previous step is homogeneous of degree n. Thus the $f_p'^{(n)}$ do form a basis for an $(n + 1)$-dimensional representation $\mathbf{U}'^{(n)}$ of $\mathbf{u}$. It is not unitary as it stands, however, except for $n = 1$. To make it unitary, we must insert a normalizing factor $\sqrt{p!\,(n-p)!}$ in the denominator of $f_p'^{(n)}$, and we denote the normalized function simply $f_p{}^{(n)}$. This normalization factor will be justified shortly. Also, to prepare for the connection with the rotation group, we now change notation by setting

$$n = 2j \qquad p = j + m$$

The normalized basis functions are then given by the symmetrical form

$$f_m{}^{(j)} = \frac{\xi_1{}^{j+m}\xi_2{}^{j-m}}{\sqrt{(j+m)!\,(j-m)!}} \qquad m = -j, \ldots, j-1, j \qquad (5\text{-}29)$$

We may now work out the unitary-representation matrices $U^{(j)}{}_{m'm}$ explicitly by simply using the binomial expansion. Thus

$$P(\mathbf{u})f_m{}^{(j)} = \frac{(a\xi_1 - b^*\xi_2)^{j+m}(b\xi_1 + a^*\xi_2)^{j-m}}{\sqrt{(j+m)!\,(j-m)!}}$$

$$= \sum_\kappa \frac{(a\xi_1)^{j+m-\kappa}(-b^*\xi_2)^\kappa (j+m)!}{\sqrt{(j+m)!}\,\kappa!\,(j+m-\kappa)!} \sum_{\kappa'} \frac{(b\xi_1)^{j-m-\kappa'}(a^*\xi_2)^{\kappa'}(j-m)!}{\sqrt{(j-m)!}\,(j-m-\kappa')!\,\kappa'!}$$

Note that the general term contains $\xi_1{}^{2j-(\kappa+\kappa')}\xi_2{}^{\kappa+\kappa'} \sim f_{j-\kappa-\kappa'}^{(j)}$. Thus, if we set $\kappa' = j - m' - \kappa$, this is the $f_{m'}{}^{(j)}$ term in the sum. Eliminating the index κ' in favor of m', and collecting factors, we obtain

$$P(\mathbf{u})f_m^{(j)} = \sum_{m'} f_{m'}{}^{(j)} U^{(j)}(\mathbf{u})_{m'm} \qquad (5\text{-}30a)$$

where

$$U^{(j)}(\mathbf{u})_{m'm} = \sum_\kappa \frac{(-1)^\kappa \sqrt{(j+m)!\,(j-m)!\,(j+m')!\,(j-m')!}}{\kappa!\,(j+m-\kappa)!\,(j-\kappa-m')!\,(\kappa+m'-m)!}$$

$$\times\, a^{j+m-\kappa}(a^*)^{j-\kappa-m'}b^{\kappa+m'-m}(b^*)^\kappa \qquad (5\text{-}30b)$$

The summation runs over all values of κ for which the denominator is finite. For example, if $m' = j$, the denominator is infinite unless $\kappa = 0$. [This is so because the factorial of a negative integer is taken to be infinite, as follows from the relation between the factorial and gamma functions, namely, $n! = \Gamma(n+1)$.] Thus in this case there is only one term in the sum, and we find

$$U^{(j)}(\mathbf{u})_{jm} = \sqrt{\frac{(2j)!}{(j+m)!\,(j-m)!}}\, a^{j+m}b^{j-m} \qquad (5\text{-}31)$$

The matrices defined in (5-30) as j takes on all integral and half-integral values include all the distinct, irreducible, unitary representations of the two-dimensional unitary group. We now set about verifying this statement.

First, we show that the factorial denominator in (5-29) is in fact the proper normalizing factor. We use the generalized Unsöld theorem (Sec. 4-9), and consider the sum $\sum_m |f_m^{(j)}|^2$, which must be invariant if the $f_m^{(j)}$ are (normalized) basis functions for a unitary representation.

$$\sum_{m=-j}^{j} |f_m^{(j)}|^2 = \sum_m \frac{|\xi_1^2|^{j+m} |\xi_2^2|^{j-m}}{(j+m)!\,(j-m)!} = \frac{(|\xi_1^2| + |\xi_2^2|)^{2j}}{(2j)!}$$

where the last step follows by the binomial theorem. But $|\xi_1^2| + |\xi_2^2|$ is invariant under all the operations $\mathbf{u}$, since ξ_1 and ξ_2 are by definition basis functions of a unitary representation, and so Unsöld's theorem applies to them. But any power of an invariant is also invariant. Thus $\sum_m |f_m^{(j)}|^2$ is invariant as required. Evidently this would not have been true without the factorial denominator, which allowed the application of the binomial theorem.

Next we prove the *irreducibility* of these representations by use of Schur's lemma (Sec. 3-2). We start by restricting attention to diagonal matrices $\mathbf{u}$. Then $b = 0$, $|a| = 1$, and we may set $a = e^{i\alpha/2}$. The only nonvanishing terms in (5-30) then arise when $\kappa = 0$ and $m' = m$. In this case (5-30b) simplifies to

$$U_{m'm}^{(j)}(\alpha) = \delta_{m'm}(e^{i\alpha/2})^{j+m}(e^{-i\alpha/2})^{j-m} = \delta_{m'm}e^{im\alpha} \tag{5-32}$$

Now, following Schur's lemma, we assume that there exists a matrix $\mathbf{M}$ which commutes with *all* $\mathbf{U}$, and hence certainly with all members of this restricted set. Then

$$\mathbf{MU} = \mathbf{UM}$$

$$M_{mm'}U_{m'm'} = U_{mm}M_{mm'}$$

$$M_{mm'}(e^{im\alpha} - e^{im'\alpha}) = 0$$

Therefore $M_{mm'} = 0$ for $m \neq m'$, and $\mathbf{M}$ must be diagonal. Now we consider a general matrix $\mathbf{u}$ which produces off-diagonal as well as diagonal terms in $\mathbf{U}$, as is evident from (5-30b). Then $\mathbf{MU} = \mathbf{UM}$ implies

$$M_{mm}U_{mm'} = U_{mm'}M_{m'm'}$$

and since $U_{mm'} \neq 0$ by hypothesis, $M_{mm} = M_{m'm'}$, and $\mathbf{M}$ must be not only diagonal but a constant matrix. From Schur's lemma it then follows that the matrices $\mathbf{U}^{(j)}$ form irreducible representations of the unitary group.

Finally, we show the *completeness* of this set of representations. This follows from the orthogonality of characters. We may restrict our attention to diagonal $\mathbf{u}$ since all $\mathbf{u}$ are in the same class with some diagonal $\mathbf{u}'$ related to it by a unitary transformation. From (5-32), we have for a diagonal matrix

$$\chi^{(j)}(\alpha) = \sum_m U_{mm}^{(j)}(\alpha) = \sum_{m=-j}^{j} e^{im\alpha} \tag{5-33}$$

where now j can take on integral or half-integral values. Any other distinct representation would need to have a character orthogonal [with some suitable weighting function $g(\alpha)$] to all such $\chi^{(j)}$ and therefore to all $\chi^{(j)} - \chi^{(j-1)} = 2 \cos j\alpha$. But since we now include half integral j as well as integral j, these functions form a complete set in the region from 0 to $\pm 2\pi$ over which α can roam in forming all distinct $\mathbf{u}$. Thus no other representations are possible. Note that the half-integral j values are required to extend the region of completeness for a cosine series from $\pm \pi$ to $\pm 2\pi$.

5-6 Representation of the Rotation Group by Representations of the Unitary Group

Having worked out the homomorphism between the rotation group and the unitary group, and the complete set of irreducible representations of the latter, we can now write out the irreducible representations of the rotation group directly. The simplest one is the trivial case $\mathbf{U}^{(0)}(\mathbf{u}) = 1$, which corresponds to $\mathbf{D}^0(\alpha, \beta, \gamma) = 1$. This merely states that a quantum state with no angular momentum (an S state) is invariant under all rotations.

Next we have $\mathbf{U}^{(\frac{1}{2})}(\mathbf{u})$, which is just $\mathbf{u}$ itself. The $\mathbf{u}$ matrix which induces a coordinate transformation characterized by the Eulerian angles α, β, γ is given by (5-27). Hence (5-27) must be equivalent to the two-dimensional representation of the rotation group based on angular-momentum eigenfunctions for $j = \frac{1}{2}$. In fact, by working through the Pauli matrices, we have developed precisely the representation matrices based on $\varphi_{\frac{1}{2}}^{(\frac{1}{2})}$ and $\varphi_{-\frac{1}{2}}^{(\frac{1}{2})}$ for the first and second rows, respectively. Even the phases are conventional. Only one caution is required. We have developed the correspondence between $\mathbf{u}(\alpha, \beta, \gamma)$ and $\mathbf{R}[\mathbf{u}(\alpha, \beta, \gamma)]$ for positive rotations of *coordinate axes* about successively rotated axes. Thus, to represent positive rotations of the contours of the functions about fixed axes, which is our conventional interpretation, we must use

$$\mathbf{R}^{-1}[\mathbf{u}(\alpha, \beta, \gamma)] = \mathbf{R}[\mathbf{u}(-\gamma, -\beta, -\alpha)].$$

Accordingly, we obtain

$$\mathbf{D}^{(\frac{1}{2})}(\alpha, \beta, \gamma) = \begin{pmatrix} e^{-i(\alpha+\gamma)/2} \cos\dfrac{\beta}{2} & -e^{-i(\alpha-\gamma)/2} \sin\dfrac{\beta}{2} \\[2ex] e^{i(\alpha-\gamma)/2} \sin\dfrac{\beta}{2} & e^{i(\alpha+\gamma)/2} \cos\dfrac{\beta}{2} \end{pmatrix} \blacktriangleright \tag{5-34}$$

This matrix tells how the two eigenfunctions of spin $\frac{1}{2}$ transform under an arbitrary rotation, specified by the Eulerian angles. As noted earlier, we see from (5-34) that one of these half-integral angular-momentum eigenfunctions transforms into its negative under a rotation by 2π in any direction, and we must consider rotations of 4π before we find the identity. These eigenfunctions are called *spinors* to distinguish their unusual rotational-transformation properties from those of vectors. If we accept this essential difference, we can consider (5-34) to form a single-valued representation for rotations up to 4π. This was the approach used in Sec. 4-7 in treating the crystal double groups. No physical nonsense can arise from the sign change under rotation by 2π, since all physically significant quantities are matrix elements of the sort

$$\int \psi_i^* F \psi_j \, d\tau$$

which will be unchanged by a simultaneous sign change in ψ_i and ψ_j. (The matrix element will be *zero* if one of the functions has half integral j and the other has integral j.)

We can obtain all the higher-dimensional representations of the rotation group by using the higher-dimensional representations (5-30) of the unitary group and the connection rule

$$\mathbf{D}^{(j)}(\alpha, \beta, \gamma) = \mathbf{U}^{(j)}[\mathbf{u}(-\gamma, -\beta, -\alpha)]$$

That is, we set

$$a = e^{-i(\alpha+\gamma)/2} \cos \frac{\beta}{2}$$

$$b = -e^{-i(\alpha-\gamma)/2} \sin \frac{\beta}{2}$$

in (5-30). If this is done, we obtain

$$D^{(j)}(\alpha, \beta, \gamma)_{m'm}$$
$$= e^{-im'\alpha} e^{-im\gamma} \sum_\kappa \frac{(-1)^\kappa \sqrt{(j+m)!\,(j-m)!\,(j+m')!\,(j-m')!}}{\kappa!\,(j+m-\kappa)!\,(j-m'-\kappa)!\,(\kappa+m'-m)!}$$
$$\times \left(\cos \frac{\beta}{2}\right)^{2j-2\kappa-m'+m} \left(-\sin \frac{\beta}{2}\right)^{2\kappa+m'-m} \blacktriangleright \quad (5\text{-}35)$$

as the general expression for the representation of the transformation properties of angular-momentum eigenfunctions under rotation through the Eulerian angles.

We have already considered the special cases $\mathbf{D}^{(0)}$ and $\mathbf{D}^{(\frac{1}{2})}$. Now let us examine the explicit form for $j = 1$ which gives the transformation properties of p functions.   Using the general formula (5-35), we find

$$\mathbf{D}^{(1)}(\alpha, \beta, \gamma) = \begin{bmatrix} e^{-i\alpha} \dfrac{1 + \cos \beta}{2} e^{-i\gamma} & -e^{-i\alpha} \dfrac{\sin \beta}{\sqrt{2}} & e^{-i\alpha} \dfrac{1 - \cos \beta}{2} e^{i\gamma} \\[2ex] \dfrac{\sin \beta}{\sqrt{2}} e^{-i\gamma} & \cos \beta & -\dfrac{\sin \beta}{\sqrt{2}} e^{i\gamma} \\[2ex] e^{i\alpha} \dfrac{1 - \cos \beta}{2} e^{-i\gamma} & e^{i\alpha} \dfrac{\sin \beta}{\sqrt{2}} & e^{i\alpha} \dfrac{1 + \cos \beta}{2} e^{i\gamma} \end{bmatrix}$$

$$(5\text{-}36)$$

In this result we note that the half angles have all disappeared.   This shows that $\mathbf{D}^{(1)}$ is invariant under rotations of 2π and hence is single-valued, as it should be for integral j.   We can see that this property is general by inspecting the general form (5-30b).   Going from $\mathbf{u}$ to $-\mathbf{u}$ introduces a factor

$$(-1)^{j+m-\kappa}(-1)^{j-\kappa-m'}(-1)^{\kappa+m'-m}(-1)^{\kappa} = (-1)^{2j}$$

into $\mathbf{U}^{(j)}$ and hence $\mathbf{D}^{(j)}$ but gives the same rotation of coordinates $\mathbf{R}(\mathbf{u})$. Thus, if j is half-integral, we have an ambiguity of sign but none appears if j is integral.

5-7   Application of the Rotation-representation Matrices

First, let us review the conventions about sense and order of the rotations under which the $\mathbf{D}^{(j)}(\alpha, \beta, \gamma)$ may be used.   In all cases, we assume the Condon and Shortley phase convention for the angular-momentum eigenfunctions, namely, $\Theta_l^{|m|} = (-1)^m \Theta_l^{-|m|}$, so that $\varphi_{-m}{}^{(j)} = (-1)^m \varphi_m{}^{(j)*}$. We always assume right-handed coordinate systems, and rotations are considered positive if they are of the sense which would advance a right-hand screw in the positive direction along the axis.   In these terms, the fundamental case for which the $D^{(j)}(\alpha, \beta, \gamma)$ were derived is that in which α, β, γ are positive rotations of the *functions* with respect to a permanently fixed coordinate system, with the rotation by γ being performed first.   As we noted in Sec. 5-3, we could obtain the same rotation of a function if we imagined a coordinate system to be attached to it and then made successive rotations about these rotated axes in the reverse order, so that the α rotation was made first.   Even in this case, though, the rotated function is still considered to be represented in terms of basis functions set up with respect to the permanent space-fixed axes.

An alternative view can be adopted, in which the function under operation is considered permanently *fixed* in space and the *axes* (and basis functions) are considered to be rotated.   To apply our formulas to this viewpoint, we note that, if the same rotation were applied to both the axes and the function being transformed, clearly the relative position would be unchanged and so the transformation matrix would simply be the identity.   That is,

$$\mathbf{D}^{(j)}_{\text{axes}}(\alpha, \beta, \gamma) \, \mathbf{D}^{(j)}_{\text{funct}}(\alpha, \beta, \gamma) = \mathbf{E}$$

Thus $\mathbf{D}^{(j)}_{\text{funct}}(\alpha, \beta, \gamma) = \mathbf{D}^{(j)}_{\text{axes}}(\alpha, \beta, \gamma)^{-1} = \mathbf{D}^{(j)}_{\text{axes}}(-\gamma, -\beta, -\alpha)$

That is, we can use the same $\mathbf{D}^{(j)}(\alpha, \beta, \gamma)$ in the second viewpoint to represent rotations of *axes* by $-\alpha, -\beta, -\gamma$ in the reverse order to that used if the function were being rotated.   For example, if we rotated axes about space-fixed axes, we would have to make negative rotations by α, β, γ, with the α rotation being performed first.

The conclusions are summarized in Table 5-1.

Table 5-1

Eulerian-angle conventions

Rotated	Sense	With respect to	Order of operators
a. Function	+	Fixed axes	$P(\alpha)P(\beta)P(\gamma)$
b. Function	+	Rotated axes	$P(\gamma)P(\beta)P(\alpha)$
c. Axes	−	Fixed axes	$P(\gamma)P(\beta)P(\alpha)$
d. Axes	−	Rotated axes	$P(\alpha)P(\beta)P(\gamma)$

Application to combined basis functions.   In many applications, such as to directed-valence orbitals, one is interested in knowing the effect of a rotation on some wavefunction which is a linear combination of several basis functions.   In this case, if

$$\psi = \sum_m \varphi_m{}^{(j)} c_m \equiv \boldsymbol{\varphi}^{(j)}\mathbf{c}$$

$$P(\alpha, \beta, \gamma)\psi = \sum_m c_m P(\alpha, \beta, \gamma)\varphi_m{}^{(j)}$$

$$= \sum_m c_m \sum_{m'} \varphi_{m'}{}^{(j)} D^{(j)}(\alpha, \beta, \gamma)_{m'm}$$

$$= \sum_{m'} \varphi_{m'}{}^{(j)}[\mathbf{D}^{(j)}(\alpha, \beta, \gamma)\mathbf{c}]_{m'} = \boldsymbol{\varphi}^{(j)}\mathbf{c}' \qquad (5\text{-}37)$$

where $\mathbf{c}$ is the column vector composed of the coefficients c_m.   Thus multiplying the representation matrix $\mathbf{D}$ into the column vector $\mathbf{c}$ yields the column vector $\mathbf{c}'$ of the new coefficients after the transformation.

As a simple example, consider the p functions arising from $j = 1$.   The basis functions are

$$\varphi_1 = -\sqrt{\frac{3}{8\pi}} \sin \theta \, e^{i\varphi} = -\sqrt{\frac{3}{8\pi}} \frac{x + iy}{r}$$

$$\varphi_0 = \sqrt{\frac{3}{4\pi}} \cos \theta = \sqrt{\frac{3}{4\pi}} \frac{z}{r} \qquad\qquad (5\text{-}38)$$

$$\varphi_{-1} = \sqrt{\frac{3}{8\pi}} \sin \theta \, e^{-i\varphi} = \sqrt{\frac{3}{8\pi}} \frac{x - iy}{r}$$

We can form directed orbitals along the three axes by taking the three combinations

$$\psi_x = \frac{1}{\sqrt{2}}\,(\varphi_{-1} - \varphi_1)$$

$$\psi_y = \frac{i}{\sqrt{2}}\,(\varphi_{-1} + \varphi_1) \tag{5-39}$$

$$\psi_z = \varphi_0$$

As we noted earlier, these three p functions transform under rotations like the three components of an ordinary vector. As an illustration, now consider the change in these functions produced by a rotation through the Eulerian angles $\alpha = 0$, $\beta = \pi/2$, $\gamma = \pi/4$. Then, from (5-36),

$$\mathbf{D}^{(1)}\left(0, \frac{\pi}{2}, \frac{\pi}{4}\right) = \begin{bmatrix} \dfrac{1-i}{2\sqrt{2}} & -\dfrac{1}{\sqrt{2}} & \dfrac{1+i}{2\sqrt{2}} \\[2ex] \dfrac{1-i}{2} & 0 & -\dfrac{(1+i)}{2} \\[2ex] \dfrac{1-i}{2\sqrt{2}} & \dfrac{1}{\sqrt{2}} & \dfrac{1+i}{2\sqrt{2}} \end{bmatrix}$$

Multiplying this into the column vectors representing ψ_x, ψ_y, and ψ_z, we find

$$P\psi_x = \mathbf{D}^{(1)} \begin{pmatrix} -\dfrac{1}{\sqrt{2}} \\[2ex] 0 \\[2ex] \dfrac{1}{\sqrt{2}} \end{pmatrix} = \begin{pmatrix} \dfrac{i}{2} \\[2ex] -\dfrac{1}{\sqrt{2}} \\[2ex] \dfrac{i}{2} \end{pmatrix} = \frac{1}{\sqrt{2}}\,(\psi_y - \psi_z)$$

$$P\psi_y = \mathbf{D}^{(1)} \begin{pmatrix} \dfrac{i}{\sqrt{2}} \\[2ex] 0 \\[2ex] \dfrac{i}{\sqrt{2}} \end{pmatrix} = \begin{pmatrix} \dfrac{i}{2} \\[2ex] \dfrac{1}{\sqrt{2}} \\[2ex] \dfrac{i}{2} \end{pmatrix} = \frac{1}{\sqrt{2}}\,(\psi_y + \psi_z)$$

$$P\psi_z = \mathbf{D}^{(1)} \begin{pmatrix} 0 \\[2ex] 1 \\[2ex] 0 \end{pmatrix} = \begin{pmatrix} -\dfrac{1}{\sqrt{2}} \\[2ex] 0 \\[2ex] \dfrac{1}{\sqrt{2}} \end{pmatrix} = \psi_x$$

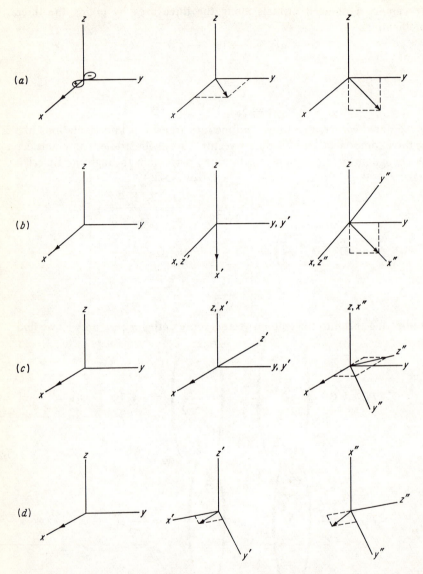

Fig. 5-4. *Successive rotations applied to ψ_x (a p function oriented along the x axis) for Euler angles $\alpha = 0$, $\beta = \pi/2$, $\gamma = \pi/4$, as discussed in text. As illustrated in the first diagram, the arrow represents the direction of the positive lobe of the wavefunction. Cases (a) to (d) correspond to the four rows of Table 5-1. All four viewpoints lead to $(\psi_y - \psi_z)/\sqrt{2}$, as found in the text, but (a) and (d) are simplest to follow and are the most commonly used interpretations.*

Since these three directed p functions can be represented by unit vectors along the axes, it is easy to follow through the rotations and verify by geometrical inspection that the results found here correctly describe the effect of the rotations taken in any of the four interpretations listed in Table 5-1. The successive rotations are illustrated in Fig. 5-4. This inspection procedure would not be possible for higher j values, however, and so one must rely on the $\mathbf{D}^{(j)}$ matrices.

5-8 Vector Model for Addition of Angular Momenta

In this section we give the group-theoretical basis for the familiar vector-model rule of elementary atomic physics. This rule states that, when two angular momenta j_1 and j_2 are coupled together vectorially, the magnitude J of the resultant angular momentum $\mathbf{J}$ may have any of the integrally spaced values between $|j_1 - j_2|$ and $j_1 + j_2$.

Consider two systems of angular momentum j_1 and j_2. Each could be a single electron on the same atom provided that we take care of the Pauli principle by a separate consideration. In a central field, they will have $(2j_1 + 1)$- and $(2j_2 + 1)$-fold degeneracies, there being that many basis functions in the irreducible representations of the full rotation group. If we now combine the systems mentally, there will be $(2j_1 + 1) \times (2j_2 + 1)$ basis functions of the sort $u_{m_1}(1)v_{m_2}(2)$, and these will form a basis for a reducible representation of the rotation group which is the direct product of the two irreducible representations. With interactions between the systems allowed for, the various irreducible representations obtained by decomposing the direct-product representation can have different energies. We now seek the general rule governing this decomposition.

If we make a rotation R,

$$P_R u_{m_1}(1) = \sum_{m_1'} u_{m_1'}(1) D^{(j_1)}(R)_{m_1' m_1}$$

$$P_R v_{m_2}(2) = \sum_{m_2'} v_{m_2'}(2) D^{(j_2)}(R)_{m_2' m_2}$$

and hence

$$P_R u_{m_1}(1) v_{m_2}(2) = \sum_{m_1', m_2'} u_{m_1'}(1) v_{m_2'}(2) D^{(j_1)}(R)_{m_1' m_1} D^{(j_2)}(R)_{m_2' m_2}$$

$$= \sum_{m_1', m_2'} u_{m_1'}(1) v_{m_2'}(2) \Delta(R)_{m_1' m_2', m_1 m_2}$$

where $\Delta(R)$ is the direct product

$$\Delta(R) = \mathbf{D}^{(j_1)}(R) \times \mathbf{D}^{(j_2)}(R)$$

To decompose $\Delta(R)$, we use the character rule

$$\chi^{(\Delta)}(R) = \chi^{(j_1)}(R)\chi^{(j_2)}(R) = \sum_J a_{j_1 j_2 J} \chi^{(J)}(R)$$

Our formula (3-20) for $a_{j_1 j_2 J}$ is not very useful because there are an infinite number of classes. However, we can easily carry out the decomposition by inspection. Contracting $a_{j_1 j_2 J}$ to a_J, we require that

$$\chi(\alpha)^{(j_1)} \chi(\alpha)^{(j_2)} = \sum_{m_1=-j_1}^{j_1} e^{-im_1\alpha} \sum_{m_2=-j_2}^{j_2} e^{-im_2\alpha} = \sum_{m_1 m_2} e^{-i(m_1+m_2)\alpha}$$

$$= \sum_{J=0}^{\infty} a_J \sum_{M=-J}^{J} e^{-iM\alpha}$$

Clearly $|M| \le j_1 + j_2$. Thus the a_J vanish for $J > j_1 + j_2$. There is one term for $|M| = j_1 + j_2$; so $a_{j_1+j_2} = 1$. There will be two terms with $M = j_1 + j_2 - 1$, namely, $(j_1 - 1) + j_2$ and $j_1 + (j_2 - 1)$. One of these is already accounted for by the sum with $a_{j_1+j_2}$, leaving one for the $(j_1 + j_2 - 1)$ sum. Continuing in this way, we find that all nonzero a_J are unity, so that we can write

$$\chi^{(j_1)}(\alpha) \chi^{(j_2)}(\alpha) = \sum_{J=|j_1-j_2|}^{j_1+j_2} \sum_{M=-J}^{J} e^{-iM\alpha} = \sum_{J=|j_1-j_2|}^{j_1+j_2} \chi^{(J)}(\alpha)$$

Hence it follows that

$$D^{(j_1)} \times D^{(j_2)} = \sum_{J=|j_1-j_2|}^{j_1+j_2} D^{(J)} = D^{(j_1+j_2)} + D^{(j_1+j_2-1)} + \cdots + D^{(|j_1-j_2|)}$$

(5-40)

That is, when two systems with angular momenta j_1 and j_2 are combined, the resulting state may have any of the angular momenta allowed by the elementary vector model.

Since the basis functions of each $D^{(J)}$ transform among themselves under a rotation of the total system, they will still be degenerate even in the case of interaction between the two systems provided that the Hamiltonian is invariant under such overall rotations. For example, if we consider only the orbital angular momenta, we may combine two electrons in an atom with angular momenta l_1 and l_2 to get $L = |l_1 - l_2|, \ldots, l_1 + l_2$. Now the Hamiltonian is not invariant under a rotation of the coordinates of one electron, since the Coulomb repulsion depends on the relative position of the two electrons. However, it *is* invariant under a joint rotation of both electrons. Hence $\mathbf{L} = \mathbf{l}_1 + \mathbf{l}_2$ is conserved, and L and M_L are good quantum numbers for representation labels. It should be noted, however, that one must *not* try to apply this result if, for example, one is combining angular momenta of two systems with centers at separate points in space if the interaction depends on the direction of the axis between them. In this case, the Hamiltonian is clearly not invariant under even a joint rotation of coordinates, unless of course one rotates the axis also. But then the total angular momentum would have to include that of the framework separating the two centers as well as that of the two subsystems. To be specific, an *isotropic-exchange coupling* $H' \sim \mathbf{j}_1 \cdot \mathbf{j}_2$ between two spins will leave $\mathbf{J} = \mathbf{j}_1 + \mathbf{j}_2$ as a good quantum number, but *anisotropic exchange* or a simple dipole-dipole coupling will not, because in these cases there is another term

$$\sim (\mathbf{r}_{12} \cdot \mathbf{j}_1)(\mathbf{r}_{12} \cdot \mathbf{j}_2)$$

which depends on the direction of the separation vector $\mathbf{r}_{12}$.

5-9   The Wigner or Clebsch-Gordan Coefficients

The foregoing vector-model proof has shown that

$$D^{(j_1)} \times D^{(j_2)} = \sum_{J=|j_1-j_2|}^{j_1+j_2} D^{(J)}$$

This means that a similarity transformation must exist which reduces $\mathbf{D}^{(j_1)} \times \mathbf{D}^{(j_2)}$ to the block form

$$\mathbf{M}(R) = \begin{bmatrix} \mathbf{D}^{|j_1-j_2|}(R) & 0 & \cdots & 0 \\ 0 & \mathbf{D}^{|j_1-j_2|+1}(R) & \cdots & 0 \\ \multicolumn{4}{c}{\dotfill} \\ 0 & 0 & \cdots & \mathbf{D}^{(j_1+j_2)}(R) \end{bmatrix}$$

Expressed more compactly,

$$\mathbf{D}^{(j_1)}(R) \times \mathbf{D}^{(j_2)}(R) = \mathbf{A}^{-1}\mathbf{M}(R)\mathbf{A} \qquad M(R)_{J'M',JM} = \delta_{J'J}D^{(J)}{}_{M'M} \qquad (5\text{-}41a)$$

where $\mathbf{A}$ is as yet an undetermined unitary matrix. This matrix $\mathbf{A}$ must be "square in the broader sense." That is, $\mathbf{A}$ has an equal number of rows and columns—$(2j_1 + 1)(2j_2 + 1) = \sum_J (2J + 1)$—but the rows are labeled by JM, whereas the columns are labeled by $m_1 m_2$. A typical element from the matrix equation above is

$$D^{(j_1)}(R)_{m_1'm_1}D^{(j_2)}(R)_{m_2'm_2} = \sum_{J,M,M'} A^{-1}_{m_1'm_2',JM'}D^{(J)}(R)_{M'M}A_{JM,m_1m_2} \qquad (5\text{-}41b)$$

The matrix $\mathbf{A}$ is useful because it prescribes the linear combinations of the $u_{m_1}v_{m_2}$ functions which are to be taken so as to diagonalize the total angular momentum J and its z component M, namely,

$$\psi_M{}^J = \sum_{m_1m_2} u_{m_1}v_{m_2}(A^{-1})_{m_1m_2,JM} \qquad (5\text{-}42a)$$

and conversely

$$u_{m_1}v_{m_2} = \sum_{JM} \psi_M{}^J A_{JM,m_1m_2} \qquad (5\text{-}42b)$$

Of course, since $\mathbf{A}$ is unitary, $\mathbf{A}^{-1} = \mathbf{A}^\dagger = \tilde{\mathbf{A}}^*$. In fact, we shall find a matrix $\mathbf{A}$ which is real. Therefore $\mathbf{A}^{-1} = \tilde{\mathbf{A}}$.

First, we confirm that if $\mathbf{A}$ satisfies (5-41) then these linear combinations do in fact transform according to $\mathbf{D}^{(J)}$, this being the only real requirement for the identification of $\psi_M{}^J$. We consider a rotation R to be applied to (5-42a),

$$
\begin{aligned}
P_R\psi_M{}^J &= \sum_{m_1 m_2} P_R u_{m_1} P_R v_{m_2} (A^{-1})_{m_1 m_2, JM} \\
&= \sum_{m_1 m_2, m_1' m_2'} u_{m_1'} v_{m_2'} D^{(j_1)}(R)_{m_1' m_1} D^{(j_2)}(R)_{m_2' m_2} (A^{-1})_{m_1 m_2, JM} \\
&= \sum_{m_1 m_2, m_1' m_2', J'M'} \psi_{M'}{}^{J'} A_{J'M', m_1' m_2'} D^{(j_1)}(R)_{m_1' m_1} \\
&\quad \times D^{(j_2)}(R)_{m_2' m_2} (A^{-1})_{m_1 m_2, JM} \\
&= \sum_{J'M'} \psi_{M'}{}^{J'} [A(\mathbf{D}^{(j_1)} \times \mathbf{D}^{(j_2)})A^{-1}]_{J'M', JM} \\
&= \sum_{J'M'} \psi_{M'}{}^{J'} [\mathbf{M}(\mathbf{R})]_{J'M', JM} \\
&= \sum_{J'M'} \psi_{M'}{}^{J'} D^{(J)}(R)_{M'M} \delta_{J'J} \\
&= \sum_{M'} \psi_{M'}{}^{J} D^{(J)}(R)_{M'M}
\end{aligned}
$$

Thus we have verified that the matrices $\mathbf{A}^{-1}$ give the linear combinations of $u_{m_1} v_{m_2}$ which transform properly under a joint rotation of the axes of both the systems being combined. If such a rotation leaves the Hamiltonian invariant even in the presence of interaction, these $\psi_M{}^J$ are the "proper linear combinations" for the combined system. This suggests the utility of the matrix of coefficients in practical problems where we are combining such angular momenta as $\mathbf{L} + \mathbf{S} = \mathbf{J}$ (the classic example), $\mathbf{j}_1 + \mathbf{j}_2 = \mathbf{J}$ (j-j coupling), $\mathbf{I} + \mathbf{J} = \mathbf{F}$ (hyperfine coupling), and even isotopic spin in nuclear theory. This last application may seem a bit surprising at first, since isotopic spin is not an angular momentum at all. However, since our group-theoretical development was based only on rotational invariance, it is clearly not restricted to rotations in ordinary space. It works equally well in isotopic-spin space.

Determination of coefficients. The Wigner coefficients which we seek are simply the elements of the matrix $\mathbf{A}$ with the properties which we have just sketched. Let us now proceed to determine this matrix explicitly. We start by considering the effect of a rotation by α about the z axis. We can express this in two different ways,

$$
P_R \psi_M{}^J = e^{-iM\alpha} \psi_M{}^J = e^{-iM\alpha} \sum_{m_1, m_2} u_{m_1} v_{m_2} A^{-1}_{m_1 m_2, JM}
$$

or

$$
P_R \sum_{m_1, m_2} u_{m_1} v_{m_2} A^{-1}_{m_1 m_2, JM} = \sum_{m_1, m_2} e^{-im_1\alpha} u_{m_1} e^{-im_2\alpha} v_{m_2} A^{-1}_{m_1 m_2, JM}
$$

Since the $u_{m_1} v_{m_2}$ form an orthonormal set, the coefficient of each term in the sum must be the same as the corresponding one in the other (equivalent) sum. This will be true only if the only nonvanishing $A^{-1}_{m_1 m_2, JM}$ occur when

$$
M = m_1 + m_2
$$

This restriction expresses the conservation of J_z, though it was obtained from a consideration of transformation properties. We can now condense our notation slightly, by setting

$$A_{JM,m_1 m_2} = a_{Jm_1 m_2}\delta_{M,m_1+m_2}$$

In terms of **a**, we may express (5-41b) as

$$D^{(j_1)}(R)_{m_1'm_1}D^{(j_2)}(R)_{m_2'm_2} = \sum_{J=|j_1-j_2|}^{j_1+j_2} a^{-1}_{m_1'm_2'J}D^{(J)}(R)_{m_1'+m_2',m_1+m_2}a_{Jm_1 m_2} \quad (5\text{-}43)$$

Note that the multiple summation has been reduced to only a single sum over J.

The next step is to note that **A** is fixed by our conditions only within multiplication by a diagonal matrix of the form $V_{J'M',JM} = w_J\delta_{J'J}\delta_{M'M}$, where $|w_J| = 1$. We see this by noting that a **V** of this form commutes with $\mathbf{M}(R)$ and hence that $(\mathbf{VA})^{-1}\mathbf{M}(\mathbf{VA}) = \mathbf{A}^{-1}\mathbf{V}^{-1}\mathbf{MVA} = \mathbf{A}^{-1}\mathbf{MV}^{-1}\mathbf{VA} = \mathbf{A}^{-1}\mathbf{MA}$. This degree of freedom allows us an arbitrary phase choice for each value of J in **A**. We use this freedom to choose $A_{J,j_1-j_2;j_1,-j_2} = a_{J,j_1,-j_2} = |a_{J,j_1,-j_2}|$ so that these elements are real and positive.

Now we proceed by multiplying through (5-43) by $D^{(J')}(R)^*_{m_1'+m_2',m_1+m_2}$, integrating over R, and using the orthogonality theorem. By using the unitarity of **A**, this can be written

$$\int D^{(j_1)}(R)_{m_1'm_1}D^{(j_2)}(R)_{m_2'm_2}D^{(J')}(R)^*_{m_1'+m_2',m_1+m_2}\,dR$$

$$= \sum_J a^*_{Jm_1'm_2'}a_{Jm_1 m_2}\int D^{(J)}(R)_{m_1'+m_2',m_1+m_2}D^{(J')}(R)^*_{m_1'+m_2',m_1+m_2}\,dR \quad (5\text{-}44)$$

By orthogonality, the right member vanishes unless $J = J'$. This reduces the sum to a single term, and we now change notation by dropping the prime on J. The right member is then $a^*_{Jm_1'm_2'}a_{Jm_1 m_2}[\int dR/(2J+1)]$.

Now make the special choice $m_1' = j_1$, $m_2' = -j_2$. Then our phase convention assures us that $a_{Jm_1'm_2'}$ is real and positive. Further, in these special cases, our general formula (5-35) for $\mathbf{D}^{(j)}(R)$ reduces to a single term, as illustrated for $\mathbf{U}^{(j)}(\mathbf{u})$ in (5-31). Upon taking advantage of this simplification, but using the general formula for $D^{(J)}(R)^*$, the left member of (5-44) becomes

$$\left(\begin{array}{c}2j_1\\j_1-m_1\end{array}\right)^{\frac{1}{2}}\left(\begin{array}{c}2j_2\\j_2-m_2\end{array}\right)^{\frac{1}{2}}$$

$$\times\sum_\kappa\frac{(-1)^{\kappa+j_2+m_2}[(J+m_1+m_2)!\,(J-m_1-m_2)!}{(J-j_1+j_2-\kappa)!\,(J+m_1+m_2-\kappa)!\,\kappa!\,(\kappa+j_1-j_2-m_1-m_2)!}$$
$$\qquad\qquad\qquad\qquad\qquad{}\times(J+j_1-j_2)!\,(J-j_1+j_2)!]^{\frac{1}{2}}$$

$$\times\int\left(\cos\frac{\beta}{2}\right)^{2(J+j_2+m_1-\kappa)}\left(\sin\frac{\beta}{2}\right)^{2(j_1-m_1+\kappa)}dR \quad (5\text{-}45)$$

[The notation $\binom{n}{m}$ signifies the binomial coefficient $n!/m!\,(n-m)!$.] This integral may be related to the normalization integral $\int dR$ by converting it to the form of a normalization integral for the simple special case $(m' = j)$,

$$\int |D^{(j)}(R)_{jm}|^2 \, dR = \binom{2j}{j-m} \int \left(\cos\frac{\beta}{2}\right)^{2(j+m)} \left(\sin\frac{\beta}{2}\right)^{2(j-m)} dR \equiv \frac{\int dR}{2j+1}$$

To make the connection, set

$$j + m = J + j_2 + m_1 - \kappa \qquad\qquad j = \frac{J + j_1 + j_2}{2}$$

or

$$j - m = j_1 - m_1 + \kappa \qquad\qquad m = \frac{J + j_2 - j_1 + 2m_1 - 2\kappa}{2}$$

Then the integral in (5-45) is just

$$\frac{\int dR}{J + j_1 + j_2 + 1} \frac{1}{\binom{2j}{j-m}} = \frac{\int dR}{J + j_1 + j_2 + 1} \frac{(J + j_2 + m_1 - \kappa)!\,(j_1 - m_1 + \kappa)!}{(J + j_1 + j_2)!}$$

$$= \frac{(J + j_2 + m_1 - \kappa)!\,(j_1 - m_1 + \kappa)!}{(J + j_1 + j_2 + 1)!} \int dR$$

Inserting this in (5-45) and then (5-45) into (5-44), canceling the $\int dR$, and multiplying through by $(2J + 1)$ leads to

$$a^*_{J,j_1,-j_2} a_{Jm_1 m_2}$$

$$= (2J+1) \sum_\kappa \frac{(-1)^{\kappa + j_2 + m_2}[(2j_1)!\,(2j_2)!\,(J + m_1 + m_2)!\,(J - m_1 - m_2)!}{(J + j_1 + j_2 + 1)!\,[(j_1 - m_1)!\,(j_1 + m_1)!\,(j_2 - m_2)!\,(j_2 + m_2)!]^{\frac{1}{2}}}$$

$$\times \frac{(J + j_2 + m_1 - \kappa)!\,(j_1 - m_1 + \kappa)!}{(J - j_1 + j_2 - \kappa)!\,(J + m_1 + m_2 - \kappa)!\,\kappa!\,(\kappa + j_1 - j_2 - m_1 - m_2)!}$$

$$\tag{5-46}$$

Now we can immediately determine $a_{J,j_1,-j_2}$ by setting $m_1 = j_1$ and $m_2 = -j_2$ and then taking the positive square root of the right member of (5-46). After an algebraic reduction given by Wigner,[1] we obtain

$$a_{J,j_1,-j_2} = \left[\frac{(2J + 1)(2j_1)!\,(2j_2)!}{(J + j_1 + j_2 + 1)!\,(j_1 + j_2 - J)!}\right]^{\frac{1}{2}}$$

[1] E. Wigner, "Group Theory," p. 194, Academic Press, Inc., New York, 1959.

Factoring this out of (5-46) and introducing a more complete notation to indicate all the angular momenta involved, we have our final result,

$$A_{m_1 m_2 M}^{j_1 j_2 J} = a_{J m_1 m_2}^{(j_1 j_2)}$$

$$= \frac{[(J + j_1 - j_2)!\,(J - j_1 + j_2)!\,(j_1 + j_2 - J)!\,(J + m_1 + m_2)!\,(J - m_1 - m_2)!]^{\frac{1}{2}}}{[(J + j_1 + j_2 + 1)!\,(j_1 - m_1)!\,(j_1 + m_1)!\,(j_2 - m_2)!\,(j_2 + m_2)!]^{\frac{1}{2}}}$$

$$\times \sum_{\kappa} \frac{(-1)^{\kappa + j_2 + m_2}\sqrt{(2J + 1)}(J + j_2 + m_1 - \kappa)!\,(j_1 - m_1 + \kappa)!}{(J - j_1 + j_2 - \kappa)!\,(J + m_1 + m_2 - \kappa)!\,\kappa!\,(\kappa + j_1 - j_2 - m_1 - m_2)!}$$

▶ (5-47)

In this equation the summation extends over all values of κ for which the factorials in the denominator do not become infinite. (Recall that $0! = 1$, but factorials of negative integers are infinite.) Note that all the coefficients are real, so that $\mathbf{A}^{-1} = \tilde{\mathbf{A}}$. Though hard to see from this form, the coefficient is unchanged (except for a possible sign change) when we interchange (j_1, m_1) with (j_2, m_2) in the formula. Such symmetry is clearly necessary if the formula is to be correct.

In the case $m_1 + m_2 = M = J$, only the term with $\kappa = J - j_1 + j_2$ survives in (5-47), and we obtain the relatively simple result

$$a_{J, m_1, J - m_1} = (-1)^{j_1 - m_1}$$

$$\times \frac{[(2J + 1)!\,(j_1 + j_2 - J)!\,(j_1 + m_1)!\,(J + j_2 - m_1)!]^{\frac{1}{2}}}{[(J + j_1 + j_2 + 1)!\,(J + j_1 - j_2)!\,(J - j_1 + j_2)!\,(j_1 - m_1)!\,(j_2 - J + m_1)!]^{\frac{1}{2}}}$$

(5-48)

We note that this is positive whenever $m_1 = j_1$, as chosen in our convention. In particular, consider the "stretched case," where $J = j_1 + j_2$ and $M = J$. Then $m_1 = j_1$, $m_2 = j_2$, and (5-48) reduces to unity, as it must since this is the only way by which $M = j_1 + j_2$ can be obtained.

We have now finished deriving explicit formulas for the famous Wigner or Clebsch-Gordan coefficients. They are the coefficients for the combination of subsystem angular-momentum eigenfunctions to form eigenfunctions of the total angular momentum according to the formula

$$\psi_M^{\,J} = \sum_m a_{J, m, M - m}^{(j_1 j_2)} u_m^{(j_1)} v_{M - m}^{(j_2)} = \sum_m A_{m, M - m, M}^{j_1 j_2 J} u_m^{(j_1)} v_{M - m}^{(j_2)} \qquad ▶ (5-49)$$

Since $M = m_1 + m_2$ in all cases, we shall often abbreviate the latter notation from $A_{m_1 m_2 M}^{j_1 j_2 J}$ to $A_{m_1 m_2}^{j_1 j_2 J}$.

5-10 Notation, Tabulations, and Symmetry Properties of the Wigner Coefficients

The formula (5-47) is very cumbersome to use, though completely general and explicit. Alternative methods exist for finding the coefficients in specific cases by use of recursion formulas. These methods have been described by various authors, and many different systems of notation have been developed. In this section, we survey this literature to make it accessible for detailed use.

One of the most useful single tabulations is that of Condon and Shortley.[1] They tabulate explicit formulas for $j_2 = \frac{1}{2}$, 1, $\frac{3}{2}$, and 2 for general values of all the other parameters. [In their notation, $A^{j_1 j_2 j_3}_{m_1 m_2 m_3} = a^{(j_1 j_2)}_{j_3 m_1 m_2} = (j_1 j_2 m_1 m_2 | j_1 j_2 j_3 m_3)$]. These results handle all problems in which the smaller of the two angular momenta to be added does not exceed 2. A similar table for $j_2 = \frac{5}{2}$ has been given by Saito and Morita,[2] and one for $j_2 = 3$ has been given by Falkoff, Colladay, and Sells.[3] A decimal tabulation has been given by Simon[4] for cases in which all angular momenta are less than $\frac{9}{2}$. More convenient for many purposes are the nondecimal tables giving the coefficients as square roots of rational fractions, due to Cohen. These are reproduced as an appendix in Heine's book[5] for all cases where $j_1 + j_2 \leqslant 4$.

The number of coefficients which need to be tabulated is reduced by many symmetry relations which govern the effect of permutation of indices and changes of sign. These results are also useful for extracting results of a general nature in the consideration of particular problems. These symmetry properties are extensively discussed in the books of Rose, of Edmonds, and of Rotenberg et al.[6] To facilitate the discussion of these symmetry properties, symmetrized vector-coupling coefficients have been introduced by many authors. These are based on the idea that, by reversing the sense of the resultant $\mathbf{j} = \mathbf{j}_3$, one has $\mathbf{j}_1 + \mathbf{j}_2 + \mathbf{j}_3 = 0$, in which all three j's enter equivalently. The first of these symmetrized forms seems to have been the V coefficients of Racah.[7] These are related to the Wigner or Clebsch-Gordan coefficients by the relation

$$V(j_1 j_2 j_3, m_1 m_2 m_3) = (-1)^{j_3 - m_3}(2j_3 + 1)^{-\frac{1}{2}} A^{j_1 j_2 j_3}_{m_1 m_2 - m_3} \tag{5-50}$$

More recently, Wigner introduced his so-called "3-j symbol,"[8] which is related to the V coefficients by a simple phase change, namely,

$$\begin{pmatrix} j_1 & j_2 & j_3 \\ m_1 & m_2 & m_3 \end{pmatrix} = (-1)^{j_3 + j_2 - j_1} V(j_1 j_2 j_3, m_1 m_2 m_3) \tag{5-51}$$

[1] Condon and Shortley, op. cit., pp. 76–77.

[2] R. Saito and M. Morita, Progr. Theoret. Phys. (Kyoto), 13, 540 (1955).

[3] D. L. Falkoff, G. S. Colladay, and R. E. Sells, Can. J. Phys., 30, 253 (1952).

[4] A. Simon, Oak Ridge Natl. Lab. Rept. ORNL-1718 (1954).

[5] V. Heine, "Group Theory," Pergamon Press, New York, 1960.

[6] M. E. Rose, op. cit., chap. 3; A. R. Edmonds, "Angular Momentum in Quantum Mechanics," chap. 3, Princeton University Press, Princeton, N.J., 1957; M. Rotenberg, R. Bivins, N. Metropolis, and J. K. Wooten, Jr., "The 3-j and 6-j Symbols," Technology Press, Cambridge, Mass., 1959.

[7] G. Racah, Phys. Rev., 62, 438 (1942).

[8] See, for example, Wigner, op. cit., chap. 24.

This 3-j symbol is equal to Schwinger's[1] X coefficient $X(j_1 j_2 j_3, m_1 m_2 m_3)$. The advantage of these symmetrized coefficients is that the introduction of the statistical weight factor $(2j_3 + 1)^{-\frac{1}{2}}$ removes an asymmetry between j_3 and j_1, j_2. The symmetry properties of the 3-j symbols (or the X coefficients) are as follows:

Equality under cyclic permutation of columns

$$\begin{pmatrix} j_1 & j_2 & j_3 \\ m_1 & m_2 & m_3 \end{pmatrix} = \begin{pmatrix} j_2 & j_3 & j_1 \\ m_2 & m_3 & m_1 \end{pmatrix} = \begin{pmatrix} j_3 & j_1 & j_2 \\ m_3 & m_1 & m_2 \end{pmatrix} \tag{5-52a}$$

Possible sign change under interchange of two columns

$$\begin{pmatrix} j_2 & j_1 & j_3 \\ m_2 & m_1 & m_3 \end{pmatrix} = \begin{pmatrix} j_1 & j_3 & j_2 \\ m_1 & m_3 & m_2 \end{pmatrix} = \begin{pmatrix} j_3 & j_2 & j_1 \\ m_3 & m_2 & m_1 \end{pmatrix}$$

$$= (-1)^{j_1+j_2+j_3} \begin{pmatrix} j_1 & j_2 & j_3 \\ m_1 & m_2 & m_3 \end{pmatrix} \tag{5-52b}$$

Possible sign change on replacement of m by $-m$

$$\begin{pmatrix} j_1 & j_2 & j_3 \\ -m_1 & -m_2 & -m_3 \end{pmatrix} = (-1)^{j_1+j_2+j_3} \begin{pmatrix} j_1 & j_2 & j_3 \\ m_1 & m_2 & m_3 \end{pmatrix} \tag{5-52c}$$

From the relations (5-50) and (5-51) analogous results can be obtained from (5-52) for the somewhat less symmetrical V coefficients and for the

Table 5-2

Wigner coefficients for $j_2 = \frac{1}{2}$

$j_2 = \frac{1}{2}$	$m_2 = \frac{1}{2}$	$m_2 = -\frac{1}{2}$
$j = j_1 + \frac{1}{2}$	$\left(\dfrac{j_1 + m + \frac{1}{2}}{2j_1 + 1}\right)^{\frac{1}{2}}$	$\left(\dfrac{j_1 - m + \frac{1}{2}}{2j_1 + 1}\right)^{\frac{1}{2}}$
$j = j_1 - \frac{1}{2}$	$-\left(\dfrac{j_1 - m + \frac{1}{2}}{2j_1 + 1}\right)^{\frac{1}{2}}$	$\left(\dfrac{j_1 + m + \frac{1}{2}}{2j_1 + 1}\right)^{\frac{1}{2}}$

ordinary unsymmetrized Wigner or Clebsch-Gordan coefficients. An exhaustive numerical tabulation of these 3-j symbols, covering all cases for $j_1, j_2, j_3 \leqslant 8$, has been given by Rotenberg et al. in the book cited above.

We conclude this section by quoting the explicit form of the Wigner coefficients (in the ordinary unsymmetrized form) for some simple cases. The case of $j = \frac{1}{2}$ is particularly important because it gives the formulas for combining the spin angular momentum of an electron with its orbital angular momentum. The coefficients for this case are given in Table 5-2.

[1] J. Schwinger, On Angular Momentum, *USAEC Rept.* NYO-3071 (1952).

For example, then,

$$\psi_m^{(l+\frac{1}{2})} = \left(\frac{l+m+\frac{1}{2}}{2l+1}\right)^{\frac{1}{2}} u_{m-\frac{1}{2}}^{(l)} v_{\frac{1}{2}}^{(\frac{1}{2})} + \left(\frac{l-m+\frac{1}{2}}{2l+1}\right)^{\frac{1}{2}} u_{m+\frac{1}{2}}^{(l)} v_{-\frac{1}{2}}^{(\frac{1}{2})}$$

Specializing further, if $l = 1$, we have the states

$$\psi_{\frac{3}{2}}^{(\frac{3}{2})} = u_1^{(1)} v_{\frac{1}{2}}^{(\frac{1}{2})}$$

$$\psi_{\frac{1}{2}}^{(\frac{3}{2})} = \sqrt{\tfrac{2}{3}} u_0^{(1)} v_{\frac{1}{2}}^{(\frac{1}{2})} + \sqrt{\tfrac{1}{3}} u_1^{(1)} v_{-\frac{1}{2}}^{(\frac{1}{2})}$$

and similar expressions for $\psi_{-\frac{1}{2}}^{(\frac{3}{2})}$ and $\psi_{-\frac{3}{2}}^{(\frac{3}{2})}$.

Table 5-3

The Wigner coefficients for $j_2 = 1$

$j_2 = 1$	$m_2 = 1$	$m_2 = 0$	$m_2 = -1$
$j = j_1 + 1$	$\left[\dfrac{(j_1 + m)(j_1 + m + 1)}{(2j_1 + 1)(2j_1 + 2)}\right]^{\frac{1}{2}}$	$\left[\dfrac{(j_1 - m + 1)(j_1 + m + 1)}{(2j_1 + 1)(j_1 + 1)}\right]^{\frac{1}{2}}$	$\left[\dfrac{(j_1 - m)(j_1 - m + 1)}{(2j_1 + 1)(2j_1 + 2)}\right]^{\frac{1}{2}}$
$j = j_1$	$-\left[\dfrac{(j_1 + m)(j_1 - m + 1)}{2j_1(j_1 + 1)}\right]^{\frac{1}{2}}$	$\dfrac{m}{[j_1(j_1 + 1)]^{\frac{1}{2}}}$	$\left[\dfrac{(j_1 - m)(j_1 + m + 1)}{2j_1(j_1 + 1)}\right]^{\frac{1}{2}}$
$j = j_1 - 1$	$\left[\dfrac{(j_1 - m)(j_1 - m + 1)}{2j_1(2j_1 + 1)}\right]^{\frac{1}{2}}$	$-\left[\dfrac{(j_1 - m)(j_1 + m)}{j_1(2j_1 + 1)}\right]^{\frac{1}{2}}$	$\left[\dfrac{(j_1 + m + 1)(j_1 + m)}{2j_1(2j_1 + 1)}\right]^{\frac{1}{2}}$

The coefficients for the case $j_2 = 1$ are given in Table 5-3. We shall find these coefficients very useful in the next sections in conjunction with the Wigner-Eckart theorem when applied to vector operators.

5-11 Tensor Operators

Scalar operators. A scalar S is defined as a quantity which is invariant under all rotations. Hence, we can apply the general theorem (Sec. 4-9) on matrix elements of invariant operators to conclude that there will be no matrix elements of a scalar operator connecting functions belonging to different irreducible representations (or different rows of the same one) of the full rotation group. For example, there are no matrix elements of the energy between states with different J or M in an atom, because the Hamiltonian is invariant under rotation and hence is a scalar. Moreover, all states within one irreducible representation have the same diagonal matrix elements (of energy, in our example). Expressed symbolically,

$$S_{N'j'm',Njm} \equiv (\psi_{N'j'm'}, S\psi_{Njm}) \equiv (N'j'm' |S| Njm)$$

$$= \delta_{jj'}\delta_{mm'}(N'j\|S\| Nj) \tag{5-53}$$

where we have used N to indicate all other quantum numbers which are irrelevant to the discussion of rotational properties. The quantity $(N'j\|S\|Nj)$ is a *reduced matrix element* which may depend on N, N', and j but is independent of m.

We now proceed to work out generalizations of this theorem for operators with more complicated rotational properties.

Vector operators. To obtain a general definition of the transformed operator O' related to the operator O by a coordinate transformation $\mathbf{R}$, we require that

$$O'\psi' = (O\psi)'$$

where $\psi' = P_R\psi$ is the transformed wavefunction. Applying this, we have

$$O'P_R\psi = P_RO\psi$$

But the relation must hold for arbitrary ψ. Hence, we have the operator equality

$$O'P_R = P_RO$$

or

$$O' = P_ROP_R^{-1} \tag{5-54}$$

An operator $\mathbf{V}$ is said to be a vector operator if it has three components (V_x, V_y, V_z) which transform like (x, y, z) under a coordinate transformation $\mathbf{R}$. That is,

$$\mathbf{V}' = P_R\mathbf{V}P_R^{-1} = \mathbf{R}^{-1}\mathbf{V} \tag{5-55}$$

where $\mathbf{R}^{-1}$ enters because of our convention that $P_Rf(\mathbf{x}) = f(\mathbf{R}^{-1}\mathbf{x})$. Expressing a component of this equation in cartesian form, we have

$$V_i' = P_RV_iP_R^{-1} = \sum_j (R^{-1})_{ij}V_j = \sum_j V_jR_{ji} \tag{5-56}$$

since for a real orthogonal transformation $\mathbf{R}^{-1} = \tilde{\mathbf{R}}$.

To illustrate these rather formal rules, consider the example $\mathbf{V} = \nabla$, and let $\mathbf{R}$ represent a rotation by $\pi/2$ about z. Then

$$\mathbf{R} = \begin{pmatrix} 0 & 1 & 0 \\ -1 & 0 & 0 \\ 0 & 0 & 1 \end{pmatrix}$$

and accordingly $P_Rf(x, y, z) = f(-y, x, z)$ and $P_R^{-1}f(x, y, z) = f(y, -x, z)$. Hence

$$P_RV_xP_R^{-1}\psi(x, y, z) = P_R\frac{\partial}{\partial x}\psi(y, -x, z) = \frac{\partial}{\partial(-y)}\psi(x, y, z) = -V_y\psi(x, y, z)$$

which is just what is required by the definition of a vector operator.

Although these cartesian coordinates are satisfactory here, it is often more convenient to introduce *spherical components* based on the rotational eigenfunctions. These will be much easier to generalize further. Conforming to our standard phase conventions, we define

$$V_1 = -\frac{V_x + iV_y}{\sqrt{2}} \equiv \frac{-V_+}{\sqrt{2}}$$

$$V_0 = V_z \tag{5-57}$$

$$V_{-1} = \frac{V_x - iV_y}{\sqrt{2}} \equiv \frac{V_-}{\sqrt{2}}$$

One must be careful in expressing a vector in terms of these components because

$$\mathbf{V} = V_x\mathbf{u}_x + V_y\mathbf{u}_y + V_z\mathbf{u}_z = -V_{-1}\mathbf{u}_1 + V_0\mathbf{u}_0 - V_1\mathbf{u}_{-1} = \sum_{\mu=-1}^{1}(-1)^\mu V_{-\mu}\mathbf{u}_\mu \tag{5-58}$$

where the $\mathbf{u}$'s are unit vectors and $\mathbf{u}_{\pm 1} = \mp(\mathbf{u}_x \pm i\mathbf{u}_y)/\sqrt{2}$, $\mathbf{u}_0 = \mathbf{u}_z$. The orthonormality relation of these new unit vectors is $\mathbf{u}_\mu \cdot \mathbf{u}_{\mu'} = \delta_{\mu,-\mu'}(-1)^\mu$, as is easily verified from their definition. Therefore the ordinary scalar product of two vectors has the form

$$\mathbf{V} \cdot \mathbf{W} = \sum_{\mu\mu'}(-1)^{\mu+\mu'} V_{-\mu}W_{-\mu'}\mathbf{u}_\mu \cdot \mathbf{u}_{\mu'}$$

$$= \sum_{\mu}(-1)^\mu V_{-\mu}W_\mu \tag{5-59}$$

If $\mathbf{V}$ has real cartesian components, this is equivalent to $\sum_\mu V_\mu^* W_\mu$, our definition of the Hermitian scalar product.

Since these new spherical components have been set up to have the same structure as the angular-momentum eigenfunctions for $j = 1$, they must have the same transformation properties, namely,

$$P_R V_\mu P_R^{-1} = \sum_{\mu'} V_{\mu'} D^{(1)}(R)_{\mu'\mu} \tag{5-60}$$

Tensor operators. The spherical vectors of the previous section are readily generalized to *spherical tensors* or *irreducible tensors* of arbitrary rank. We define an irreducible-tensor operator of the ωth rank to be an operator with $(2\omega + 1)$ components $T_\mu^{(\omega)}$ which transform under rotations according to

$$P_R T_\mu^{(\omega)} P_R^{-1} = \sum_{\mu'} T_{\mu'}^{(\omega)} D^{(\omega)}(R)_{\mu'\mu} \tag{5-61}$$

These irreducible tensors have considerable advantages over ordinary cartesian tensors. For example, a cartesian tensor of rank 2 has $3^2 = 9$ components $T_{ij}(i, j = x, y, z)$. The component T_{xy} is a quantity with the same transformation properties as the product of the coordinates xy. If a rotation of coordinates is carried out, these T_{ij} are transformed into new components T'_{ij} which are linear combinations of the old according to

$$T'_{ij} = \sum_{kl} T_{kl} R_{ki} R_{lj} \tag{5-62}$$

where R_{ki} is the matrix of direction cosines specifying the corresponding coordinate transformation. Hence, if we treat the double subscript ij as a single index α, the $T_{ij} = T_\alpha$ form a basis for a nine-dimensional representation of the rotation group, $T'_\alpha = \sum_\beta T_\beta \Gamma(R)_{\beta\alpha}$ with $\Gamma(R)_{\beta\alpha} = \Gamma(R)_{kl,ij} = R_{ki} R_{lj}$, or $\mathbf{\Gamma}(R) = \mathbf{R} \times \mathbf{R}$. Since we have already seen that the coordinates (x, y, z) transform according to $D^{(1)}$ (within a unitary transformation), evidently $R \times R = D^{(1)} \times D^{(1)} = D^{(0)} + D^{(1)} + D^{(2)}$ by the vector-model principle. Thus, under the group of rotations, the representation based on the nine components of a second-rank cartesian tensor may be decomposed into simpler one, three, and five-dimensional representations having the transformation properties of $D^{(0)}$, $D^{(1)}$, and $D^{(2)}$. These are examples of the irreducible tensors introduced at the beginning of this section. In general, a cartesian tensor of rank r may be decomposed into irreducible tensors whose ranks may be determined by successive applications of the vector-model principle. For example, a third-rank tensor T_{ijk} has 27 components. It may be decomposed as follows,

$$D^{(1)} \times D^{(1)} \times D^{(1)} = D^{(1)} \times [D^{(0)} + D^{(1)} + D^{(2)}]$$

$$= D^{(1)} + [D^{(0)} + D^{(1)} + D^{(2)}] + [D^{(1)} + D^{(2)} + D^{(3)}]$$

$$= D^{(0)} + 3D^{(1)} + 2D^{(2)} + D^{(3)}$$

into seven distinct irreducible tensors of ranks 0 through 3. Evidently there will always be just one irreducible tensor of rank equal to that of the cartesian tensor being reduced.

Let us return to the decomposition of the cartesian second-rank tensor to see the nature of the irreducible tensors in more detail. First, let us note that the trace of a tensor

$$T = \operatorname{Tr} \mathbf{T} = \sum_i T_{ii}$$

is a scalar, or a quantity invariant under any unitary transformation such as a rotation of axes. Hence it forms a basis for $D^{(0)}$. Next, consider the three antisymmetric combinations obtained from

$$A_z = \tfrac{1}{2}(T_{xy} - T_{yx})$$

by cyclic permutation. Note that this A_z will transform under rotation like $(V_x U_y - V_y U_x)$, where $\mathbf{V}$ and $\mathbf{U}$ are arbitrary vectors. But this expression is just the z component of the vector product $\mathbf{V} \times \mathbf{U}$, which transforms as a vector under rotation (but as an axial vector under inversion since $\mathbf{V}$ and $\mathbf{U}$ are polar vectors). Hence the elements A_x, A_y, and A_z form a basis for a three-dimensional irreducible representation equivalent to $D^{(1)}$. Note that if the tensor $\mathbf{T}$ were antisymmetric (or "skew-symmetric"), it would be completely specified by these three elements, and would have the form

$$\begin{pmatrix} 0 & A_z & -A_y \\ -A_z & 0 & A_x \\ A_y & -A_x & 0 \end{pmatrix}$$

Since the elements of such an antisymmetric tensor transform under rotations like those of a vector but go into themselves under inversion of coordinates, such a tensor is the prototype form of an axial vector.

Finally, we find the basis of the five-dimensional irreducible representation by forming the traceless symmetric tensor $\mathbf{S}$ with elements

$$S_{ij} = \tfrac{1}{2}(T_{ij} + T_{ji}) - \tfrac{1}{3}T$$

This has six elements, but only two of the diagonal ones are independent since the trace is zero. Thus one could choose five elements which contain all the information, and these would serve as the basis of a five-dimensional *irreducible* representation. A convenient choice of five elements in terms of cartesian coordinates is the set suggested by the five d functions we used earlier in discussing crystal-field splittings, namely, S_{xy}, S_{yz}, S_{zx}, $S_{xx} - S_{yy}$, $2S_{zz} - S_{xx} - S_{yy}$. However, to use the $\mathbf{D}^{(2)}$ matrices to represent the transformation properties, we must use the five functions with the same symmetry as the angular-momentum eigenfunctions $\varphi_m{}^{(2)}$. These differ from the above-mentioned set (after normalization) by only a unitary transformation.

As a specific example, consider the electric field gradient tensor, whose elements are

$$\varphi_{ij} = \frac{\partial^2 \varphi}{\partial x_i\, \partial x_j} = \frac{-\partial E_i}{\partial x_j} = \frac{-\partial E_j}{\partial x_i}$$

The invariant is $\sum_i \varphi_{ii} = \nabla^2 \varphi = -4\pi\rho$, which certainly is independent of orientation of the axes. The antisymmetric tensor vanishes because, e.g.,

$$A_z = \frac{1}{2}\left[\frac{\partial^2 \varphi}{\partial x\, \partial y} - \frac{\partial^2 \varphi}{\partial y\, \partial x}\right] = 0$$

The remaining five elements could be taken to be φ_{xy}, φ_{yz}, φ_{zx}, $\varphi_{xx} - \varphi_{yy}$, $2\varphi_{zz} - \varphi_{xx} - \varphi_{yy}$. These elements give all the information required to calculate the quadrupole coupling, for example.

It having been seen how the cartesian tensors are built up by considering the transformation properties of products of cartesian coordinates, it is not surprising that we can take the same approach for the spherical tensors.

That is, if we have two vectors with spherical components $V_{\pm 1,0}$ and $U_{\pm 1,0}$, then we can consider the nine cross products as forming a basis for a reducible representation of the rotation group. By the same argument given above, this decomposes into $D^{(0)}$, $D^{(1)}$, and $D^{(2)}$. We get the invariant from the scalar product

$$T_0^0 = \sum_\mu (-1)^\mu V_\mu U_{-\mu} \qquad (5\text{-}63a)$$

the vector from such antisymmetric products as

$$T_{\pm 1}^{(1)} \sim (V_{\pm 1} U_0 - V_0 U_{\pm 1}) \qquad T_0^{(1)} \sim (V_1 U_{-1} - V_{-1} U_1) \qquad (5\text{-}63b)$$

and the second-rank tensor from the symmetric combinations, i.e.,

$$T_{\pm 2}^{(2)} \sim V_{\pm 1} U_{\pm 1} \qquad T_{\pm 1}^{(2)} \sim (V_{\pm 1} U_0 + V_0 U_{\pm 1})$$
$$T_0^{(2)} \sim [2V_0 U_0 + (V_1 U_{-1} + V_{-1} U_1)] \qquad (5\text{-}63c)$$

These linear combinations are specified by the Wigner coefficients for combining $j_1 = 1$ and $j_2 = 1$ to form $J = 0$, 1, and 2, respectively. This technique can be extended indefinitely to work out possible structures for the higher-rank spherical tensors.

Two spherical tensors may be multiplied and contracted according to the rule

$$T_M^{(\Omega)}(A_1, A_2) = \sum_{\mu_1} a_{\Omega, \mu_1, M - \mu_1}^{(\omega_1 \omega_2)} T_{\mu_1}^{(\omega_1)}(A_1) T_{M - \mu_1}^{(\omega_2)}(A_2) \qquad (5\text{-}64)$$

where a is the appropriate Wigner coefficient. This formula is exactly analogous to the one for combining angular-momentum eigenfunctions, and it may be proved in exactly the same way by using only the transformation properties of the T_μ under rotations. From this formula we see that, when tensors are multiplied, we combine the ranks according to the vector model. A particularly important example is the formation of a tensor of zero rank (a scalar invariant) by the combination of two tensors of the same rank. In this case the Wigner coefficient is just $(-1)^{\omega_1 - \mu_1}(2\omega_1 + 1)^{-\frac{1}{2}}$. Dropping the factors which are independent of μ in $T_0^{(0)}$, we have the invariant

$$\mathcal{I}(A_1, A_2) = \sum_\mu (-1)^\mu T_\mu^{(\omega)}(A_1) T_{-\mu}^{(\omega)}(A_2) \qquad (5\text{-}65)$$

This invariant is often called the *scalar product of the two tensors*. We note that it reduces to the definition of the scalar product for three-dimensional vectors as found previously. Such forms play a prominent role in the expression of any rotationally invariant Hamiltonian which depends on two types of variables.

An example of considerable importance is interaction of two dipoles. The classical expression

$$H = \frac{\boldsymbol{\mu}_1 \cdot \boldsymbol{\mu}_2}{r_{12}^3} - \frac{3(\boldsymbol{\mu}_1 \cdot \mathbf{r}_{12})(\boldsymbol{\mu}_2 \cdot \mathbf{r}_{12})}{r_{12}^5} \qquad (5\text{-}65a)$$

can be expanded into the following form (with $\mu = \gamma \hbar \mathbf{I}$),

$$H = \gamma^2 \hbar^2 r_{12}^{-3}(A + B + C + D + E + F)$$

where
$$A = I_z^{(1)} I_z^{(2)}(3 \cos^2 \theta_{12} - 1)$$

$$B = -\tfrac{1}{4}(I_-^{(1)} I_+^{(2)} + I_+^{(1)} I_-^{(2)})(3 \cos^2 \theta_{12} - 1)$$

$$C = \tfrac{3}{2}(I_+^{(1)} I_z^{(2)} + I_z^{(1)} I_+^{(2)}) \sin \theta_{12} \cos \theta_{12} e^{-i\varphi_{12}} \qquad (5\text{-}65b)$$

$$D = \tfrac{3}{2}(I_-^{(1)} I_z^{(2)} + I_z^{(1)} I_-^{(2)}) \sin \theta_{12} \cos \theta_{12} e^{i\varphi_{12}}$$

$$E = \tfrac{3}{4} I_+^{(1)} I_+^{(2)} \sin^2 \theta_{12} e^{-2i\varphi_{12}}$$

$$F = \tfrac{3}{4} I_-^{(1)} I_-^{(2)} \sin^2 \theta_{12} e^{2i\varphi_{12}}$$

In these expressions θ_{12} and φ_{12} are the angles specifying the orientation of the vector $\mathbf{r}_{12}$. The terms A and B both belong to $\mu = 0$, as is clear from the angular dependence $(3 \cos^2 \theta_{12} - 1)$. If the spin parts of A and B are summed together, they equal

$$\tfrac{1}{2}[2I_z^{(1)} I_z^{(2)} - I_x^{(1)} I_x^{(2)} - I_y^{(1)} I_y^{(2)}] = \tfrac{1}{2}[2I_0^{(1)} I_0^{(2)} + I_1^{(1)} I_{-1}^{(2)} + I_{-1}^{(1)} I_1^{(2)}]$$

in terms of cartesian or spherical components. In either form the expression is easily recognized to have the characteristic transformation property of $T_0^{(2)}$. C and D belong to $\mu = \pm 1$, and E and F belong to $\mu = \pm 2$. In all cases the factor involving the spins $\mathbf{I}^{(1)}$ and $\mathbf{I}^{(2)}$ is of the form given in (5-63c). On the other hand, the factor involving θ_{12} and φ_{12} is simply a spherical harmonic of order 2. Decomposition of H in this manner gives a clear physical picture of the interaction in terms of spin flips produced by a time variation of the spatial coordinates θ_{12} and φ_{12}.

Another example having the same structure could be set up representing the interaction of a nuclear quadrupole-moment tensor with the electric field gradient tensor which we discussed earlier in the section. This example is deferred to a later chapter, however.

We conclude this section by pointing out that Racah defines these same irreducible operators by their commutation relations with angular-momentum operators.

$$[J_\pm, T_\mu^{(\omega)}] = [\omega(\omega + 1) - \mu(\mu \pm 1)]^{\frac{1}{2}} T_{\mu \pm 1}^{(\omega)} \qquad (5\text{-}66a)$$

and
$$[J_z, T_\mu^{(\omega)}] = \mu T_\mu^{(\omega)} \qquad (5\text{-}66b)$$

The angular-momentum operators themselves satisfy these rules for $\omega = 1$. Since we have shown that the rotational transformation properties of angular-momentum operators are equivalent to the commutation relations, it is not surprising that Racah's definition of a general spherical tensor operator is equivalent to our definition through transformation properties. This equivalence may be verified by consideration of infinitesimal rotations.

5-12 The Wigner-Eckart Theorem

This very useful theorem provides the generalization of the matrix-element theorem for scalar operators to the higher tensor operators. We consider a general matrix element, apply a unitary transformation P_R (which does not change the value of the matrix element), and obtain

$$(N'j'm' |T_\mu^{(\omega)}| Njm)$$

$$= (P_R\psi_{N'j'm'}, P_R T_\mu^{(\omega)} P_R^{-1} P_R \psi_{Njm})$$

$$= \left(\sum_{m''} \psi_{N'j'm''} D^{(j')}(R)_{m''m'}, \sum_{\mu'} T_{\mu'}^{(\omega)} D^{(\omega)}(R)_{\mu'\mu} \sum_{m''} \psi_{Njm''} D^{(j)}(R)_{m''m} \right)$$

$$= \sum_{\mu'm''m'''} D^{(j')}(R)_{m''m'}^* D^{(\omega)}(R)_{\mu'\mu} D^{(j)}(R)_{m'''m}(N'j'm'' |T_{\mu'}^{(\omega)}| Njm''')$$

Now, in the case of the scalar operator, $\mathbf{D}^{(\omega)} = \mathbf{D}^{(0)} = 1$, and we could integrate over R and apply the orthogonality theorem for representations to obtain the result. To handle the general case of a product of three representations, we just combine the last two by use of the Wigner coefficients using (5-41).

$$D^{(\omega)}(R)_{\mu'\mu} D^{(j)}(R)_{m''m} = \sum_{J=|j-\omega|}^{j+\omega} A_{m''\mu'}^{j\omega J} D^{(J)}(R)_{m''+\mu',m+\mu} A_{m\mu}^{j\omega J}$$

In this we have used the fact that $\mathbf{A}$ is real. Inserting this expression above, we have

$$(N'j'm' |T_\mu^{(\omega)}| Njm) = \sum_{J\mu'm''m'''} A_{m''\mu'}^{j\omega J} A_{m\mu}^{j\omega J} D^{(j')}(R)_{m''m'}^* D^{(J)}(R)_{m''+\mu',m+\mu}$$

$$\times (N'j'm'' |T_{\mu'}^{(\omega)}| Njm''')$$

If we integrate over R and divide by the volume of group-element space $\int dR$, the orthonormality of the $\mathbf{D}$'s leads to

$$(N'j'm' |T_\mu^{(\omega)}| Njm) = \sum_{J=|j-\omega|}^{j+\omega} \sum_{\mu'm''m'''} A_{m''\mu'}^{j\omega J} A_{m\mu}^{j\omega J} \frac{\delta_{Jj'}\delta_{m''+\mu',m'}\delta_{m+\mu,m'}}{2j'+1}$$

$$\times (N'j'm'' |T_{\mu'}^{(\omega)}| Njm''')$$

Note that the matrix element vanishes unless $|j - \omega| < j' < j + \omega$. Even if this inequality is satisfied, there is only one term in the sum on J, namely, $J = j'$. We may rewrite this nonvanishing term as

$$(N'j'm' |T_\mu^{(\omega)}| Njm)$$

$$= A_{m\mu}^{j\omega j'} \delta_{m',m+\mu} \sum_{\mu'm''m'''} \frac{A_{m''\mu'}^{j\omega j'} \delta_{m''+\mu',m''}(N'j'm'' |T_{\mu'}^{(\omega)}| Njm''')}{2j'+1}$$

We see that the sum is independent of m, m', μ and depends only on N, N', j, j'. Thus we may denote it by $(N'j' \| T^{(\omega)} \| Nj)$ and then obtain our theorem

$$(N'j'm' \,|\, T_\mu^{(\omega)} \,|\, Njm) = A_{m\mu}^{j\omega j'} \delta_{m', m+\mu} (N'j' \| T^{(\omega)} \| Nj)$$

$$|j - \omega| \leqslant j' \leqslant (j + \omega) \qquad \blacktriangleright \text{(5-67)}$$

The quantity $(N'j' \| T^{(\omega)} \| Nj)$ is called the *reduced matrix element* of $T^{(\omega)}$.

This result reduces the amount of calculation required to handle a problem, since the tabulated Wigner coefficients give the ratios of matrix elements of all components of a tensor operator between all states of given initial and final angular momentum. Thus one need calculate only one such matrix element to determine the reduced matrix element. This is usually done for the simple choice $\mu = 0$, $m = m'$. It should accordingly be emphasized that the *form* of the result (5-67) is all that is essential from this development. The reduced matrix element is almost invariably determined by using (5-67) together with one explicitly calculated matrix element. The summation expression for the reduced matrix element given above from the derivation is of no use for calculation.

This theorem also gives us two very general *selection rules* on matrix elements,

$$|\Delta j| = |j - j'| \leqslant \omega$$
$$\Delta m = m' - m = \mu \leqslant \omega \qquad \blacktriangleright \text{(5-68)}$$

For example, for dipole transitions, T is a vector operator, and so $\omega = 1$. Therefore $\Delta j = \pm 1$, 0. For circularly polarized radiation about the z axis, $\mu = \pm 1 = \Delta m$. For radiation polarized along the quantization axis, $\mu = 0 = \Delta m$.

Matrix elements of vector operators. The most important single example of the use of the Wigner-Eckart theorem (5-67) is the determination of the matrix elements of a vector operator. (These results are also obtained by Condon and Shortley[1] by using the commutation rules.) Suppressing the N, N' notation (although it is still implicit) and using the tabulated Wigner coefficients (although absorbing some normalization constants into the reduced matrix elements), we have the following nonvanishing elements:

$$(J + 1, M \pm 1 \,|\, V_\pm \,|\, JM) = \mp (J + 1 \| V \| J)[(J \pm M + 1)(J \pm M + 2)]^{\frac{1}{2}}$$

$$(J + 1, M \,|\, V_z \,|\, JM) = (J + 1 \| V \| J)[(J + M + 1)(J - M + 1)]^{\frac{1}{2}}$$

$$(J, M \pm 1 \,|\, V_\pm \,|\, JM) = (J \| V \| J)[(J \mp M)(J \pm M + 1)]^{\frac{1}{2}}$$

$$\equiv (J \| V \| J)(J, M \pm 1 \,|\, J_\pm \,|\, JM) \qquad \blacktriangleright \text{(5-69)}$$

$$(JM \,|\, V_z \,|\, JM) = (J \| V \| J)M \equiv (J \| V \| J)(JM \,|\, J_z \,|\, JM)$$

$$(J - 1, M \pm 1 \,|\, V_\pm \,|\, JM) = \pm (J - 1 \| V \| J)[(J \mp M)(J \mp M - 1)]^{\frac{1}{2}}$$

$$(J - 1, M \,|\, V_z \,|\, JM) = (J - 1 \| V \| J)[(J + M)(J - M)]^{\frac{1}{2}}$$

[1] Condon and Shortley, *op. cit.*, p. 63.

Note that we give the elements for $V_{\pm} = V_x \pm iV_y = \mp\sqrt{2}V_{\pm 1}$ since these are more commonly used than the spherical components $V_{\pm 1}$ themselves. Also note that, by absorbing some common factors from the Wigner coefficients into the reduced matrix elements, we have changed the normalization of the reduced matrix elements in (5-69) compared with those in (5-67). By using these results we can calculate, for example, the relative intensities of various Zeeman components $M \to M'$ of a given transition $J \to J'$ for any polarization. We also note in general that, if $J = J' = 0$, all matrix elements of V vanish. Thus we have our final selection rule, namely, $J = 0 \to 0$ *transitions are forbidden*. [This result could have been anticipated from the vector-model result that, when one combines $\omega = 1$ (for the dipole transition) with $J = 0$, one obtains *only* $J = 1$ for the final state.] Finally, we note that the matrix elements diagonal in J are proportional to the corresponding matrix elements of $\mathbf{J}$ itself. We shall make much use of this fact in later chapters by using $\mathbf{J}$ as an operator replacement for other *vector* operators. Apart from the proportionality constant $(J \| V \| J)$, we see that this is a rigorous replacement for matrix elements diagonal in J. For our applications, this fact is the most important single result of this section.

5-13   The Racah Coefficients

The Wigner or Clebsch-Gordan coefficients which we have discussed provide the unitary transformation for coupling two angular momenta j_1, j_2 together to form a resultant j. It often occurs that one must couple *three* angular momenta together. This can be done by successive use of the Wigner coefficients, but it is worth working out a number of general properties once and for all. The Racah coefficients relate the various coupling schemes which may be used in coupling $\mathbf{j}_1 + \mathbf{j}_2 + \mathbf{j}_3 = \mathbf{j}$. In this section and the following one, we shall survey these coefficients and their application.

Given $\mathbf{j}_1$, $\mathbf{j}_2$, and $\mathbf{j}_3$, we can have six simultaneously diagonalized operators, for example, $\mathbf{j}_1{}^2$, $\mathbf{j}_2{}^2$, $\mathbf{j}_3{}^2$, j_{1z}, j_{2z}, j_{3z}. However, we are more interested in representations which diagonalize $\mathbf{j}^2$ and j_z as well as the magnitudes j_1, j_2, j_3. In this case the sixth diagonal operator is $\mathbf{j}_{\text{int}}^2$, the square of an intermediate angular momentum formed by compounding two of the three. This $\mathbf{j}_{\text{int}}$ can be chosen in three ways. We consider the two,

$$\mathbf{j}_{\text{int}} = \mathbf{j}' = \mathbf{j}_1 + \mathbf{j}_2 \qquad \mathbf{j}_{\text{int}} = \mathbf{j}'' = \mathbf{j}_2 + \mathbf{j}_3$$

The relation between either of these and the third possibility could be obtained by a suitable renumbering. Now a unitary transformation $\mathbf{R}$ must exist between the eigenfunctions of the basis which diagonalizes j' and that which diagonalizes j''. Suppressing the quantum numbers j_1, j_2, j_3, which carry throughout, we write this relation as

$$\psi_m{}^j(j') = \sum_{j''} \psi_m{}^j(j'') R_{j''j'} \tag{5-70}$$

The Racah coefficients W are then defined by the following relation:

$$R_{j''j'} = [(2j'' + 1)(2j' + 1)]^{\frac{1}{2}} W(j_1 j_2 j j_3; j'j'') \tag{5-71}$$

The square root is introduced here, just as with Racah's V coefficients, to enhance the symmetry properties.

We now use the Wigner coefficients to work out some properties of the W coefficients. As before, we denote the Wigner coefficient for combining $\mathbf{j_1} + \mathbf{j_2} = \mathbf{j}$ by $A_{m_1 m_2}^{j_1 j_2 j}$. It is, of course, understood that $m = m_1 + m_2$. Then

$$\psi_m{}^{j'} = \sum_{m_1} A_{m_1, m'-m_1}^{j_1 j_2 j'} \psi_{m_1}{}^{j_1} \psi_{m'-m_1}{}^{j_2}$$

Using these, we can build up

$$\psi_m{}^j(j') = \sum_{m'} A_{m', m-m'}^{j' j_3 j} \psi_{m-m'}^{j_3} \sum_{m_1} A_{m_1, m'-m_1}^{j_1 j_2 j'} \psi_{m_1}{}^{j_1} \psi_{m'-m_1}{}^{j_2}$$

In a similar manner

$$\psi_m{}^j(j'') = \sum_{m''} A_{m-m'', m''}^{j_1 j'' j} \psi_{m-m''}^{j_1} \sum_{m_2} A_{m_2, m''-m_2}^{j_2 j_3 j''} \psi_{m_2}{}^{j_2} \psi_{m''-m_2}^{j_3}$$

Inserting these in (5-70) and equating coefficients of $\psi_{\mu_1}{}^{j_1} \psi_{\mu_2}{}^{j_2} \psi_{\mu_3}{}^{j_3}$ on both sides (noting that $\mu_1 + \mu_2 + \mu_3 = m$), we obtain

$$A_{\mu_1 \mu_2}^{j_1 j_2 j'} A_{\mu_1+\mu_2, \mu_3}^{j' j_3 j} = \sum_{j''} R_{j''j'} A_{\mu_2 \mu_3}^{j_2 j_3 j''} A_{\mu_1, \mu_2+\mu_3}^{j_1 j'' j} \tag{5-72}$$

We can now transform this into more useful forms by taking advantage of the unitary nature of the A's. For example, if we multiply by $A_{\mu_2 \mu_3}^{j_2 j_3 j''}$ for some particular j'' and sum over μ_2, holding $\mu_2 + \mu_3$ fixed, we obtain

$$\sum_{\mu_2} A_{\mu_1 \mu_2}^{j_1 j_2 j'} A_{\mu_1+\mu_2, \mu_3}^{j' j_3 j} A_{\mu_2 \mu_3}^{j_2 j_3 j''} = R_{j''j'} A_{\mu_1, \mu_2+\mu_3}^{j_1 j'' j} \tag{5-73}$$

If we now multiply through by $A_{\mu_1, \mu_2+\mu_3}^{j_1 j'' j}$ and sum on μ_1, holding

$$\mu_1 + \mu_2 + \mu_3 = \text{constant},$$

we obtain

$$\sum_{\mu_1} \sum_{\mu_2} A_{\mu_1 \mu_2}^{j_1 j_2 j'} A_{\mu_1+\mu_2, \mu_3}^{j' j_3 j} A_{\mu_2 \mu_3}^{j_2 j_3 j''} A_{\mu_1, \mu_2+\mu_3}^{j_1 j'' j} = R_{j''j'} \tag{5-74}$$

Combining (5-74) with (5-71) provides an explicit definition of the Racah coefficients in terms of the Wigner coefficients which we have determined previously. We note that the Racah coefficients are independent of all magnetic quantum numbers. We see that they cannot depend on m by inspection of (5-70), since application of J_+ or J_- to (5-70) changes the value of m while multiplying both sides by the same factor. Hence $R_{j''j'}$ must be independent of m.

In discussing the properties of the W coefficients it is customary to lighten the notation by replacing $(j_1 j_2 j j_3; j'j'')$ by $(abcd; ef)$ and also replacing $m_1 m_2 m m_3$ by $\alpha\beta\gamma\delta$, so that $\gamma = \alpha + \beta + \delta$. In this notation, the relations (5-74) and (5-71) lead to

$$W(abcd; ef) = [(2e + 1)(2f + 1)]^{-\frac{1}{2}} \sum_{\alpha,\beta} A^{abe}_{\alpha\beta} A^{edc}_{\alpha+\beta,\delta} A^{bdf}_{\beta\delta} A^{afc}_{\alpha,\beta+\delta} \qquad (5\text{-}75)$$

which explicitly defines the Racah coefficients. Racah was able to carry out the indicated summation and obtain the following expression:

$$W(abcd; ef) = \Delta(abe)\Delta(cde)\Delta(acf)\Delta(bdf)$$

$$\times \sum_{\kappa} \frac{(-1)^{\kappa+a+b+c+d}(\kappa + 1)!}{(\kappa - a - b - e)! \, (\kappa - c - d - e)!} \qquad (5\text{-}76)$$

$$\times (\kappa - a - c - f)! \, (\kappa - b - d - f)! \, (a + b + c + d - \kappa)!$$

$$\times (a + d + e + f - \kappa)! \, (b + c + e + f - \kappa)!$$

In this, the "triangle coefficients" are defined by

$$\Delta(abc) = \left[\frac{(a + b - c)! \, (a - b + c)! \, (-a + b + c)!}{(a + b + c + 1)!} \right]^{\frac{1}{2}}$$

This expression is finite only if a, b, c satisfy the "triangular condition" that $\mathbf{a} + \mathbf{b} = \mathbf{c}$ be allowed by the vector model. That is, the difference of any two must not exceed the third. Thus we expect the selection rule that $W(abcd; ef)$ cannot occur unless (abe), (edc), (bdf), and (afc) satisfy the triangular condition. In fact detailed consideration shows that there will be an infinite denominator (the factorial of a negative integer) for all terms in the sum in (5-76) if the triangular conditions are not satisfied. Thus these coefficients have a value of zero. This is reasonable, since otherwise we would have nonphysical coupling arrangements. The situation can be visualized graphically by a vector-model quadrilateral in which the two intermediate angular momenta form the diagonals. Such a quadrilateral is shown in Fig. 5-5.

Properties of the Racah coefficients. The roles of the six angular momenta in $W(abcd; ef)$ may be interchanged in 24 ways while the triangular relations are still preserved. The only effect of such permutations is a possible phase change. Detailed consideration shows that the following permutations leave the coefficient unchanged:

$$(abcd; ef)(badc; ef)(cdab; ef)(dcba; ef)(acbd; fe)(cadb; fe)(bdac; fe)(dbca; fe)$$

$$(5\text{-}77a)$$

The following permutations give $(-1)^{b+c-e-f}W(abcd;ef)$:

$$(aefd;bc)(eadf;bc)(fdae;bc)(dfea;bc)(afed;cb)(fade;cb)(edaf;cb)(defa;cb)$$
$$(5\text{-}77b)$$

The following give $(-1)^{a+d-e-f}W(abcd;ef)$:

$$(ebcf;ad)(befc;ad)(cfeb;ad)(fcbe;ad)(ecbf;da)(cefb;da)(bfec;da)(fbce;da)$$
$$(5\text{-}77c)$$

By use of these relations, any index can be moved to any position for the purpose of carrying out calculations.

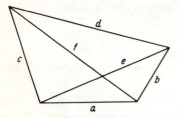

Fig. 5-5. *Vector-model quadrilateral showing the relation of the six coupled angular momenta involved in a Racah coefficient.*

Whenever the area of one of the triangles vanishes, i.e., when one member of a triad in (5-76) equals the sum of the other two, the sum reduces to a single term. For example, if $e = a + b$, then both $(\kappa - a - b - c - d)!$ and $(a + b + c + d - \kappa)!$ appear in the denominator. The only value of κ which keeps these both finite is $\kappa = a + b + c + d$, in which case both are $0! = 1$. In this way we obtain

$$W(abcd; a + b, f)$$

$$= \left[\frac{(2a)!\,(2b)!\,(a+b+c-d)!\,(a+b+c+d+1)!\,(a+b+d-c)!}{(2a+2b+1)!\,(a+c+f+1)!\,(c+d-a-b)!\,(b+f+d+1)!} \right.$$

$$\times \left. \frac{(c+f-a)!\,(d+f-b)!}{(a+c-f)!\,(a+f-c)!\,(b+d-f)!\,(b+f-d)!} \right]^{\frac{1}{2}} \quad (5\text{-}78)$$

By use of the symmetry relations of the preceding paragraph, (5-78) may be used in general when any of the four triangles has vanishing area.

Another simple special case occurs when one of the arguments of W vanishes. This causes the quadrilateral to reduce to a triangle. A simple reduction using the symmetry properties leads to

$$W(abcd; 0f) = \frac{(-1)^{f-b-d}\delta_{ab}\delta_{cd}}{[(2b+1)(2d+1)]^{\frac{1}{2}}} \quad (5\text{-}79)$$

Since the $R_{j''j'} = R_{fe}$ form a unitary transformation, we have the orthonormality relations

$$\sum_e R_{fe}R_{ge} = \delta_{fg}$$

By using the definition (5-71), this implies the following orthonormality property of the Racah coefficients:

$$\sum_e (2e + 1)(2f + 1)W(abcd; ef)W(abcd; eg) = \delta_{fg} \qquad (5\text{-}80)$$

Another useful sum over W coefficients is Racah's sum rule,

$$\sum_e (2e + 1)(-1)^{a+b-e}W(abcd; ef)W(bacd; eg) = W(agfb; cd) \quad (5\text{-}81)$$

Obviously these coefficients are even more awkward to handle algebraically than the Wigner coefficients. Considerable effort has been expended in preparing helpful tables. A useful source of algebraic expressions for $e = \frac{1}{2}$, 1, $\frac{3}{2}$, and 2 is the review article of Biedenharn, Blatt, and Rose.[1] Extensive numerical tables, covering all cases for angular momenta up to 8 in steps of $\frac{1}{2}$, are given in the book of Rotenberg et al.[2] Actually, they tabulate the so-called $6\text{-}j$ coefficients, related to Racah's W coefficients by a phase factor, namely,

$$W(abcd; ef) = (-1)^{a+b+c+d} \begin{Bmatrix} a & b & e \\ d & c & f \end{Bmatrix}$$

so as to have improved symmetry characteristics.

5-14 Application of Racah Coefficients

The Racah coefficients have proved useful in an extremely wide range of applications. The unifying characteristic is the importance of states of definite angular momentum. Examples of these areas are atomic and molecular spectra, angular correlation of nuclear radiation, angular distributions in nuclear reactions, and multipole-field expansions.

An important type of application arises in evaluating matrix elements of a rotationally invariant operator, such as a term in a Hamiltonian, which is expressed as the contraction of two irreducible tensors of rank L. We may denote this as follows:

$$T^{(L)}(1) \cdot T^{(L)}(2) = \sum_M (-1)^M T_M^{(L)}(1) T_{-M}^{(L)}(2)$$

This represents an interaction between two independent subsystems 1 and 2, each characterized by its own angular momentum. For example, if this represented the spin-orbit coupling, system 1 is the orbital motion with angular momentum $j_1 = l$ and system 2 is the spin with $j_2 = s$. Regardless of the value of l, though, these two tensors are both of rank $L = 1$, since all angular momenta are (axial) vector operators.

[1] L. C. Biedenharn, J. M. Blatt, and M. E. Rose, *Revs. Modern Phys.*, **24**, 249 (1952).
[2] Rotenberg, Bivins, Metropolis, and Wooten, *op. cit.*

In general we are interested in matrix elements of the type

$$(j_1'j_2'j'm' | T^{(L)}(1) \cdot T^{(L)}(2)| j_1 j_2 jm)$$

We can express the initial and final states in terms of decoupled states, using the Wigner coefficients,

$$\psi_m{}^j = \sum_{m_1} \psi_{m_1}{}^{j_1} \psi_{m-m_1}{}^{j_2} A_{m_1, m-m_1}^{j_1 j_2 j}$$

$$\psi_{m'}{}^{j'} = \sum_{m_1'} \psi_{m_1'}{}^{j_1'} \psi_{m'-m_1'}{}^{j_2'} A_{m_1', m'-m_1'}^{j_1' j_2' j'}$$

By using these, the matrix element becomes

$$\sum_{M m_1' m_1} (-1)^M A_{m_1', m'-m_1'}^{j_1' j_2' j'} A_{m_1, m-m_1}^{j_1 j_2 j} (j_1' m_1' | T_M{}^{(L)}(1)| j_1 m_1)$$
$$\times (j_2', m'-m_1' | T_{-M}{}^{(L)}(2)| j_2, m-m_1)$$

if we take advantage of the chance of factoring the matrix element which arises because the two factors operate in separate decoupled spaces. We may now apply the Wigner-Eckart theorem to each of these factored matrix elements. For example,

$$(j_1' m_1' | T_M{}^{(L)}(1)| j_1 m_1) = (j_1' \| T^{(L)}(1)\| j_1) A_{m_1 M}^{j_1 L j_1'} \delta_{m_1', M+m_1}$$

We may also apply the Wigner-Eckart theorem to the entire matrix element $T^{(L)}(1) \cdot T^{(L)}(2)$. Since it is an invariant (tensor of zero rank), we have

$$(j'm' | T^{(L)}(1) \cdot T^{(L)}(2)| jm) = A_{m0}^{j0j'} \delta_{m'm}(j \| T^{(L)}(1) \cdot T^{(L)}(2)\| j')$$
$$= \delta_{j'j} \delta_{m'm}(j \| T^{(L)}(1) \cdot T^{(L)}(2)\| j) \tag{5-82}$$

Thus, as expected, the matrix element vanishes unless $j' = j$ and $m' = m$. Now, if we combine the pieces set up above, taking advantage of the fact that $M = m_1' - m_1$ to eliminate one summation, we have

$$(j_1'j_2'j'm' | T^{(L)}(1) \cdot T^{(L)}(2)| j_1 j_2 jm) = \delta_{m'm} \delta_{j'j}(j_1' \| T^{(L)}(1)\| j_1)(j_2' \| T^{(L)}(2)\| j_2)$$
$$\times \sum_{m_1 m_1'} (-1)^{m_1' - m_1} A_{m_1, m-m_1}^{j_1 j_2 j} A_{m-m_1, m_1-m_1'}^{j_2 L j_2'} A_{m_1, m_1'}^{j_1 L j_1'} A_{m_1', m'-m_1'}^{j_1' j_2' j'}$$

By suitable juggling of indices and use of the symmetry properties of the Wigner coefficients, this sum can be brought into the form (5-75) defining the Racah coefficients. In this way we can express the matrix element in terms of a single Racah coefficient, which can be found in the tables. The final result is

$$(j_1'j_2'j'm' | T^{(L)}(1) \cdot T^{(L)}(2)| j_1 j_2 jm) = \delta_{m'm} \delta_{j'j}(-1)^{j_1' + j_2 - j} W(j_1 j_2 j_1' j_2'; jL)$$
$$\times [(2j_1' + 1)(2j_2' + 1)]^{\frac{1}{2}}(j_1' \| T^{(L)}(1)\| j_1)(j_2' \| T^{(L)}(2)\| j_2) \tag{5-83}$$

In view of (5-82), we see that the Racah coefficient can be considered to express the reduced matrix element of the contraction $T^{(L)}(1) \cdot T^{(L)}(2)$ in terms of the reduced matrix elements of $T^{(L)}(1)$ in the $j_1 m_1$ representation and of $T^{(L)}(2)$ in the $j_2 m_2$ representation. Since these are relatively easy to calculate, this is a considerable simplification.

Another important result which may be obtained by similar manipulation[1] is an expression for the matrix elements of one of the tensors $T^{(L)}(1)$, which operates on only part of the system, in the coupled representation which diagonalizes the *total* angular momentum. This result is an example of the general Wigner-Eckart theorem;

$$(j_1' j_2' j' m' | T_M^{(L)}(1) | j_1 j_2 j m) = A_{mM}^{jLj'} \delta_{m', m+M} (j_1' j_2' j' \| T^{(L)}(1) \| j_1 j_2 j) \quad (5\text{-}84a)$$

but in this case the Racah coefficients explicitly relate the reduced matrix element with that in the decoupled representation, which is relatively easy to calculate.

$$(j_1' j_2' j' \| T^{(L)}(1) \| j_1 j_2 j) = \delta_{j_2' j_2} (-1)^{j_2 + L - j_1 - j'} [(2j_1' + 1)(2j + 1)]^{\frac{1}{2}}$$
$$\times W(j_1 j j_1' j'; j_2 L)(j_1' \| T^{(L)}(1) \| j_1) \quad (5\text{-}84b)$$

5-15 The Rotation-Inversion Group

We have now completed our study of the group of all proper rotations. This group represents a large part of the symmetry of many physical systems. However, as we noted in the previous chapter, in many cases we also have symmetry under inversion as well as under proper rotations. Operations containing the inversion as well as a rotation are referred to as *improper* rotations. They are quite distinct from the proper rotations and have the property of turning a right-handed into a left-handed coordinate system, for example. Clearly no proper rotation could accomplish this, since all proper rotations can be built up out of infinitesimal rotations. The combined group formed as the direct product of the group of the inversion with the pure rotation group is called the *rotation-inversion group*, the *rotation-reflection group*, the *full orthogonal group* (as opposed to the orthogonal group with determinant $+1$, which gives the pure rotations), or, in the German literature, the *Drehspiegelungsgruppe*.

The coordinate transformation matrix of the inversion operator i in x, y, z coordinate space is simply

$$\mathbf{i} = \begin{pmatrix} -1 & 0 & 0 \\ 0 & -1 & 0 \\ 0 & 0 & -1 \end{pmatrix}$$

[1] See, for example, Rose's book, *op. cit.*, p. 118.

which changes $\mathbf{r}$ into $-\mathbf{r}$. Obviously this matrix commutes with all the 3×3 rotation matrices. Thus we may form the direct-product group in the simple manner indicated in Sec. 3-9. As noted there, this leads to two irreducible representations of the direct-product group for each one of the proper-rotation group. The *positive*, or *even*, one of these is just the same for proper and improper rotations, whereas the *negative*, or *odd*, one has all signs changed for improper rotations. That is,

$$D_+^{(l)}(iR) = D_+^{(l)}(R)$$
$$D_-^{(l)}(iR) = -D_-^{(l)}(R)$$

(5-85)

We have used l rather than j to label the representations here because these simple considerations are valid only for the single-valued representations arising from integral angular momenta. As noted in Sec. 4-10, for a single-particle orbital angular-momentum eigenfunction, the parity is determined by l, through the transformation properties of the spherical harmonics, to be $(-)^l$. If we combine a number of particles, there is no longer any such unique relation with the total orbital angular momentum L. Rather, the parity is given by $\prod_k (-)^{l_k}$, where l_k is the orbital angular momentum of the kth particle.

If we now introduce spin, this has no effect on these parity-build-up considerations. The spin, being an angular momentum, is an axial vector, with sense invariant under inversion. The intrinsic parity, which determines the change of sign of the spin *functions* (as opposed to the spin direction) under inversion, is arbitrary so long as no transmutation of particles is considered. We may take it to be even for electrons. Thus the parity of an atomic state is still determined only by the orbital part of the wavefunction. Concerning the representations for half-integral angular momenta, we see that we have the same double-valuedness as in the pure rotation group discussed earlier. This ambiguity of sign of the representation matrices, even for proper rotations, obscures the simple distinction of even and odd representations indicated in (5-85). However, by definition, this relation still must be valid if we restrict the rotation R to the unit operation so that only the inversion operation is involved.

It might appear that in the double-valued representations parity loses its significance because of the intrinsic sign ambiguity arising because of the sign change under a rotation by 2π. This is, of course, not true, as may be illustrated by considering the application of parity considerations in selection rules for dipole transitions. Since both initial- and final-state wavefunctions enter the matrix element together with the electric-dipole operator, a rotation by 2π leaves the matrix element unchanged whether the system has integral or half-integral angular momentum. However, inversion leaves it unchanged, and hence allowed to be nonvanishing, *only* if the initial and final states belong to representations of opposite parity.

5-16 Time-reversal Symmetry

The inclusion of inversion with the spatial rotations as a symmetry operation suggests through the relativistic equivalence of space and time that we also include a time-reversal operation T, which takes t into $-t$. Application of this operation classically requires reversing all velocities and letting time proceed in the reverse direction so that a system runs back through its past history. That this is a classically allowed symmetry follows from the fact that $\mathbf{F} = m\mathbf{a}$ involves only a *second* derivative with respect to time. A similar situation holds in quantum mechanics provided that H is real, as it is expected to be, since it represents an energy. If this is so, and if we momentarily neglect spin, T simply transforms ψ into ψ^*. We can see this by considering the time-dependent Schrödinger equation

$$H\psi = i\hbar \frac{\partial \psi}{\partial t}$$

If we take the complex conjugate and use the reality of H, we obtain

$$H\psi^* = i\hbar \frac{\partial \psi^*}{\partial(-t)}$$

Thus the state ψ^* will evolve into $+t$ in exactly the same way as ψ would have evolved in the $-t$ direction. Since probability density is proportional to $|\psi|^2$, it is unaffected by the operation. However, the momentum distribution, being proportional to the imaginary part of $\psi^* \nabla \psi$, is reversed, as it should be. If we are considering a stationary state, clearly its energy is unchanged, since $H\psi = E\psi$ implies $H\psi^* = E\psi^*$. There will or will not be a degeneracy associated with this symmetry, depending on whether the set of ψ^* is, or is not, linearly independent of the set of ψ corresponding to a given energy. In these considerations it must be kept in mind that any magnetic field present in H must be reversed in the time-reversal operation. This is taken care of automatically if the field is produced by part of the quantum-mechanical system under consideration, because a reversal of momenta reverses the field. However, if one is considering a system immersed in an external field, it must be explicitly reversed. For example, the wavefunction with $L_z = 1$, namely, $\psi \sim e^{i\varphi}$, goes into $\psi^* = e^{-i\varphi}$, with $L_z = -1$. In the absence of a magnetic field, this must have the same energy. In the presence of a field, the energy will remain the same, provided that the field is reversed when the transformation is made. If we had considered only an internal magnetic-dipolar interaction between two atoms, then T would have reversed both magnetic moments and the energy would be automatically unchanged.

It is important to note that T is not a linear and unitary operator as have been all our previous transformations. This is evident because $T(c\psi) = c^*(T\psi)$, whereas, under a unitary transformation, linearity would have required $T(c\psi) = c(T\psi)$. An operator with this property is said to be *antilinear*, and if it also conserves the magnitude of the scalar product so that

$$|(T\psi, T\varphi)| = |(\psi, \varphi)|$$

then it is called *antiunitary*.

Determination of the time-inversion operator. When we take spin into account, T is not simply the operation of complex conjugation as it is for a simple Schrödinger wavefunction as considered above. However, it *is* an antiunitary operator. This can be seen by noting that the time dependence of any energy eigenfunction is $e^{-iE_n t/\hbar}$, whether or not spin is involved. Hence, if several eigenfunctions of different energy are superposed, the time-dependent coefficient $c_n(t)$ must go into $c_n^*(t)$ if the direction of time is reversed. An antiunitary operator will make this transformation.

Now, the product of any two antiunitary operators, such as T and complex conjugation K, is evidently a unitary operator U.

$$TK = U$$

But in that case

$$T = UK \tag{5-86}$$

since K^2 is the identity. This result shows that the most general form T can take is the product of a unitary operator and the operation of complex conjugation. We can narrow this down by using the physical fact that $T^2\psi$ can differ from ψ by at most a constant factor of unit magnitude. Thus, in matrix notation,

$$\mathbf{T}^2 = \mathbf{UKUK} = \mathbf{UU}^* = c\mathbf{E} \tag{5-87}$$

where $\mathbf{E}$ is the unit matrix. Using the unitarity of $\mathbf{U}$, we have

$$\mathbf{U} = c(\mathbf{U}^*)^{-1} = c\tilde{\mathbf{U}}$$

Taking the transpose,

$$\tilde{\mathbf{U}} = c\mathbf{U} = c^2\tilde{\mathbf{U}}$$

Thus $c^2 = 1$, $c = \pm 1$, and the unitary matrix $\mathbf{U}$ must be either completely symmetric or antisymmetric on taking the transpose. Returning to (5-87), we see that it becomes

$$\mathbf{T}^2 = \pm\mathbf{E} \tag{5-88}$$

so that no possibility exists except that $T^2\psi = \pm\psi$. Clearly, when $T = K$ as for simple Schrödinger functions, $T^2 = E$. However, when there is an odd number of particles with half-integral spin, it turns out that the lower sign will hold.

To see that this is the case, we note that simple physical operators can be divided into two classes, those containing even powers of time, and those containing odd powers. The first class contains ordinary coordinate operators; the second contains linear and angular-momentum operators. Operators of the first sort will commute with T, for example,

$$T\mathbf{q} = \mathbf{q}T \tag{5-89a}$$

whereas those of the second sort will *anticommute*, for example,

$$T\mathbf{p} = -\mathbf{p}T \tag{5-89b}$$

Using the explicit forms $\mathbf{p} = (\hbar/i)\boldsymbol{\nabla}$ and $T = UK$, we can readily see that (5-89) implies that U commutes with multiplication by, and differentiation with respect to, the spatial coordinates and hence that U can have an effect only on the spin variables. We now determine U by observing that the spin components of each particle obey (5-89b). For example,

$$T\mathbf{s}_{ix} = -\mathbf{s}_{ix}T \tag{5-90}$$

where $i = 1, 2, \ldots, n$ runs over all particles. But the matrices of s_{ix} are all real. Hence

$$T\mathbf{s}_{ix} = UK\mathbf{s}_{ix} = U\mathbf{s}_{ix}K$$

By using (5-90), this must equal $-\mathbf{s}_{ix}UK$. Hence U anticommutes with $\mathbf{s}_{ix}$. A similar argument holds for $\mathbf{s}_{iz}$, which is also real. However, since $\mathbf{s}_{iy}$ is purely imaginary, this argument shows that U *commutes* with $\mathbf{s}_{iy}$. These commutation properties of U immediately suggest that U be chosen as a simple product of Pauli matrices,

$$U = \sigma_{1y}\sigma_{2y} \cdots \sigma_{ny} \tag{5-91}$$

which satisfies all the requirements, including normalization. In fact it can be shown that, apart from a phase factor, this is a unique solution. Hence, we conclude that

$$T = \sigma_{1y}\sigma_{2y} \cdots \sigma_{ny}K \tag{5-92}$$

If we allow this operator to act on a state function, we have

$$T\psi(\mathbf{r}_1, \mu_1, \ldots, \mathbf{r}_n, \mu_n) = (-i)^{2(\mu_1 + \cdots + \mu_n)}\psi(\mathbf{r}_1, -\mu_1, \ldots, \mathbf{r}_n, -\mu_n)^* \tag{5-93}$$

where μ_i is here used to denote the quantized z component ($\pm\frac{1}{2}$) of the ith spin. From this result it is evident that, if n is even, ($\mu_1 + \cdots + \mu_n$) is integral and hence $T^2\psi = \psi$. However, if n is odd, we see that $T^2\psi = -\psi$, as indicated above.

Kramers' theorem. From this result we can immediately obtain the celebrated *Kramers' theorem*.[1] This states that all energy levels of a system containing an odd number of electrons must be at least doubly degenerate regardless of how low the symmetry is, provided that there are no magnetic fields present to remove time-reversal symmetry. This theorem follows from the fact that, if $T^2\psi = -\psi$, then $T\psi$ is orthogonal to ψ. Since $T\psi$ must also

---

[1] H. A. Kramers, *Koninkl. Ned. Akad. Wetenschap., Proc.*, **33**, 959 (1930).

have the same energy as ψ, by time-reversal symmetry, at least this double degeneracy must exist. The orthogonality is proved by noting first that, for any functions ψ and φ, we have

$$(T\psi, T\varphi) = (UK\psi, UK\varphi) = (K\psi, K\varphi) = (\psi, \varphi)^* = (\varphi, \psi) \qquad (5\text{-}94)$$

Hence $(T\psi, \psi) = (T\psi, T^2\psi) = (T\psi, -\psi) = -(T\psi, \psi)$

and $(T\psi, \psi)$ must equal zero. Evidently the proof fails if there is an even number of electrons, so that $T^2\psi = \psi$, and in fact in this case no degeneracy need exist unless there is some spatial symmetry.

Representations and extra degeneracies. Because of the antiunitary nature of T, it is not possible to find a matrix representation in the usual sense for the group containing T and all the other antiunitary operators TR_i formed as products of T with ordinary unitary operators representing spatial coordinate transformations. However, we can still define matrices $D(TR_i)$ by

$$TR_i\psi_\kappa = \sum_\lambda \psi_\lambda D(TR_i)_{\lambda\kappa} \qquad (5\text{-}95)$$

because the transformed eigenfunction must still be expressible as a linear combination of the degenerate set of partners ψ_λ. Moreover, these matrices are unitary, on the assumption that the ψ_λ are chosen to be orthogonal. The proof goes exactly as the analogous one in Sec. 3-6, by using (5-94). The difficulty that arises is that, e.g.,

$$D(TR_iR_j) = D(TR_i)D(R_j)^* \neq D(TR_i)D(R_j) \qquad (5\text{-}96)$$

This difficulty is fundamental and ineradicable. Accordingly, Wigner[1] introduces the nomenclature *corepresentations* for these matrix systems which arise when a group containing antiunitary as well as unitary operations is under consideration.

Without going into this problem in detail, we can make considerable progress by noting that given a set of degenerate partners $\psi_\kappa^{(j)}$, arising from spatial symmetry, which form a basis for a representation $D^{(j)}(R)$, we can form a new set of degenerate functions $T\psi_\kappa^{(j)} = \bar\psi_\kappa^{(j)}$. Under spatial transformations, these new functions transform according to $D^{(j)}(R)^*$. This is readily seen by considering

$$R\bar\psi_\kappa^{(j)} = RT\psi_\kappa^{(j)} = TR\psi_\kappa^{(j)} = T\sum_\lambda \psi_\lambda^{(j)} D^{(j)}(R)_{\lambda\kappa}$$

$$= \sum_\lambda (T\psi_\lambda^{(j)}) D^{(j)}(R)_{\lambda\kappa}^*$$

$$= \sum_\lambda \bar\psi_\lambda^{(j)} D^{(j)}(R)_{\lambda\kappa}^* \qquad (5\text{-}97)$$

where we have used the fact that time reversal commutes with any ordinary coordinate transformation.

[1] Wigner, *op. cit.*, chap. 26.

The question then arises whether or not any new degeneracy is brought in by the time-reversal symmetry, i.e., whether the $\bar{\psi}_\lambda{}^{(j)}$ are linearly independent of the $\psi_\lambda{}^{(j)}$. Wigner has shown[1] that this question can be resolved by consideration of the representation matrices $\mathbf{D}$ and $\mathbf{D}^*$ (for the operations R alone, without considering the antiunitary operations). Three cases arise:

    a. $\mathbf{D}$ and $\mathbf{D}^*$ are equivalent to the same real irreducible representation.
    b. $\mathbf{D}$ and $\mathbf{D}^*$ are inequivalent.
    c. $\mathbf{D}$ and $\mathbf{D}^*$ are equivalent but cannot be made real.

The results also depend on whether we have integral or half-integral spin. We consider first the case of integral spin, so that $T^2 = +1$.

In case *a*, we can imagine that a suitable transformation has been carried out so that $\mathbf{D}(R) = \mathbf{D}(R)^* =$ real for both the ψ_λ and $\bar{\psi}_\lambda$. The same $D(R)$ will also hold for the real functions $u_\lambda = (\psi_\lambda + \bar{\psi}_\lambda)$ and $v_\lambda = i(\psi_\lambda - \bar{\psi}_\lambda)$. But because of their reality, no symmetry operation, *including T*, will mix any of the u's with the v's. Thus the u's or v's alone each form a basis for the same representation of the total symmetry. Since the two sets have the same energy, they must be identical, barring *accidental* degeneracy. Thus, in this case, no new degeneracy is introduced by time reversal symmetry.

In case *b*, since $\mathbf{D}$ and $\mathbf{D}^*$ are inequivalent representations, the $\bar{\psi}_\lambda$ and ψ_μ are all mutually orthogonal; yet both sets are required together to represent the full symmetry of the group. Hence the degeneracy is doubled by the presence of time-reversal symmetry.

In case *c*, since $\mathbf{D}$ and $\mathbf{D}^*$ cannot be made real, it follows that the ψ_λ and $\bar{\psi}_\lambda$ cannot be made real, since if they were,

$$D(R)_{\mu\lambda} = (\psi_\mu, P_R\psi_\lambda)$$

would also be real for all members of the group of spatial operations. But if the ψ_λ and $\bar{\psi}_\lambda$ cannot be made real, then both these independent functions are required in the basis for the full group including T and the degeneracy is doubled.

Now, let us consider the case of half-integral spin, where $T^2 = -1$. In case *a*, if we proceed as above, the proof fails because, with $T^2 = -1$,

$$Tu_\lambda = iv_\lambda \qquad Tv_\lambda = -iu_\lambda$$

so that time reversal mixes the u's and v's just as effectively as it does the ψ's and $\bar{\psi}$'s. Thus, both are required, and the degeneracy is doubled. For case *b*, the argument goes through as above, and the degeneracy is doubled. For case *c*, the argument is rather more lengthy, and we shall simply state Wigner's result that there is no extra degeneracy.

[1] E. P. Wigner, *Göttingen Nachrichten, Math.-Physik. Kl.*, p. 546, 1932.

The above results are made easier to apply by using a criterion due to Frobenius and Schur[1] which allows cases *a*, *b*, and *c* to be distinguished with only the character table and the diagonal elements of the group-multiplication table known.   The test is as follows: Form the sum of the characters of the squares of all *h* group elements.   This sum has the value *h*, 0, or $-h$, according as we are dealing with cases *a*, *b*, or *c*.   That is,

$$\sum_R \chi^{(j)}(R^2) = \begin{cases} h & \text{case } a \\ 0 & \text{case } b \\ -h & \text{case } c \end{cases} \qquad (5\text{-}98)$$

We prove this result using the great orthogonality theorem, Eq. (3-8).   In all cases,

$$\sum_R \chi^{(j)}(R^2) = \sum_R \sum_{\mu,\lambda} D^{(j)}(R)_{\mu\lambda} D^{(j)}(R)_{\lambda\mu}$$

In case *a*, where $\mathbf{D}^{(j)} = \mathbf{D}^{(j)*}$, this becomes, by using (3-8),

$$\sum_R \chi^{(j)}(R^2) = \sum_{\mu,\lambda} \sum_R D^{(j)}(R)^*_{\mu\lambda} D^{(j)}(R)_{\lambda\mu} = \sum_{\mu\lambda} \delta_{\mu\lambda} \frac{h}{l_j} = h$$

In case *b*, where $D^{(j)}$ and $D^{(j)*}$ are inequivalent, we can set $D^{(j)} = D^{(k)*}$, where $D^{(k)}$ is not related to $D^{(j)}$.   But then the orthogonality theorem leads to zero.   In case *c*, we have $\mathbf{D}^* = \mathbf{SDS}^{-1}$.   Using this, with the orthogonality theorem, leads to

$$\sum_R \chi^{(j)}(R^2) = \frac{h}{l_j} \text{Tr} (\mathbf{S}^*\mathbf{S})$$

Table 5-4

Effect of time-reversal symmetry on degeneracy

Case	Frobenius-Schur test $\sum_R \chi(R^2)$	Relation between D and D*	Integral spin	Half-integral spin
a	h	D, D* can be made real and equal	No extra degeneracy	Doubled degeneracy
b	0	D, D* are inequivalent	Doubled degeneracy	Doubled degeneracy
c	−h	D, D* are equivalent, but cannot be made real	Doubled degeneracy	No extra degeneracy

[1] Frobenius and Schur, *Berl. Ber.*, p. 186, 1906.

It can be shown that $\mathbf{S}$ must be antisymmetric ($\mathbf{S} = -\widetilde{\mathbf{S}}$) if $\mathbf{D}$ cannot be made real. Then $\mathbf{S}^* = -\mathbf{S}^{-1}$, and $\mathrm{Tr}\,(\mathbf{S}^*\mathbf{S}) = -l_j$. Thus

$$\sum_R \chi^{(j)}(R^2) = -h$$

as was to be shown.

The results of this section are summarized in Table 5-4. It might be remarked that time-reversal symmetry gives no additional degeneracy at all when the symmetry is the full rotation group. However, it becomes an important symmetry property in systems such as molecules or crystals with relatively low symmetry. For example, in the point group C_3, we have the character table

C_3	E	C_3	$C_3{}^2$	
A	1	1	1	
$E\Big\{$	1	ω	ω^2	$\omega = e^{2\pi i/3}$
	1	ω^2	ω	

In the absence of time-reversal symmetry the two lower lines would form distinct one-dimensional representations, with $\sum_R \chi(R^2) = 1 + \omega + \omega^2 = 0$.

However, since time reversal exists, they must be degenerate and hence they are conventionally lumped together as a two-dimensional representation E, despite the fact that no spatial operator in the group C_3 will transform a function from one line to the other.

5-17   More General Invariances

The rotation group, plus space and time inversion, is as much of the atomic symmetry as we shall consider in any detail. Further symmetries exist, however, and we shall now sketch the general situation of which these particular symmetries form an important part. The interest in this question arises from the intimate relation between symmetry properties and conservation laws. If the Hamiltonian of a system is invariant under a set of mutually commuting coordinate transformations R_i, this means that we can have simultaneously sharp values for the conjugate momenta. For example, we have already noted that the linear momentum $\mathbf{p}$ is best defined as that quantity which is "conserved," is a "constant of the motion," or is a "good quantum number" if H has translational invariance. Similarly, invariance under rotation leads to the conservation of angular momentum, invariance under translation in time to energy conservation, invariance under space inversion to parity conservation, and so on. Thus it is worthwhile to seek out the full symmetry of the problem.

In addition to the coordinate transformations mentioned above, the theory of relativity states that all nonaccelerated frames of reference are equivalent. In other words, H must be invariant under any proper Lorentz transformation, these forming the *Lorentz group*. Such a transformation can be viewed geometrically as a rotation through a complex angle in the four-dimensional space with coordinates $(x, y, z, \tau = ict)$. For example, if there is a relative velocity $v = \beta c$ along x, then the rotation is in the $x\tau$ plane and the transformation equations are

$$x' = \gamma x - i\beta\gamma\tau = x \cos\theta - \tau \sin\theta$$

$$y' = y$$

$$z' = z \tag{5-99}$$

$$\tau' = \gamma\tau + i\beta\gamma x = x \sin\theta + \tau \cos\theta$$

where $\gamma = (1 - \beta^2)^{-\frac{1}{2}}$ and $\tan\theta = i\beta$, or $\theta = i \tanh^{-1}\beta$. Such a rotation leaves the length $(x^2 + y^2 + z^2 - c^2t^2)^{\frac{1}{2}}$ invariant. If we combine this Lorentz group with the translations in space and time, we have what is known as the *proper Poincaré group*. If, in addition, we add the space and time inversions, we have the *unrestricted Poincaré group*.

To conclude the list of possible invariances, we mention charge conjugation C. This operation replaces each elementary particle by its antiparticle but leaves the space-time degrees of freedom unchanged. All the operators i, C, T commute with each other and with the Lorentz group $\mathscr{L}$. Thus we can represent the presumed total symmetry (apart from the translations, which just assure momentum and energy conservation) as the direct product of the continuous group $\mathscr{L}$ with the discrete Abelian group E, i, C, T, iC, iT, CT, iCT.

In discussing invariances of this sort, it is worthwhile to distinguish between two different levels of generality. As long as the discussion is confined to some specific concrete Hamiltonian, one can make definite statements. For example, if all the terms representing both internal interactions and externally applied fields have inversion symmetry, parity is rigorously a good quantum number in the formal problem to be solved. If one then applies a perturbation such as an electric field which destroys the inversion symmetry, parity fails as a quantum number and matrix elements coupling odd and even states exist. The extent of the breakdown will be determined by the relative strength of the nonsymmetric interaction and the energetic separation of states of opposite parity under the symmetric part of the Hamiltonian. If these should lie close together or actually be degenerate, then the breakdown of parity would be large. This accounts for the first-order Stark effect in the $n = 2$ state of hydrogen, in the approximation that the $2s$ and $2p$ levels are treated as degenerate.

The basic questions of recent interest are on the more general level.   In this case, the symmetry operation is applied to all of nature.   This eliminates all "external" interactions and leaves only the "internal" interactions specified by fundamental physical laws.   Then we are asking: (1) Would a possible physical world with the same physical laws be the result of inverting all coordinates through some arbitrary origin, thus transforming a world of right-handed systems into left-handed ones?   (2) Would a possible world result if all velocities were reversed (including currents producing magnetic fields, etc.) and time were allowed to develop in the reverse sense?   (3) Would a possible world result if all particles were transformed to antiparticles?

Until 1956, it had been thought likely that the answers to all these questions were affirmative.   In that year, however, Lee and Yang noted[1] that there was no good evidence for parity conservation in the nuclear processes involving weak interactions, and in fact some evidence against it.   They proposed further experimental tests, which, when carried out,[2] showed conclusively that parity is not conserved in the β decay of nuclei nor in the decay of π mesons.   Thus we are forced to conclude that, in addition to the usual scalar terms in the Hamiltonian, there must also be pseudoscalar terms which change sign even when *all* coordinates are inverted.   The experimental situation is perhaps most easily understood in the β-decay experiment of Wu et al.   The crucial experimentally observed fact is that the direction of a polar vector (the velocity **v** of the emitted β particles) is determined by the sense of an axial vector (the magnetic field **H**, which polarizes the nuclei).   Since inversion leaves **H** unchanged, while reversing **v**, the fact is inconsistent with universal inversion symmetry.

It has been suggested[3] that the situation might best be handled by dropping inversion and charge conjugation as separate symmetry operators, retaining only their product iC as a form of generalized inversion.   In the experiment cited, this would now cause **H** to reverse, since positrons would now be carrying the current in the magnet, and the antinuclei would emit anti-β particles in the same direction as the normal particles had been before.   This would remove the contradiction.   The directions of the various vectors involved are illustrated schematically in Fig. 5-6.   This suggestion is welcome because of the Schwinger-Lüders-Pauli theorem.[4]   This shows that, under very general assumptions, the product iCT should be a valid symmetry operation in any local field theory.   Thus, if T is an allowed operation, iC must also be one.   Since the reduced group E, iC, T, iCT is isomorphic with the group E,

[1] T. D. Lee and C. N. Yang, *Phys. Rev.*, **104**, 254 (1956).

[2] C. S. Wu, E. Ambler, R. W. Hayward, D. D. Hoppes, and R. P. Hudson, *Phys. Rev.*, **105**, 1413 (1957); R. L. Garwin, L. M. Lederman, and M. Weinrich, *Phys. Rev.*, **105**, 1415 (1957).

[3] E. P. Wigner, *Revs. Modern Phys.*, **29**, 255 (1957).

[4] See, for example, G. Lüders, *Kgl. Danske Videnskab. Selskab Mat.-fys. Medd.*, **28** (5), 1954.

i, T, iT, it forms a group isomorphic with the unrestricted Poincaré group, when combined with the Lorentz group and the translations.   However, the apparent necessity of joining C to i to form an allowed symmetry operator is initially a bit unsettling because of all the evidence in favor of the simple

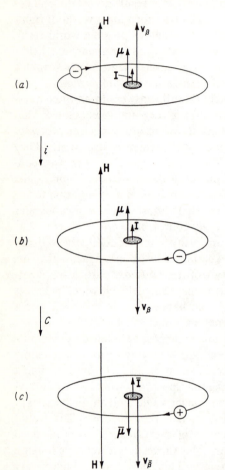

Fig. 5-6.   *Schematic diagram of β-decay experiment of Wu et al., showing effect of successive application of i and C operators.   Figure shows current circulating in magnet to produce field* **H** *which orients magnetic moment* μ *of nucleus, with associated angular momentum* **I**, *and also the velocity* v_β *of the emitted β- particle. Arrangement (b) is not observed, though (a) is.   Hence i is not a symmetry operation.*

i operator.   This situation illustrates the fact that great and effective use can be made of symmetry properties which are not ultimately accurate for *all* types of forces.   So long as the property holds for all the forces playing a significant role in a process, the corresponding conservation law will be well obeyed.   In general, as one considers weaker and weaker forces, the residual symmetry group will tend to decrease.   In the last analysis, only experimental investigation can provide us with the knowledge of what symmetry actually exists in nature.

EXERCISES

5-1   Use the fact that we can write $J = \dfrac{\hbar}{i} \dfrac{\partial}{\partial \varphi}$, where φ is the angle of rotation about the axis of J, to show that

$$P_\theta \psi(\varphi) \equiv e^{-i\theta J/\hbar} \psi(\varphi) = \psi(\varphi - \theta)$$

5-2   Simplify the general formula for $D^{(j)}_{m'm}(\alpha, \beta, \gamma)$ to that for $D^{(j)}_{m'j}(\alpha, \beta, \gamma)$. Using this simplified formula, work out all the coefficients $D^{(2)}_{m'2}(\alpha, \beta, \gamma)$.

5-3   When one measures the component of a magnetic moment along a magnetic field in a Stern-Gerlach type of experiment, one forces a system to give a sharp value to the component of angular momentum in that direction. The probability of finding any given value of m is just $|c_m|^2$, where $\psi = \sum_m c_m \psi_m^{(j)}$, and the z axis is the axis of the field. Assume that the fields in two successive experiments on the same atoms are inclined at an angle θ. (*a*) What is the probability as a function of θ that an atom which was in state m in the first experiment will be found in state m' in the second experiment for arbitrary j? (You may assume that the wavefunction does not change in passing from one experiment to the next.) (*b*) Tabulate explicit answers for the cases $j = 0, \frac{1}{2}$, and 1. (*c*) Finally, specialize these matrices for $\theta = \pi/2$.

5-4   A spinor is said to point in the z direction if $\psi = e^{i\delta} u_+$, where $u_+ = \begin{pmatrix} 1 \\ 0 \end{pmatrix}$ and δ is any real number, so that J_z is certainly $+\frac{1}{2}$. (*a*) Using this definition and the representation $D^{(\frac{1}{2})}$ of the rotation group, in what direction is the spinor $\psi' = a_- u_- + a_+ u_+ = \begin{pmatrix} a_+ \\ a_- \end{pmatrix}$ pointing? ($|a_+|^2 + |a_-|^2 = 1$.) Specify the direction by giving the polar and azimuthal angles θ and φ. (*b*) What are a_+ and a_- for a spinor pointing in the x direction and the y direction? Remember that a_+ and a_- are determined only within a common phase factor.

5-5   Write down the specific transformation matrices for single rotations through angle θ of spinors ($j = \frac{1}{2}$) about the y and z axes. Work out the transformation matrix for a rotation θ about the x axis, using the general result for $D^{(\frac{1}{2})}(\alpha, \beta, \gamma)$. Using these results, note that rotating a *spinor* u_+ by π about x and about y leads to spinors of different phase. Use $D^{(1)}(\alpha, \beta, \gamma)$ to show that rotating a z-directed *vector* $\psi_z = \psi (j = 1, m = 0)$ in the same way leads to no phase difference.

5-6   Use the Wigner coefficients to express the wavefunction of an atomic state with $L = 2, S = 2, J = 0$ in terms of products of orbital and spin functions having $L = 2$ and $S = 2$, respectively. What is the probability that $S_z = 2$ in this state?

5-7   Consider the state $l = 1, s = \frac{1}{2}, j = \frac{1}{2}, j_z = m$. (*a*) Show that the charge distribution is spherically symmetric. (*b*) Find the spatial distribution of spin direction for the state $m = +\frac{1}{2}$. (Spin direction is defined as the direction ζ along which S_ζ is certainly $+\frac{1}{2}$.) Make a simple sketch showing your results in the xz plane.

5-8   The Landé g factor of a state is defined as

$$g = \frac{\mu_z}{\beta M} = \frac{(JM \,|2S_z + L_z|\, JM)}{(JM \,|S_z + L_z|\, JM)} = \frac{(JM \,|2S_z + L_z|\, JM)}{M} = 1 + \frac{(JM \,|S_z|\, JM)}{M}$$

Using the Wigner coefficients to express ψ_{JM} in terms of eigenfunctions of S_z and L_z, evaluate g for the case $S = \frac{1}{2}$ and general L. Show that your result is in agreement with that calculated for this case using the vector model of elementary physics. [See Eq. (6-65).]

5-9 (a) Calculate the relative intensities of all possible π ($\Delta M = 0$) and σ ($\Delta M = \pm 1$) Zeeman components (in a weak magnetic field) for the transition between a $^2S_{\frac{1}{2}}$ and a $^2P_{\frac{3}{2}}$ state. In this calculation assume linear polarization along x and z for σ and π transitions, respectively. (b) Using the g factors computed in Exercise 5-8 to determine the energy differences, prepare a diagram showing the splitting and intensity pattern which would be observed. Label all the lines, and distinguish the σ and π transitions.

5-10 Consider the many-electron electronic spin-spin interaction. Note that in general this can be decomposed into a sum of irreducible tensor operators as indicated for the case of two spins in Eq. (5-65). Consider a system perturbed from spherical symmetry to only axial symmetry, which is in a Σ state. That is, the orbital wavefunction is invariant under rotation of coordinates about the axis. If the system has an electronic spin S, which is an approximately good quantum number because the axial perturbation is small compared with energy differences to states of different S, use the Wigner-Eckart theorem to show that the spin-spin interaction produces energy splittings within this state (i.e., diagonal in S) equivalent to those produced by the effective Hamiltonian

$$H = D[3S_z^2 - S(S + 1)] \qquad D = \text{constant}$$

operating on the total-spin state functions. This effect gives rise to the fine-structure splitting of the ^6S Mn^{++} paramagnetic-resonance spectrum in an axial crystal field. It also produces the large term ($\sim 2 \text{ cm}^{-1}$) coupling the spin to the figure axis in the $^3\Sigma$ ground state of O_2. (HINT: This problem requires some physical thought, but not a great deal of mathematical manipulation.)

5-11 Use the Wigner-Eckart theorem and the Racah coefficients to work out an explicit expression for the matrix elements of the form

$$(j_1, j_2, J - 1, M \,|j_{1z}|\, j_1, j_2, J, M)$$

where j_1 and j_2 have been coupled to form J and $J - 1$ in the initial and final states, respectively.

5-12 (a) Classify the following as to whether they transform as (polar) vectors or as pseudovectors (axial vectors), i.e., whether they do or do not change sign under the inversion operator ($\mathbf{r}' = -\mathbf{r}$): velocity, momentum, grad, angular momentum, force, electric field, magnetic field, magnetic vector potential, magnetic moment. (b) Show that the Hamiltonian for an atom is invariant under inversion when it includes the Coulomb interaction among electrons and nucleus, spin-orbit coupling (taken in the form $\lambda \, \Sigma \, \mathbf{l}_i \cdot \mathbf{s}_i$), spin-spin coupling [as in Eq. (5-65)], and the interaction with a uniform magnetic field. Hence deduce that parity is an exact quantum number for describing the state of an atom under these conditions. (c) Would this conclusion be true if the spin-orbit coupling took the form of $\lambda \sum_i \mathbf{r}_i \cdot \mathbf{s}_i$? Explain.

5-13   Classify the operators listed in Exercise 5-12*a* according to whether they are even or odd under time inversion (including time inversion of sources of any fields).

5-14   The $2\ {}^2S_{\frac{1}{2}}$ state of atomic H is metastable since dipole transitions to the ground state ($1\ {}^2S_{\frac{1}{2}}$) are forbidden by the parity rule.   How intense must a perturbing static electric field be to reduce the lifetime of the metastable state to 10^{-3} sec?   [HELPFUL INFORMATION: In the absence of external fields the $2\ {}^2S_{\frac{1}{2}}$ state lies 1,058 Mc/sec above the $2\ {}^2P_{\frac{1}{2}}$ state because of the Lamb shift.   For simplicity, you may ignore the nearby $2\ {}^2P_{\frac{3}{2}}$ state.   The general expression for transition probability per unit time[1] via electric dipole transitions is

$$w = \frac{4e^2 \omega_{kn}{}^3 \left| \mathbf{r}_{kn} \right|^2}{3\hbar c^3}$$

This effect of an electric field is called "quenching" of metastables.]

REFERENCES

CONDON, E. U., and G. H. SHORTLEY: "The Theory of Atomic Spectra," chap. III, Cambridge University Press, New York, 1951.

EDMONDS, A. R.: "Angular Momentum in Quantum Mechanics," Princeton University Press, Princeton, N.J., 1957.

FANO, U., and G. RACAH: "Irreducible Tensorial Sets," Academic Press Inc., New York, 1959.

ROSE, M. E.: "Elementary Theory of Angular Momentum," John Wiley & Sons, Inc., New York, 1957.

ROTENBERG, M., R. BIVINS, N. METROPOLIS, and J. K. WOOTEN, Jr.: "The 3-*j* and 6-*j* Symbols," Technology Press, Cambridge, Mass., 1959.

WIGNER, E. P.: "Group Theory and Its Application to the Quantum Mechanics of Atomic Spectra," Academic Press Inc., New York, 1959.

[1] L. I. Schiff, "Quantum Mechanics," 2d ed., p. 261, McGraw-Hill Book Company, Inc., New York, 1955.

6 QUANTUM MECHANICS OF ATOMS

We now turn back from the development of group-theoretical methods to our basic objective, namely, the calculation of the properties of atoms, molecules, and solids by quantum-mechanical means. Henceforth the group theory will be brought to bear when helpful, but our central objective is to formulate a quantitative understanding of these physical systems rather than simply to illustrate one particular theoretical tool.

Although there are similarities in the theory of atoms and of molecules, the calculations in the two cases differ sufficiently so that we shall first concentrate on the simpler atomic problem. Here we have a single massive positive nucleus providing a central force center. This results in a Hamiltonian with overall spherical symmetry, which allows great simplification. Before plunging into detailed calculations, let us briefly review the elements of atomic structure so that we may see just what we are trying to calculate more quantitatively.

6-1 Review of Elementary Atomic Structure and Nomenclature

To specify the quantum state of an atom, we first specify the *configuration*. This tells us which one-electron central-field eigenstates (labeled by n and l) are occupied. For example, the ground state of the silicon atom ($Z = 14$) has the configuration $1s^2 2s^2 2p^6 3s^2 3p^2$. Since the electrons interact among themselves, the energy of the atom is not determined by this much information

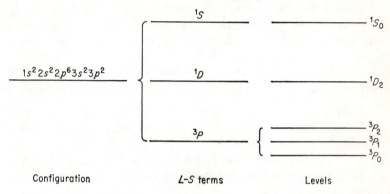

Configuration   L-S terms   Levels

Fig. 6-1. *Hierarchy of levels for the ground configuration of silicon.*

alone. We must also know how the various one-electron functions are coupled together to form eigenfunctions for the entire atom. For this purpose we can neglect the closed shells, since there is no freedom left if all states are occupied. Thus in our example, we concentrate on the $3p^2$ part. Applying our vector-model result to the orbital eigenfunctions,

$$D^{(1)} \times D^{(1)} = D^{(0)} + D^{(1)} + D^{(2)} = S + P + D$$

while for the spins

$$D^{(\frac{1}{2})} \times D^{(\frac{1}{2})} = D^{(0)} + D^{(1)}$$

Thus two p electrons can form the following *L-S terms*: ^1S, ^1P, ^1D, ^3S, ^3P, ^3D. Later we shall see how the Pauli principle excludes all except ^3P, ^1S, ^1D if the electrons are *equivalent* (same n and l), as in this example. The Wigner coefficients could be used to express the eigenfunctions of L, M_L in terms of those of l_1, m_1 and l_2, m_2, for example. These *L-S* terms differ in energy by the order of an intra-atomic Coulomb energy, i.e., typically a few electron volts.

Finally we introduce the small magnetic energies referred to as the spin-orbit coupling and decompose the *L-S* terms into individual *levels* of definite total angular momentum J. The various levels of a given term form its *multiplet structure*. The intervals in the multiplet are of order $(v/c)^2 \sim (Z_{\text{eff}}/137)^2 (l/n^2)$ less than the Coulomb energies, namely, of order 1 to 1,000 cm^{-1}. This hierarchy of level structure may be depicted schematically as shown in Fig. 6-1. The individual levels are $(2J + 1)$-fold degenerate, and these sets of degenerate functions form bases for irreducible representations of the full rotation group of the whole system as a unit. Since the Hamiltonian is rigorously invariant under such rotations, these degeneracies are exact to all orders of approximation so long as no external fields are applied to disturb the isotropy of space. In other words, J is rigorously a good quantum number. This is *not* true of the one-electron quantum numbers which enter the configuration labels, since the Hamiltonian is not invariant under a rotation of the coordinates of one particle. In fact, the Wigner coefficients prescribe what linear combinations of one-electron functions form these proper eigenfunctions. In the L-S coupling approximation, this prescription would be carried out in stages: first coupling $\mathbf{l}_1 + \mathbf{l}_2 = \mathbf{L}$ and $\mathbf{s}_1 + \mathbf{s}_2 = \mathbf{S}$; then coupling $\mathbf{L} + \mathbf{S} = \mathbf{J}$.

The three *Hund rules* determine the ground state of an atom whose configuration is given. These rules dictate that to find the ground state:

1. Choose the maximum value of S consistent with the Pauli principle.

2. Choose the maximum value of L consistent with the Pauli principle and rule 1.

3a. If the shell is less than half full, choose $J = J_{\min} = |L - S|$.

3b. If the shell is more than half full, choose $J = J_{\max} = L + S$.

Applying these rules to the Si example, we find that the ground state is $^3P_0$, as shown in the diagram above. Although we shall examine these results more closely later on, it is probably worth emphasizing right here that, despite its appearance, rule 1 has nothing to do with a magnetic coupling between spins. Rather it arises from the antisymmetry of the wavefunction. This keeps electrons with parallel spins out of each other's way, thus lowering the *electrostatic-repulsion energy*. Only rule 3 is based on magnetic energies.

If we consider heavy atoms where Z is high, v/c becomes larger and the magnetic forces become much more important. In this limiting case, the concept of L-S terms is not very useful. Instead we couple the spin of each electron to its own orbital motion forming a resultant j_i. We then specify a level by quoting all the j_i's and the total J which they form by vector addition. Of course J is a rigorous quantum number in this *j-j coupling* scheme, just as with L-S coupling. We shall confine our attention largely to the L-S coupling case, which pertains to the more common light atoms.

If high accuracy is required, one cannot ignore the admixture of configurations other than the nominal one by the nonspherical part of the one-electron Hamiltonian. This part arises, for example, when electrons other than the one under consideration are present outside closed shells. This admixture is called *configuration interaction*, and it is usually ignored in approximate work. If it is taken into account, we are at least assured that complete accuracy can be obtained if enough configurations are included, because the set of all configurations forms a complete orthogonal set. Convergence in this method of *superposition of configurations* is often rather slow, however.

6-2 The Hamiltonian

We shall usually confine our attention to the approximate Hamiltonian

$$H = \sum_{i=1}^{N} \left[\frac{\mathbf{p}_i^2}{2m} - \frac{Ze^2}{r_i} + \xi(r_i)\mathbf{l}_i \cdot \mathbf{s}_i \right] + \sum_{i>j} \frac{e^2}{r_{ij}} \tag{6-1}$$

As will be discussed later, the spin-orbit coupling may be handled approximately by the term in $\xi(r) = (2m^2c^2r)^{-1}(\partial U/\partial r)$, where $U(r)$ is a spherically symmetric approximation to the potential inside the atom. We neglect all other magnetic and relativistic effects, hyperfine structure, etc., for the present. In fact, even (6-1) is too difficult to handle all at once; so we separate a simple spherically symmetric term H_0 and treat the rest as a perturbation H'. Thus

$$H_0 = \sum_i \left[\frac{\mathbf{p}_i^2}{2m} + U(r_i) \right]$$

$$\tag{6-2}$$

and $$H' = \sum_i \left[\sum_{j>i} \frac{e^2}{r_{ij}} - \frac{Ze^2}{r_i} - U(r_i) \right] + \sum_i \xi(r_i)\mathbf{l}_i \cdot \mathbf{s}_i$$

The physical idea of H_0 is that we consider each electron to move in $U(r_i)$, the Coulomb field of the nucleus as "screened" by the other electrons. For very small r, $U(r) \approx -Ze^2/r$, but for large r, the potential is much weaker, more like simply $-e^2/r$ in a neutral atom. By restricting U to the spherical form, we assure the separation of variables, and each electron will have a well-defined angular momentum, as assumed in the simple configuration labels. All that remains to be found is the radial dependence of the individual eigenfunctions. After this has been done, one introduces the first part of H' to split apart the L-S terms and then the second part to produce the multiplet structure.

6-3 Approximate Eigenfunctions

Simple approximate eigenfunctions. The simplest approximation is to assume *hydrogenic functions*. These differ from the hydrogen eigenfunctions only by the replacement of the nuclear charge by $(Z - s)$, where s is the screening charge of the inner electrons. Slater[1] has proposed a simple rule for choosing s. He assumes that each electron in the same shell as the one

---

[1] J. C. Slater, *Phys. Rev.*, **36**, 57 (1930).

under consideration contributes 0.35 to s; each electron in the next inner shell contributes 0.85; each electron in still lower shells contributes a full 1.00 to the shielding. This leads to fair results but suffers from the inconvenience that the functions for the various shells are not orthogonal since they are solutions for Hamiltonians with different potentials.

A more systematic approach for the light atoms is to choose a simple form for each radial function and then to choose the best values for parameters entering the function by the variational method of minimizing the total energy. The method was introduced by Morse, Young, and Haurwitz,[1] extended by Duncanson and Coulson,[2] and recently carried further by machine computation by Tubis.[3] The assumed functions are

$$u(1s) = \left(\frac{u^3 a^3}{\pi}\right)^{\frac{1}{2}} e^{-uar}$$

$$u(2s) = \left(\frac{u^5}{3\pi N}\right)^{\frac{1}{2}}\left[re^{-ur} - 3\left(\frac{A}{u}\right)e^{-ubr}\right]$$

$$u(2p_0) = \left(\frac{u^5 c^5}{\pi}\right)^{\frac{1}{2}} r\cos\theta e^{-ucr}$$

$$u(2p_{\pm 1}) = \left(\frac{u^5 c^5}{2\pi}\right)^{\frac{1}{2}} r\sin\theta e^{\pm i\varphi} e^{-ucr}$$

(6-3)

where $A = (a + b)^3/(1 + a)^4$ is chosen for orthogonality of $u(1s)$ and $u(2s)$. The normalization constant N has the value $N = 1 - [48A/(1 + b)^4] + 3A^2/b^2$. Within this framework, the values of the parameters a, b, c, and u are determined and tabulated for all atoms up to 10 electrons. These calculations are actually made including the nonspherical part of the Hamiltonian, so that the L-S terms have distinct energies. Results are usually quoted for two low-lying L-S terms. These functions are quite useful for semiquantitative calculations.

More accurate variational eigenfunctions. We have just mentioned the use of a variational method to obtain approximate eigenfunctions. This method is based on the fact that the Schrödinger equation is the Euler equation which results when the calculus of variations is applied to finding the function ψ which minimizes $E = \int \psi^* H \psi \, d\tau$ subject to the constraint that the normalization integral $\int \psi^* \psi \, d\tau$ be held constant.[4] Therefore the exact solution of the Schrödinger equation for the ground state is precisely

[1] P. M. Morse, L. A. Young, and E. Haurwitz, *Phys. Rev.*, **48**, 948 (1935).
[2] W. E. Duncanson and C. A. Coulson, *Proc. Roy. Soc. Edinburgh*, **62**, 37 (1944).
[3] A. Tubis, *Phys. Rev.*, **102**, 1049 (1956).
[4] See, for example, H. Margenau and G. M. Murphy, "The Mathematics of Physics and Chemistry," 1st ed., p. 208, D. Van Nostrand Company, Inc., Princeton, N.J., 1943.

the function giving the absolute minimum energy.   In any case, since the energy is stationary with respect to small changes in ψ about any eigen-function, quite good energy eigenvalues may be obtained by using only moderately good approximate eigenfunctions.   Moreover, we always get an upper bound for the ground-state energy and a simple analytic criterion for choosing the "best" approximate ψ within a class of functions such as those considered above.   To get better eigenfunctions, one then must generalize the class of functions to which the variation method is applied by increasing the number of free parameters.

This approach has been carried particularly far for the normal state of He, which is the simplest atomic problem which is not soluble exactly. Hylleraas[1] carried this through using about 10 parameters and including terms depending explicitly on r_{12} to represent the correlation of the two electrons.   He managed to get an energy 24 cm^{-1} *below* the experimental energy of $-198,310.5$ cm^{-1}.   This apparent contradiction of the variational theorem is a consequence of neglect of relativistic and other higher cor-rections in the Hamiltonian.   When these are inserted, the agreement is satisfactory.   More recently, Kinoshita[2] has carried out a 39-parameter variational calculation of the nonrelativistic problem and made more care-ful allowance for the relativistic and quantum electrodynamic (Lamb-shift) effects.   The most exhaustive calculation is that of Pekeris,[3] who used more than 200 terms and obtained an accuracy estimated to be 0.01 cm^{-1}, an order of magnitude better than the experimental value.   This makes it pretty clear that no important errors remain in the Hamiltonian and that excellent results can be obtained by this method if one pushes it far enough.   The method is not very practical for the heavier atoms with more electrons, however.

Hartree functions.   A very reasonable type of approximate function was chosen by Hartree, namely,

$$\psi(\mathbf{r}_1, s_1, \ldots, \mathbf{r}_N, s_N) = u_1(\mathbf{r}_1)\chi_1(s_1) \cdots u_N(\mathbf{r}_N)\chi_N(s_N) \tag{6-4}$$

In this, the $u_i(\mathbf{r}_i)$ are one-electron orbital functions.   The $\chi_i(s_i)$ are either $\alpha(s_i)$ or $\beta(s_i)$, the conventional spin function α referring to $m_s = +\frac{1}{2}$ and β to $m_s = -\frac{1}{2}$.   Such a simple product function is the simplest conceivable function expressing the idea that the state is characterized by a configuration specified by the one-electron u_i functions.   Clearly such a function is only approximate because, for one thing, it completely omits all *correlation* of the coordinates of the electrons; i.e., the probability distribution of r_1 is in-dependent of where $r_2 \cdots r_N$ might be at the time.   In fact, the e^2/r_{ij} Coulomb repulsion will tend to keep the electrons away from each other in the true

[1] E. A. Hylleraas, *Z. Physik*, **54**, 347 (1929).
[2] T. Kinoshita, *Phys. Rev.*, **105**, 1490 (1957).
[3] C. L. Pekeris, *Phys. Rev.*, **112**, 1649 (1958).

wavefunction. Since no allowance is made for that here, the energy of the Hartree function must exceed the correct (and minimum) energy. However, within the framework of all functions which can be written as simple products, there should be a "best" one characterized by a set of functions $u_1 \cdots u_N$ which together give the lowest energy. These best functions are the so-called "Hartree functions," which have been calculated numerically for many atoms. Since the entire functional form is adjustable, better results should be possible than with a restricted analytic approximation to $u_i(r)$.

These functions are found by iteration. First one guesses a set of functions $u_1 \cdots u_N$. Then one calculates the spherically averaged potential U_i seen by each electron by averaging the potential produced by the *other* electrons ($j \neq i$) and the nucleus.

$$U_i(r_i) = \frac{-Ze^2}{r_i} + \left[\sum_{j \neq i} \frac{e^2}{r_{ij}} \right]_{\text{av.}} \qquad (6\text{-}5a)$$

The Schrödinger equation containing these potentials,

$$\left[\frac{\mathbf{p}_i{}^2}{2m} + U_i(r_i) \right] u_i = E_i u_i \qquad (6\text{-}5b)$$

is solved by numerical integration. The resulting set of u_i will not in general agree with the assumed one. On the basis of this new set, another guess is made, and the calculation is repeated over and over until *self-consistency* is reached. One can show that these self-consistent solutions to the set of coupled integrodifferential equations described above (the "Hartree equations") also satisfy the variational requirement of lowest energy. The self-consistent potential $U_i(r_i)$ for the valence electrons is the proper potential to use in calculating the spin-orbit interaction function $\xi(r)$ if it is desired to calculate multiplet splittings. The Hartree functions for a large number of atoms have been calculated and tabulated by various authors. A review of the method, together with an index to the early calculations, has been given by Hartree.[1] Another listing is given in the Landolt-Börnstein tables (1951 edition), and Slater[2] has given a bibliography of calculations up to 1958. These indices also include reference to the more accurate Hartree-Fock functions, which we consider next.

Antisymmetric functions. Although the Hartree functions are good, they suffer from a serious group-theoretical objection. All electrons are identical or dynamically equivalent. Therefore $u_2(1)u_1(2)$ must be just as good a wavefunction as $u_1(1)u_2(2)$, and it must have the same energy. Thus any

[1] D. R. Hartree, *Repts. Progr. Phys.*, **11**, 113 (1946–1947). See also Hartree's book, "The Calculation of Atomic Structures," J. Wiley & Sons, Inc., New York, 1957, which lists work up to 1956.

[2] J. C. Slater, "Quantum Theory of Atomic Structure," vol. I, app. 16, McGraw-Hill Book Company, Inc., New York, 1960.

linear combination would serve. In general we might expect that we could replace the simple product function by

$$\psi = \sum_p c_p P[u_1(1)u_2(2) \cdots u_N(N)]$$

where P permutes the order of the electrons and c_p gives the weighting of the Pth permutation.

Because the electrons are in fact identical, no physically observable quantity (matrix element) can depend on the numbering. That is,

$$\bar{f} = \int \psi^*(1 \cdots N) f \psi(1 \cdots N) \, d\tau_1 \cdots d\tau_N$$

must be unchanged for an arbitrary exchange of r_i and r_k for an arbitrary function f. This is possible only if $P_{ik}\psi = e^{i\delta}\psi$, so that $(P_{ik}\psi)^*(P_{ik}\psi) = \psi^*\psi$. There is a further restriction imposed by the fact that $P_{ik}{}^2\psi = \psi$, which must be true since the second interchange simply restores the original order. These two requirements may be satisfied only if we have

$$P_{ik}\psi = (\pm 1)\psi \tag{6-6}$$

That is, the wave function is allowed to belong only to these two one-dimensional representations of the permutation group—the *symmetric* and the *antisymmetric*. Experimental evidence (Pauli exclusion principle, etc.) shows that electrons, protons, and all other half-integral spin particles (fermions) belong to the antisymmetric representation; on the other hand, all particles with integral spin (bosons) belong to the symmetric representation.

From the above discussion it follows that any correct electronic eigenfunction must be antisymmetric under interchange of electrons. The simple Hartree functions lack this property. However, we can antisymmetrize them or any other functions with the antisymmetrizing operator A. This operator is defined by

$$\Psi = A\psi = (N!)^{-\frac{1}{2}} \sum_P (-1)^p P\psi(r_1, \ldots, r_N) \tag{6-7}$$

where p is the number of binary permutations in the permutation P. This function Ψ has the required antisymmetric property, since interchanging i, j changes all the p's by 1, thus changing the sign of $A\psi$. The factor $(N!)^{-\frac{1}{2}}$ preserves normalization if ψ is a product of orthonormal u_i's. Note that, applied to such a product, this operator is equivalent to forming the determinant

$$\Psi = (N!)^{-\frac{1}{2}} \begin{vmatrix} u_1(1) & u_2(1) & \cdots & u_N(1) \\ u_1(2) & u_2(2) & \cdots & u_N(2) \\ \cdots & \cdots & \cdots & \cdots \\ u_1(N) & \cdots & \cdots & u_N(N) \end{vmatrix} \tag{6-8}$$

[The notation $u_1(1)$ now includes both space and spin parts.]   These "Slater determinants" are the simplest generalization of the simple product function to have the proper symmetry under permutation of identical particles. Accordingly, they play a leading role in quantum mechanical calculations.

We can draw two simple but important inferences directly from this determinantal form of electronic wavefunctions:

1. If $u_i = u_j$, two columns are identical and the determinant vanishes. This gives us the elementary form of the *Pauli principle* forbidding more than one electron in any given eigenstate.

2. If $(r_i, s_i) = (r_j, s_j)$, there are two identical rows and again the determinant vanishes.   This shows that two electrons with parallel spins are never at the same place in real space and, by continuity, seldom very near to each other.   This fact introduces a *kinematic correlation* of electrons with *parallel spin* which reduces the Coulomb-repulsion energy.   Since this energy lowering is a consequence of the antisymmetry which is required by the "exchange degeneracy," this physical correlation turns out to be the basis for the intra-atomic "exchange energy" which causes parallel spins (maximum value of S) to lower the energy as stated in the first *Hund rule*.

Hartree-Fock functions.   If we now seek the best functions which can be written as a single determinant of this sort, the variational method again leads to a set of coupled integrodifferential equations for the single-electron functions u_i.   These more complicated equations are called the *Hartree-Fock equations*, and they were first found independently by Fock and by Slater in putting Hartree's rather intuitive method on a sounder mathematical footing.   These equations are also solved numerically by iteration to self-consistency.   In recent years, Hartree-Fock functions have usually been calculated rather than the simpler Hartree functions, though the results are not radically different.   They *are* better, however, and have the practical advantage that the various u_i are rigorously orthogonal to each other.   This was not true for the Hartree functions, since in that case a different potential is used for each subshell of the atom.   Actually, the Hartree-Fock functions also have a different effective potential for each u_i.   However, it turns out that the exchange terms introduced by the antisymmetry assure orthogonality.   We defer our derivation and discussion of the Hartree-Fock equations until after the next section, in which we develop the tools needed for all calculations with determinantal wavefunctions.

6-4   Calculation of Matrix Elements between Determinantal Wavefunctions

We shall require many matrix elements of various operators between determinants of orthonormal functions.   All those of most interest are either one- or two-electron operators. This allows us to reduce the whole procedure to a few rules, which we now derive.

Matrix elements of $F = \sum_i f(r_i)$. The simplest type of operator is a sum of identical one-electron operators, each operating on the coordinates of one electron. Examples are $T = (-\hbar^2/2m)\sum_i \nabla_i^2$, $V_Z = -\sum_i Ze^2/r_i$, and $\mu_e = e\sum_i r_i$. We are interested in the matrix element

$$(A \,|F|\, B) = \int \cdots \int \Psi^*(A)F\Psi(B)\, d\tau_1 \cdots d\tau_N \tag{6-9}$$

where $\Psi(A)$ is the antisymmetrized product (sometimes abbreviated AP)

$$\Psi(A) = (N!)^{-\frac{1}{2}}\sum_P (-1)^P P\psi(A) = A\{a_1(1)a_2(2)\cdots a_N(N)\} \tag{6-10}$$

We have now condensed our notation by writing a_i to mean the spin-orbital function characterized by the set of quantum numbers a_i. Expanding the matrix element yields

$$(A \,|F|\, B) = (N!)^{-1}\sum_i\sum_P\sum_{P'} (-1)^{p+p'}\int \cdots \int P\psi^*(A)f(i)P'\psi(B)\, d\tau_1 \cdots d\tau_N \tag{6-11}$$

We now choose to view the permutation operator P as operating on the quantum-number indices a_i rather than on the coordinates r_i, the two being quite equivalent. For example, if $P = P_{12}$,

$$P\psi(A) = P[a_1(1)a_2(2)a_3(3)\cdots]$$
$$= Pa_1(1)Pa_2(2)Pa_3(3)\cdots = a_2(1)a_1(2)a_3(3)\cdots$$

Thus the integral above becomes

$$\int Pa_1(1)P'b_1(1)\, d\tau_1 \cdots \int Pa_i(i)f(i)P'b_i(i)\, d\tau_i \cdots \int Pa_N(N)P'b_N(N)\, d\tau_N$$

Each of these integral factors (except the ith) is an orthonormality integral which will vanish unless $Pa_k = P'b_k$, etc. Therefore at least $(N-1)$ of the set of N spin-orbital functions in A and in B must be the same. This gives us the selection rule

$$(A \,|F|\, B) = 0 \tag{6-12}$$

if A, B differ by more than one pair of occupied functions.

Now, let a_k and b_l be the two functions which differ in the two sets of functions A and B. The subscripts k, l will in general differ because to get consistent results we must establish a fixed ordering convention. Our convention for atoms is to order the nl values in the conventional order for specifying a configuration. Within a subshell, the order is in decreasing order of m_l values and, for each m_l, α before β in the spin functions. (Of course, in molecular and solid-state problems, some other convenient convention is chosen.) We can then form a set B',

$$B' = a_1 a_2 \cdots a_{k-1} b_l a_{k+1} \cdots a_N$$

which will *not* in general be in conventional order, but in which all the a_i functions are in the same position as in A. Now $\Psi(B') = \pm\Psi(B)$, the minus sign applying if B' differs from B by an odd number of interchanges. Then $(A |F| B) = \pm (A |F| B')$, the sign carrying through from the result of reordering.

We may calculate $(A |F| B')$ quite simply. In order that the product of orthonormality integrals shall not vanish, the permutations P and P' must be the same so as to keep corresponding functions together. Thus the $(-1)^{p+p'} = +1$, and all terms come in positively. Further, to get a non-zero contribution, P must be such as to bring the function a_k to the ith position. For a given i, there are $(N-1)!$ ways of doing this. But this can be done for each i; so the result must be multiplied by N. The $N(N-1)!/N!$ factors cancel, leaving

$$(A |F| B) = \pm \int a_k^*(1) f(1) b_l(1) \, d\tau = \pm(a_k |f| b_l) \qquad \blacktriangleright \text{ (6-13)}$$

where the $\pm$ sign is determined by whether an even or odd permutation is required to bring b_l into register with a_k, starting in the standard order.

The argument is similar for the diagonal elements. However, since the two sets are now identical, no $\pm$ signs enter from rearrangements. P must always equal P'. Thus

$$(A |F| A) = (N!)^{-1} \sum_i \sum_P \int P a_i^*(i) f(i) P a_i(i) \, d\tau_i$$

For each of the N values of i, there are $(N-1)!$ ways of permuting the indices not equal to i when each of the N a_k's is moved into the ith position. Hence we obtain

$$(A |F| A) = \sum_{i=1}^N \int a_i^*(1) f(1) a_i(1) \, d\tau_1 = \sum_{i=1}^N (a_i |f| a_i) \qquad \blacktriangleright \text{ (6-14)}$$

This result states the very reasonable conclusion that the expectation value of F in the many-electron state A is simply the sum of the expectation values of f over the occcupied orbitals a_i.

Matrix elements of $G = \sum_{i>j} g(i,j)$. Since we can assume that any terms of the sort $g(i, i)$ are handled by the rules just derived for functions of the type F, we lose no generality by excluding the diagonal terms in this two-electron operator G. Typical examples of operators of the type G are the Coulomb energy $\sum\limits_{i>j} e^2/r_{ij}$ and the dipole-dipole energy $\sum\limits_{i>j} r_{ij}^{-3}[\boldsymbol{\mu}_i \cdot \boldsymbol{\mu}_j - 3(\mathbf{r}_{ij} \cdot \boldsymbol{\mu}_i)$ $(\mathbf{r}_{ij} \cdot \boldsymbol{\mu}_j)/r_{ij}^2]$. For such operators, we have

$$(A \,|G\,|B) = (N!)^{-1} \sum_{i>j} \sum_{P'P} (-1)^{p+p'} \iint Pa_i^*(i)Pa_j^*(j)g(i, j)P'b_i(i)P'b_j(j)$$
$$\times \, d\tau_i \, d\tau_j \, \delta(Pa_1, P'b_1) \cdots \delta(Pa_N, P'b_N) \tag{6-15}$$

where the product of δ functions arises from the orthonormality integrals and omits a factor only for i and j. In this case, the selection rule is that

$$(A \,|G|\, B) = 0 \tag{6-16}$$

if A, B differ by more than two pairs of functions. We then have to consider three possible cases: A, B differing by 2, 1, or 0 pairs of functions.

Considering first the case where A and B differ by two pairs of functions, we can rearrange B to B', as we did above, so that all the functions b_i which equal some a_i are paired off. This leaves the unmatched functions a_k and a_l paired off with the unmatched functions b_m and b_n. These can be paired either as $(k, l; m, n)$ or $(k, l; n, m)$, with the number of binary permutations, and hence the sign, changing from one to the other. By following a term-counting scheme like that given above for F, we then are led to the result

$$(A \,|G|\, B) = \pm \left\{ \iint a_k^*(1)a_l^*(2)g(1, 2)b_m(1)b_n(2) \, d\tau_1 \, d\tau_2 \right.$$
$$\left. - \iint a_k^*(1)a_l^*(2)g(1, 2)b_n(1)b_m(2) \, d\tau_1 \, d\tau_2 \right\}$$
$$= \pm\{(a_k a_l \,|g|\, b_m b_n) - (a_k a_l \,|g|\, b_n b_m)\} \quad \blacktriangleright \tag{6-17}$$

The $\pm$ sign is determined by the permutation required to bring B into B' with the pairing $(k, l; m, n)$.

By similar arguments, when A and B differ by only one pair $b_l \neq a_k$, we have

$$(A \,|G|\, B) = \pm \sum_t \left\{ \iint a_k^*(1)a_t^*(2)g(1, 2)b_l(1)a_t(2) \, d\tau_1 \, d\tau_2 \right.$$
$$\left. - \iint a_k^*(1)a_t^*(2)g(1, 2)a_t(1)b_l(2) \, d\tau_1 \, d\tau_2 \right\}$$
$$= \pm \sum_t \{(a_k a_t \,|g|\, b_l a_t) - (a_k a_t \,|g|\, a_t b_l)\} \quad \blacktriangleright \tag{6-18}$$

where t runs over the $(N - 1)$ sets of functions common to A and B. Finally, the diagonal elements are given by

$$(A \,|G|\, A) = \sum_{k>t=1}^{N} \left\{ \iint a_k^*(1)a_t^*(2)g(1, 2)a_k(1)a_t(2) \, d\tau_1 \, d\tau_2 \right.$$
$$\left. - \iint a_k^*(1)a_t^*(2)g(1, 2)a_t(1)a_k(2) \, d\tau_1 \, d\tau_2 \right\}$$
$$= \sum_{k>t} \{(a_k a_t \,|g|\, a_k a_t) - (a_k a_t \,|g|\, a_t a_k)\} \quad \blacktriangleright \tag{6-19}$$

The integrals with positive sign are called *direct integrals* and the ones with negative sign are called *exchange integrals* because the functions a_t and a_k are exchanged between initial and final states in the integral. The presence of the exchange integrals is strictly a consequence of our use of an antisymmetrized form. They would not occur, for example, if we evaluated $(A \,|G|\, A)$ using the simple product function $\psi(A)$. In general, all proper exchange terms arise from this antisymmetry.

If $g(i, j)$ is independent of spin, we can carry out the "integration" over spin coordinates implied by the integral signs in a trivial manner. In the case of the off-diagonal elements, inspection of (6-17) and (6-18) shows that $(A \,|G|\, B)$ will vanish between states differing in $M_S = \Sigma m_s$, because in each case m_s must match for each electron. This is as it must be, since if G is spin-independent, it must commute with S_z. In the diagonal element $(A|\, G|\, A)$, we get simply a factor of unity in the direct integral and a factor of $\delta(m_s{}^k, m_s{}^t)$ in the exchange integral. Thus there are exchange integrals of a spin-*independent* operator only for electrons of like spin. This introduces an energy which formally depends on the relative orientation of spins, and hence it plays a central role in magnetic problems, despite the fact that the fundamental interaction is usually the electrostatic energy e^2/r_{ij}, which has nothing magnetic about it. This term in the energy, proportional to $-\mathbf{s}_k \cdot \mathbf{s}_t$, is the formal basis for the Heisenberg–Dirac–Van Vleck[1] *vector model*, which has often been used in rather phenomenological treatments of magnetic problems. Stated more precisely, if we consider two electrons occupying two orthogonal orbitals a and b, one can readily show that the energies of singlet and triplet states are $E_0 + K_{12} \pm J_{12}$, where E_0 is the energy in the absence of any interaction between the electrons, $K_{12} = (ab \,|g|\, ab)$, and $J_{12} = (ab \,|g|\, ba)$. This can be written as $E_0 + K_{12} - \frac{1}{2}J_{12} - 2J_{12}\mathbf{s}_1 \cdot \mathbf{s}_2$. The spin-dependent part is then generalized to

$$-2 \sum_{k>l} J_{kl}\mathbf{s}_k \cdot \mathbf{s}_l \qquad (6\text{-}20)$$

Note that, if $J > 0$, spins tend to align parallel to give a lower energy. In fact, if $g(i, j) = e^2/r_{ij}$, the exchange integral can be proved to be positive definite.[2] This provides the quantum-mechanical basis for the Hund rule that the L-S term with maximum S has the lowest energy, and we see that it is based on electrostatic, not magnetic, interactions.

This result (6-20) also forms the basis of Heisenberg's theory of ferromagnetism of solids. However, since antiferromagnetic and diamagnetic materials are more common than ferromagnetic ones, it is clear that it must also be possible for J to be negative if (6-20) is to hold. As stated above,

[1] J. H. Van Vleck, "Theory of Electric and Magnetic Susceptibilities," chap. 12, Oxford University Press, New York, 1932.

[2] J. C. Slater, "Quantum Theory of Atomic Structure," vol. I, app. 19, McGraw-Hill Book Company, Inc., New York, 1960.

this is not possible with J defined as a proper exchange integral. If we combine all terms which depend on $\mathbf{s}_1 \cdot \mathbf{s}_2$, including those arising from nonorthogonality of the functions a, b as well as the proper exchange integral, either sign becomes possible. However, in this case J_{12} has become difficult to calculate, and in many cases it is really little more than a disposable parameter in a phenomenological treatment. Despite this fact, such treatments are often of great value in accounting for the behavior of physical systems in which *ab initio* calculations are much too difficult to be possible.[1] Further discussion of exchange integrals will be given in Sec. 7-6 on the Heitler-London Method.

6-5 Hartree-Fock Method

Our program is to use the variational theorem to find the best set of N one-electron orbitals to use in a determinantal wavefunction. This is done by minimizing the diagonal energy matrix element while constraining the functions to stay orthonormal. For definiteness, we take an atomic Hamiltonian, excluding small magnetic-energy terms like the spin-orbit coupling, namely,

$$H = \sum_i \left[\frac{\mathbf{p}_i{}^2}{2m} - \frac{Ze^2}{r_i} \right] + \sum_{i>j} \frac{e^2}{r_{ij}}$$

It should be emphasized, however, that the method is in principle equally applicable to molecular and solid-state problems if Ze^2/r is replaced by a suitably generalized many-center potential. We use the results of the last section, denoting the various orbitals (including spin) by u_i. This gives us

$$\int \Psi^* H \Psi \, d\tau = \sum_i \int u_i^*(1) \left[-\frac{\hbar^2 \nabla_1{}^2}{2m} - \frac{Ze^2}{r_1} \right] u_i(1) \, d\tau_1 + \sum_{i>j} \iint u_i^*(1) u_j^*(2) \left(\frac{e^2}{r_{12}} \right)$$

$$\times \, u_i(1) u_j(2) \, d\tau_1 \, d\tau_2 - \sum_{i>j,\|} \iint u_i^*(1) u_j^*(2) \left(\frac{e^2}{r_{12}} \right) u_j(1) u_i(2) \, d\tau_1 \, d\tau_2$$

$$(6\text{-}21)$$

The last sum is to be taken only over pairs of electrons with parallel spins, because they are the exchange integrals and e^2/r_{12} is independent of spin. Now we imagine one of the $u_i(r)$ to be varied by a small amount $\delta u_i(r)$; this of course produces a like change δu_i^* in u_i^*. One can show that all terms arising from the variation of u_i^* are complex conjugates of those arising from the variation in u_i. This is elementary for the potential-energy terms but requires an application of Green's theorem for the term in ∇^2. Thus, if we make the energy insensitive in first order to change in u_i^* alone, we have already satisfied our requirement for an extremal energy. Since we also

[1] It is perhaps symbolic of the gulf between these two approaches that in the *ab initio* literature J and K are commonly used for the direct and exchange integrals, respectively, whereas in the "vector-model" literature, the usage is usually the reverse.

wish to preserve the normalization of the u_i, we add the Lagrange multiplier $(-\epsilon_{ii})$ times the normalization integral $\int u_i^*(1)u_i(1)\, d\tau_1$ to our energy integral before performing the variation. Further, to preserve orthogonalization, we add the orthogonality integrals $\int u_i^*(1)u_j(1)\, d\tau_1$ with Lagrange multipliers $-\epsilon_{ij}$. By the usual method in the calculus of variations, we then argue that, if the sum of integrals is to vanish for an arbitrary δu_i^*, the coefficient of δu_i^* must be everywhere zero. This leads directly to the Hartree-Fock equation,

$$\left[-\frac{\hbar^2 \nabla_1^2}{2m} - \frac{Ze^2}{r_1}\right]u_i(1) + \sum_{j \neq i}\left[\int u_j^*(2)\left(\frac{e^2}{r_{12}}\right)u_j(2)\, d\tau_2\right]u_i(1)$$

$$- \sum_{j \neq i, \parallel}\left[\int u_j^*(2)\left(\frac{e^2}{r_{12}}\right)u_i(2)\, d\tau_2\right]u_j(1) = \sum_j \epsilon_{ij}u_j(1) \qquad \blacktriangleright \quad (6\text{-}22)$$

The terms in the first line are exactly those which we had in the simple Hartree case: the kinetic energy, the potential energy due to the nucleus, and the Coulomb repulsion averaged over the positions of the other electrons. The exchange integrals do not appear in the Hartree equations because they arise only with antisymmetrized functions. Also, since the Hartree functions are not orthogonal, we have no terms in the right member except $\epsilon_{ii}u_i(1)$, because the other Lagrange multipliers would not appear. In this way, we see how the Hartree equations (6-5) follow from a variational approach if we neglect antisymmetry.

Let us now clarify these extra terms which distinguish the Hartree-Fock from the Hartree equations. First, we note that only the best *determinant* is fixed by the variational principle. Since a determinant is unchanged by a unitary transformation, the u_i are *not* completely determined. This freedom can be used to throw ϵ_{ij} into diagonal form, so that we again have only the single term $\epsilon_{ii}u_i \equiv E_i u_i$ appearing in Eq. (6-22). We shall assume that this simplification has in fact been carried through.

To give more physical insight into the remaining sum of exchange integrals which arises from the antisymmetry, we follow a method suggested by Slater.[1] First, he observes that we can remove the restriction that $j \neq i$ in both the summations, since the two integrals added in this way just cancel each other. Then, if we multiply and divide the exchange term by $u_i^*(1)u_i(1)$, we have the following revised form of the Hartree-Fock equations:

$$\left[-\frac{\hbar^2 \nabla_1^2}{2m} - \frac{Ze^2}{r_1}\right]u_i(1) + \left[\sum_j \int u_j^*(2)\left(\frac{e^2}{r_{12}}\right)u_j(2)\, d\tau_2\right]u_i(1)$$

$$- \left[\sum_{j \parallel i} \frac{\int u_i^*(1)u_j^*(2)(e^2/r_{12})u_j(1)u_i(2)\, d\tau_2}{u_i^*(1)u_i(1)}\right]u_i(1) = E_i u_i(1) \quad (6\text{-}23)$$

[1] J. C. Slater, *Phys. Rev.*, **81**, 385 (1951).

This now looks exactly like $Hu_i = E_i u_i$. The effective Hamiltonian now includes the kinetic energy, nuclear attraction, the repulsion of *all* electrons (including the one we are interested in), and the last term. Since we can be sure the electron really exerts no force on itself, this last term must somehow take care of that fact. The last term is negative and subtracts the repulsion of a charge density at $\mathbf{r}_2$ of magnitude (apart from a factor e)

$$\rho_{ex}^{(i)}(1, 2) = \sum_{j \parallel i} \frac{u_i^*(1)u_j^*(2)u_j(1)u_i(2)}{u_i^*(1)u_i(1)} \tag{6-24}$$

We call this the *exchange charge density* seen by the electron in the ith function when it is at $\mathbf{r}_1$. Integration over $\mathbf{r}_2$, upon using the orthogonality of u_i and u_j for $i \neq j$, shows that this exchange charge density contains exactly one electron. Further, at $\mathbf{r}_2 = \mathbf{r}_1$, $\rho_{ex}^{(i)}(1, 1) = \sum_{j \parallel i} |u_j(1)|^2$. That is, it exactly subtracts out the charge density of all electrons with spin parallel to that of the ith function at $\mathbf{r}_1$. This is consistent with our earlier remark that the antisymmetry guaranteed that electrons with parallel spin were never at the same point in space. The exact shape of $\rho_{ex}(1, 2)$ will depend on the details of the u's. However, to a crude approximation, it will have the effect of removing all the parallel spin charge from a sphere centered at $\mathbf{r}_1$ and of radius R such that

$$\int \rho_{ex} \, d\tau \approx \frac{4\pi}{3} \sum_{j \parallel i} |u_j(1)|^2 R^3 = 1 \tag{6-25}$$

More pictorially, each electron moves in the field of the nucleus, the electrons of opposite spin, and those of parallel spin outside an "exchange hole," or "Fermi hole" (of radius R), which follows around wherever it goes.

Slater has used this physical picture as the basis for a simplified form of the Hartree-Fock equations. In this he forms a suitable *average* exchange charge density for electrons of each spin and then calculates *all* his eigenfunctions in the resulting two potentials, which are, of course, identical if there are no unpaired spins in the system. This eliminates the complications which arise from having each electron move in a different potential. The approximation is particularly useful in handling excited states, for excited orbitals can be calculated which are automatically orthogonal to the normal ones because they arise from the same Hamiltonian.

It should be noted that the effective Hartree-Fock potential *does* depend on the spin orientation, unless the atom has only paired spins throughout. This has important consequences for the wave functions in magnetic atoms. In particular, it produces a difference in the orbital functions associated with up and down spins, even for inner shells where both are occupied. To treat this accurately, one must resort to the *unrestricted Hartree-Fock method*,[1] in which the usual simplifying restriction that the orbital functions

[1] G. W. Pratt, Jr., *Phys. Rev.*, **102**, 1303 (1956).

depend only on n, l, m_l (not m_s) is dropped. This doubles the number of functions to be determined and hence is not ordinarily used except in treating effects depending directly on the *difference* between the up- and down-spin functions.

An important example of such an effect is in the hyperfine structure coupling in Mn^{++}. This ion has a nominal $1s^2 2s^2 2p^6 3s^2 3p^6 3d^5$ configuration. Since all the inner subshells are complete, the magnetic properties of the ion arise from the five d electrons whose spins are parallel in accordance with the Hund rule. As we shall see in Sec. 6-11, the dominant hyperfine coupling term is proportional to the density of unpaired spins at the nucleus. Since a d function must vanish as r^2 as $r \to 0$, there is no direct coupling to the $3d$ electrons and there would be no "contact" hyperfine splitting in the restricted Hartree-Fock approximation. Actually, a very sizable splitting is observed. This has been computed with fair success by Wood and Pratt[1] and by Heine[2] by use of the unrestricted Hartree-Fock method with the Slater approximation for the exchange potential. They find that the $1s$ and $2s$ orbitals for spins parallel to the $3d$ spins are pulled outward by the attractive exchange potential, leaving a negative net spin density at the nucleus. The effect on the $3s$ orbitals, which are peaked at about the same radius as the $3d$, is of opposite sign but less in magnitude. The resultant effect is of the correct sign and order of magnitude (about 1 per cent unbalance between spin densities), whereas previous calculations[3] using an apparently more restricted approach had yielded a result which was an order of magnitude too small.

6-6   Calculation of L-S-term Energies

Degeneracy within a configuration. If the configuration consists entirely of closed shells, there is only one possible determinantal eigenfunction. For example, if we have six p electrons, the wavefunction must be

$$\Psi = A\{p_1\alpha(1)p_1\beta(2)p_0\alpha(3)p_0\beta(4)p_{-1}\alpha(5)p_{-1}\beta(6)\} \tag{6-26a}$$

where A is our antisymmetrizing operator defined in (6-7). We shall often condense this notation still further to write

$$\Psi = (1^+ \ 1^- \ 0^+ \ 0^- \ -1^+ \ -1^-) \tag{6-26b}$$

No other assignment of six p electrons would be consistent with the exclusion principle.

[1] J. H. Wood and G. W. Pratt, *Phys. Rev.*, **107**, 995 (1957).
[2] V. Heine, *Phys. Rev.*, **107**, 1002 (1957).
[3] A. Abragam, J. Horowitz, and M. H. L. Pryce, *Proc. Roy. Soc.* (*London*), **A230**, 169 (1955).

If a shell is not full, we have many possibilities—all belonging to the same configuration. For example, if we have two inequivalent p electrons, such as $2p3p$, 36 possible determinants can be formed, since each electron has its choice of six states. Even if we consider two *equivalent* p electrons, we have $\binom{6}{2} = 15$ possible determinants. All these determinants must have the same energy in the purely central-field approximation. Therefore, when we introduce the interelectronic perturbation

$$H' = \sum_{i>j} \frac{e^2}{r_{ij}} - \sum_{i} \left[U(r_i) + \frac{Ze^2}{r_i} \right]$$

we must solve the secular equation

$$|(H^0 + H')_{ij} - E\delta_{ij}| = 0 \tag{6-27}$$

connecting these states exactly—that is, without any perturbation-theory approximation—because all the diagonal elements $H_{ii}{}^0$ are the same. Obviously this would involve an immense amount of labor if no means were available to help factor the secular equation. Fortunately, the group-theoretical properties of angular-momentum operators provide a large amount of initial factorization.

Factorization by angular momentum. The method is based on the fact that the approximate Hamiltonian H is invariant, not only under an overall rotation of coordinates, but also under separate rotation of spatial and spin coordinates, so long as we are neglecting spin-orbit coupling. In other words, H commutes, not only with $\mathbf{J}$, but also with $\mathbf{L}$ and $\mathbf{S}$. We may then apply our matrix-element theorem (Sec. 4-9 or 5-11) for invariant (scalar) operators to state that there can be no matrix elements of H between states with different L, S, or J or different M_L, M_S, or M_J, since these quantum numbers label the irreducible representations and their rows. Thus we can factor the secular equation by choosing determinants, or linear combinations of determinants, which are already angular-momentum eigenfunctions.

We are limited in the extent we can go in this direction by the fact that all angular momenta that commute with H do not commute with each other. We can simultaneously diagonalize four angular-momentum operators. The most useful sets have the good quantum numbers

$$LSM_LM_S \quad \text{and} \quad LSJM$$

It is easiest to start with the LSM_LM_S sets, since we are using one-electron functions characterized by m_l and m_s. Thus we automatically give "sharp" values to $M_L = \Sigma m_l$ and $M_S = \Sigma m_s$. Then, if we sort out states by their M_L and M_S values, we shall already have achieved a great amount of factorization. As an example, we consider our np^2 example of two equivalent p electrons. We may then classify all the 15 determinants consistent with the exclusion principle as in Table 6-1. Evidently only the upper left quadrant of this table need be written out, since the others can be obtained by suitable systematic sign changes. Since there are no matrix elements between states of different M_L or M_S, inspection of this table shows that our 15×15 secular equation will factor down to eight 1×1's, two 2×2's, and one 3×3, by $M_L M_S$ factorization alone.

Table 6-1

Possible np^2 determinants

M_L	M_S		
	1	0	−1
2		(1^+1^-)	
1	(1^+0^+)	$(1^+0^-)(1^-0^+)$	(1^-0^-)
0	$(1^+ -1^+)$	$(1^+ -1^-)(1^- -1^+)(0^+0^-)$	$(1^- -1^-)$
−1	$(0^+ -1^+)$	$(0^+ -1^-)(0^- -1^+)$	$(0^- -1^-)$
−2		$(-1^+ -1^-)$	

To proceed further, we must inquire after the L and S values as well. For example, the state (1^+1^-) must have $L \geq 2$, since $M_L = 2$. But in combining two systems with $l = 1$, the vector model allows only $L = 0$, 1, 2; therefore, it must be a D state $(L = 2)$. Since only one state with $M_L = 2$ exists, it must be part of a ^1D term, since for ^3D we would have to have two others with $M_S = \pm 1$. These are forbidden by the exclusion principle, however, since they would require double occupancies like (1^+1^+). Now that we know that a ^1D term is present, we must also have the other four "partners" in the five-dimensional representation. This accounts for 1 state out of each of the $M_S = 0$ column. The remaining 10 states are easily seen to be those required for a $^3P(M_L = \pm 1, 0; M_S = \pm 1, 0)$ and a $^1S(M_L = M_S = 0)$ term. It is in this way that we obtain the result that the terms allowed by the Pauli principle in the np^2 configuration are ^1D, ^3P, and ^1S.

At this time we might note that only for the cases in Table 6-1 where there is only one determinant are these determinants also eigenfunctions of L and S. In the other cases, each determinant is a mixture of the various L-S terms represented therein. It is easily possible to find the linear combinations within a set which also diagonalize L and S. For example, by using

$$L_-\Psi(L = 2, M_L = 2, S = 0) = L_-(1^+1^-)$$

$$= \sqrt{L(L+1) - 2 \cdot 1}\,\Psi(L = 2, M_L = 1, S = 0)$$

and the methods developed in Sec. 6-8, we find that

$$\psi(L = 2, M_L = 1, S = 0) = \frac{[(1^+0^-) - (1^-0^+)]}{\sqrt{2}}$$

In this way, we could find linear combinations of determinants which would completely diagonalize the Hamiltonian in this example. However, another method, due to Slater, is still more efficient in obtaining the energy eigenvalues which we seek.

Slater sum-rule method. This method is based on the fact that the energy of an L-S term is independent of M_L and M_S. Thus we have only three *distinct* eigenvalues to find from our 15×15 problem, namely, the energies of the ^1D, ^3P, and ^1S terms. We can calculate these energies for whatever M_L and M_S are most convenient. When possible we choose M_L and M_S so that there is only one determinant in the set, assuring us that we are dealing with an eigenfunction of L and S, that is, a pure term. For example,

$$E(^1D) = (1^+1^- |H| 1^+1^-)$$

and
$$E(^3P) = (1^+0^+ |H| 1^+0^+) \tag{6-28a}$$

All that remains to find is $E(^1S)$. This will be one of the three roots of the 3×3 secular equation obtained from the three determinants with $M_L = M_S = 0$. But we already know the other two roots, $E(^1D)$ and $E(^3P)$, and by the invariance of traces under unitary transformations, we know that the sum of the three roots is just the sum of the three diagonal energies. Thus we can express our unknown energy somewhat symbolically as

$$E(^1S) = Tr\,(3 \times 3) - E(^1D) - E(^3P)$$
$$= \langle 1^+ -1^- \rangle + \langle 1^- -1^+ \rangle + \langle 0^+0^- \rangle - \langle 1^+1^- \rangle - \langle 1^+0^+ \rangle \tag{6-28b}$$

where in the last line we have abbreviated the notation for diagonal matrix elements still further.

Thus we have succeeded in finding all the distinct eigenvalues of a 15×15 secular equation by calculating only five diagonal matrix elements. This trick is of very general utility and handles all cases until we have two distinct terms with the same L-S values arising from the same configuration. In this case, one must actually solve a small secular equation, because rotational symmetry is no longer enough to determine the eigenfunctions completely.

6-7 Evaluation of Matrix Elements of the Energy

Let us examine these diagonal matrix elements to which we have reduced the problem. They are matrix elements of

$$H = \sum_i \left[\frac{\mathbf{p}_i^2}{2m} - \frac{Ze^2}{r_i} \right] + \sum_{i>j} \frac{e^2}{r_{ij}}$$

Now, the one-electron terms will contribute equally to all diagonal elements since all come from the same configuration. Thus these terms only shift the energetic center of gravity for the configuration and have no effect on the splitting apart of the terms. This term structure is determined wholly by $Q \equiv \sum_{i>j} e^2/r_{ij}$.

Applying our general results for operators of the type G, we have

$$(A \,|Q|\, A) = \sum_{a>b} \left[\left(ab \left| \frac{e^2}{r_{12}} \right| ab \right) - \left(ab \left| \frac{e^2}{r_{12}} \right| ba \right) \right] \qquad (6\text{-}29)$$

This sum runs over all pairs of electrons and may be broken up into three parts: (1) pairs in closed shells, (2) pairs with one electron in a closed shell, and (3) pairs with neither electron in a closed shell. Since closed shells have spherical symmetry (Unsöld's theorem), we do not expect terms like (1) and (2) to split the terms when summed over all occupied states. Rather, they represent the "shielding" effect of the inner closed shells upon the potential experienced by the valence electrons. This fact may be proved rigorously[1], but since it seems physically evident as well, we shall omit the formal proof here. This leaves the entire splitting to be determined by the part

$$(A \,|Q|\, A) = \sum_{a>b=1}^{N'} [J(a, b) - K(a, b)] \qquad (6\text{-}30)$$

where N' includes only the electrons not in closed shells and J and K are the conventional symbols for the direct and exchange integrals of the Coulomb interaction. This substantiates our earlier statement that the energy differences among the L-S terms belonging to a given configuration arise from the Coulomb repulsion.

Calculation of matrix elements of the electrostatic interaction. Rather than calculate simply the diagonal elements required in using the Slater sum rule, we shall set up the machinery for handling the most general case that can arise. This has the form

$$(ab \,|q|\, cd) = \iint a^*(1)b^*(2) \left(\frac{e^2}{r_{12}} \right) c(1)d(2) \, d\tau_1 \, d\tau_2 \qquad (6\text{-}31)$$

where typically

$$a(1) = \left(\frac{1}{r_1} \right) R_{n^a l^a}(r_1) \Theta_{l^a m_l^a}(\theta_1) \Phi_{m_l^a}(\varphi_1) \chi_{m_s^a}(s_1) \qquad (6\text{-}32a)$$

For ease in writing, we shall write this

$$a(1) = \left(\frac{1}{r_1} \right) R_1(n^a l^a) \Theta_1(l^a m_l^a) \Phi_1(m_l^a) \chi_1(m_s^a) \qquad (6\text{-}32b)$$

[1] Condon and Shortley, *op. cit.*, p. 182.

The integral is then a sixfold integral over spatial coordinates and a two-fold summation over spin coordinates. The spin sum gives zero unless $m_s{}^a = m_s{}^c$ and $m_s{}^b = m_s{}^d$. If these hold, the sum yields unity, and we have only the spatial integrals to deal with.

To proceed, we expand $1/r_{12}$ in the conventional way,

$$\frac{1}{r_{12}} = \frac{1}{(r_1{}^2 + r_2{}^2 - 2r_1 r_2 \cos \omega)^{\frac{1}{2}}} = \sum_{k=0}^{\infty} \frac{r_<{}^k}{r_>{}^{k+1}} P_k (\cos \omega) \qquad (6\text{-}33)$$

where $r_<$ and $r_>$ are the lesser and greater of $|r_1|$, $|r_2|$ and ω is the angle between $\mathbf{r}_1$ and $\mathbf{r}_2$. Next, we expand $P_k (\cos \omega)$, using the spherical harmonic addition formula,

$$P_k(\cos \omega) = \frac{4\pi}{2k + 1} \sum_{m=-k}^{k} \Theta_1(km)\Theta_2(km)\Phi_1(m)\Phi_2^*(m) \qquad (6\text{-}34a)$$

where

$$\cos \omega = \cos \theta_1 \cos \theta_2 + \sin \theta_1 \sin \theta_2 \cos (\varphi_1 - \varphi_2) \qquad (6\text{-}34b)$$

Integrating the resulting expression over φ_1 and φ_2 merely gives δ functions which assure conservation of L_z (together with a normalization factor of $\dfrac{1}{2\pi}$). Thus the matrix element vanishes unless

$$m = m_l{}^a - m_l{}^c = m_l{}^d - m_l{}^b$$

or

$$m_l{}^a + m_l{}^b = m_l{}^c + m_l{}^d$$

This is as it must be, since e^2/r_{12} is invariant under a general rotation of co-ordinates of *all* particles. The remaining integral is

$$(ab \,|q|\, cd) = 2e^2 \sum_{k=0}^{\infty} \frac{1}{2k + 1} \int_0^{\infty} \int \frac{r_<{}^k}{r_>{}^{k+1}} R_1(n^a l^a) R_2(n^b l^b) R_1(n^c l^c) R_2(n^d l^d) \, dr_1 \, dr_2$$

$$\times \int_0^{\pi} \int \Theta_1(l^a m_l{}^a)\Theta_2(l^b m_l{}^b)\Theta_1(l^c m_l{}^c)\Theta_2(l^d m_l{}^d)$$

$$\times \Theta_1(k, m_a - m_c)\Theta_2(k, m_l{}^d - m_l{}^b) \sin \theta_1 \sin \theta_2 \, d\theta_1 \, d\theta_2 \qquad (6\text{-}35)$$

or

$$(ab \,|q|\, cd) = \delta(m_s{}^a, m_s{}^c)\delta(m_s{}^b, m_s{}^d)\delta(m_l{}^a + m_l{}^b, m_l{}^c + m_l{}^d) \sum_{k=0}^{\infty} c_k(l^a m_l{}^a, l^c m_l{}^c)$$

$$\times c^k(l^d m_l{}^d, l^b m_l{}^b) R^k(n^a l^a n^b l^b, n^c l^c n^d l^d) \qquad \blacktriangleright \quad (6\text{-}36)$$

The coefficients c^k are integrals of products of three Legendre functions. These have been worked out by Gaunt and tabulated once and for all.[1] The formal definition is

$$c^k(lm_l, l'm_l') = \sqrt{\frac{2}{2k + 1}} \int_0^{\pi} \Theta(k, m_l - m_l')\Theta(lm_l)\Theta(l'm_l') \sin \theta \, d\theta \qquad (6\text{-}37)$$

---

[1] J. A. Gaunt, *Trans. Roy. Soc. (London)*, **A228**, 195 (1928). A more accessible table of results may be found in Condon and Shortley, *op. cit.*, pp. 178–179; a table adequate for *s*, *p*, *d* electrons is given in Appendix C of this book.

We note that $c^k(lm_l, l'm_l') = (-1)^{m_l-m_l'}c^k(l'm_l', lm_l)$ follows from our standard phase choice for the $\Theta(lm_l)$. Actually these Gaunt coefficients c^k can be readily expressed in terms of Wigner coefficients, and in fact this entire development could be carried out within the framework of the tensor-operator formalism which we developed in the previous chapter. We shall not do this here, however, but shall simply state that the connection is

$$c^k(lm_l, l'm_l') = \left[\frac{(2l'+1)}{(2l+1)}\right]^{\frac{1}{2}} A_{000}^{kl'l} A_{m_l-m_l',m_l',m_l}^{kl'l} \qquad (6\text{-}37a)$$

where the $A_{m_1m_2m_3}^{j_1j_2j_3}$ are the Wigner coefficients. The other quantity, R^k, is a radial integral which depends on the particular approximate form for the radial wave function. The definition is just

$$R^k(n^al^an^bl^b, n^cl^cn^dl^d) = e^2 \int_0^\infty \int \frac{r_<^k}{r_>^{k+1}} R_1(n^al^a)R_2(n^bl^b)R_1(n^cl^c)R_2(n^dl^d) \, dr_1 \, dr_2$$

$$(6\text{-}38)$$

These radial integrals may have to be evaluated numerically if we use Hartree functions, for example. Luckily, in most cases of interest, only a few terms in the sum on k in (6-36) are nonzero, because of the properties of the c^k. That is, $c^k(l, l') = 0$ unless k satisfies the vector-model condition $|l - l'| \leqslant k \leqslant l + l'$ and unless further $k + l + l' =$ even. This limitation on the number of k values allows these remaining integrals R^k to be used as parameters in describing the term structure, and interesting results for *ratios* of term separations may often be deduced by using only the c^k, which are tabulated.

Diagonal elements. Since only the diagonal elements are needed for using the Slater method, we now specialize our results for this case. We have

$$(A \,|Q|\, A) = \sum_{a>b} [J(a, b) - K(a, b)] \equiv \sum_{a>b} [(ab \,|q|\, ab) - (ab \,|q|\, ba)]$$

Because of the symmetry which exists when only two distinct orbitals (a, b) are involved, we can condense the notation and reduce our general expression to

$$J(a, b) = \sum_{k=0}^\infty a^k(l^am_l^a, l^bm_l^b)F^k(n^al^a, n^bl^b) \qquad (6\text{-}39a)$$

and

$$K(a, b) = \delta(m_s^a, m_s^b) \sum_{k=0}^\infty b^k(l^am_l^a, l^bm_l^b)G^k(n^al^a, n^bl^b) \qquad (6\text{-}39b)$$

where we have introduced the new notations

$$a^k(l^am_l^a, l^bm_l^b) = c^k(l^am_l^a, l^am_l^a)c^k(l^bm_l^b, l^bm_l^b) \qquad (6\text{-}40a)$$

$$b^k(l^am_l^a, l^bm_l^b) = [c^k(l^am_l^a, l^bm_l^b)]^2 \qquad (6\text{-}40b)$$

and

$$F^k(n^al^a, n^bl^b) = R^k(n^al^an^bl^b, n^al^an^bl^b) \qquad (6\text{-}41a)$$

$$G^k(n^al^a, n^bl^b) = R^k(n^al^an^bl^b, n^bl^bn^al^a) \qquad (6\text{-}41b)$$

These coefficients a^k and b^k have also been tabulated.[1] We note that $F^k = G^k$ if the electrons are equivalent.

Example (np^2). As shown in (6-28), in order to get the L-S-term energies, we need the diagonal energies $\langle 1^+1^- \rangle$, $\langle 1^+0^+ \rangle$, $\langle 1^+ -1^- \rangle$, $\langle 1^- -1^+ \rangle$, and $\langle 0^+0^- \rangle$. Apart from the common additive constant coming from the closed shells,

$$\langle 1^+1^- \rangle = J(1^+1^-) - K(1^+1^-)$$

$$= \sum_k a^k(p, 1, p, 1)F^k(np, np) - 0$$

Inspecting the tables of a^k coefficients, we find that the only nonvanishing coefficients are for $k = 0, 2$. Thus

$$\langle 1^+1^- \rangle = F^0(np, np) + (\tfrac{1}{25})F^2(np, np) = F_0(np, np) + F_2(np, np)$$

where we have indicated the absorption of numerical constants into the F^k by dropping the superscript to a subscript. Similarly,

$$\langle 1^+0^+ \rangle = J(1^+0^+) - K(1^+0^+)$$

$$= \sum a^k(p, 1, p, 0)F^k(np, np) - \sum b^k(p, 1, p, 0)G^k(np, np)$$

$$= F^0 - \frac{2F^2}{25} - \frac{3F^2}{25}$$

$$\equiv F_0 - 5F_2$$

Continuing, we obtain

$$\langle 1^+ -1^- \rangle = F_0 + F_2$$
$$\langle 1^- -1^+ \rangle = F_0 + F_2$$
$$\langle 0^+0^- \rangle = F_0 + 4F_2$$

Substituting the results back into the results of the Slater sum rule (6-28), we have the term energies (apart from a common additive constant) given by

$$E(^1S) = F_0 + 10F_2$$
$$E(^1D) = F_0 + F_2 \tag{6-42}$$
$$E(^3P) = F_0 - 5F_2$$

Inspection of the definition of the F_k shows that they are all positive definite integrals. Thus our results are consistent with Hund's rules, which require the ^3P level to lie lowest. We note that the remaining terms also lie in the order suggested by the rule that maximum L lies lowest. In general, however, one can rely on the rules to give only the one lowest term correctly.

---

[1] See Condon and Shortley, *op. cit.*, p. 180; an abbreviated table is given in Appendix C of this book.

To calculate the absolute values of these energies, we would need to have good radial functions to evaluate the radial integrals F_k. However, we can test the theory without knowing F_k by comparing ratios. For example,

$$\frac{E(^1S) - E(^1D)}{E(^1D) - E(^3P)} = \frac{9F_2}{6F_2} = \frac{3}{2}$$

The result should be independent of F_2 to the accuracy of our single-configuration approximation. Experimental values for various np^2 atoms are 1.13, 1.14, 1.14, 1.48, 1.50, 1.39. The departures from 1.50 are largely due to the admixture of $np(n+1)p$ configurations. The general order-of-magnitude agreement is not bad, though, and the terms at least fall in the proper order. For simplicity we shall usually be satisfied to use these single-configuration approximations for the L-S terms as the starting point for considering fine and hyperfine structures. Before proceeding to that, however, let us examine the nature of these L-S eigen*functions* a bit more directly.

6-8 Eigenfunctions and Angular-momentum Operations

In Sec. 6-6, we commented on the fact that only in simple cases are the individual determinants eigenfunctions of L, S as well as M_L, M_S. For example, (1^+1^-) has $L = 2$, $M_L = 2$, $S = M_S = 0$. However, by operating on this determinant with L_-, we can find the combinations of determinants which form the partners of this function with $M_L = 1, 0, -1, -2$. Such operations can be considered simply as applications of our general formulas for computing matrix elements of operators of the type F or G. Thus, for example,

$$L_-\Psi_A = \sum_B \Psi_B(B\,|L_-|\,A) \tag{6-43}$$

where $L_- = \sum_i L_-^i$. However, these angular-momentum operations occur so frequently that it is worthwhile to work out explicit results.

L_z **and** S_z. These operators are simple to handle since we have built our determinants out of eigenfunctions of L_z^i and S_z^i. Thus, when we operate with

$$L_z = \sum_{i=1}^N L_z^i = L_z^1 + L_z^2 + L_z^3 + \cdots + L_z^N$$

we obtain N determinants, each identical to the original one except for a factor m_l^i. Summing these, we have

$$L_z\Psi_A = \sum_i (m_l^i)\Psi_A = M_L\Psi_A \tag{6-44a}$$

Similarly,

$$S_z\Psi_A = \sum_i (m_s^i)\Psi_A = M_S\Psi_A \tag{6-44b}$$

and these determinants are shown to be eigenfunctions of L_z and S_z, as we have known all along.

$L_\pm$ **and** $S_\pm$. These are a bit more complicated, but using (6-13) it is easy to see that this works out to

$$L_\pm\Psi(n^1l^1m_l{}^1m_s{}^1;\ldots)$$
$$= \sum_i [l^i(l^i+1) - m_l{}^i(m_l{}^i \pm 1)]^{\frac{1}{2}}\Psi(\ldots; n^il^i, m_l{}^i \pm 1, m_s{}^i;\ldots) \quad (6\text{-}45a)$$

and

$$S_\pm\Psi(n^1l^1m_l{}^1m_s{}^1;\ldots)$$
$$= \sum_i [(\tfrac{3}{4}) - m_s{}^i(m_s{}^i \pm 1)]^{\frac{1}{2}}\Psi(\ldots; n^il^im_l{}^i,m_s{}^i \pm 1;\ldots) \quad (6\text{-}45b)$$

[Unlike (6-13), no $\pm$ signs appear here because we have not brought the resulting determinants to standard order.] Note that the only nonzero value of $[(\tfrac{3}{4}) - m_s{}^i(m_s{}^i \pm 1)]$ is 1. This occurs when $m_s{}^i = \mp\tfrac{1}{2}$. Also note that any Ψ's which are formally generated but which are inconsistent with the Pauli principle because of double occupancy of a spin-orbital function will vanish because of the antisymmetry of the determinants.

$\mathbf{L}^2$ **and** $\mathbf{S}^2$. We expand this as

$$\mathbf{L}^2 = (\sum_i \mathbf{L}^i)^2 = \sum_i (\mathbf{L}^i)^2 + 2\sum_{i>j} \mathbf{L}^i \cdot \mathbf{L}^j$$
$$= \sum_i l^i(l^i+1) + 2\sum_{i>j} L_z{}^iL_z{}^j + \sum_{i>j}(L_+{}^iL_-{}^j + L_-{}^iL_+{}^j)$$

The first two sums give diagonal matrix elements; the last one gives off-diagonal ones. The total effect of the operator is as follows:

$$\mathbf{L}^2\Psi(n^1l^1m_l{}^1m_s{}^1;\ldots)$$
$$= [\sum_i l^i(l^i+1) + 2\sum_{i>j}m_l{}^im_l{}^j]\Psi(n^1l^1m_l{}^1m_s{}^1;\ldots)$$
$$+ \sum_{i\neq j}[l^i(l^i+1) - m_l{}^i(m_l{}^i+1)]^{\frac{1}{2}}[l^j(l^j+1) - m_l{}^j(m_l{}^j-1)]^{\frac{1}{2}}$$
$$\times \Psi(\ldots; n^i, l^i, m_l{}^i+1, m_s{}^i;\ldots; n^j, l^j, m_l{}^j-1, m_s{}^j;\ldots) \quad (6\text{-}46a)$$

Similarly, we have for the spin operators

$$\mathbf{S}^2\Psi(n^1l^1m_l{}^1m_s{}^1;\ldots) = \left[\frac{3N}{4} + 2\sum_{i>j}m_s{}^im_s{}^j\right]\Psi(n^1l^1m_l{}^1m_s{}^1;\ldots)$$
$$+ \sum_{i\neq j}[(\tfrac{3}{4}) - m_s{}^i(m_s{}^i+1)]^{\frac{1}{2}}[(\tfrac{3}{4}) - m_s{}^j(m_s{}^j-1)]^{\frac{1}{2}}$$
$$\times \Psi(\ldots; n^il^im_l{}^i, m_s{}^i+1;\ldots; n^jl^jm_s{}^j, m_s{}^j-1;\ldots)$$

We can simplify the first term by noting that

$$2\sum_{i>j}m_s{}^im_s{}^j = \sum_i m_s{}^i\sum_{j\neq i}m_s{}^j = \sum_i m_s{}^i(M_S - m_s{}^i)$$
$$= M_S{}^2 - \sum_i(m_s{}^i)^2 = M_S{}^2 - \frac{N}{4}$$

Thus the first coefficient is $[M_S^2 + N/2]$. Since the only nonzero value for the square roots is unity, the second term also can be simplified. If any spatial orbital is occupied by two electrons with paired spins, it appears in the second term with spins reversed, which produces just a sign change because of the antisymmetry. This term cancels the contribution in the first term of all paired spins. This leaves us with the result

$$\mathbf{S}^2 \Psi_A = \left[M_S^2 + \frac{\text{number of unpaired spins}}{2} \right] \Psi_A + \sum \Psi'_{A'} \qquad (6\text{-}46b)$$

where Ψ'_A are all the determinants which differ from Ψ_A by a single interchange of spin orientations between distinct spatial orbitals.

Let us now apply this machinery to our np^2 example. Starting with $\Psi(LM_LSM_S) = \Psi(2200) = (1^+1^-)$, we have

$$L_- \Psi(2200) = \sqrt{2 \cdot 3 - 2 \cdot 1}\, \Psi(2100) = 2\Psi(2100)$$

and

$$L_- \Psi(2200) = L_-(1^+1^-) = \sqrt{1 \cdot 2 - 1 \cdot 0}\,(0^+1^-) + \sqrt{1 \cdot 2 - 1 \cdot 0}\,(1^+0^-)$$

$$= \sqrt{2}[(0^+1^-) + (1^+0^-)]$$

Comparing, we have

$$\Psi(2100) = \frac{1}{\sqrt{2}}\,[(0^+1^-) + (1^+0^-)]$$

or, if the orbitals are returned to standard order,

$$\Psi(2100) = \frac{1}{\sqrt{2}}\,[(1^+0^-) - (1^-0^+)]$$

If we operated on this linear combination with $\mathbf{L}^2$, we could verify directly that it is in fact an eigenfunction satisfying

$$\mathbf{L}^2 \frac{(1^+0^-) - (1^-0^+)}{\sqrt{2}} = 2 \cdot 3\, \frac{(1^+0^-) - (1^-0^+)}{\sqrt{2}}$$

The ease with which these matrix elements and eigenfunctions are found indicates another useful technique for reducing secular equations. We could, for example, easily compute matrix elements of $\mathbf{L}^2$ within the 3×3 $M_L = M_S = 0$ block which arises in the np^2 problem. If we diagonalized this $\mathbf{L}^2$ matrix by solving the simple secular equation, we would have found the three linear combinations belonging to ^1S, ^1D, ^3P, respectively, corresponding to eigenvalues 0, $2 \cdot 3$, $1 \cdot 2$. Exactly these linear combinations also must diagonalize the 3×3 matrix of the *Hamiltonian*, whose matrix elements are in general much harder to find. Although in the present case this method would be a bit more difficult than Slater's sum rule, it does play a very useful role in molecular calculations. In that case, one is often interested only in finding an $S = 0$ ground state. By first diagonalizing S^2, one can find all the linear combinations of determinants which need to be considered in the final secular equation connecting several configurations.

6-9 Calculation of Fine Structure

We now take the eigenfunctions and eigenvalues found for the L-S terms as the starting point for introducing the *magnetic* or *relativistic* perturbations. These energies are of order $(v/c)^2$ smaller than the term separations; so for the lighter atoms it is satisfactory to consider the fine structure of each L-S term separately by perturbation theory. Because the various $M_L M_S$ sub-states of a term are degenerate, it is first necessary to organize linear combinations of them which are proper zero-order eigenfunctions such that there are no off-diagonal elements of the perturbation between them. This presents no difficulty, because we know that the total Hamiltonian including these spin-orbit coupling terms is invariant under rotations of the *total* system. Thus, according to the group theory, there are no matrix elements of H between states of different J and M_J. But the Wigner coefficients prescribe exactly what linear combinations of $M_L M_S$ states form the eigenfunctions of J, M_J. Thus they provide just the zero-order linear combinations which we require. Actually, we can usually get the eigen*values* by tricks which make even the Wigner coefficients unnecessary.

One-electron case. To get a start, we consider the simple case of a single electron moving in the spherical potential outside closed shells. The magnetic-perturbation energy of the spin can be expressed as

$$H' = -\mathbf{\mu}_S \cdot \mathbf{H}_{\text{eff}} = \frac{e\hbar}{mc} \mathbf{s} \cdot \left(\mathbf{E} \times \frac{\mathbf{v}}{c} \right) \qquad (6\text{-}47)$$

where we have used the Lorentz transformation to find the effective magnetic field seen by the spin as it moves through the central electric field. This field may be expressed by

$$\mathbf{E} = -\frac{\partial V}{\partial r} \mathbf{r}_0 \equiv \frac{1}{e} \frac{\partial U}{\partial r} \mathbf{r}_0$$

where $\mathbf{r}_0$ is a radial unit vector. Substituting this above, we have

$$H' = \frac{\hbar}{mc^2} \frac{\partial U}{\partial r} \mathbf{s} \cdot (\mathbf{r}_0 \times \mathbf{v}) = \frac{\hbar}{rm^2c^2} \frac{\partial U}{\partial r} \mathbf{s} \cdot (\mathbf{r} \times \mathbf{p}) = \frac{\hbar^2}{m^2c^2} \left(\frac{1}{r} \frac{\partial U}{\partial r} \right) \mathbf{l} \cdot \mathbf{s}$$

This simplified treatment has failed to introduce the factor of $\frac{1}{2}$ which results from the Thomas precession[1] of relativistic kinematics or from a

---

[1] L. H. Thomas, *Nature*, **117**, 514 (1926).

straightforward reduction of the Dirac equation to a two-component form. Inserting this correction factor, we obtain

$$H' = \xi(r)\mathbf{l} \cdot \mathbf{s} \qquad \xi(r) = \frac{\hbar^2}{2m^2c^2} \frac{1}{r} \frac{\partial U}{\partial r} \qquad (6\text{-}48)$$

(Note that we take $\mathbf{l}$ and $\mathbf{s}$ in units of $\hbar$ making them dimensionless quantities in these equations.)

When we deal with the many-electron problem, we shall simply generalize this to $H' = \sum_i \xi(r_i)\mathbf{l}_i \cdot \mathbf{s}_i$. This includes only the interaction of each spin with its own orbit. This is the dominant term since the central field is stronger than the interelectronic interaction. We thereby neglect the orbit-orbit, spin–other–orbit, and spin-spin terms. The effect of this neglect has been considered by various authors.[1]

Matrix elements. In evaluating matrix elements of H', we take advantage of the fact that it is a product of three operators, each operating on completely independent coordinates of the electron. $\xi(r)$ operates on the radial coordinate, $\mathbf{l}$ on angular coordinates of the orbital position, and $\mathbf{s}$ on the spin coordinates. This enables us to factor the matrix element as follows:

$$(n'l'm_l'm_s' \,|H'|\, nlm_lm_s) = (n'l' \,|\xi(r)|\, nl)(l'm_l' \,|\mathbf{l}|\, lm_l) \cdot (m_s' \,|\mathbf{s}|\, m_s) \quad (6\text{-}49)$$

Except for the radial integral, the other factors are elementary matrix elements which can be written down immediately. Since $\mathbf{l}$ is diagonal in l, $l' = l$. Noting that $\mathbf{l} \cdot \mathbf{s}$ is a scalar operator, we know that $\Delta M = (m_l' + m_s') - (m_l + m_s)$ must be zero. Since $\mathbf{l}$ and $\mathbf{s}$ separately are vector operators, they have $\Delta m_l = \pm 1, 0$ and $\Delta m_s = \mp 1, 0$ selection rules. Thus we conclude that l and $m = m_l + m_s$ are good quantum numbers but m_l and m_s are not. Instead, we know by our general arguments of rotational invariance that $|\mathbf{j}| = |\mathbf{l} + \mathbf{s}|$ is now a good quantum number. This fact enables us to get our answer by a short cut requiring almost no calculation.

We note that

$$\mathbf{j}^2 = (\mathbf{l} + \mathbf{s})^2 = \mathbf{l}^2 + \mathbf{s}^2 + 2\mathbf{l} \cdot \mathbf{s}$$

Therefore

$$\mathbf{l} \cdot \mathbf{s} = \frac{\mathbf{j}^2 - \mathbf{l}^2 - \mathbf{s}^2}{2} \qquad (6\text{-}50)$$

This is a quantum-mechanical operator equation. Since, in the representation based on the actual eigenfunctions fixed by the Wigner coefficients, $\mathbf{j}^2$, $\mathbf{l}^2$, $\mathbf{s}^2$ are all diagonal, $\mathbf{l} \cdot \mathbf{s}$ also has only diagonal elements in this representation and these can be written down by inspection. Thus we conclude that

$$E(nljm) = E_0(nl) + \zeta(nl)(ljm \,|\mathbf{l} \cdot \mathbf{s}|\, ljm)$$

---

[1] For example, see R. E. Trees, *Phys. Rev.*, **82**, 683 (1951); H. H. Marvin, *Phys. Rev.*, **71**, 102 (1947); R. G. Breene, Jr., *Phys. Rev.*, **119**, 1615 (1960).

or $\qquad E(nljm) = E_0(nl) + \dfrac{\zeta(nl)}{2} [j(j+1) - l(l+1) - s(s+1)]$ $\qquad$ (6-51)

where $\zeta(nl) = \displaystyle\int_0^\infty R_{nl}^2(r)\xi(r)\,dr$ is the appropriate radial integral.

Now, for a single electron, we have $s = \frac{1}{2}$ and $j = l \pm \frac{1}{2}$. Inserting these values in (6-51) leads to

$$E(nljm) = E_0(nl) + \zeta(nl) \begin{cases} \dfrac{l}{2} \\[2mm] \dfrac{-(l+1)}{2} \end{cases} \quad \text{for } j = \begin{cases} l + \dfrac{1}{2} \\[2mm] l - \dfrac{1}{2} \end{cases} \qquad (6\text{-}52)$$

For all reasonable potentials, $\partial U/\partial r > 0$; so $\zeta > 0$. Thus *the state of lower j lies lowest, as required by the Hund rule.* This fact could, of course, be confirmed by an elementary consideration of the direction of $\mathbf{H}_{\text{eff}}$ relative to the orientation of the orbit. The doublet separation $(l + \frac{1}{2})\zeta(nl)$ is clearly of magnetic origin and smaller than electronic energies by a factor of order α^2. For example, the sodium D lines at 17,000 cm^{-1} result from the transition $3p \to 3s$. The $3p$ term is split into $^2P_{\frac{3}{2}}$ and $^2P_{\frac{1}{2}}$ with a separation $3\zeta/2$ of 17 cm^{-1}. This splitting is mirrored in the splitting of the optical transition into a doublet with this separation, which amounts to 0.1 per cent in this case.

Many-electron case. Here we take our perturbing term to be

$$H' = \sum_i \xi(r_i)\mathbf{l}_i \cdot \mathbf{s}_i$$

The presence of more than one electron complicates the calculation of matrix elements. This occurs because $\mathbf{l}_i$ and $\mathbf{s}_i$ are now only ordinary vector operators with respect to rotations of space and spin coordinates, whereas in the one-electron case they were the total angular momentum corresponding to these respective rotations. Thus, whereas $\mathbf{l}$ and $\mathbf{s}$ were diagonal in l and s for one electron, the Wigner-Eckart theorem (Sec. 5-12) tells us that $\mathbf{l}_i$ and $\mathbf{s}_i$ will have matrix elements of the type $\Delta L = \pm 1, \Delta S = \pm 1$ as well as $\Delta L = \Delta S = 0$, in the general case. However, as long as ζ is small compared with term separations, the simple L-S coupling scheme is still a good approximation. Thus we shall ignore the effects of admixtures of other L and S values by the spin-orbit coupling, and consider only the matrix elements within a given term, i.e., those elements which are diagonal in L and S.

For these, we can take advantage of the fact, noted in Sec. 5-12, that the matrix elements of a vector operator are proportional to those of the corresponding angular momentum. Noting that $\xi(r)$ is a scalar, we may include it with no change in transformation properties. Thus, using n to denote all quantum numbers specifying the configuration, we have

$$(nLM'_LSM'_S |\xi(r_i)\mathbf{l}_i \cdot \mathbf{s}_i| nLM_LSM_S)$$

$$= (nLM'_L |\xi(r_i)\mathbf{l}_i| nLM_L) \cdot (nSM'_S| \mathbf{s}_i| nSM_S)$$

$$= (nL \|\xi(r_i)l_i\| nL)(nS \|s_i\| nS)(LM'_L |\mathbf{L}| LM_L) \cdot (SM'_S |\mathbf{S}| SM_S)$$

If we now sum over all the electrons, we have

$$(nLM'_LSM'_S |H'| nLM_LSM_S) = \zeta(nLS)(LM'_LSM'_S |\mathbf{L} \cdot \mathbf{S}| LM_LSM_S) \quad (6\text{-}53)$$

where
$$\zeta(nLS) = \sum_i (nL \|\xi(r_i)l_i\| nL)(nS \|s_i\| nS)$$

Thus, *within an L-S term*, the matrix elements of $\sum_i \xi(r_i)\mathbf{l}_i \cdot \mathbf{s}_i$ are proportional to those of $\mathbf{L} \cdot \mathbf{S}$. This is an operator equation, since we have shown it to be true for a general element (within the term). Since operator equations are invariant under a unitary transformation, this result also holds in the *LSJM* representation based on the proper zero-order eigenfunctions. Thus we can write down our energy eigenvalues by inspection, just as in the one-electron case, obtaining

$$E(nLSJM) = E_0(nLS) + \zeta(nLS)\frac{J(J + 1) - L(L + 1) - S(S + 1)}{2} \quad \blacktriangleright (6\text{-}54)$$

The result is exact for the Hamiltonian used, within the approximation that we neglect the effects of coupling to other terms. These effects can be taken into account by using the Racah coefficients to handle the terms off-diagonal in L and S. In fact Racah was led to develop his methods from consideration of problems of this type.

Landé interval rule. For the time being, $\zeta(nLS)$ is considered merely as a parameter which will be evaluated shortly. The theory can be tested in its present form by considering ratios of separations of levels. The separation of Jth and $(J - 1)$st level is

$$E(nLSJM) - E(nLS, J - 1, M) = \left(\frac{\zeta}{2}\right)[J(J + 1) - (J - 1)J] = \zeta(nLS)J$$

$$(6\text{-}55)$$

Thus the separations of adjacent multiplet levels should be proportional to the higher of the J values concerned. This is the *Landé interval rule*.

As an example showing how well this holds, we cite some data on the $3d^6 4s^2 \, {}^5D$ term of iron in Table 6-2. If the rule were rigorously true, the entries in the last column would all be equal. The deviations represent the effect of the smaller interaction terms, which we have omitted.

One can readily see that the center of gravity of the levels of a multiplet [each level being weighted with its statistical weight $(2J + 1)$] is exactly where it would be in the absence of spin-orbit effects. This follows from the invariance of the diagonal sum and the fact that the diagonal values of H' in an LSM_LM_S representation are ζM_LM_S. Thus

$$H'_{av} = [\zeta/(2L + 1)(2S + 1)] \sum_{-L}^{L} M_L \sum_{-S}^{S} M_S = 0$$

This fact was used to determine suitable average energies for terms in checking the predictions on the L-S-term separations in Sec. 6-7.

Table 6-2

Multiplet intervals in the ^5D terms of iron

Level	Energy, cm^{-1}	Interval	$\zeta \equiv interval/J$
$^5D_4$	0.0	-415.9	-103.9
$^5D_3$	415.9	-288.1	-96.1
$^5D_2$	704.0	-184.1	-92.1
$^5D_1$	888.1	-89.9	-89.9
$^5D_0$	978.1		

Absolute term intervals. The Landé interval rule was obtained effortlessly by group theory. However, to evaluate actual magnitudes of energy splittings, we always must evaluate some parameter or radial integral about which the group theory gives no information. We now seek to evaluate the parameter $\zeta(nLS)$, which gives the scale factor for the fine structure of a given L-S term. We can reduce this parameter to simple integrals by a method similar to the Slater-sum-rule method for calculating term separations.

The method is based on comparing two alternative expressions for the same matrix element. For a diagonal element, our foregoing discussion allows us to write

$$(nLSM_LM_S \mid \sum_i \xi(r_i)\mathbf{l}_i \cdot \mathbf{s}_i \mid nLSM_LM_S) = \zeta(nLS)(LSM_LM_S \mid \mathbf{L} \cdot \mathbf{S} \mid LSM_LM_S)$$

$$= \zeta(nLS)M_LM_S \qquad (6\text{-}56a)$$

But, by our rule for evaluating diagonal elements of operators of type $F = \sum_i f_i$ over determinantal wave functions, we also know that this matrix element must have the value

$$\sum_i (a^i \mid \xi(r)\mathbf{l} \cdot \mathbf{s} \mid a^i) = \sum_i \zeta(n^il^i)m_l{}^im_s{}^i \qquad (6\text{-}56b)$$

where $\zeta(nl)$ are the same integrals as those defined in handling the one-electron problem. This allows a straightforward evaluation of $\zeta(nLS)$ as long as there is only a single determinant having the given M_L, M_S values.

For example, in our np^2 case, the only $M_L = M_S = 1$ determinant is (1^+0^+). This belongs to the ^3P term. Applying the equality above,

$$\zeta(np^2, {}^3P) = \frac{1}{M_L M_S} \sum_i \zeta(n^i l^i) m_l{}^i m_s{}^i = \zeta(np) \cdot 1 \cdot \frac{1}{2} + 0 = \frac{\zeta(np)}{2}$$

If there is more than one determinant from the configuration under consideration having given M_L, M_S, we extend this method, using the invariance of diagonal sums under a unitary transformation, to write

$$M_L M_S \sum_{\text{terms}} \zeta(nLS) = \sum_{\text{det}} \left[\sum_i \zeta(n^i l^i) m_l{}^i m_s{}^i \right] \tag{6-57}$$

The sums over both terms and determinants are to run over the given $M_L M_S$ block of the table (such as Table 6-1) pertaining to the given configuration. By a stepwise procedure like that in Slater's sum-rule method, the ζ values for all terms can be found. Our np^2 example is too simple to provide a useful example, since the only multiple entries in Table 6-1 are in $M_S = 0$ blocks. However, if we consider two *inequivalent* p electrons $(npn'p)$, we can set up an illustration. The determinant (1^+1^+) belongs to ^3D with $M_L = 2$, $M_S = 1$. Using the simple method, we obtain $\zeta(npn'p\ {}^3D) = \frac{1}{4}[\zeta(np) + \zeta(n'p)]$. Now, proceeding to the $M_L = M_S = 1$ block, we find two determinants: (1^+0^+) and (0^+1^+). Suitable combinations of these belong to the terms ^3D and ^3P. Using (6-57), we equate

$$\zeta(npn'p\ {}^3D) + \zeta(npn'p\ {}^3P) = \tfrac{1}{2}\zeta(np) + \tfrac{1}{2}\zeta(n'p)$$

By subtraction, we then find

$$\zeta(npn'p\ {}^3P) = \zeta(npn'p\ {}^3D) = \frac{[\zeta(np) + \zeta(n'p)]}{4} \tag{6-58}$$

As an experimental test, consider the $2p3p$ configuration of atomic carbon. There it is found that $\zeta(^3D) = 11.2$, $10.6\ \text{cm}^{-1}$ and $\zeta(^3P) = 10.2$, $12.5\ \text{cm}^{-1}$, the alternative values being evaluated from different multiplet intervals. We see that in fact $\zeta(^3D) \simeq \zeta(^3P)$, and we also note that the interval rule is quite well satisfied.

Configurations of the form nl^x and Hund's rules. An important special case in atomic structure is that in which all electrons not in closed shells belong to a single subshell and we are interested in the ground term. According to the first two Hund rules, this term will have the largest values of S and L consistent with the exclusion principle. For such states the block with $M_L = L$ and $M_S = S$ always contains but a single determinant, and we may evaluate $\zeta(nLS)$ very easily. We distinguish two cases.

1. Shell less than half full ■ Here all the spins *can* be parallel, and in the ground state they *will* be, by Hund's first rule. For example, for $3d^4$, the ground term is ^5D, and we have $(2^+1^+0^+ -1^+)$ for the determinant with $M_L = L = 2$, $M_S = S = 2$.   In general, we have from (6-56)

$$M_L M_S \zeta(nLS) = \zeta(nl) \sum_i m_l{}^i m_s{}^i$$

In our special case this becomes

$$LS\zeta(nLS) = \zeta(nl) \cdot \frac{1}{2} \cdot \sum_i m_l{}^i = \frac{\zeta(nl)M_L}{2} = \frac{\zeta(nl)L}{2}$$

Therefore $$\zeta(nl^x, LS) = \frac{\zeta(nl)}{2S}   \qquad x < 2l + 1 \qquad (6\text{-}59a)$$

  2. *Shell more than half full*   ■   In this case, some spins must be reversed to satisfy the Pauli principle.   For example, in the $3d^6\,^5D$ ground term,

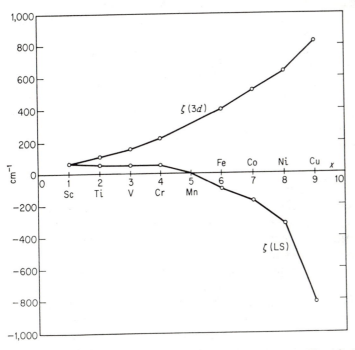

Fig. 6-2. *Spin-orbit parameters in $3d^x$ transition elements.   The splitting parameters $\zeta(LS)$ are computed as indicated in Table 6-2 and averaged over the various splittings. The data used are for the $3d^x4s^2$ configurations of the neutral atoms.   (From Charlotte E. Moore, "Atomic Energy Levels," Natl. Bur. Standards, Circ. 467, vols. I and II, 1949 and 1952.   A very similar figure appears in Condon and Shortley.)*

$(2^+1^+0^+ -1^+ -2^+2^-)$ is the determinant for the $M_L = M_S = 2$ state. In general, we have for this type of case,

$$M_L M_S \zeta(nLS) = LS\zeta(nLS) = \zeta(nl) \left\{ \frac{1}{2} \sum_{-l}^{l} m_l{}^i - \frac{1}{2}M_L \right\} = \frac{-\zeta(nl)L}{2}$$

Thus
$$\zeta(nl^x, LS) = \frac{-\zeta(nl)}{2S} \qquad x > 2l+1 \qquad (6\text{-}59b)$$

Comparing (6-59a) and (6-59b), we see that the sign of the effective spin-orbit parameter $\zeta(nl^x, LS)$ changes from a positive to a negative value if the shell is more than half full. This causes the state of lowest energy to change from minimum J to maximum J in this case, as required by the third Hund rule. In case $x = 2l + 1$, so that the shell is exactly half full, $S = (2l + 1)/2$ but $L = 0$. Thus $J = S = (2l + 1)/2$ only, and there is no multiplet structure. Hence no rule is required for this case.

Example: $3d^x$. In Fig. 6-2, we plot the values of $\zeta(3d)$ and $\zeta(LS)$ versus x for the atoms of the iron transition series. Note that $\zeta(LS)$ is initially nearly constant, then changes sign and increases rapidly. On the other hand, $\zeta(3d)$ stays positive and increases steadily in magnitude with Z and hence $(\partial U/\partial r)$, as expected from (6-48).

This situation is of particular importance in understanding electronic-spin resonance experiments on iron-group salts. In this case, the g factor is shifted from the free-electron value (2.0023) by second-order effects of the spin-orbit coupling proportional to $-\zeta(LS)$. Thus for the first half of the shell (for example, Cr, V) $g < 2$, but it is close to 2, since ζ is small. For the second half of the shell (for example, Fe, Ni, Cu) g exceeds 2.0023 by an amount which increases rapidly with $\zeta(LS)$.

6-10   Zeeman Effect

With the inclusion of the relativistic, or spin-orbit, coupling terms in the Hamiltonian in the previous section, we have reduced the atomic degeneracy to the full extent consistent with the rotational invariance of the problem. This invariance requires all $M(\equiv M_J)$ states belonging to a given fine-structure level (J) to be degenerate, because there are rotational-symmetry operations that take states with one M value into states with another. This residual degeneracy is completely lifted by applying a magnetic field **H**, which destroys the isotropy of space by introducing a preferred direction. The remaining symmetry is only the group of two-dimensional rotations about **H**, which, being Abelian, has only one-dimensional representations. Hence all degeneracy is lifted, and, as discussed in Sec. 3-7, states are characterized by their M value, such that rotation by an angle φ about **H** multiplies the function by $e^{-iM\varphi}$. We now calculate the magnitude of the splitting of the energy levels produced by the applied field.

The perturbing Hamiltonian for a magnetic field **H** in the z direction is

$$H_Z = -\mathbf{\mu} \cdot \mathbf{H} = \beta(L_z + g_s S_z)\mathsf{H} \tag{6-60}$$

where $\mathbf{\mu}$ is the magnetic moment of the atom, β is the Bohr magneton, L and S are the orbital and spin angular momenta in units of $\hbar$, and g_s is the anomalous g value of the electron spin. According to the usual derivation from the Dirac equation,[1] g_s is exactly 2, as had been found empirically from spectra. However, in 1948, Schwinger[2] showed that, to first order in $\alpha = e^2/\hbar c \approx \frac{1}{137}$, g_s should have the value $2(1 + \alpha/2\pi) = 2.0023$. This small correction arises from quantum electrodynamic effects which enter from treating the electromagnetic field in vacuum as a quantum-mechanical system with which the electron interacts. This calculation was carried to higher order by Karplus and Kroll,[3] who obtained $g_s = 2.002291$. This calculation was later corrected by Sommerfield,[4] who found $g_s = 2.002319$. Precise atomic-beam experiments confirm this value quite well. Actually, many other relativistic corrections enter when the electron is moving in an atom, but these require more specific individual treatment for each atom.[5] Since these corrections are outside the scope of our treatment, we shall simply take g_s as a given quantity, slightly larger than 2. We shall also neglect the terms in the Hamiltonian which involve the square of the magnetic vector potential, since these are ordinarily negligible compared with the linear terms.

Weak-field limit. For weak magnetic fields, the matrix elements of H_Z are so small compared with the separation between multiplet levels with different values of J that we can restrict our attention to matrix elements diagonal in J. Noting that both **L** and **S** transform as vectors, we may use the Wigner-Eckart theorem, and in particular the vector matrix elements of (5-69), to write that

$$(JM' |\mathbf{L}| JM) = (J \|L\| J)(JM' |\mathbf{J}| JM)$$

and

$$(JM' |\mathbf{S}| JM) = (J \|S\| J)(JM' |\mathbf{J}| JM) \tag{6-61}$$

We can evaluate the reduced matrix elements by a simple argument. First, noting that $\mathbf{L} + \mathbf{S} = \mathbf{J}$, we see that $(J \|L\| J) + (J \|S\| J) = 1$. For convenience in writing, let $(J \|S\| J) = \alpha$. Then $(J \|L\| J) = 1 - \alpha$. Now note that $\mathbf{L}^2 = (\mathbf{J} - \mathbf{S})^2 = \mathbf{J}^2 + \mathbf{S}^2 - 2\mathbf{J} \cdot \mathbf{S}$. If we take the $|LSJM)$ diagonal element of this equation, we have

$$L(L + 1) = J(J + 1) + S(S + 1) - 2\alpha J(J + 1)$$

[1] P. A. M. Dirac, "The Principles of Quantum Mechanics," 3d ed., chap. XI, Oxford University Press, New York, 1947.

[2] J. Schwinger, *Phys. Rev.*, **73**, 416 (1948); **76**, 790 (1949).

[3] R. Karplus and N. M. Kroll, *Phys. Rev.*, **77**, 536 (1950).

[4] C. M. Sommerfield, *Phys. Rev.*, **107**, 328 (1957).

[5] See, for example, A. Abragam and J. H. Van Vleck, *Phys. Rev.*, **92**, 1448 (1953).

Solving for α, we obtain

$$\alpha = (J \parallel S \parallel J) = \frac{J(J+1) + S(S+1) - L(L+1)}{2J(J+1)}$$

$$(6\text{-}62)$$

and $\quad 1 - \alpha = (J \parallel L \parallel J) = \frac{J(J+1) - S(S+1) + L(L+1)}{2J(J+1)}$$

By using (6-61) and (6-62), (6-60) becomes

$$H_Z = [(1 - \alpha) + g_s\alpha]\beta J_z H \equiv g_J \beta J_z H \qquad (6\text{-}63)$$

which defines the g_J factor. Since $J_z = M$ is diagonal in our representation, (6-63) directly gives the eigenvalues of the magnetic perturbation.

$$E_M = g_J \beta M H \qquad (6\text{-}64)$$

These are to be added to the field-free energy level. Note that, if g_s is set equal to exactly 2, g_J becomes $(1 + \alpha)$. This is the usual expression due to Landé, namely,

$$g_J = 1 + \frac{J(J+1) + S(S+1) - L(L+1)}{2J(J+1)} \qquad (6\text{-}65)$$

Though usually obtained in elementary courses by nonrigorous means, we see that this is a rigorous result of our group theory, provided that we take $g_s \equiv 2$.

Strong-field limit (Paschen-Back effect). Since **L** and **S** are vector operators, they have $\Delta J = \pm 1$ matrix elements in addition to the $\Delta J = 0$ elements used above. In very strong fields $[\beta H \gg \zeta(LS)]$, these matrix elements are more important energetically than the spin-orbit coupling which couples **L** and **S** to form **J**, giving the fine-structure splittings. In this case, we abandon the $|LSJM\rangle$ basis functions, suitable to field free space, and revert to the $|LM_LSM_S\rangle$ functions. The latter are suitable zero-order functions for a strong field which "decouples" **L** from **S**, leaving both precessing about **H**. The diagonal matrix elements of H_Z can be written down directly as

$$H_Z = \beta H(M_L + g_s M_S) \qquad (6\text{-}66)$$

This is the first-order magnetic energy. However, we must remember that the spin-orbit coupling is still present. From (6-53), it can be written

$$H_{S\text{-}O} = \zeta \mathbf{L} \cdot \mathbf{S} = \zeta[L_z S_z + \tfrac{1}{2}(L_+ S_- + L_- S_+)] \qquad (6\text{-}67)$$

Adding the diagonal terms of (6-67) to (6-66), we have for the first-order energy

$$E(M_L, M_S) = \zeta M_L M_S + \beta H(M_L + g_s M_S) \qquad (6\text{-}68)$$

This will describe the very strong field limit (complete Paschen-Back effect).

Typically fields of 10^6 gauss are required to reach this limit, because $\zeta \sim 100$ cm^{-1}. Thus this limit can be observed only in relatively light atoms, where ζ is small.

Intermediate-field region (incomplete Paschen-Back effect). When $\beta H \sim \zeta$, neither limiting approximation is valid, and we must deal with off-diagonal as well as diagonal matrix elements, regardless of our choice of quantization scheme. Clearly $M = M_L + M_S$ must remain a rigorous quantum number throughout, because of rotational invariance about the field direction. However, if we use the $M_L M_S$ quantization scheme, for example, though H_Z is diagonal, $H_{S\text{-}O}$ has elements of the form $\Delta M_L = \pm 1$, $\Delta M_S = \mp 1$, as well as the diagonal elements used above. From (6-67), we see that these elements are

$$(M_L \pm 1, M_S \mp 1 \,|H_{S\text{-}O}|\, M_L, M_S)$$

$$= \frac{\zeta}{2} [L(L + 1) - M_L(M_L \pm 1)]^{\frac{1}{2}} [S(S + 1) - M_S(M_S \mp 1)]^{\frac{1}{2}} \quad (6\text{-}69)$$

Using them in second-order perturbation theory, we can extend the usefulness of the high-field limit downward.

On the other hand, if we use JM quantization, $H_{S\text{-}O}$ is diagonal and $H_Z = \beta H[J_z + (g_s - 1)S_z]$ has a diagonal part (in J_z) and a part (in S_z) which has $\Delta J = \pm 1$ matrix elements as well as diagonal ones. These may be found by specializing (5-84), which involves a simple application of Racah coefficients (see Exercise 5-11). Using these elements, perturbation theory will give the departures from a linear Zeeman effect which set in when the Zeeman splitting ($\sim \beta H$) begins to become comparable with the multiplet interval ($\sim \zeta$).

When more accurate solutions are required than can be obtained by perturbation theory, a secular equation must be solved exactly for the eigenvalues. Note that, since $M = J_z$ remains a valid quantum number (representation label) throughout, the secular equation will be between only the zero-order states having a given M value. This will at most be the number of levels in the multiplet, namely, $(2S + 1)$ or $(2L + 1)$, whichever is smaller. If we have $S = \frac{1}{2}$, there is never more than a 2×2 to solve and we may solve the problem algebraically. For illustration, consider the states with $M = M_L' + \frac{1}{2}$. The Hamiltonian matrix connecting the zero-order states with $(M_L, M_S) = (M_L', \frac{1}{2})$ and $(M_L' + 1, -\frac{1}{2})$ is

$$\begin{bmatrix} \dfrac{\zeta M_L'}{2} + \beta H\left(M_L' + \dfrac{g_s}{2}\right) & \dfrac{\zeta}{2}[L(L+1) - M_L'(M_L' + 1)]^{\frac{1}{2}} \\[3ex] \dfrac{\zeta}{2}[L(L+1) - M_L'(M_L' + 1)]^{\frac{1}{2}} & \dfrac{-\zeta(M_L' + 1)}{2} + \beta H\left(M_L' + 1 - \dfrac{g_s}{2}\right) \end{bmatrix}$$

$$(6\text{-}70)$$

The solutions of the secular equation are then [after substituting M for $(M'_L + \frac{1}{2})$]

$$E_{\pm} = \beta H M - \frac{\zeta}{4} \pm \frac{\zeta}{2} \left\{ \left(L + \frac{1}{2} \right)^2 + \frac{2\beta H M}{\zeta} (g_s - 1) + \frac{\beta^2 H^2}{\zeta^2} (g_s - 1)^2 \right\}^{\frac{1}{2}}$$

(6-71)

It is easily verified that this expression reduces correctly to the limiting values derived earlier, both when $\beta H \gg \zeta$ and when $\beta H \ll \zeta$. To illustrate

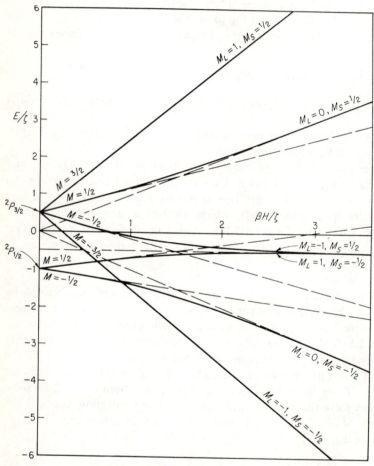

Fig. 6-3. *Transition to Paschen-Back region for* ^2P *term (schematic).*

the transition from the weak to strong-field limit, in Fig. 6-3 we graphically display the result (6-71) for the special case of $L = 1$. In other words, we show the splitting of the levels of a ^2P state in a magnetic field. The dashed lines indicate the extrapolation of the low-field linear Zeeman effect which would occur if the Paschen-Back effect did not set in.   Dashed lines are also used to indicate the extrapolation of the high-field limit down to zero field.

Spectra.   The energy-level structure having been worked out, the observed spectrum is determined provided that one knows the selection rules. These are fixed by the matrix elements of vector operators for the usual dipole transitions and were discussed in Sec. 5-12.   The results are that $\Delta M = 0$ for radiation polarized along **H** and $\Delta M = \pm 1$ for radiation polarized perpendicular to **H**.   The latter two can be separated by using circularly polarized radiation in the $+$ and $-$ senses.

If a transition is induced between the Zeeman-split sublevels of a single multiplet level, it will be a *magnetic*-dipole transition, because the parities of initial and final states are the same.   In this case, it is the polarization of the high-frequency magnetic field that enters in the selection rule (and the frequency will typically be in the microwave range).   However, if the transition is an ordinary optical one between levels of different parity, the selection rules are governed by the polarization of the optical-frequency electric field and the effect of the magnetic field is only to produce a small splitting of the line into closely spaced components.   The latter case is, of course, what is commonly referred to as the Zeeman effect.

6-11   Magnetic Hyperfine Structure

According to our group theory, in the absence of external fields, we expect J to be a rigorous quantum number and the degeneracy of the $(2J + 1)$ states in a level to be absolute.   Yet when atomic spectra are observed, one often finds that the fine-structure lines split even further into *hyperfine structure* (h.f.s.).   Since group theory allows no such splitting from the electronic degrees of freedom, it must result from an additional degree of freedom.   Pauli (1924) suggested that, if the nucleus had a spin, it could provide this.   The extra structure is then associated with the new irreducible representations of the full rotation group based on the direct product of nuclear and electronic eigenfunctions.

The energetic interaction which produces the splitting can be either electric or magnetic, the most important terms being the electric-quadrupole and magnetic-dipole interactions.   As is well known, parity is conserved to a very high order under all the strong nuclear interactions, and therefore for most purposes parity is a good nuclear quantum number.   This prevents the nucleus from having a permanent (i.e., diagonal) electric-dipole moment, because matrix elements of such a polar vector must vanish between states of the same parity.   The same argument excludes all the odd electric multipoles.   (The parity-violating weak interactions produce electric-dipole moments far too small to be detected.)   Since the magnetic-dipole moment is an axial vector, there is no rule against the existence of permanent magnetic-dipole moments.

The parity considerations also allow magnetic-octupole, -32-pole, etc., moments. However, the magnitude of the effect falls rapidly with increasing multipole order so that only relatively recently has even an octupole effect been measured.[1] We shall confine our attention to the dipole interaction, where the Hamiltonian is simply

$$H_m = -\boldsymbol{\mu}_I \cdot \mathbf{H}_{\text{eff}} = -\boldsymbol{\mu}_I \cdot (\mathbf{H}_0 + \mathbf{H}_J) \tag{6-72}$$

where $\mathbf{H}_0$ is any externally applied field and $\mathbf{H}_J$ is the field at the nucleus produced by the electronic motion of the atom indicated by its angular momentum $\mathbf{J}$.

It is convenient to express the nuclear dipole moment $\boldsymbol{\mu}_I$ as

$$\boldsymbol{\mu}_I = \left(\frac{\mu}{I}\right)\mathbf{I}$$

where μ and I are the tabulated moment and spin of the nucleus. The nuclear moment μ can be determined by measuring the energy-level separation in a known external field $\mathbf{H}_0$, provided that we can make a suitable allowance for $\mathbf{H}_J$. Now, since $\mathbf{H}_J$ must transform as a vector under rotations of the electronic wavefunction, its matrix elements can be found by using the Wigner-Eckart theorem. In particular, since the h.f.s. energies are so small compared with the fine-structure energies separating states of different J, we usually need consider only elements of $\mathbf{H}_J$ diagonal in J. But, as we have noted before, such matrix elements are proportional to those of $\mathbf{J}$ itself. Thus we have

$$(JM' |\mathbf{H}_J| JM) = A(JM' |\mathbf{J}| JM)$$

We evaluate A by postmultiplying both sides by the matrix operator $\mathbf{J}$. This leads to

$$A = \frac{(JM |\mathbf{H}_J \cdot \mathbf{J}| JM)}{(JM |\mathbf{J} \cdot \mathbf{J}| JM)} = \frac{(JM |\mathbf{H}_J \cdot \mathbf{J}| JM)}{J(J+1)}$$

Therefore, in the absence of an external field, (6-72) becomes

$$H_m = a\mathbf{I} \cdot \mathbf{J} \qquad \blacktriangleright \tag{6-73a}$$

with

$$a = -\frac{\mu_I(JM |\mathbf{H}_J \cdot \mathbf{J}| JM)}{IJ(J+1)} \tag{6-73b}$$

With this result in hand, we have reduced the magnetic h.f.s. problem to a form identical with the spin-orbit interaction, which we have already

---

[1] V. Jaccarino, J. G. King, R. A. Satten, and H. H Stroke, *Phys. Rev.*, **94**, 1798 (1954).

solved.   If we define a total angular momentum $\mathbf{F} = \mathbf{I} + \mathbf{J}$, we have

$$\mathbf{I} \cdot \mathbf{J} = \tfrac{1}{2}(\mathbf{F}^2 - \mathbf{I}^2 - \mathbf{J}^2)$$

and hence

$$E(nLSJIF) = E_0(nLSJI) + \left(\frac{a}{2}\right)[F(F+1) - J(J+1) - I(I+1)] \quad (6\text{-}74)$$

The Landé interval rule becomes

$$E(F) - E(F-1) = aF \qquad (6\text{-}75)$$

This should hold with excellent accuracy because I and J are *very* near to rigorous quantum numbers.   The breakdown is proportional to $(a/\Delta E)$, where ΔE is the energy separation from a state of different J (or I) which is coupled to the state under consideration by $\Delta J = \pm 1$ (or $\Delta I = \pm 1$) matrix elements of the complete Hamiltonian.   Since $a \sim 0.1$ cm^{-1}, while $\Delta E \sim \zeta(LS) \sim 100$ cm^{-1}, this error should be very small.   Since the ΔE for a change of I is in the nuclear energy (millions of electron volts) range, such effects are completely negligible.   Any deviation from the Landé interval rule (after correcting for the small admixtures of states with $J \pm 1$) must be ascribed to the higher multipoles, principally the electric-quadrupole interaction.

Calculation of a for non-S-states.   In order to use this result to determine μ, we must be able to calculate a from the atomic wavefunction.   The most obvious contribution to $\mathbf{H}_J$, and hence a, is the magnetic field produced by the orbital currents about the nucleus,

$$\mathbf{H}_L = \sum_i \frac{e}{c} \frac{\mathbf{v}_i \times \mathbf{r}_i}{r_i^3} = \frac{e}{mc} \sum_i \frac{\mathbf{p}_i \times \mathbf{r}_i}{r_i^3} = -2\beta \sum_i \frac{\mathbf{l}_i}{r_i^3} \qquad (6\text{-}76)$$

where we measure $\mathbf{l}_i$ in units of $\hbar$ and β is the Bohr magneton.

The dipole-dipole interaction between the nuclear moment and the electronic spins may be expressed as

$$H_{dd} = -\boldsymbol{\mu}_I \cdot \mathbf{H}_S \qquad (6\text{-}77a)$$

where

$$\mathbf{H}_S = 2\beta \sum_i \frac{1}{r_i^3}\left[\mathbf{s}_i - 3\frac{(\mathbf{r}_i \cdot \mathbf{s}_i)\mathbf{r}_i}{r_i^2}\right] \qquad (6\text{-}77b)$$

Since both $\mathbf{r}_i$ and $\mathbf{s}_i$ transform as vectors under rotations of all electronic coordinates, $\mathbf{H}_S$ will do so also and hence will have its matrix elements diagonal in J proportional to those of $\mathbf{J}$, as required.

Combining the orbital and spin parts, we have

$$\mathbf{H}_J = -2\beta \sum_i \frac{1}{r_i^3}\left[\mathbf{l}_i - \mathbf{s}_i + \frac{3(\mathbf{r}_i \cdot \mathbf{s}_i)\mathbf{r}_i}{r_i^2}\right] \qquad \blacktriangleright (6\text{-}78)$$

Specialization for single-electron outside closed shells. When there is but one electron not in a filled subshell, the only noncanceling terms in (6-78) arise from it. We may also drop the distinction between $\mathbf{l}_i$ and $\mathbf{L}$, for example. Thus, (6-78) becomes

$$\mathbf{H}_J = -\frac{2\beta}{r^3}\left[\mathbf{L} - \mathbf{S} + \frac{3(\mathbf{r} \cdot \mathbf{S})\mathbf{r}}{r^2}\right] \tag{6-79}$$

To obtain a from (6-73b), we must evaluate $\mathbf{H}_J \cdot \mathbf{J}$. This is

$$\mathbf{H}_J \cdot \mathbf{J} = \frac{-2\beta}{r^3}\left[\mathbf{L} - \mathbf{S} + \frac{3(\mathbf{r} \cdot \mathbf{S})\mathbf{r}}{r^2}\right] \cdot (\mathbf{L} + \mathbf{S})$$

Multiplying this out, and noting that $\mathbf{r} \cdot \mathbf{L} = \mathbf{r} \cdot (\mathbf{r} \times \mathbf{p}) = 0$, we obtain

$$-\frac{2\beta}{r^3}\left[\mathbf{L}^2 - \mathbf{S}^2 + \frac{3(\mathbf{r} \cdot \mathbf{S})^2}{r^2}\right]$$

If we now expand the last term in cartesian coordinates and use the properties of $S = \frac{1}{2}$ spin matrices that $S_x^2 = S_y^2 = S_z^2 = S(S + 1)/3 = \frac{1}{4}$ and that $S_i S_j = -S_j S_i$, we find that it gives $S(S + 1)$, just canceling the second term. Thus we obtain $\mathbf{H}_J \cdot \mathbf{J} = -2\beta \mathbf{L}^2/r^3$, and, from (6-73b),

$$a = 2\beta \frac{\mu_I}{I}\left\langle\frac{1}{r^3}\right\rangle_{\text{av}} \frac{L(L + 1)}{J(J + 1)} \qquad \blacktriangleright \ (6\text{-}80)$$

In a hydrogenic atom, $\langle 1/r^3 \rangle_{\text{av}}$ can be expressed as

$$\left\langle\frac{1}{r^3}\right\rangle_{\text{av}} = \frac{1}{a_0^3}\frac{Z^3}{n^3 L(L + 1)(L + \frac{1}{2})} \tag{6-81}$$

where $a_0 = \hbar^2/me^2$ is the first Bohr radius. This leads to an $\mathbf{H}_J$ of the order of 10^5 gauss for the lowest states of hydrogen, and the a value is

$$a_H = \frac{2\beta}{a_0^3}\left(\frac{\mu_I}{I}\right)\frac{Z^3}{n^3(L + \frac{1}{2})J(J + 1)} \tag{6-82}$$

For atoms with an inner core of filled shells, (6-82) is not valid though it can be used for estimation if a suitable effective Z is inserted. In principle, one should simply insert $\langle 1/r^3 \rangle_{\text{av}}$, as calculated from a Hartree-Fock radial function, into (6-80). In practice, such functions are not always available, nor are they very accurate. Hence, it is frequently more reliable to infer $\langle 1/r^3 \rangle_{\text{av}}$ from the experimentally measured fine-structure splitting. As noted in (6-51),

$$\zeta = \langle \xi(r) \rangle_{\text{av}} = \frac{\hbar^2}{2m^2c^2}\left\langle\frac{1}{r}\frac{dU}{dr}\right\rangle_{\text{av}}$$

Thus, if we define a $Z_p(r)$ which reproduces the potential by

$$U(r) = \frac{-Z_p(r)e^2}{r}$$

then the field is reproduced by a function $Z_f(r)$ defined by

$$\frac{dU}{dr} = \frac{e^2}{r^2}\left[Z_p(r) - r\frac{dZ_p(r)}{dr}\right] \equiv \frac{Z_f(r)e^2}{r^2}$$

and
$$\zeta = \frac{\hbar^2}{2m^2c^2}\left\langle\frac{Z_f(r)e^2}{r^3}\right\rangle_{av} = 2\beta^2\left\langle\frac{Z_f(r)}{r^3}\right\rangle_{av} \tag{6-83}$$

We can now express a in terms of the fine-structure splitting between $J = L \pm \frac{1}{2}$, namely, $\delta = (L + \frac{1}{2})\zeta$, as follows

$$\frac{a}{\delta} = \frac{\mu_I}{I\beta}\frac{L(L+1)}{(L+\frac{1}{2})J(J+1)}\frac{1}{Z_i} \tag{6-84a}$$

where
$$Z_i = \frac{\langle Z_f(r)/r^3\rangle_{av}}{\langle 1/r^3\rangle_{av}} \tag{6-84b}$$

Since the $1/r^3$ weighting factor emphasizes the interior of the atom, Z_i is an "interior" effective Z value, which is known semiempirically and from approximate calculations to be of the order of $(Z - 4)$ for p electrons.[1] From the form of (6-84) we see clearly that hyperfine-structure splittings are smaller than fine-structure splittings by a factor of the order of $(\mu_I/\beta) \sim (m/M) \sim 10^{-3}$.

Actually, there are sizable relativistic corrections to the result (6-84) which may amount to a factor of 2 or so in the heavy atoms. These were treated by Breit and by Racah in the early days of the subject.[2] Such calculations have been reviewed extensively by Kopfermann.[3]

Calculation of a for S-states. In an S-state, $L = 0$, and $J = S$. In this case, (6-80) becomes indeterminate, since $\langle 1/r^3\rangle$ becomes infinite, whereas $L(L + 1)$ goes to zero. In fact, if we consider only the terms treated above, $\mathbf{H}_J$ is actually zero. $\mathbf{H}_L$ is obviously zero, since $L = 0$, and $\mathbf{H}_S$ [as given by (6-77b)] can be shown to be zero for the spherical distribution of spin density which exists in an S state. However, a closer scrutiny of the problem shows that, in addition to (6-77b), there is a so-called "contact" hyperfine interaction which arises whenever there is a nonzero density of electronic magnetization inside the nucleus itself. This interaction was first worked

---

[1] S. Goudsmit, *Phys. Rev.*, **43**, 636 (1933); E. Fermi and E. Segrè, *Z. Physik*, **82**, 729 (1933).

[2] G. Breit, *Phys. Rev.*, **38**, 463 (1931); G. Racah, *Z. Physik*, **71**, 431 (1931).

[3] H. Kopfermann, "Nuclear Moments," chap. III, Academic Press Inc., New York, 1958.

out by Fermi[1] in 1930, using the Dirac equation to handle the relativistic quantum mechanics of an electron moving very near (or inside) the nucleus. More recently, it has been pointed out[2] that the existence of this contact interaction can be understood completely classically by using ordinary magnetostatics. Because of the greater clarity and simplicity of the classical argument, we shall be content to restrict our attention to it.

The simplest model for the nuclear-dipole moment would be a rotating spherical shell of radius r_0 with uniform surface-charge density. From classical magnetostatics, this will produce a pure dipole field outside the shell and a uniform field inside the shell of magnitude

$$\mathbf{B}_{\text{in}} = \frac{2}{r_0^3} \boldsymbol{\mu} = \frac{8\pi}{3V} \boldsymbol{\mu}$$

where V is the volume of the nuclear sphere. The magnetic interaction energy of this shell with the electronic magnetic-moment density $\mathbf{M}(\mathbf{r})$ can be written as

$$E = -\int \mathbf{B}(\mathbf{r}) \cdot \mathbf{M}(\mathbf{r}) \, d\tau$$

Since $\mathbf{M}(\mathbf{r})$ is spherically symmetric in an S state, the integral of the product of $\mathbf{M}(\mathbf{r})$ with the dipole field outside r_0 vanishes by symmetry. This leaves only the integral inside the nuclear sphere, namely,

$$E = -\frac{8\pi}{3V} \boldsymbol{\mu} \cdot \mathbf{M}(0) \cdot V$$

where $\mathbf{M}(0)$ is the density of electronic magnetization inside the nucleus, which can be assumed constant over such a small volume. Since $\mathbf{M}(0) = -g_s\beta \sum_i |\psi_i(0)|^2 \mathbf{s}_i$, where $|\psi_i(0)|^2$ is the probability density of the ith electron at the nucleus, we see that we can write the contact interaction energy operator as

$$H_c = \frac{8\pi}{3} g_s\beta \sum_i \delta(\mathbf{r}_i)\mathbf{s}_i \cdot \boldsymbol{\mu} \qquad \blacktriangleright \quad (6\text{-}85)$$

or, for the case of a single unpaired electron,

$$H_c = a_c \mathbf{I} \cdot \mathbf{S} \qquad (6\text{-}86a)$$

with

$$a_c = \left(\frac{\mu_I}{I}\right) \frac{8\pi}{3} g_s\beta \, |\psi(0)|^2 \qquad (6\text{-}86b)$$

[1] E. Fermi, *Z. Physik*, **60**, 320 (1930).

[2] See, for example, R. A. Ferrell, *Am. J. Phys.*, **28**, 484 (1960). Although this reference is almost certainly not the first to present this viewpoint, it is one of the more accessible. We should point out that the contact term can also be derived along with the usual dipole interaction (6–77b) if one calculates the field of the spin dipole moment $\mathbf{M}_s$ as $\mathbf{B}_s = \text{curl } \mathbf{A}_s$, with $\mathbf{A}_s = c^{-1} \int \mathbf{j}_s(\mathbf{r}')/|\mathbf{r} - \mathbf{r}'| \, d\tau'$ and $\mathbf{j}_s = c \text{ curl } \mathbf{M}_s$. With proper attention to vector-analysis identities, this procedure leads to both the ordinary dipole interaction and the contact interaction. The central point is that $\nabla^2(1/r) = -4\pi\delta(r)$. Because of the more physical and less mathematical nature of the argument given in the text, we have given it preference here.

Since the size of the nuclear sphere has dropped out, we can immediately generalize this argument to an arbitrary spherically symmetric distribution of nuclear magnetism. Because of the spherical symmetry of the contacting electrons, only the spherical part of the nuclear moment contributes. Hence, the result (6-85) or (6-86) should be generally valid, as is in fact the case.

It is a curious fact that for hydrogen, $|\psi(0)|^2$ has such a value that the a calculated by this formula agrees exactly with the value obtained if L is set equal to zero in our previously derived formula (6-82) for a_H. This apparent accident has been clarified by Satten,[1] who has shown that (6-86) can be obtained directly from (6-80) by a limiting process in which L is formally allowed to approach zero continuously. Since $\langle 1/r^3 \rangle \sim L^{-1}$, this leads to a finite value in (6-80) and detailed calculation shows that the numerical coefficient is exactly that found by Fermi. This consideration makes it reasonable that S-electron h.f.s. coupling is generally stronger than that of other electrons, as is implied by the factor $(L + \tfrac{1}{2})^{-1}$ in (6-82). Naturally this contact h.f.s. term arises only with S electrons because $\psi(0) = 0$ for $L > 0$.

Addition of external magnetic field. Any externally applied field $\mathbf{H}_0$ will interact with the magnetic moment both of the electrons and of the nucleus. Thus, the complete Hamiltonian is

$$H = a\mathbf{I} \cdot \mathbf{J} - \left(\frac{\mu_J}{J}\right)\mathbf{J} \cdot \mathbf{H}_0 - \left(\frac{\mu_I}{I}\right)\mathbf{I} \cdot \mathbf{H}_0 \qquad (6\text{-}87)$$

where (μ_J/J) is usually written $- g_J\beta$, g_J being the Landé g factor of the electronic state as given in (6-65). As with the ordinary Zeeman effect, the solution is particularly simple in either the weak or the strong-field limit. In fact, we may directly transcribe the results of Sec. 6-10, with obvious changes in notation and allowance for the fact that neither (μ_J/J) nor (μ_I/I) has any universal value.

In the weak-field limit $(\mu_J \mathsf{H}_0 \ll a)$, we take diagonal elements in an $|IJFM\rangle$ representation and obtain

$$E(IJFM) = \left(\frac{a}{2}\right)[F(F + 1) - I(I + 1) - J(J + 1)]$$

$$- \frac{M\mathsf{H}_0}{2F(F + 1)}\left\{ \left(\frac{\mu_J}{J}\right)[F(F + 1) - I(I + 1) + J(J + 1)] \right.$$

$$\left. + \left(\frac{\mu_I}{I}\right)[F(F + 1) + I(I + 1) - J(J + 1)] \right\} \quad (6\text{-}88)$$

This formula describes how the $(2F + 1)$-fold degenerate levels split into $(2F + 1)$ equally spaced sublevels, with separation increasing linearly with

[1] R. A. Satten, unpublished.

H_0. Note that, since $\mu_I \sim (m/M)\mu_J$, the term in (μ_I/I) can be neglected except when very high accuracy is required. On the other hand, if the experiment is intended to measure μ_I from this term, an overall accuracy of a part per million is required to provide reasonable accuracy for μ_I.

In the strong-field, or Back-Goudsmit, limit, $\mu_J H \gg a$, and the $\Delta F = \pm 1$ off-diagonal elements of $\mathbf{J} \cdot \mathbf{H}_0$ destroy F as a useful zero-order quantum number. Instead, we choose a decoupled representation which diagonalizes

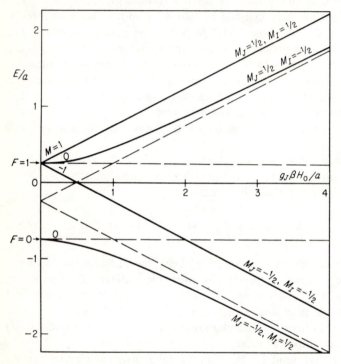

Fig. 6-4. *Application of Breit-Rabi formula for $I = J = \frac{1}{2}$.*

M_J and M_I, and hence the Zeeman energy, but leaves $\mathbf{I} \cdot \mathbf{J}$ partly off diagonal. The diagonal energy in the $|IJM_IM_J\rangle$ representation forms the strong-field limit to the energy expression, namely,

$$E(IJM_IM_J) = aM_JM_I - H_0\left[\left(\frac{\mu_J}{J}\right)M_J + \left(\frac{\mu_I}{I}\right)M_I\right] \qquad (6\text{-}89)$$

For intermediate field strengths, one must solve the secular equation more exactly, the effect of off-diagonal terms being included just as discussed in Sec. 6-10. In the simple case of $J = \frac{1}{2}$ and arbitrary I, the exact solution given there as Eq. (6-71) may be applied. In this connection, it is referred to as the *Breit-Rabi formula*[1] and is usually quoted in the form:

$$E(IJFM) = -\frac{\Delta W}{2(2I + 1)} - \frac{\mu_I}{I}H_0 M \pm \frac{\Delta W}{2}\left[1 + \frac{4Mx}{2I + 1} + x^2\right]^{\frac{1}{2}} \blacktriangleright \quad (6\text{-}90)$$

where $\Delta W = (a/2)(2I + 1)$ is the field-free h.f.s. splitting and

$$x = [-(\mu_J/J) + (\mu_I/I)](H_0/\Delta W)$$

is a dimensionless measure of the strength of the applied field. This formula plays a central role in the interpretation of atomic-beam experiments.

The results are sketched in Fig. 6-4 for the simplest case, namely, $I = J = \frac{1}{2}$, $F = 0, 1$. This applies to the ground state of atomic hydrogen, for example. Note that the $M = \pm 1$ levels diverge linearly at all fields because they are always pure states with $M_J = M_I = \pm\frac{1}{2}$, respectively. The two $M = 0$ levels, however, are composed of $M_J = -M_I = \pm\frac{1}{2}$ in a linear combination which depends on H_0. At $H_0 = 0$, they are mixed equally, giving no resultant magnetic moment, and hence $(\partial E/\partial H_0)_0 = 0$. As H_0 increases, the off-diagonal matrix elements between them change these proportions in such a way as to give the characteristic "repulsion" of two energy levels with the same group-theoretical symmetry.

In concluding this section, we might emphasize that, because typically $a \sim 10^{-3}\zeta$, the values of H_0 giving the Back-Goudsmit decoupling of F into I and J occur at a field of order 10^3 gauss, whereas the Paschen-Back decoupling of J into L and S occurs only for $H_0 \sim 10^6$ gauss. Thus one does not usually need to treat the intermediate case of both decoupling simultaneously.

6-12 Electric Hyperfine Structure

The term in the energy which we seek here is the electric-quadrupole inter-action. This has usually been deduced by elementary but tedious means. Instead of this, we shall take advantage of our development of Wigner and Racah coefficients to give a more elegant treatment of the electrostatic interaction between the nuclear and electronic charge. In our treatment we shall neglect any electronic charge which lies within the nuclear radius. As long as we consider only atoms, only S electrons penetrate the nucleus. Since they have a spherically symmetric distribution, they give no quadrupole coupling and so no error results from their omission. Under this as-sumption, we always have $r_< = r_n$ and $r_> = r_e$. Therefore the expansion of $|\mathbf{r}_n - \mathbf{r}_e|^{-1}$ in terms of the Legendre polynomials following (6-33) is simply

$$\frac{1}{r_{ne}} = \frac{1}{r_e}\sum_L \left(\frac{r_n}{r_e}\right)^L P_L(\cos \omega_{en}) \qquad (6\text{-}91)$$

[1] G. Breit and I. Rabi, *Phys. Rev.*, **38**, 2082 (1931).

where $\mathbf{r}_n$ and $\mathbf{r}_e$ are measured from the center of the nucleus to an element of nuclear and electronic charge, respectively, and ω_{en} is the angle between the directions ω_e and ω_n of $\mathbf{r}_e$ and $\mathbf{r}_n$. Introducing the spherical harmonic-addition formula (6-34) and normalized spherical harmonics $Y_M{}^L$, we have

$$\frac{1}{r_{ne}} = \frac{1}{r_e} \sum_L \left(\frac{r_n}{r_e}\right)^L \left(\frac{4\pi}{2L+1}\right) \sum_{M=-L}^{L} (-1)^M Y_M{}^L(\omega_n) Y_{-M}{}^L(\omega_e) \qquad (6\text{-}92)$$

We note that this interaction automatically breaks up into a sum of scalar operators, each of which is the contraction of two irreducible tensor operators of rank L. We now examine the first few terms.

For $L = 0$, we have simply $1/r_e$, the ordinary Coulomb monopole term. If we integrate this interaction over the nuclear- and electronic-charge densities ρ_n and ρ_e, we find its contribution to the system energy to be

$$\left[\int \rho_n(r_n)\, d\tau_n\right] \cdot \left[\int \frac{\rho_e(r_e)\, d\tau_e}{r_e}\right] = [Ze] \cdot \int \frac{\rho_e(r_e)\, d\tau_e}{r_e} \qquad (6\text{-}93)$$

For $L = 1$, we have three terms which can be rearranged to correspond to the three cartesian components. The energy is the scalar product of the two vectors,

$$\left[\int \rho_n(r_n)\mathbf{r}_n\, d\tau_n\right] \cdot \left[\int \rho_e(r_e)\mathbf{r}_e/r_e{}^3\, d\tau_e\right] = -\boldsymbol{\mu}_e \cdot \mathbf{E} \qquad (6\text{-}94)$$

where $\boldsymbol{\mu}_e$ is the nuclear electric-dipole moment and $\mathbf{E}$ is the electric field at the nucleus produced by the electronic-charge cloud. This term vanishes both because $\boldsymbol{\mu}_e = 0$ by parity conservation and because $\mathbf{E} = 0$, if the nucleus is in a position of equilibrium.

The term of interest is the quadrupole term, $L = 2$. This can be expressed as the product of two cartesian dyadics $\frac{1}{6}\mathbf{Q} : \nabla\mathbf{E}$, but it is much simpler to leave it in its spherical form. The matrix elements which we desire can then be written down immediately by use of our equation (5-83), namely,

$$(IJFM|\, H_Q\, |IJFM) = \frac{4\pi}{5}\, (IJFM\, |\sum_M (-1)^M r_n{}^2 Y_M{}^2(\omega_n) r_e{}^{-3} Y_{-M}{}^2(\omega_e)|\, IJFM)$$

$$= \frac{4\pi}{5}\, (-1)^{I+J-F}[(2I+1)(2J+1)]^{\frac{1}{2}}(I\| r_n{}^2 Y^2(\omega_n) \|I)$$

$$\times (J\| r_e{}^{-3} Y^2(\omega_e) \|J)W(IJIJ; F2) \qquad (6\text{-}95)$$

The entire dependence on F is in the Racah coefficient. Upon using a symmetry property of the W coefficient, $W(IJIJ; F2) = W(IIJJ; 2F)$. This second form is of a type given in the tables of Biedenharn, Blatt, and Rose.[1] It is

$$W(IIJJ; 2F) = 6(-1)^{F-I-J} \left[\frac{(2J-2)!\,(2I-2)!}{(2J+3)!\,(2I+3)!}\right]^{\frac{1}{2}}$$

$$\times [C(C+1) - \tfrac{4}{3}I(I+1)J(J+1)] \qquad (6\text{-}96)$$

[1] L. C. Biedenharn, J. M. Blatt, and M. E. Rose, *Revs. Modern Phys.*, **24**, 256 (1952).

where $C = F(F+1) - I(I+1) - J(J+1)$. Clearly C is the diagonal matrix element of $2\mathbf{I} \cdot \mathbf{J}$. Thus we have shown that the energy eigenvalues (for given I and J) are proportional to $[3(\mathbf{I} \cdot \mathbf{J})^2 + \frac{3}{2}(\mathbf{I} \cdot \mathbf{J}) - \mathbf{I}^2\mathbf{J}^2]$, the result usually obtained by the elementary methods.

Although the answer obtained above is complete in so far as the dependence on F is concerned, it is worthwhile giving some attention to clarifying the significance of the reduced matrix elements so that we have a detailed connection with the conventional results. That is, we wish to express $(I\| r_n^2 Y^2(\omega_n) \|I)$ in terms of the nuclear-quadrupole moment Q and $(J\| r_e^{-3} Y^2(\omega_e) \|J)$ in terms of the field-gradient coupling constant q_J. From its definition in the Wigner-Eckart theorem, we have

$$(I\| r^2 Y^2(\omega_n) \|I) = \frac{(I, M_I + \mu| r^2 Y_\mu^2(\omega_n) |IM_I)}{A_{M_I\mu}^{I2I}} \qquad (6\text{-}97)$$

This ratio can be evaluated for any values of μ and M_I, but it is simplest to choose $\mu = 0$ and $M_I = I$. Then (from the tables in Condon and Shortley)

$$A_{I0}^{I2I} = \frac{3M_I^2 - I(I+1)}{[(2I-1)I(I+1)(2I+3)]^{\frac{1}{2}}} = \left[\frac{I(2I-1)}{(I+1)(2I+3)}\right]^{\frac{1}{2}} \qquad (6\text{-}98)$$

Now, the nuclear-quadrupole moment Q is conventionally *defined* by

$$eQ = \int [\rho_n(r_n)]_{M_I=I} (3z_n^2 - r_n^2) \, d\tau_n = \left(\frac{16\pi}{5}\right)^{\frac{1}{2}} \int \rho(r_n) r_n^2 Y_0^2(\omega_n) \, d\tau_n \qquad (6\text{-}99)$$

Q has the dimensions of area, and it is a measure of the elongation along z of the nuclear charge when the nuclear spin is aligned as nearly along the z axis as possible. The indicated integration is equivalent to taking the $|II)$ diagonal matrix element. Thus, combining our results, we have

$$eQ = (II| \left(\frac{16\pi}{5}\right)^{\frac{1}{2}} r_n^2 Y_0^2(\omega_n) |II) = \left[\frac{16\pi}{5} \frac{I(2I-1)}{(I+1)(2I+3)}\right]^{\frac{1}{2}} (I\| r_n^2 Y^2(\omega_n) \|I) \qquad (6\text{-}100)$$

We note that $Q = 0$ for $I = 0$ or $I = \frac{1}{2}$. Thus a nucleus can have no quadrupole moment as defined here unless its spin exceeds $\frac{1}{2}$.

In a similar manner, the field-gradient parameter in the electronic state J is defined by

$$eq_J = \left(\frac{\partial^2 V}{\partial z^2}\right)_{r=0, M_J=J} = \int [\rho_e(r_e)]_{M_J=J} \frac{3z_e^2 - r_e^2}{r_e^5} \, d\tau_e$$

$$= \left[\frac{16\pi}{5} \frac{J(2J-1)}{(J+1)(2J+3)}\right]^{\frac{1}{2}} (J\| r_e^{-3} Y^2(\omega_e) \|J) \qquad (6\text{-}101)$$

From this, we note that q_J vanishes unless $J > \frac{1}{2}$. Thus quadrupole effects can be observed only if both I and J exceed $\frac{1}{2}$.

Finally, we use these relations to eliminate the reduced matrix elements in our energy expression (6-95), using q_J and Q as parameters instead. After canceling common factors, we obtain

$$(IJFM| \, H_Q \, |IJFM) = \frac{e^2 q_J Q}{2I(2I-1)J(2J-1)} \left\{ \tfrac{3}{4}C(C+1) - I(I+1)J(J+1) \right\}$$

▶ (6-102)

with $C = F(F+1) - I(I+1) - J(J+1)$. This is the form in which the result is usually quoted, though often with C written as $2\mathbf{I} \cdot \mathbf{J}$.

For simple molecular problems, one can work out the explicit dependence of q_J on the rotational state and include it in the formula. We shall not go into any derivations[1] at this point but shall simply quote the result for the case in which the nucleus under consideration lies on the axis of symmetry of a symmetrical top molecule, i.e., one with two equal moments of inertia. In this case

$$q_J = \frac{3K^2 - J(J+1)}{(J+1)(2J+3)} q_0 \tag{6-103}$$

where q_0 is computed with respect to the molecular symmetry axis (rather than with respect to an axis fixed in space) and K is the quantized component of $\mathbf{J}$ along that axis. Note that $q_J \to q_0$ if $K = J \to \infty$, as must be true classically in this limit. Also note that, in a linear molecule, $\mathbf{J}$ must be perpendicular to the axis, giving $K = 0$. Hence in that case, $q_J = -Jq_0/(2J+3)$.

Returning to the atomic case, since I and J must exceed $\tfrac{1}{2}$ for quadrupole effects, there is always a magnetic-dipole interaction as well. Thus the total h.f.s. Hamiltonian will be

$$H_{hfs} = H_m + H_Q$$

Because of overall rotational invariance, however, F must be a good quantum number even in this case. In fact, the energy levels are readily written down by simply summing the contributions of H_m and H_Q to the diagonal matrix element of H_{hfs} in an $|IJFM)$ representation. Thus

$$E(IJFM) = \frac{a}{2}C + \frac{e^2 q_J Q}{2I(2I-1)J(2J-1)} \left\{ \tfrac{3}{4}C(C+1) - I(I+1)J(J+1) \right\}$$

(6-104)

where again $C = 2\mathbf{I} \cdot \mathbf{J} = F(F+1) - I(I+1) - J(J+1)$.

If we now add a weak magnetic field, we can proceed exactly as before, since our zero-order eigenfunctions are unchanged. Thus we simply add to (6-104) the term in H_0 given in (6-88) to describe the linear Zeeman-effect

---

[1] See, for example, C. H. Townes and A. L. Schawlow, "Microwave Spectroscopy," chap. 6, McGraw-Hill Book Company, Inc., New York, 1955; N. F. Ramsey, "Nuclear Moments," p. 19, John Wiley & Sons, Inc., New York, 1953.

region. In the other extreme, the strong-field region, we take $|IJM_IM_J)$ quantization and add to (6-89) the diagonal elements of H_Q in this representation. These are readily obtained from (6-95) by using the Wigner-Eckart theorem on the electronic and nuclear operators separately. Thus

$$(IJM_IM_J| \, H_Q \, |IJM_IM_J) = \frac{4\pi}{5} \, (I\| \, r_n^2 \, Y^2(\omega_n) \, \|I)(J\| \, r_e^{-3} \, Y^2(\omega_e) \, \|J) A_{M_I,0}^{I2I} A_{M_J,0}^{J2J}$$

$$(6\text{-}105)$$

Using (6-100) and (6-101) to express the reduced matrix elements in terms of eQ and eq_J, and using the fact that

$$A_{M_I,0}^{I2I} = \frac{3M_I^2 - I(I+1)}{[(2I-1)I(I+1)(2I+3)]^{\frac{1}{2}}}$$

(and similarly for $A_{M_J,0}^{J2J}$), we find that

$$(IJM_IM_J| \, H_Q \, |IJM_IM_J) = \frac{e^2Qq_J}{4} \, \frac{[3M_I^2 - I(I+1)]}{I(2I-1)} \, \frac{[3M_J^2 - J(J+1)]}{J(2J-1)}$$

$$(6\text{-}106)$$

This is the term to be added to (6-89) to get the contribution of H_Q to the energy in the strong-field limit.

In the intermediate-field region, we need off-diagonal matrix elements as well. These are obtained by the same procedure used in (6-105), except that in the $(M_I + \mu, M_J - \mu| \, H_Q \, |M_IM_J)$ matrix element, we have a factor of $A_{M_I,\mu}^{I2I} A_{M_J,-\mu}^{J2J}$. The formula in (6-105) is the specialization of this result for $\mu = 0$. Given all the matrix elements, perturbation theory or more exact procedures can be used to compute the energy levels.

Further refinements. Our treatment so far has indicated the main features in the calculation of hyperfine structure. However, to get very high accuracy a number of corrections must be considered. One arises because of the existence of $\Delta J = \pm 1$ matrix elements in H_m and of $\Delta J = \pm 1, \pm 2$ matrix elements in H_Q, in addition to the $\Delta J = 0$ elements which we have used above. These off-diagonal elements will be particularly important if the fine-structure interval is relatively small, rather than immense in comparison with h.f.s. energies. They can be calculated from generalizations of the procedures given above, but of course new parameters, such as $q_{JJ'}$, are introduced, which must be estimated theoretically if correction for these effects is to be made.

Another sort of correction, which fortunately does not change the form of the results but does interfere with a simple interpretation of the numerical values, is the so-called "Sternheimer[1] shielding" or "antishielding

[1] R. M. Sternheimer, *Phys. Rev.*, **80**, 102 (1950); **84**, 244 (1951); **86**, 316 (1952); **95**, 736 (1954).

effect." Sternheimer has shown that the exterior field of a nuclear-quadru-pole moment produces a significant polarization of the core electrons of an atom. This can have the effect of either increasing or decreasing the actual magnitude of e^2qQ from that obtained by considering only unperturbed valence electrons as producing a field acting on the bare nucleus. These corrections can be sizable and are hard to evaluate reliably. Thus abso-lute values of Q are hard to obtain with confidence even though the un-perturbed electronic wavefunction may be reasonably well known.

After making the small corrections for off-diagonal elements mentioned above, the field-free h.f.s. spectrum should be rigorously fitted by adjustment of two parameters: a for the magnetic dipole and e^2qQ for the electric quadrupole. If this is not possible, one is led to attribute the discrepancy to the effect of a nuclear-octupole moment, the next nonvanishing nuclear moment. Such moments have now been measured for several nuclei, but the interaction cannot possibly be observed unless both I and J are $\frac{3}{2}$ or greater. This can be seen in an elementary way by noting that, if either I or J is less than $\frac{3}{2}$, there will be at most two energy-level differences to fit, and this can certainly be done with two parameters. The same conclusion could, of course, be reached by inspecting the Wigner coefficients. Although we shall not carry the problem further here, extensive discussions of calculations, including those with the higher nuclear moments, are available in the litera-ture.[1]

EXERCISES

6-1 The wavefunction of a helium atom is reasonably approximated as a product of two hydrogenic 1s functions $Ne^{-\alpha r}$, where α is a parameter to be deter-mined by the variational principle and N is a normalizing constant. Find N, and then compute the potential experienced by one electron in the field of the nucleus and the other electron. Show that it is given by

$$U(r) = -\frac{e^2}{r}\{1 + (1 + \alpha r)e^{-2\alpha r}\}$$

6-2 (a) Using the potential found in Exercise 6-1, calculate the total energy of an electron with wavefunction $N'e^{-\alpha'r}$. (b) Set up an expression to find the value of α' giving the lowest energy for given α. Then, set $\alpha = \alpha'$ as required by sym-metry, solve for α in units of $(1/a_0)$, and find the corresponding binding energy of a He 1s electron. Compare with the experimental ionization potential of 1.81 rydbergs. (c) Note that you get a different, and incorrect, answer, if you set $\alpha = \alpha'$ *before* applying the variational principle to the one-electron energy. Explain this discrepancy. (d) Calculate the total binding energy of the two-electron system with respect to the energy of separated particles. Note that this is *not*

[1] See, for example, C. Schwartz, *Phys. Rev.*, **105**, 173 (1957).

twice the value found in (b). Since the binding energy of an atom with $Z = 2$ with one electron in a $1s$ orbital is rigorously 4 rydbergs, evaluate the binding energy of the second electron by subtracting the 4 rydbergs from the two-electron binding energy. Compare with the result of (b). The near equality of these results illustrates Koopmans' theorem[1], namely, that the ionization potential differs only in second order from the Hartree one-electron energy parameter. In fact, helium provides a very severe test of the theorem, since the removal of one electron makes its largest change when it is removed from a two-electron system.

6-3 (a) Consider a lithium atom ($Z = 3$) with a $1s^2$ core and one valence electron. As suggested by Exercises 6-1 and 6-2, assume that the core charge can be roughly approximated by that of hydrogenic $1s$ functions, but with an effective $(Z - s) = 2$, so that $\rho/e = (2a^3/\pi)e^{-2ar}$, with $a = 2/a_0$. Use Poisson's equation of electrostatics to calculate the potential as a function of r, and show that it is given by

$$U = \frac{-e^2}{r} - (Z - 1)\left(\frac{e^2}{r}\right)(1 + ar)e^{-2ar}$$

(b) From the potential found in (a), compute and plot $Z_p(r)$ and $Z_f(r)$, the effective nuclear charges for potential and field, as defined in conjunction with Eq. (6-83).

6-4 Approximate the excited-state $2p$ orbital for the neutral lithium atom of Exercise 6-3 by the expression

$$\psi_{2p}(r, \theta, \varphi) = Are^{-br}Y_l{}^m(\theta, \varphi)$$

where A is a normalizing constant and b is an adjustable parameter. Taking the potential found in Exercise 6-3a, find the best value of b by the variational method. (You will have to use successive approximations, but convergence is rapid.) For this best value, what is the binding energy of the $2p$ electron? (The experimental value is 0.260 rydberg.)

6-5 Using the potential and ψ_{2p} found in Exercises 6-3 and 6-4, calculate the spin-orbit parameter $\zeta(2p)$, and thence the splitting between the $^2P_{\frac{1}{2}}$ and $^2P_{\frac{3}{2}}$ states. (The experimental splitting is 0.34 cm^{-1}.)

6-6 Compute the value of $\langle 1/r^3 \rangle$ for ψ_{2p} as given in Exercise 6-4. From this value and the results of Exercise 6-5, compute Z_i as defined in (6-84b). [The usual procedure is actually the reverse—namely, one takes ζ from experiment, estimates Z_i, and then infers $\langle 1/r^3 \rangle$ for the interpretation of h.f.s. data.]

In Exercises 6-7 to 6-10 consider the nd^2 configuration.

6-7 By setting up an M_L, M_S table of determinants as in Table 6-1, work out the terms allowed by the exclusion principle. . Note that the table need be set up only for $M_L \geq 0$, $M_S \geq 0$, since the rest follows from symmetry.

6-8 (a) Express the energies of these allowed terms in terms of diagonal matrix elements between determinantal functions. (Use Slater-sum-rule method.)
(b) Reduce these in terms of the standard radial integrals $F_0(nd^2)$, $F_2(nd^2)$, $F_4(nd^2)$.
(c) Find the condition which these F's must satisfy if the Hund rules for determining the ground term are to be satisfied.

[1] *Physica,* **1,** 104 (1933).

6-9 (a) Work out the eigenfunctions of the $M_L = 3$, $M_S = 0$ states belonging to the ^1G and ^3F terms as linear combinations of the simple determinants. (b) Using the tables of Wigner coefficients, find the $M_J = 3$ eigenfunctions of both the $J = 3$ and $J = 4$ levels of the ^3F term, again as linear combinations of simple determinants.

6-10 Find the spin-orbit coupling parameter $\zeta(LS)$ for all nd^2 L-S terms in terms of the parameter

$$\zeta(nd) = (nd|\,\xi(r)\,|nd) = \int_0^\infty R_{nl}{}^2(r)\xi(r)\,dr$$

where $\psi(r, \theta, \varphi) = r^{-1}R_{nl}(r)Y_l{}^m(\theta, \varphi)$ and $\int_0^\infty R_{nl}{}^2(r)\,dr = 1$

6-11 Assume that one has three orthogonal orbital wavefunctions a, b, and c, each occupied by one electron. This system will have a spin degeneracy of $2^3 = 8$, because we can associate either α or β spin functions with each spatial orbital. (a) Enumerate the possible determinantal states. (b) For the usual spin-free approximate Hamiltonian, eigenvalues of S_z and $\mathbf{S}^2$ are good representation labels. What eigenvalues of these operators occur in this exercise, and how frequently does each appear? (c) Find orthonormal combinations of the above determinants which diagonalize these spin operators. Note that your results are not unique for any symmetries appearing more than once. In that case a secular equation based on the Hamiltonian is required to obtain the proper eigenfunctions. How large is the maximum secular problem remaining after applying the spin symmetry considerations?

6-12 If an atom is in a $2p^2$ $^3P_0$ state, to which of the following states is an electric-dipole transition allowed? Explain in each case. (a) $2p3d\,^3D_2$, (b) $2s2p\,^3P_1$, (c) $2s3s\,^3S_1$, (d) $2s2p\,^1P_1$.

6-13 (a) Use second-order perturbation theory to compute the corrections to the low-field Zeeman-effect result (6-88). The off-diagonal matrix elements of J_z and I_z in an $|IJFM)$ representation which you require can be obtained by using (5-84) as in Exercise 5-11. (b) Show that, if you specialize your result to $J = \frac{1}{2}$, it is in agreement with the exact Breit-Rabi result to terms in x^2.

6-14 Derive the contact hyperfine interaction (6-85) in addition to the usual hyperfine coupling of dipolar form by the method indicated in the footnote preceding that equation.

REFERENCES

CONDON, E. U., and G. H. SHORTLEY: "The Theory of Atomic Spectra," chaps. 5–7, 16, Cambridge University Press, New York, 1951.

EYRING, H., J. WALTER, and G. E. KIMBALL: "Quantum Chemistry," chap. 9, John Wiley & Sons, Inc., New York, 1944.

HARTREE, D. R.: "The Calculation of Atomic Structures," John Wiley & Sons, Inc., New York, 1957.

KOPFERMANN, H.: "Nuclear Moments," chaps. 1, 3, Academic Press Inc., New York, 1958.

MOTT, N. F., and I. N. SNEDDON: "Wave Mechanics and Its Applications," chaps. 5 and 6, Oxford University Press, New York, 1948.

PAULING, L., and E. B. WILSON, JR.: "Introduction to Quantum Mechanics," chaps. 8 and 9, McGraw-Hill Book Company, Inc., New York, 1935.

RAMSEY, N. F.: "Nuclear Moments," chap. 2, John Wiley & Sons, Inc., New York, 1953.

SLATER, J. C.: "Quantum Theory of Atomic Structure," vols. I and II, McGraw-Hill Book Company, Inc., New York, 1960.

7 MOLECULAR QUANTUM MECHANICS

7-1 Born-Oppenheimer Approximation

At the foundation of all quantum-mechanical treatments of molecular problems is the fact that the great disparity of electronic and nuclear masses allows the problems of nuclear and electronic motion largely to be separated. This was shown in 1927 by Born and Oppenheimer in their classic paper.[1] In this section we review the results of this treatment. We take as our total Hamiltonian

$$H = - \sum_\alpha \frac{\hbar^2}{2M_\alpha} \nabla_\alpha^2 - \sum_i \frac{\hbar^2}{2m} \nabla_i^2 + V_{nn} + V_{ne} + V_{ee} \qquad (7\text{-}1)$$

where α runs over all the nuclei, i runs over the electrons, and the V's are potential energies depending on positions of nuclei (n) and electrons (e).

[1] M. Born and R. Oppenheimer, *Ann. Physik*, **87**, 457 (1927).

Now, if the nuclei were "clamped" in position as if infinitely massive, the Hamiltonian would be simply

$$H_e = -\frac{\hbar^2}{2m} \sum_i \mathbf{\nabla}_i^2 + V_{ee} + V_{ne} + V_{nn} \qquad (7\text{-}2)$$

and there would be electronic eigenfunctions $\psi_e(\mathbf{r}_i; \mathbf{R}_\alpha)$ which satisfy $H_e\psi_e = E_e\psi_e$. These electronic eigenfunctions and eigenvalues depend parametrically on $\mathbf{R}_\alpha$ because of the term V_{ne}. As long as we hold the nuclei clamped in place, we can consider the term V_{nn} as a simple additive constant in the Hamiltonian. This will have no effect on the ψ_e but will add directly to the eigenvalue E_e. We may now write the total Hamiltonian as

$$H = H_e + H_n = H_e - \sum_\alpha \frac{\hbar^2}{2M_\alpha} \mathbf{\nabla}_\alpha^2 \qquad (7\text{-}3)$$

What Born and Oppenheimer did was to show that, because (m/M_α) is so small, we can treat H_n as a small perturbation. The suitable expansion parameter turns out to be $\kappa = (m/M)^{\frac{1}{4}}$, where M is an average nuclear mass. Clearly, κ is typically of order 0.1. Born and Oppenheimer carried out a systematic expansion of the eigenfunctions and eigenvalues in powers of κ. If one considers only small displacements from an equilibrium configuration of nuclei, the terms linear in κ vanish in defining the equilibrium configuration, the quadratic terms give the vibrational energy, the cubic terms vanish, and the quartic terms give the rotational energy, and some corrections to the vibrational energy. It also turns out that one must carry the analysis through fourth order to determine properly the complete zero-order eigenfunction. This is natural, since the eigenfunction must include the rotational as well as vibrational and electronic parts. We shall not go through this lengthy, but rather straightforward, analysis. Rather, we shall merely confirm that the results are reasonable.

In the zero-order Born-Oppenheimer approximation to the eigenfunctions (which is all that one normally uses), one writes an approximate total eigenfunction of H as a simple product,

$$\psi = \psi_e\psi_n \qquad (7\text{-}4)$$

where $\psi_e(\mathbf{r}_i; \mathbf{R}_\alpha)$ is the eigenfunction of H_e and $\psi_n(\mathbf{R}_\alpha)$ depends only on $\mathbf{R}_\alpha$. We now test this approximation by trying to satisfy the equation

$$H\psi = (H_e + H_n)\psi_e\psi_n = E\psi_e\psi_n \qquad (7\text{-}5)$$

To carry out the indicated operations, we note that

$$\mathbf{\nabla}_\alpha^2\psi_e\psi_n = \mathbf{\nabla}_\alpha \cdot \mathbf{\nabla}_\alpha\psi_e\psi_n = \psi_e\mathbf{\nabla}_\alpha^2\psi_n + 2\mathbf{\nabla}_\alpha\psi_e \cdot \mathbf{\nabla}_\alpha\psi_n + \psi_n\mathbf{\nabla}_\alpha^2\psi_e$$

and

$$\mathbf{\nabla}_i^2\psi_e\psi_n = \psi_n\mathbf{\nabla}_i^2\psi_e$$

since ψ_n is independent of $\mathbf{r}_i$ but ψ_e depends on $\mathbf{R}_\alpha$ parametrically. Thus, we have

$$
\begin{aligned}
H\psi &= (H_e + H_n)\psi_e\psi_n \\
&= -\left\{ \sum_\alpha \frac{\hbar^2}{2M_\alpha} \boldsymbol{\nabla}_\alpha \psi_e \cdot \boldsymbol{\nabla}_\alpha \psi_n + \sum_\alpha \frac{\hbar^2}{2M_\alpha} \psi_n \boldsymbol{\nabla}_\alpha^2 \psi_e \right\} - \psi_e \sum_\alpha \frac{\hbar^2}{2M_\alpha} \boldsymbol{\nabla}_\alpha^2 \psi_n \\
&\quad - \psi_n \sum_i \frac{\hbar^2}{2m} \boldsymbol{\nabla}_i^2 \psi_e + (V_{nn} + V_{ne} + V_{ee})\psi_e\psi_n \\
&= -\{\quad\} + \psi_n H_e \psi_e + \psi_e H_n \psi_n \\
&= -\{\quad\} + E_e(\mathbf{R}_\alpha)\psi_e\psi_n + \psi_e H_n \psi_n
\end{aligned}
\tag{7-6}
$$

Now, if we neglect the term in braces, we can set $H\psi = E\psi$, factor out ψ_e, since it is no longer operated on, and obtain

$$
[H_n + E_e(\mathbf{R}_\alpha)]\psi_n = E\psi_n
\tag{7-7}
$$

This equation serves as the Schrödinger equation for ψ_n. It says simply that the nuclei move in an effective potential which is the electronic energy for the relevant state [including $V_{nn}(\mathbf{R}_\alpha)$], as a function of the internuclear distances. This separation of the problem is physically very reasonable. The light electrons orbit around so rapidly compared with the motion of the heavy nuclei that we can approximately solve for their motion by considering the nuclei to be fixed. We then include the energy of this electronic motion in the effective potential for the quasi-static nuclear motion.

To test the validity of the approximation, we now examine the size of the term in braces which we had to drop to obtain our simple results. This term might be viewed as a perturbation H' which will admix other product functions $\psi_e'\psi_n'$ in forming the true eigenfunction. So long as the matrix elements of H' are small compared with the zero-order energy differences, the approximation will be good. We can check this by comparing the terms in braces in (7-6) with the nuclear kinetic energy term, the smallest term not in braces. The ratios of these terms are

$$
\frac{\boldsymbol{\nabla}_\alpha \psi_e \cdot \boldsymbol{\nabla}_\alpha \psi_n}{\psi_e \boldsymbol{\nabla}_\alpha^2 \psi_n} \quad \text{and} \quad \frac{\psi_n \boldsymbol{\nabla}_\alpha^2 \psi_e}{\psi_e \boldsymbol{\nabla}_\alpha^2 \psi_n}
$$

These ratios are small to the extent that ψ_e depends less rapidly on $\mathbf{R}_\alpha$ than does ψ_n. Now, ψ_n goes rapidly to zero if $\mathbf{R}_\alpha$ is displaced by a distance of the order of the amplitude of the zero-point motion, whereas ψ_e is so extended that it changes significantly only when $\mathbf{R}_\alpha$ is displaced by a distance of the order of internuclear distances. The ratio of zero-point amplitude to electronic extension is readily shown to be of order $\kappa = (m/M)^{\frac{1}{4}}$, by an argument based on the observation that in the harmonic-oscillator approximation the restoring force for a given displacement is of the same order for electrons and nuclei but that the masses differ in the ratio m/M. Thus we see that the neglected terms would enter only in higher-order terms in the expansion parameter κ. This justifies our neglect of them as our usual starting approximation. Serious errors arise only in symmetrical molecules, where some of the zero-order eigenfunctions are actually degenerate. Then these neglected terms are able to exert decisive influence in splitting the degeneracy and determining the symmetries of the actual eigenfunctions. However, because this complication introduces only a small secular equation, we shall defer discussion of such situations until specific examples occur in our systematic treatment. This systematic approach commences with a discussion of the methods of solving the "clamped-nuclei" electronic problem. The solutions of this will provide not only the electronic energy levels but also the effective potential for vibrational motion.

7-2 Simple Electronic Eigenfunctions

The simplest molecules are those with only two centers. Since these already introduce most of the effects which make the molecular problem differ from the atomic one, while remaining relatively simple to think about, we shall concentrate quite a bit of attention on them.

As soon as there is more than one center, we loose the spherical symmetry which played such a vital role in our atomic calculations. This destroys the usefulness of $\mathbf{L}^2$ and (except in linear molecules) L_z for factoring secular equations. Also, it makes the Hartree-Fock equations, which apply just as well in principle, hopeless to solve rigorously, because our unknown eigenfunctions can no longer be expressed in terms of a single variable, $|r|$, as in the atomic case.

Our best hope in general is to set up a series of orthonormal orbitals having the symmetry of whatever molecule we are dealing with, calculate matrix elements of H_e between them, and solve the secular equation. This will give a set of *molecular orbitals* (MO's), which are then used somewhat as the atomic orbitals were used to build up atomic configurations. For this procedure to be practical, the starting functions (*symmetry orbitals*) must be well chosen to speed convergence. This precludes using atomic functions about one center but suggests the use of *linear combinations* of *atomic orbitals* (LCAO's) centered on the various nuclei. Such LCAO MO's form the basis for a large share of current calculations on molecular structure. Another possible method for diatomic molecules is the use of MO's which really have the two centers "built in." The exact eigenfunctions for H_2^+, which can be found by separation of variables in elliptic coordinates, are the prototypes for such functions, just as the H atomic eigenfunctions are the foundation for approximations suitable for atomic problems.

Hydrogen-molecule ion, H_2^+.   Since an exact solution[1] is available for the hydrogen-molecule ion, it is valuable as a practice problem for trying out approximation methods which can be carried over to the less simple molecules.

The simplest approximation to the ground state which we could set up would be an MOLCAO

$$\psi = c_a a + c_b b \tag{7-8}$$

where a and b are ordinary hydrogen $1s$ wavefunctions centered on protons a and b, the two centers of the problem.   Symmetry requires that the two

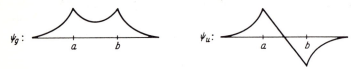

Fig. 7-1.   *Even and odd molecular orbitals for the hydrogen-molecule ion, H_2^+.*

centers are equivalent.   Therefore $|c_a| = |c_b|$, and $c_b = \pm c_a$.   We must be a little careful of normalization, because the functions a and b are not orthogonal.   The normalization requirement is

$$1 = \int \psi^* \psi \, d\tau = \int (c_a^* a^* \pm c_b^* b^*)(c_a a \pm c_a b) \, d\tau$$

$$= 2 |c_a|^2 \pm 2 |c_a|^2 \int a^* b \, d\tau \tag{7-9}$$

since a and b are normalized and real.   The integral

$$S_{ab} = \int a^* b \, d\tau \tag{7-10}$$

is called the *overlap integral* between the orbitals a and b.   It is a measure of the degree of nonorthogonality.   We can solve the normalization equation for $|c_a|$ and obtain the following two approximate molecular orbitals, taking c_a to be real for simplicity:

$$\psi_g = \frac{1}{\sqrt{2}\sqrt{1 + S_{ab}}} (a + b) \qquad \psi_u = \frac{1}{\sqrt{2}\sqrt{1 - S_{ab}}} (a - b) \tag{7-11}$$

Of these, ψ_g is obviously even on inversion through the midpoint of a and b, whereas ψ_u is odd.   The qualitative forms of these wavefunctions are indicated in Fig. 7-1.   Clearly these two wavefunctions will have different

[1] E. Teller, *Z. Physik*, **61**, 458 (1930); D. Bates, K. Ledsham, and A. L. Stewart, *Phil. Trans. Roy. Soc. (London)*, **246**, 215 (1953).

energies, ψ_g lying below ψ_u. The odd function, ψ_u, would appear to have more energy for two reasons: (1) There is less charge in the central region, which enjoys the attractive potential of both nuclei and correspondingly more on the ends, where it feels only one nucleus. (2) The odd function has an extra node, and hence more kinetic energy associated with slope of the wavefunction. Although argument 2 has a certain heuristic appeal and is often quoted, it cannot really be valid, because it is inconsistent with the virial theorem.[1] For the equilibrium nuclear configuration and for the exact eigenfunction, this theorem requires that the kinetic energy be exactly half the magnitude of the (negative) potential energy. Accordingly, the kinetic energy must cancel half the binding energy, and for maximum binding there will be a *maximum* kinetic energy. Thus, in fact, argument 1 must be relied on to account for the difference in energy of the two functions.

If the binding energy of ψ_g is computed as a function of internuclear distance $R_{ab} = R$, one finds a curve of the sort shown in Fig. 7-2. The

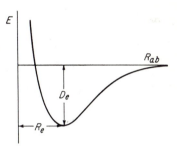

Fig. 7-2. *Dependence of molecular binding on internuclear distance.*

maximum binding (lowest energy) occurs at $R_e = 1.32$ A, where the binding energy D_e is 1.76 ev. Since the correct values, as found by experiment or from the exact solution, are $R_e = 1.06$ A and $D_e = 2.79$ ev, this leaves much room for improvement. One very simple but effective improvement[2] is to introduce an effective nuclear charge $Z'e$ into the hydrogenic functions. This allows adjustment of the radial scale of the atomic orbitals for each value of internuclear distance R. In this way one obtains the proper value $R_e = 1.06$ A and improves D_e to 2.25 ev, cutting the error in half. At R_e, $Z' = 1.228$, which lies between the values corresponding to counting one and both hydrogen nuclei, as might be expected. If one further modifies the atomic orbitals by polarizing them along the axis (by adding some p_σ character to the s functions), the binding energy can be improved to 2.71 ev. Finally, a simple analytic variation function, due to James,[3] can be

[1] For a discussion of the implications of the virial theorem in quantum mechanics, see P. O. Löwdin, *J. Mol. Spectroscopy*, **3**, 46 (1959).

[2] B. N. Finkelstein and G. E. Horowitz, *Z. Physik*, **48**, 118 (1928).

[3] H. M. James, *J. Chem. Phys.*, **3**, 7 (1935).

set up in elliptical coordinates, namely,

$$\psi = e^{-\delta\mu}(1 + c\nu^2)$$

where the elliptic coordinates are defined by $\mu = (r_a + r_b)/R_{ab}$ and $\nu = (r_a - r_b)/R_{ab}$. This function gives 2.772 ev, which is really quite near the true value. This list of results should provide an idea of the accuracy which results from various types of approximations. Further details may be found in various texts[1] as well as in the original literature.

7-3   Irreducible Representations for Linear Molecules

Homonuclear diatomic molecules $D_{\infty h}$. The notation D_{∞} means that rotations C_φ through all possible angles φ about the symmetry (z) axis are symmetry operations and that twofold rotations C_2' can be made about any axis in the bisecting plane perpendicular to the z axis. The subscript h implies the existence of the horizontal (xy) reflection plane with associated reflection σ_h. Combining this reflection with a 180° rotation about z gives the inversion i. Combining i with a C_2' gives a reflection in a vertical plane, σ_v, which is usually preferred for symmetry classification. The irreducible representations and states are labeled in a way which indicates the transformation properties under C_φ, i, and (for Σ states) σ_v. The character table is given here, in slightly different form from that given in the Appendix:

$D_{\infty h}$	E	$2C_\varphi$	σ_v	i	$2iC_\varphi$	$i\sigma_v = C_2'$
Σ_g^+	1	1	1	1	1	1
Σ_u^+	1	1	1	−1	−1	−1
Σ_g^-	1	1	−1	1	1	−1
Σ_u^-	1	1	−1	−1	−1	1
Π_g	2	$2\cos\varphi$	0	2	$2\cos\varphi$	0
Π_u	2	$2\cos\varphi$	0	−2	$-2\cos\varphi$	0
Δ_g	2	$2\cos2\varphi$	0	2	$2\cos2\varphi$	0
Δ_u	2	$2\cos2\varphi$	0	−2	$-2\cos2\varphi$	0
...	...	...	...	...	...	...

The primary quantum number in the representation label is $|L_z| = |\Lambda|$. This is indicated by a Greek capital letter in analogy with the atomic scheme ($S, P, D, \ldots$). Thus we have $\Sigma, \Pi, \Delta, \ldots$ for $|\Lambda| = 0, 1, 2, \ldots$. Small letters ($\sigma, \pi, \delta, \ldots$) are used if one is classifying a single electron state, again

---

[1] See, for example, H. Eyring, J. Walter, and G. E. Kimball, "Quantum Chemistry," chap. XI, John Wiley & Sons, Inc., New York, 1944; and L. Pauling and E. B. Wilson, Jr., "Introduction to Quantum Mechanics," chap. XII, McGraw-Hill Book Company, Inc., New York, 1935.

in analogy with the atomic case. Except for Σ states, which must be non-degenerate, all these states are doubly degenerate since the two senses of rotation have the same energy. The basis functions transform like $e^{i\Lambda\varphi} = e^{\pm i|\Lambda|\varphi}$ under rotations. The character under C_φ for all representations follows immediately from this knowledge. The inversion character merely depends on the label, g or u, and the number of functions in the basis set. The $\pm$ label on the Σ states tells the character under σ_v, that is, whether or not the wavefunction changes sign when reflected in a plane through the internuclear axis. We note that the characters of all the two-dimensional representations under σ_v are zero. This follows from the fact that

$$e^{\pm i|\Lambda|\varphi} \sim (x \pm iy)^{|\Lambda|}$$

A σ_v reflection takes y into $-y$, thus interchanging the basis functions, giving a traceless matrix representation. The characters under iC_φ and $i\sigma_v = C_2'$ follow from the characters under C_φ and σ_v since i has a diagonal matrix representation.

These group-theoretical quantum numbers do not completely specify a state. We must give other information, equivalent to the configuration description in the atomic case, which specifies the nature of all occupied states, including principal quantum numbers as well as symmetry. The two conventional ways of specifying a one-electron state consist in giving the appropriate atomic quantum numbers in the limits (1) separated atoms and (2) united atom. For example, the H_2^+ ground state, ψ_g, would be labeled (1) $\sigma_g 1s$ or (2) $1s\sigma_g$. On the other hand, the ψ_u excited state would be labeled (1) $\sigma_u 1s$ or (2) $2p\sigma_u$. Note that the group-theoretical labels based on absolute symmetry considerations always are the same in either limiting description. In the united-atom scheme, ψ_u is called $2p\sigma_u$ because, as the two centers approach close together as if to form a united atom, the molecular orbital has the symmetry of an atomic p_z orbital about the midpoint.

Since it is relatively easy to estimate the order and position of the energy levels in the two limiting cases $R_{ab} \to 0$ and ∞, we can gain a qualitative picture of the energy-level structure at intermediate values of R_{ab} by setting up a *correlation diagram*. This correlation is based on the "no-crossing" rule on energy levels belonging to the same irreducible representation. A few of the lower levels are indicated in Fig. 7-3.

Examples. H_2 ■ Here we have only two electrons; so both can occupy the lowest σ_g orbital with spins opposed. Then $\Lambda = \sum_i \lambda_i = 0$, and we must have a Σ state. Since all orbitals are even under inversion and reflection, the state is $(\sigma_g 1s)^2$, $^1\Sigma_g^+$.

Li_2 ■ The six electrons fill the three lowest levels on the diagram (Fig. 7-3). By simple arguments, this gives a state $(\sigma_g 1s)^2(\sigma_u 1s)^2(\sigma_g 2s)^2$, $^1\Sigma_g^+$.

N_2 ■ (14 electrons). Assuming that the order of the levels at each end is reliable, we still can make an unambiguous assignment using the correlation diagram alone. Experiment confirms that in fact the ground state is $(\sigma_g 1s)^2(\sigma_u 1s)^2(\sigma_g 2s)^2(\sigma_u 2s)^2(\sigma_g 2p)^2(\pi_u 2p)^4$. Again, the completely filled "shells" lead to a $^1\Sigma_g^+$ state. Note that the $\pi_u 2p$ level holds four electrons, rather than two, because of the existence of $\Lambda = \pm 1$ degeneracy in addition to the spin degeneracy.

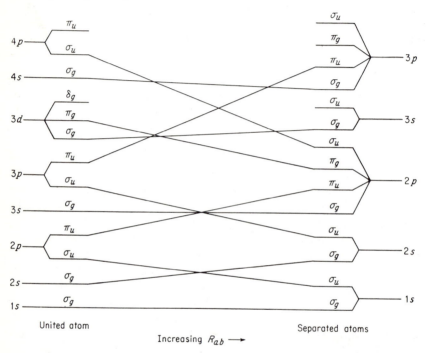

Fig. 7-3. *Correlation diagram for homonuclear diatomic molecules.*

O_2 ■ (16 electrons). If we add two more electrons to the N_2 filled shells, they go in as $(\pi_g 2p)^2$. This leaves an unfilled shell, since there is room for four $\pi_g 2p$ electrons. Now Hund's rule has a chance to work, and indeed the ground state has the two spins parallel and thus orbits necessarily opposed. This leads to $\Lambda = 0$, but $S = 1$. It is a gerade state because we are adding π_g electrons. But it is — under reflection because reflection interchanges the two orbitals, which changes the sign of the antisymmetric electronic wavefunctions. Thus the ground state of O_2 is a $^3\Sigma_g^-$ state. From the same configuration we can also form $^1\Sigma_g^+$ and $^1\Delta_g$ excited states. These three states enter into the most prominent atmospheric absorption bands of O_2.

Heteronuclear diatomic molecules $C_{\infty v}$.   In this case, we no longer have a center of symmetry, and hence we loose the parity quantum number.   The symmetry group contains only C_φ and σ_v.   The character table is

$C_{\infty v}$	E	$2C_\varphi$	σ_v
Σ^+	1	1	1
Σ^-	1	1	-1
Π	2	$2\cos\varphi$	0
Δ	2	$2\cos 2\varphi$	0
...	...	...	...

as can be seen by employing the arguments used for $D_{\infty h}$.   Since the two atoms are now inequivalent, the correlation diagram is heavily modified at the separated-atom end, which now has twice as many distinct levels, each having half the degeneracy it had before.[1]   The molecular orbitals in this case completely lose their inversion symmetry, and in general the low-lying orbitals have more charge attracted to the center with higher Z.

7-4   The Hydrogen Molecule

Since this is the simplest multielectron molecule, we shall look at it in some detail to illustrate and compare various methods.   Regardless of the method, the Hamiltonian operator is the same and can be written

$$H = -\tfrac{1}{2}\nabla_1^2 - \tfrac{1}{2}\nabla_2^2 - \frac{1}{R_{a1}} - \frac{1}{R_{a2}} - \frac{1}{R_{b1}} - \frac{1}{R_{b2}} + \frac{1}{r_{12}} + \frac{1}{R_{ab}} \quad (7\text{-}12)$$

In this, we have shifted into *atomic units*, in which length is measured in units of the Bohr radius a_0, energy in units of 2 rydbergs or e^2/a_0, charge in

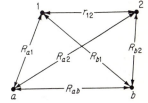

Fig. 7-4.   *Geometry of H_2 molecule.*

units of e, mass in units of m, and angular momentum in units of $\hbar$.   As indicated in Fig. 7-4, R_{a1} is the distance from nucleus a to electron 1, etc. An exact solution for eigenvalues and eigenfunctions of this Hamiltonian is no longer possible because of the r_{12}^{-1} interaction term.   Therefore we shall examine several approximation schemes.

[1] See Eyring, Walter, and Kimball, *op. cit.*, p. 210.

7-5 Molecular Orbitals

In this approximation to H_2, we describe the $^1\Sigma_g{}^+$ ground state of the system by putting both electrons in the lowest or $(\sigma_g 1s)$ MOLCAO. The wavefunction is then given by

$$\psi = N[a(1) + b(1)][a(2) + b(2)] \frac{[\alpha(1)\beta(2) - \beta(1)\alpha(2)]}{\sqrt{2}} \qquad (7\text{-}13)$$

N is a normalizing factor, and, as in Sec. 7-2, $a(1)$ is a hydrogenic $1s$ function about nucleus a, having the explicit form $\pi^{-\frac{1}{2}}e^{-R_{a1}}$ in atomic units. The last factor is the antisymmetric spin function corresponding to a singlet state. This provides the overall antisymmetry required by the Pauli principle, as will be discussed later in more detail. If the energy is calculated with this ψ, we find the maximum binding energy $D_e = 2.65$ ev occurring at $R_{ab} = R_e = 1.6a_0$. The experimental values are $D_e = 4.72$ ev and $R_e = 1.40a_0$. In seeking the origin of this error of about 2 ev in the binding energy, we should recall that in $H_2{}^+$ the simple MO ground state fell about 1 ev short of the correct value. This error was eliminated by adjusting the radial scale and introducing polarization terms which threw charge into the desirable region between the two nuclei, where it was bound by the attraction of both. Coulson[1] carried out a similar calculation for H_2 in which he determined a five-term approximation to the Hartree-Fock solution for the "best" σ_g molecular orbitals. With these MO's he increased the calculated binding energy to about 3.6 ev, or halfway to the correct value. The remaining error must lie largely in the fact that there is no interelectronic correlation in the MO wavefunction. To illustrate this, let us return to the simple LCAO MO's of Eq. (7-13) and expand the spatial part of that wavefunction to see what it is saying about the electrons. We find

$$\psi \sim a(1)a(2) + b(1)b(2) + a(1)b(2) + b(1)a(2)$$

From this, we note immediately that *ionic states*, in which both electrons are on one center [such as $a(1)\,a(2)$], are given equal weight with the *covalent* ones [such as $a(1)b(2)$], in which there is one electron on each center. Since it takes more energy to remove an electron from one center than we gain by adding it as a *second* electron to the other center, these ionic states are expected to be energetically less favorable. Therefore for best results they should appear in ψ with less weight, if at all.

We can make this improvement in several ways. The simplest is simply to drop the offending ionic terms. This leads to the *Heitler-London method*,[2] which gives complete interatomic correlation. That is, if electron 1 is on nucleus a, then 2 is surely on b. This obviously will give the correct answers

[1] C. A. Coulson, *Proc. Cambridge Phil. Soc.*, **34**, 204 (1938).
[2] W. Heitler and F. London, *Z. Physik*, **44**, 455 (1927).

as $R_{ab} \rightarrow \infty$, whereas our simple MO picture would become hopelessly bad in that limit. At the point of maximum binding, the Heitler-London method yields $R_e = 1.64a_0$ and $D_e = 3.14$ ev, which is in better agreement with experiment by about 0.5 ev than the simple MO method. By introducing a radial scale factor and polarization terms into the functions a and b, as described in connection with H_2^+, the computed binding energy can be increased to 4.02 ev, again about 0.5 ev better than the "best" MO value.

We can do a slightly better job if the ionic states are not just thrown out but rather are reduced to the optimum degree by the variational method. It is a useful illustration to view this as a simple example of *configuration interaction* within the MO picture. The only configurations which can mix in are those with the same overall group-theoretical symmetry, namely, $^1\Sigma_g^+$. The only other configuration of MO's made of $1s$ AO's having this symmetry is $(\sigma_v 1s)^2$. In our same approximation, the wavefunction of this singlet state is

$$\psi' = N'[a(1) - b(1)][a(2) - b(2)] \frac{[\alpha(1)\beta(2) - \beta(1)\alpha(2)]}{\sqrt{2}} \qquad (7\text{-}14)$$

Looking at the spatial part only, we find

$$\psi' \sim a(1)a(2) + b(1)b(2) - a(1)b(2) - b(1)a(2)$$

We note that the ionic terms and covalent terms enter ψ' with opposite sign, whereas in the $(\sigma_g 1s)^2$ configuration they entered with the same sign. Therefore, if we superpose the configurations to form a new function Φ given by

$$\Phi = c\psi + c'\psi' \qquad (7\text{-}15)$$

and vary c and c' for minimum energy, we are in effect finding the optimum admixture of ionic states. When carried out, this variational problem is identical with solving the 2×2 secular equation between ψ and ψ'. It turns out that at R_e the optimum admixture of ionic states is only a few per cent. Thus the Heitler-London approximation is close to correct in this case. The relation among the various approximations has been considered particularly carefully by Slater[1] and by Coulson and Fischer.[2]

We might summarize our conclusions by stating that, of the 2-ev error in the simple MOLCAO ground state of H_2, about 1 ev is eliminated by going to MO's which give a more favorable one-electron potential energy by shifting charge into the internuclear region and another $\frac{1}{2}$ ev can be eliminated by introducing a simple type of correlation into the wavefunction. Since the Heitler-London method starts with this correlation in, the starting

[1] J. C. Slater, *Phys. Rev.*, **35**, 509 (1930), **41**, 255 (1932); *Tech. Rept.* 3, Solid State and Molecular Theory Group, MIT, 1953; "Quantum Theory of Molecules and Solids," vol. I, McGraw-Hill Book Company, Inc., New York, 1963.

[2] C. A. Coulson and I. Fischer, *Phil. Mag.*, **40**, 386 (1949).

error is only 1.5 ev in the simple H-L method. This error can be reduced by the same 1 ev as with the MO's by choice of better atomic functions which shift charge into the favorable region.[1]

So far we have followed the traditional practice for simple problems of writing ψ as a product of a space function times a spin function. This technique becomes much less convenient when we have to deal with more than two electrons. Therefore, let us work out the connection between these forms and the determinantal technique developed for the atomic problem, which can handle any number of electrons.

A simple example is our MO ground state. There we have the determinant

$$\psi = \frac{1}{\sqrt{2}} \begin{vmatrix} \sigma_g(1)\alpha(1) & \sigma_g(2)\alpha(2) \\ \sigma_g(1)\beta(1) & \sigma_g(2)\beta(2) \end{vmatrix} \tag{7-16}$$

since both spin states of the $\sigma_g 1s$ orbital are (singly) occupied. The properties of determinants allow us to factor out factors common to all elements of a column. This gives

$$\psi = \sigma_g(1)\sigma_g(2) \frac{1}{\sqrt{2}} \begin{vmatrix} \alpha(1) & \alpha(2) \\ \beta(1) & \beta(2) \end{vmatrix}$$

or, upon inserting our MOLCAO approximation for σ_g, we have

$$\psi = \frac{[a(1) + b(1)]}{[2(1 + S)]^{\frac{1}{2}}} \cdot \frac{[a(2) + b(2)]}{[2(1 + S)]^{\frac{1}{2}}} \frac{1}{\sqrt{2}} \begin{vmatrix} \alpha(1) & \alpha(2) \\ \beta(1) & \beta(2) \end{vmatrix} \tag{7-17}$$

Expansion of the determinant yields the spin function

$$[\alpha(1)\beta(2) - \beta(1)\alpha(2)]/\sqrt{2}$$

which we stated above in (7-13).

A more illuminating example is found by considering the $\sigma_g\sigma_u$ configuration. Since the electrons are not equivalent here, the spins can be parallel, forming a triplet state. Let us consider the $M_s = 1$ state, in which both electrons must be in the spin state α. We may factor out the common spin functions and normalizing factors from each column, as follows,

$$\psi(^3\Sigma_u{}^+, M_s = 1) = \frac{1}{\sqrt{2}} \begin{vmatrix} \sigma_g(1)\alpha(1) & \sigma_g(2)\alpha(2) \\ \sigma_u(1)\alpha(1) & \sigma_u(2)\alpha(2) \end{vmatrix}$$

$$= \frac{\alpha(1)\alpha(2)}{2^{\frac{3}{2}}(1 - S^2)^{\frac{1}{2}}} \begin{vmatrix} a(1) + b(1) & a(2) + b(2) \\ a(1) - b(1) & a(2) - b(2) \end{vmatrix}$$

---

[1] The author is indebted to Dr. V. Heine for his illuminating remarks on this subject.

where again we have used the simple LCAO approximation to both MO's. Expanding the determinant, and collecting terms, we find

$$\psi(^3\Sigma_u^+, M_s = 1) = \frac{b(1)a(2) - a(1)b(2)}{[2(1 - S^2)]^{\frac{1}{2}}} \alpha(1)\alpha(2) \qquad (7\text{-}18)$$

From this we can obtain the states with $M_s = 0, -1$ by operating with S_-. They are unchanged except that the spin functions are

$$[\alpha(1)\beta(2) + \beta(1)\alpha(2)]/\sqrt{2}$$

and $\beta(1)\beta(2)$, respectively. We note that in all three cases these spin functions are symmetric under interchange of electrons 1 and 2, whereas the space function is antisymmetric. The resulting overall antisymmetry is an automatic consequence of using a determinantal form. We also note that the space function includes no ionic states. Therefore the MO and H-L methods are completely equivalent for this $^3\Sigma_u$ state.

If we had considered the $M_S = 0$ state of this $\sigma_g\sigma_u$ configuration, we could have written down the two determinants $(\sigma_g^+\sigma_u^-)$ and $(\sigma_g^-\sigma_u^+)$. We could then have diagonalized S^2 between these two, finding that the linear combinations $[(\sigma_g^+\sigma_u^-) \pm (\sigma_g^-\sigma_u^+)]/\sqrt{2}$ corresponded to the singlet and triplet states. The triplet state would have agreed with that found in the previous paragraph. The other combination is

$$\psi(^1\Sigma_u^+) = \frac{a(1)a(2) - b(1)b(2)}{[2(1 - S^2)]^{\frac{1}{2}}} \frac{\alpha(1)\beta(2) - \beta(1)\alpha(2)}{\sqrt{2}} \qquad (7\text{-}19)$$

We note that this $^1\Sigma_u$ state is composed purely of ionic states. Hence, it will have a higher energy than $^3\Sigma_u$ states formed from the same configuration. This illustrates the fact that the Hund rule, that the state of maximum spin lies lowest, applies to molecules as well as to atoms—provided that it is possible to form a state with $S \neq 0$ from the prescribed configuration of occupied orbitals. The basic mechanism of the energy lowering is the same as in the atomic case: the antisymmetry correlates the two electrons with parallel spin so as to keep them away from each other by keeping them on different atoms. The $^3\Sigma_u$ state will still lie *above* the $^1\Sigma_g$ ground state, however, because of the excitation energy required to raise one electron from the $\sigma_g 1s$ orbital to the $\sigma_u 1s$ orbital. This does not violate the Hund rule, of course, because the rule holds only for comparisons of states formed from the same configuration of occupied one-electron orbitals.

7-6 Heitler-London Method

Although we have described the H-L method above as being a special simplified improvement of the molecular orbital method, it was in fact introduced on its own merits by Heitler and London and it continues to serve as the basis for many calculations. Therefore we now give it some detailed critical examination, paying particular attention to points which might cause confusion in its interpretation.

The H-L approximations to the $^1\Sigma_g$ and $^3\Sigma_u$ states of H_2 are

$$\psi(^1\Sigma_g) = N[a(1)b(2) + b(1)a(2)] \tag{7-20a}$$

$$\psi(^3\Sigma_u) = N'[a(1)b(2) - b(1)a(2)] \tag{7-20b}$$

These spatial functions are written down directly from a consideration of the equivalence of the two electrons and from the assumption that correlation effects will prevent both electrons from being on the same center at the same time. The spin functions do not enter naturally, as with the determinantal method, but are handled separately. This causes some difficulty in the extension of the method of many-electron systems.

The main conceptual difficulties arise from the fact that, unlike our σ_g and σ_u MO's, the basic functions here are not orthogonal. For example,

$$\iint [a(1)b(2)][b(1)a(2)] \, d\tau_1 \, d\tau_2 = \left[\int a(1)b(1) \, d\tau_1 \right]^2 = S_{ab}^2 \equiv S^2 \tag{7-21}$$

Although some writers tend to throw away these overlap integrals when they get in the way, they are *not* small. Rather, they are often of order 0.3, and, as we shall see, they are essential for the success of the method in its usual form.

To make this point more precise, we consider in detail the energy of the $^1\Sigma$ and $^3\Sigma$ states of H_2 in the H-L approximation. This is

$$E = \iint \psi^* H\psi \, d\tau_1 \, d\tau_2$$

$$= N^2 \iint [a(1)b(2) \pm b(1)a(2)]H[a(1)b(2) \pm b(1)a(2)] \, d\tau_1 \, d\tau_2 \tag{7-22}$$

where H is the same Hamiltonian written down above as (7-12) and $N^2 = \frac{1}{2}(1 \pm S^2)^{-1}$. The terms resulting from the expansion of the integrand are usually written as

$$E = 2E_{1s}(\text{H}) + \frac{J}{1 \pm S^2} \pm \frac{K}{1 \pm S^2} \tag{7-23}$$

where $\quad J = \iint a(1)b(2)\left[-\frac{1}{R_{b1}} - \frac{1}{R_{a2}} + \frac{1}{r_{12}} + \frac{1}{R_{ab}} \right]a(1)b(2) \, d\tau_1 \, d\tau_2 \tag{7-24}$

and $\quad K = \iint a(1)b(2)\left[-\frac{1}{R_{b1}} - \frac{1}{R_{a2}} + \frac{1}{r_{12}} + \frac{1}{R_{ab}} \right]b(1)a(2) \, d\tau_1 \, d\tau_2 \tag{7-25}$

In analogy to the atomic case, J is called the *direct* integral and K the *exchange* integral.

Expanding J and using symmetry, we obtain

$$J = -2 \int \frac{a^2(1)\, d\tau_1}{R_{b1}} + \frac{1}{R_{ab}} + \iint \frac{a^2(1)b^2(2)}{r_{12}} \, d\tau_1 \, d\tau_2 \qquad (7\text{-}26)$$

This will be quite small, since it represents the Coulomb energy between two overlapping, classically distinguishable, spherically symmetric, neutral atoms. The first term is the attraction of the electron on each atom for the nucleus of the other, the second is the repulsion of the nuclei, and the third is the repulsion of the two electron clouds. If we expand K in a similar way, we find

$$K = -2S \int \frac{a(1)b(1)}{R_{b1}} \, d\tau_1 + \frac{S^2}{R_{ab}} + \iint \frac{a(1)b(2)b(1)a(2)}{r_{12}} \, d\tau_1 \, d\tau_2 \qquad (7\text{-}27)$$

These terms all have to do with the *exchange-charge density* $a(1)b(1)$. This is located in the overlap region between the nuclei, where a and b are simultaneously nonzero. The total exchange charge $\int a(1)b(1) \, d\tau_1 = S$, the overlap integral. To interpret (7-27) semiclassically, we note that we must reduce the nuclear charges from e to Se. Then the first term is the attraction of the exchange charge to the two nuclei, the second is the reduced nuclear repulsion due to the reduced nuclear charge, and the third is the electrostatic self-energy of the exchange-charge cloud. If a and b were orthogonal, only the last term would survive and it would be positive, as are all proper atomic exchange integrals. However, in hydrogen and all simple covalent bonds, S is sufficiently large and the distribution of exchange charge is such that the first term dominates and K is negative.

If we now refer back to our energy expression (7-23), we see that, because $K < 0$, the singlet state $^1\Sigma_g$ lies below the triplet $^3\Sigma_u$. This is, of course, correct and is the result we obtained earlier by considering molecular orbitals. Moreover, because $|K| > |J|$, the molecular energy lies below the energy of the separated atoms and the singlet state represents a bound molecule.

It is at this point that confusion often arises because of the complete difference in significance of this exchange integral K, and the exchange integrals of atomic theory. The latter include only the self-energy term of the exchange-charge cloud. Hence they are positive definite, and Hund's rule that the state of maximum S lies lowest follows. In the H-L case, however, the lack of orthogonality makes K negative. It is this term which lowers the energy to give the strong bonding in the $^1\Sigma_g$ state and raises the energy of the $^3\Sigma_u$ state. This apparent contradiction of the Hund rule is removed in the MO approach (which has only positive exchange integrals), since the $^3\Sigma_u$ state arises from an excited *configuration* in which one electron has been promoted from a σ_g to a σ_u orbital.

From this discussion, we see that there is no absolute significance to exchange integrals or "exchange forces." All the forces are just Coulomb interactions, and the exchange-charge densities depend on the mathematical form of the wavefunction. If we express ψ in terms of orthogonal orbitals, for example, all exchange integrals are positive. With H-L orbitals, they are negative. By suitable juggling, they could even be made to vanish. Yet in all cases, if the method is carried through consistently, we get the same final answers. In view of these facts, the frequently occurring statement that, if "the" exchange integral is positive, we have ferromagnetic coupling (spins tend to be parallel), whereas if it is negative, we have antiferromagnetic coupling (spins tend to be opposed), is seen to make sense only in so far as everyone refers to a rather standard setup with nonorthogonal orbitals of the H-L sort.

Despite the number of years that have passed since Heisenberg first discussed the exchange interaction, efforts at clarifying these points still are being vigorously pressed. A particularly penetrating review has been given by Herring.[1] He shows that, although the H-L approximation reduces correctly to the separated atoms at infinite distance, it does not become asymptotically exact for singlet-triplet energy differences. In fact, when the internuclear distance exceeds about 50 atomic units, the simple H-L theory leads to the triplet lying *below* the singlet, which is simply not correct. He suggests a modified H-L method that *is* asymptotically exact at large distances. From it he finds that the coupling is always antiferromagnetic (singlet below triplet) provided that the atomic wavefunctions have no angular nodes. He is also able to justify the form of the Heisenberg–Dirac–Van Vleck vector-model coupling law (Sec. 6-4) in the asymptotic limit. For further discussion of exchange, superexchange, and covalency, the reader is referred to the literature.[2]

7-7 Orthogonal Atomic Orbitals

The above discussion suggests that it might be interesting to consider making a H-L type of calculation using atomic orbitals so modified as to be orthogonal, unlike the normal $1s$ functions of hydrogen. This technique is similar to the introduction of Wannier functions[3] in solids. In the H_2 case, by symmetry we can simply take

$$A = C_1 a + C_2 b \qquad B = C_1 b + C_2 a \qquad (7\text{-}28)$$

[1] C. Herring, *Revs. Modern Phys.*, **34**, 631 (1962).

[2] See for example: P. W. Anderson, *Phys. Rev.*, **115**, 2 (1959), and a review article, Exchange in Insulators: Superexchange, Direct Exchange, and Double Exchange, in H. Suhl and G. Rado (eds.), "Magnetism," Academic Press Inc., New York, to be published; also R. K. Nesbet, *Phys. Rev.*, **122**, 1497 (1961); A. J. Freeman and R. E. Watson, *Phys. Rev.*, **124**, 1439 (1961).

[3] G. H. Wannier, *Phys. Rev.*, **52**, 191 (1937).

with C_1 and C_2 chosen to make A and B orthonormal. This requires

$$C_1 = \frac{1}{2(1+S)^{\frac{1}{2}}} + \frac{1}{2(1-S)^{\frac{1}{2}}} \qquad C_2 = \frac{1}{2(1+S)^{\frac{1}{2}}} - \frac{1}{2(1-S)^{\frac{1}{2}}}$$

Since S is reasonably small, C_1 is nearly 1 but C_2 is a small negative number
($\sim -S/2$). The resulting functions are shown schematically in Fig. 7-5.
The positive overlap in the middle is just canceled by the negative overlap
near the centers a, b.

Now, if one calculates the diagonal energy of the H-L $^1\Sigma_g{}^+$ ground state
function built of these, namely,

$$\psi \sim A(1)B(2) + B(1)A(2) \tag{7-29}$$

one finds no binding at all! At no value of R_{ab} does the energy lie below
that of the separated atoms. To get a decent energy, we must also form

Fig. 7-5. *Schematic diagram of orthogonalized atomic orbitals for H_2.*

the other $^1\Sigma_g{}^+$ state, which will lie still higher in energy, and solve the
secular equation between them. This configuration interaction, of course,
will lead to exactly the same final functions and energies as we found using
MO's with configuration interaction or using the ordinary H-L method with
optimum admixture of ionic states. These methods differ only in their
choice of initial approximate wavefunction. Our experience with these
three methods now suggests that, for H_2 at least, the H-L method gives a
slightly better initial approximation than the MO method, whereas the orthogonal atomic-orbital method gives a hopeless initial approximation. This
latter fact points up the extreme importance of nonorthogonality for the
success of the H-L method. Clearly caution must be exercised if any of the
initial approximations is to be trusted.

7-8 Group Theory and Molecular Orbitals

In our discussion of H_2 we found it useful to note that there were no matrix
elements of H between states belonging to different irreducible representations
of the symmetry group of the molecule. This principle is even more useful in setting up MO's for the description of polyatomic molecules. If we
set up LCAO's which are symmetry orbitals belonging to definite irreducible
representations, these will factor the large secular equations which would
otherwise occur.

In some cases, the symmetry orbitals can be set up by inspection. If this fails, we can fall back on the *basis-function generating machine* (Sec. 3-8) to produce suitable orbitals, given any function F (such as an atomic orbital) which one wants to use as part of the set of symmetry orbitals. Given the complete representation matrices, symmetry orbitals belonging to particular rows of particular irreducible representations can be generated. If we are given only the characters, we can only generate symmetry orbitals belonging to a given representation. However, if we generate l suitably related independent functions of this sort, where l is the dimensionality, we can choose l orthonormal combinations of these which can serve as basis functions for the rows of a particular unitary representation. Thus it is always possible to get a large degree of help in choosing the symmetry orbitals, even with knowledge of only the characters.

Hückel theory. To see how these principles work out in practice, let us consider the example of benzene, C_6H_6, first treated by use of molecular orbitals by Hückel.[1] This has the symmetry D_{6h}. So long as we confine our attention to functions which are all even or all odd under σ_h, we need consider only the operations of D_6. (This is similar to the situation in our crystal-field work, where we could ignore the parity operator because we were dealing with functions of a definite parity.) The character table for D_6 follows:

D_6	E	C_2	$2C_3$	$2C_6$	$3C_2'$	$3C_2''$
$\Gamma_1(A_1)$	1	1	1	1	1	1
$\Gamma_2(A_2)$	1	1	1	1	-1	-1
$\Gamma_3(B_1)$	1	-1	1	-1	1	-1
$\Gamma_4(B_2)$	1	-1	1	-1	-1	1
$\Gamma_5(E_2)$	2	2	-1	-1	0	0
$\Gamma_6(E_1)$	2	-2	-1	1	0	0
Γ_S	6	0	0	0	2	0

Figure 7-6 shows the location of typical C_2' and C_2'' axes and the labeling of the six carbon sites around the ring.

Now assume, for example, that we want to set up six orthonormal symmetry orbitals built from the six carbon $2s$ orbitals. Since these six functions transform into each other under D_6, they form a basis for a 6-dimensional representation Γ_S. By applying the same elementary considerations used earlier (Sec. 4-12) in discussing the valence-bond method, we can work out the character of Γ_S under all the group operations. This character has been written in at the bottom of the preceding character table.

[1] E. Hückel, *Z. Physik*, **70**, 204 (1931).

From this character we then deduce that

$$\Gamma_S = \Gamma_1 + \Gamma_3 + \Gamma_5 + \Gamma_6 \qquad (7\text{-}30)$$

Thus we can form two nondegenerate symmetry orbitals and two pairs of doubly degenerate symmetry orbitals from our six atomic orbitals.

We next work out these symmetry orbitals explicitly. Since Γ_1 is the totally symmetric representation, the corresponding orbital can be written down by inspection as being

$$\psi(\Gamma_1) \sim [a + b + c + d + e + f]$$

Normalizing, with allowance for a nearest-neighbor overlap integral S, we have

$$\psi_1 = \psi(\Gamma_1) = [6(1 + 2S)]^{-\frac{1}{2}}(a + b + c + d + e + f) \qquad (7\text{-}31)$$

where $a, b, \ldots$ are the atomic functions located on the corresponding carbon centers.

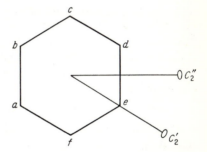

Fig. 7-6. *Geometry of benzene structure.*

The function belonging to Γ_3 is obtained by using the basis-function generating technique applied to, for example, a. Thus using (3-38),

$$
\begin{aligned}
\psi(\Gamma_3) &\sim \sum_R \chi^{(3)}(R) P_R a \\
&= 1 \cdot Ea - 1 \cdot C_2 a + 1 \sum C_3 a - 1 \sum C_6 a + 1 \sum C_2' a - 1 \sum C_2'' a \\
&= a - d + (c + e) - (b + f) + (a + e + c) - (f + b + d) \\
&= 2(a - b + c - d + e - f)
\end{aligned}
$$

Normalizing, we obtain

$$\psi_2 = \psi(\Gamma_3) = [6(1 - 2S)]^{-\frac{1}{2}}(a - b + c - d + e - f) \qquad (7\text{-}32)$$

For Γ_5, we have to deal with a two-dimensional representation. Therefore we must generate two independent functions. If we do this by operating on a and on b, we find

$$
\begin{aligned}
\psi_3' &\sim (2a + 2d - c - e - b - f) \\
\psi_4' &\sim (2b + 2e - a - c - d - f)
\end{aligned}
$$

These functions are not orthogonal as they stand. However, if one forms the sum and difference $\psi_3' \pm \psi_4'$, one obtains two orthogonal functions. (It is easy to show that the sum and difference of any two similarly normalized real functions are orthogonal.) After normalization, these orthogonal functions are

$$\psi_3 = [4(1 - S)]^{-\frac{1}{2}}(a - b + d - e) \tag{7-33a}$$

$$\psi_4 = [12(1 - S)]^{-\frac{1}{2}}(a + b - 2c + d + e - 2f) \tag{7-33b}$$

These functions then can be taken as the basis functions for the two-dimensional unitary representation Γ_5. In a similar way, we find the following basis for Γ_6:

$$\psi_5 = [4(1 + S)]^{-\frac{1}{2}}(a + b - d - e) \tag{7-34a}$$

$$\psi_6 = [12(1 + S)]^{-\frac{1}{2}}(a - b - 2c - d + e + 2f) \tag{7-34b}$$

In this simple way we have found a unitary transformation between the six atomic orbitals and six molecular-symmetry orbitals, all belonging to different rows of irreducible representations. Therefore there are no matrix elements of the Hamiltonian between them, and the 6×6 secular equation based on the six AO's has been factored completely by symmetry alone! If the matrix elements between the AO's are known, the energies of the MO's can be written down directly. For example, if we assume that the diagonal energy of the AO's is

$$(a\,|H|\,a) = (b\,|H|\,b) = \cdots = Q \tag{7-35}$$

and that only nearest-neighbor AO's have matrix elements β between them, so that

$$(a\,|H|\,b) = (b\,|H|\,c) = \cdots = (f\,|H|\,a) = \beta \tag{7-36}$$

then we have

$$W(\Gamma_1) = (\psi_1\,|H|\,\psi_1)$$
$$= [6(1 + 2S)]^{-1}(a + b + c + d + e + f|H|a + b + c + d + e + f)$$
$$= \frac{Q + 2\beta}{1 + 2S} \tag{7-37a}$$

Similarly,
$$W(\Gamma_3) = \frac{Q - 2\beta}{1 - 2S} \tag{7-37b}$$

$$W(\Gamma_5) = \frac{Q - \beta}{1 - S} \tag{7-37c}$$

$$W(\Gamma_6) = \frac{Q + \beta}{1 + S} \tag{7-37d}$$

Since the nodeless totally symmetric MO is expected to lie lowest, β is presumably negative. S is expected to be positive, since the overlapping outer parts of all the atoms have the same sign. The complete level scheme

is then as shown in Fig. 7-7. For simplicity, the figure is drawn with neglect of S compared with β/Q. Three levels are depressed below the AO
energy, and three are raised above it. If, for example, we have a single $2s$
electron contributed by each atom, the electrons will doubly occupy (two
spin orientations) the lower three levels. This total decrease in energy would
be the binding energy of the molecule arising from the $2s$ electrons. If
each atom contributed two $2s$ electrons, all levels would be filled and no

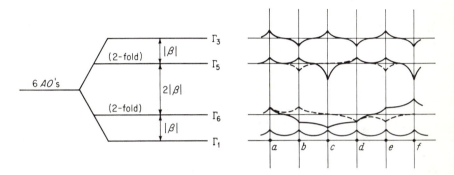

Fig. 7-7. *Molecular-orbital energy-level scheme for benzene, with overlap integrals
neglected. The form of the MO is also shown schematically in each case, a 1s atomic
orbital being used for simplicity. In case of degeneracy, the two functions are distinguished by solid and dashed curves.*

first-order contribution to binding would occur according to these results.
In fact, some repulsion would occur, as expected for closed-shell atoms.
Including higher terms in (7-37), this calculation shows an upward shift of
the center of gravity of the levels by an amount $2\,|\beta|\,S + 2QS^2 + \cdots$,
which increases with increasing overlap.

One might wonder how our Γ_6 wavefunctions display the symmetry of
the molecule, since in neither ψ_5 nor ψ_6 is there the same amount of charge
on all six equivalent centers. However, since these two functions are degen-
erate, they will be equally occupied and $|\psi_5|^2 + |\psi_6|^2$ does have equal charge
on all centers, as required by the generalized Unsöld theorem (see Sec. 4-9).

Additional insight into these wavefunctions can be obtained by noting
that the cyclic group of order 6 is a subgroup of the full symmetry of this
system. Thus the eigenfunctions of the Hamiltonian can be chosen from
those of the cyclic group, the generating element of the group being a rotation by $\pi/3$ about the center of the benzene ring. Thus by the Bloch theorem
(3-26), we must have

$$\psi_n\left(\varphi + \frac{\pi}{3}\right) = e^{i\pi n/3}\psi_n(\varphi)$$

where φ is an angle of rotation in the plane of the molecule and n takes the values 0, 1, 2, 3, 4, 5. We can construct such a function (not normalized) by taking

$$\psi_n(\mathbf{r}) = \sum_{q=1}^{6} e^{in\pi q/3} a_q(\mathbf{r}) \tag{7-38}$$

where a_q is an AO located on the qth site around the ring. These functions ψ_n represent progressive waves, traveling around the ring. Note that the functions for $n = 0$ and 3 correspond to the functions found in (7-31) and (7-32) for Γ_1 and Γ_3, respectively. The functions for $n = 2$ and 4 are degenerate, being complex conjugates of each other. Taking the real and imaginary parts separately, corresponding to standing waves, one finds the functions found in (7-33) for Γ_5. Similarly, ψ_1 and ψ_5 lead to the Γ_6 functions. Thus the eigenfunctions found earlier are closely related to Bloch functions for this cyclic "crystal" containing only six atoms.

So far we have considered LCAO symmetry orbitals derived only from the carbon $2s$ functions. Exactly the same calculations would carry through for the carbon $1s$ and hydrogen $1s$ functions. We have to be a bit more careful with the carbon $2p$ functions, since they are not invariant under rotations. The six p_R functions, directed radially, will transform exactly like the S functions under D_{6h}, and hence they will form the same kind of symmetry orbitals. The six p_π functions normal to the plane of the molecule belong to representations of the full D_{6h} group which are odd under σ_h. Hence, this type of π electron will not mix at all with those which are even under σ_h, but they will still be split into four levels: Γ_2^-, Γ_4^-, Γ_5^-, Γ_6^-. Finally, the "tangential" p_π orbitals in the plane will form a six-dimensional representation which can be reduced to Γ_2^+, Γ_4^+, Γ_5^+, Γ_6^+.

Now we are ready to put the pieces together to get the best wavefunction for the C_6H_6 molecule as a whole. We have found a set of 36 symmetry orbitals which can be formed from the 36 low-lying AO's of the types H($1s$) and C($1s$, $2s$, $2p$). These are distributed among the irreducible representations as follows:

$$\Gamma_1^+ \quad \text{and} \quad \Gamma_3^+ \text{: H}(1s)\text{C}(1s)\text{C}(2s)\text{C}(2p_R)$$

$$\Gamma_2^+ \quad \text{and} \quad \Gamma_4^+ \text{: C}(2p_T)$$

$$\Gamma_5^+ \quad \text{and} \quad \Gamma_6^+ \text{: H}(1s)\text{C}(1s)\text{C}(2s)\text{C}(2p_R)\text{C}(2p_T)$$

$$\Gamma_2^-, \Gamma_4^-, \Gamma_5^-, \quad \text{and} \quad \Gamma_6^- \text{: C}(2p_\pi)$$

Allowing for spin orientation, we thus have 72 possible states into which to put the 42 electrons of the benzene molecule. The problem is to find the 42 lowest-lying MO's into which to put the electrons to form the ground state. These "best" MO's will not in general be identical with the symmetry orbitals which we have found, but rather optimum (via the variational principle) linear combinations of those of the same symmetry. We find these optimum combinations by solving the secular equation between the various symmetry orbitals of the same symmetry. (Matrix elements are allowed between these because they belong to the same row of the same irreducible representation.) These secular equations are relatively tiny, however, and therefore easy to solve. In our example, the largest secular equations would be the 5×5's from Γ_5^+ and Γ_6^+, whereas the unfactored problem working with AO's would involve solving a 36×36! Even the 5×5 can be reduced for practical purposes to a 4×4 because the diagonal energy of the $C(1s)$ symmetry orbital will lie so far below that of the others that the slight mixing with the others can be handled by perturbation theory, if at all. In fact it is often assumed that such "inner-shell" symmetry orbitals are exact enough to be considered occupied and unperturbed and simply left out of the problem from the start.

After these small secular equations have been solved, we shall have determined the 36 MO's which are the best approximations to the true eigenfunctions which can be formed from the 36 atomic orbitals which we started with. The MO's are all orthogonal and can be treated much like the one-electron functions of atomic physics. To get the ground configuration, these are filled by starting from the lowest-lying level and working up till all the electrons are accounted for. Excitation and ionization energies are determined from the differences of one-electron MO energies. There is, however, little of the zero-order degeneracy within a configuration, which gives the L-S terms in atomic structure, because the low symmetry of the molecule severely restricts the maximum dimensionality of the irreducible representations.

Naturally, the outline given here can indicate only the barest skeleton of the modern methods for calculating molecular energy levels by the method of molecular orbitals. However, useful results in the interpretation of experimental data can be obtained even with this simple framework. For more details about realistic calculation methods, including those using the full Hartree-Fock formalism, the reader is referred to the literature.[1]

7-9 Selection Rules for Electronic Transitions

Linear molecules $C_{\infty v}$. For linear molecules, the z component of the dipole moment belongs to the Σ^+ (or A_1) representation, while the x and y components belong to the Π (or E_1) representation. Since the direct product of any representation Γ with A_1 simply reproduces Γ, the selection

[1] See, for example, C. C. J. Roothan, *Revs. Modern Phys.*, **23**, 69 (1951); A. Streitwieser, Jr., "Molecular Orbital Theory for Organic Chemists," John Wiley & Sons, Inc., New York, 1961; R. Daudel, R. Lefebvre, and C. Moser, "Quantum Chemistry," Interscience Publishers, Inc., New York, 1959.

rules for transitions induced by the z component of the dipole moment are

$$\Sigma^+ \to \Sigma^+ \qquad \Sigma^- \to \Sigma^- \qquad \Pi \to \Pi \qquad \Delta \to \Delta \qquad \text{etc.}$$

The x, y polarization case is worked out by noting that $\chi^{(\Pi)}(C_\varphi) = 2 \cos \varphi$, and in general $\chi^{(\Lambda)}(C_\varphi) \sim \cos \Lambda\varphi$. Thus, the direct-product character is proportional to $\cos \varphi \cos \Lambda\varphi$, hence to $[\cos (\Lambda + 1)\varphi + \cos (\Lambda - 1)\varphi]$, and the selection rule is $\Delta\Lambda = \pm 1$. Combining this result with that for the z component, we have the following selection rules for unpolarized radiation or randomly oriented molecules:

$$\Sigma^+ \leftrightarrow \Sigma^+, \Pi$$

$$\Sigma^- \leftrightarrow \Sigma^-, \Pi$$

$$\Pi \leftrightarrow \Sigma^+, \Sigma^-, \Pi, \Delta$$

$$\Delta \leftrightarrow \Pi, \Delta, \Phi$$

Homonuclear diatomic molecules $D_{\infty h}$. In this case, z belongs to $\Sigma_u{}^+$ and (x, y) to Π_u. Since $D_{\infty h}$ can be considered to be the direct product of $C_{\infty v}$ with the inversion, nothing new is added to the selection rules found above except that there must be a parity change in each case because the electric-dipole operator has odd parity. For example, we have $\Sigma_g{}^+ \leftrightarrow \Sigma_u{}^+, \Pi_u$ and $\Sigma_u{}^+ \leftrightarrow \Sigma_g{}^+, \Pi_g$.

Polyatomic molecules. One must find the representations to which x, y, and z belong for the symmetry group of the molecule under consideration and must use these in direct products to find the selection rules. For the example of benzene, treated in the previous section, z belongs to Γ_2 and (x, y) belong to Γ_6. Thus, by inspecting the character table, we see that transitions induced by the z component of the dipole moment are allowed for $\Gamma_1 \leftrightarrow \Gamma_2$, $\Gamma_3 \leftrightarrow \Gamma_4$, $\Gamma_5 \leftrightarrow \Gamma_5$, and $\Gamma_6 \leftrightarrow \Gamma_6$. Similarly, transitions induced by x and y components connect the following: $\Gamma_1, \Gamma_2 \leftrightarrow \Gamma_6$; $\Gamma_3, \Gamma_4 \leftrightarrow \Gamma_5$.

7-10 Vibration of Diatomic Molecules

Now that we have examined methods for handling the problems of the electronic motion in a molecule with fixed nuclear force centers, we next proceed in the Born-Oppenheimer approximation to study the nuclear motion in the effective potential provided by the electronic energy. The Hamiltonian for this stage in the problem is

$$H = \sum_\alpha \frac{P_\alpha{}^2}{2M_\alpha} + V(R_\alpha) \tag{7-39}$$

The effective potential here, $V(R_\alpha)$, is to be identified with the total electronic energy as a function of internuclear distance $E_e(R_\alpha)$, which includes, as in (7-2), the direct internuclear-repulsion energy $V_{nn}(R_\alpha)$.

To fix ideas, we again specialize to the simplest type of molecules, namely, diatomic (but not necessarily homonuclear) molecules. As is well known, the kinetic-energy term may be separated into three parts: (1) center-of-mass motion (three degrees of freedom), (2) rotation (two degrees), and (3) vibration (one degree). The terms here are listed in ascending order in energy. We now concentrate on the vibrational part. This is equivalent to the problem of a one-dimensional oscillator with *reduced mass* $\mu = M_1 M_2/(M_1 + M_2)$ and a restoring force derived from the effective potential $V(R) = E_e(R)$ obtained from the solution for the electronic energy as a function of R. This effective potential has the general form shown in Fig. 7-2.

The simplest approximation is to treat $V(R)$ as a parabola about R_e. This is quite good for small oscillations but becomes quite hopeless as one approaches dissociation. Since the solution of the harmonic oscillator is familiar from elementary quantum mechanics,[1] it will not be repeated here. Suffice it to say that the resulting energy levels are

$$E_v = (v + \tfrac{1}{2})\hbar\omega_e \tag{7-40}$$

where ω_e is the classical vibration frequency of the oscillator and $v = 0, 1, 2, \ldots$ is the vibrational quantum number. Both position and momentum operators have $\Delta v = \pm 1$ matrix-element selection rules.

The Morse approximation. The most famous and generally useful higher approximation for the molecular-vibration problem is that of Morse.[2] He sought an analytic approximation to the effective potential $V(R)$ which would have all the correct qualitative features built in. That is, he wanted a function $V(R)$ that approached a constant as $R \to \infty$, became very large (preferably infinite) as $R \to 0$, had a minimum at some equilibrium point R_e, and led to an exactly soluble Schrödinger equation with eigenvalues departing from the equally spaced harmonic-oscillator ones in the experimentally observed manner. He satisfied these requirements by the choice

$$V(R) = D_e[1 - e^{-a(R-R_e)}]^2 \tag{7-41}$$

which has the zero of energy at the minimum of the potential at R_e. As $R \to \infty$, $V(R) \to D_e$, the dissociation energy. The parameter a controls the curvature at the minimum by setting the scale of radial distance. As $R \to 0$, $V(R) \to D_e[1 - e^{aR_e}]^2 \approx D_e e^{2aR_e} \gg D_e$. In fact, $V(0)$ should be infinite because of the $Z_1 Z_2 e^2/R$ Coulomb repulsion. However, as long as $V(0) \gg D_e$, no serious error results from finiteness of $V(0)$.

[1] See, for example, L. I. Schiff, "Quantum Mechanics," 2d ed., p. 60, McGraw-Hill Book Company, Inc., New York, 1955.
[2] P. M. Morse, *Phys. Rev.*, **34**, 57 (1929).

Upon putting (7-41) into (7-39), setting up the Schrödinger equation, and separating variables, the radial equation to be solved is

$$-\frac{\hbar^2}{2\mu}\frac{d^2S}{dR^2} + \left\{\frac{J(J+1)\hbar^2}{2\mu R^2} + D_e[1 - e^{-a(R-R_e)}]^2\right\}S = ES \tag{7-42}$$

where $\psi(\mathbf{R}) = R^{-1}S(R)Y_J^M(\theta, \varphi)$ is the complete wavefunction, on the assumption that the rotational quantum number is J. The term

$$J(J+1)\hbar^2/2\mu R^2$$

is the *centrifugal potential* arising from rotation. Since it is of a smaller order of magnitude than the term in D_e, we shall neglect it for the time being, treating only the purely vibrational energy of a nonrotating vibrator. In this case, an exact solution to (7-42) may be obtained by making a change of variable by defining

$$x = e^{-a(R-R_e)} \tag{7-43}$$

leading to

$$\frac{d^2S}{dx^2} + \frac{1}{x}\frac{dS}{dx} + \frac{2\mu}{a^2\hbar^2}\left(\frac{E-D_e}{x^2} + \frac{2D_e}{x} - D_e\right)S = 0 \tag{7-44}$$

This equation can be solved in terms of the associated Laguerre polynomials (which also appear in solving the radial equation for the hydrogen atom). Since the resulting wavefunction is not of much use, it will not be quoted here. It is important to note only that the usual requirement of truncating a series to a polynomial to satisfy boundary conditions leads to a quantum condition for the energy eigenvalues. The resulting energy levels are

$$E_{\text{vib}} = \hbar\omega_e(v + \tfrac{1}{2}) - x_e\hbar\omega_e(v + \tfrac{1}{2})^2 \tag{7-45}$$

with

$$\omega_e = a\left(\frac{2D_e}{\mu}\right)^{\frac{1}{2}} \tag{7-45a}$$

and

$$x_e = \left(\frac{\hbar\omega_e}{4D_e}\right) \tag{7-45b}$$

The first term of (7-45) is obviously the ordinary harmonic-oscillator approximation, while the second gives a correction due to the "softening" of the restoring potential for large amplitudes. In fact, it was known empirically that experimental data fit the form (7-45) quite accurately, and Morse viewed this agreement as demonstrating that (7-41) did in fact give a quite good approximation to the true potential curve. This point has been verified in more detail by constructing numerically a potential function which does reproduce the experimental levels exactly and comparing it with the Morse curve.[1]

[1] See, for example, O. Klein, Z. Physik, **76**, 226 (1932); R. Rydberg, Z. Physik, **73**, 376 (1932), **80**, 514 (1933); A. L. G. Rees, Proc. Phys. Soc. (London), **59**, 998 (1947); H. M. Hulburt and J. O. Hirshfelder, J. Chem. Phys., **9**, 61 (1941).

It is interesting to note that the eigenvalues (7-45) reach a maximum of just D_e, the dissociation energy, when $(v + \frac{1}{2}) = \frac{1}{2}x_e^{-1} = 2D_e/\hbar\omega_e$. Thus, there are a finite number of discrete vibrational levels for this problem, and because of the crowding together of the upper levels, this number is just twice what it would be if the level spacing remained $\hbar\omega_e$ all the way up to the dissociation energy D_e.

Other approximations. Although the Morse approximation is probably that most often used, many others have been suggested by various authors. The first serious attempt at treating the anharmonic molecular oscillator was that of Kratzer,[1] using old quantum theory. This type of power-series approach, where one takes

$$V(R) = V(R_e) + \tfrac{1}{2}V''(R_e)(R - R_e)^2 + \tfrac{1}{6}V'''(R_e)(R - R_e)^3 + \cdots \quad (7\text{-}46)$$

was carried into wave mechanics by Dunham,[2] later simplified and extended by Sandeman.[3] This method enables one to obtain as good an approximation near R_e as desired, by simply taking more terms, and the solution is quite straightforward by using perturbation methods. However, the method works badly far from R_e, since the expansion (7-46) inevitably diverges as $R \to \infty$, rather than approaching a constant D_e.

To show how the method works, let us consider a case where only one anharmonic term is carried and take our zero of energy to be at $V(R_e)$. Then we can write

$$V(\xi) = a_2\xi^2 + a_3\xi^3 \quad (7\text{-}47)$$

where

$$\xi = \frac{(R - R_e)}{R_e}$$

is a normalized dimensionless vibrational coordinate and $a_2 = \frac{1}{2}\mu\omega_e^2R_e^2$. The matrix elements of ξ in the harmonic-oscillator approximation (neglect of a_3) are simply[4]

$$(v' \,|\xi|\, v) = \lambda[(v + 1)^{\frac{1}{2}}\delta_{v',v+1} + v^{\frac{1}{2}}\delta_{v',v-1}] \quad (7\text{-}48)$$

where

$$\lambda = \left(\frac{\hbar}{2\mu R_e^2\omega_e}\right)^{\frac{1}{2}}$$

Knowing the matrix of ξ, one can calculate the matrices of ξ^2, ξ^3, ... by simple matrix multiplication. Thus

$$(v' \,|\xi^2|\, v) = \lambda^2\{[(v + 1)(v + 2)]^{\frac{1}{2}}\delta_{v',v+2} + (2v + 1)\delta_{v',v} + [v(v - 1)]^{\frac{1}{2}}\delta_{v',v-2}\} \quad (7\text{-}49)$$

[1] A. Kratzer, *Z. Physik*, **3**, 289 (1920).
[2] J. L. Dunham, *Phys. Rev.*, **41**, 713, 721 (1932).
[3] I. Sandeman, *Proc. Roy. Soc. Edinburgh*, **60**, 210 (1940).
[4] See, for example, Schiff, *op. cit.*, p. 65.

and

$$(v' \,|\xi^3|\, v) = \lambda^3\{[(v + 1)(v + 2)(v + 3)]^{\frac{1}{2}}\delta_{v',v+3}$$

$$+ 3(v + 1)^{\frac{3}{2}}\delta_{v',v+1} + 3v^{\frac{3}{2}}\delta_{v',v-1} + [v(v - 1)(v - 2)]^{\frac{1}{2}}\delta_{v',v-3}\}. \quad (7\text{-}50)$$

Combining these with the matrix for the kinetic energy $(P^2/2\mu)$, we have

$$(v' \,|H|\, v) = (v + \tfrac{1}{2})\hbar\omega_e\delta_{v'v} + a_3(v' \,|\xi^3|\, v) \quad (7\text{-}51)$$

Calculating the anharmonic correction to second-order perturbation theory leads to

$$E(v) = (v + \tfrac{1}{2})\hbar\omega_e - \frac{30a_3{}^2\lambda^6}{\hbar\omega_e}(v + \tfrac{1}{2})^2 + \text{const} \quad (7\text{-}52)$$

Thus, to this order of perturbation theory, we find the same form of answer as found with the Morse potential, namely, (7-45). The present result focuses on the initial anharmonicity near $R = R_e$, whereas the Morse curve gives an overall approximation for the potential all the way out to dissociation.

Numerous other potentials have been proposed, with three, four, or more adjustable parameters. Each of these has its own advantages and might be considered. In particular, it should be pointed out that some of these are definitely superior in agreement with experiment to the Morse potential, which has perhaps been overworked because of its simplicity and early invention. For a critical survey of almost two dozen such functions, the reader is referred to a review article by Varshni.[1]

7-11   Normal Modes in Polyatomic Molecules

Let us consider the motion of N nuclei in a general molecule as governed by the energy function $V(\mathbf{R}_\alpha)$ determined by solution of the electronic problem. We specify the position of all the nuclei by giving $3N$ coordinates ξ_k, each measured from the equilibrium position of the relevant nucleus. Therefore the energy V will have its minimum value if all the coordinates are zero and will increase no faster than quadratically as any coordinate departs from zero. That is, there can be no linear term, since this would contradict the assumed equilibrium. However, there will certainly be anharmonic terms ($\sim\xi_k{}^3$, etc.) which need to be considered for large ξ_k. These anharmonic terms introduce coupling between the normal modes, which could be treated as correction terms. These coupling terms, for example, allow thermal equilibrium to be reached after a single mode is excited, a process which could not occur if the modes were truly independent. For the time being, however, we shall neglect these terms and shall consider the decomposition of the possible motions of the nuclei into independent normal modes of vibration. The group theory is very helpful in finding the particular combinations

[1] Y. P. Varshni, *Revs. Modern Phys.*, **29**, 664 (1957).

$Q_K = \sum_k a_{kK}\xi_k$ which describe the normal modes of a molecule of given symmetry.

A geometrical viewpoint may be helpful in an initial consideration of these modes. We view the set of $3N$ ξ_k's as the coordinates of a point in a $3N$-dimensional phase space. In the equilibrium position the phase point is at the origin. Now, for any small distortion or displacement of the molecule, there is a corresponding small displacement of the representative point in some direction from the origin to the point $(\xi_1 \cdots \xi_{3N})$. Knowing $V(\mathbf{R}_\alpha) \equiv V(\xi_k)$, we can compute the increase in potential energy for any small motion in this $3N$-dimensional space. Obviously, if we impart a uniform translation in physical space to the molecule as a whole, there is no increase in potential energy, and hence no restoring force. Similarly, a uniform rotation produces no potential energy or restoring force. Thus there are six independent $3N$-dimensional vector displacements which can be made with no restoring force. (There are only five for a linear molecule, because only two angles are required to specify the rotational orientation.) For example, a displacement of the molecule by (Δz) in the z direction would be given by a phase-space vector displacement $\boldsymbol{\xi} = (\Delta z)[\mathbf{u}_3 + \mathbf{u}_6 + \cdots + \mathbf{u}_{3N}]$ where the $\mathbf{u}_k$ are unit vectors representing displacements of individual particles along one of the cartesian axes.

We now imagine that the system has been given an arbitrary displacement from the origin in any direction orthogonal to the six (or five, for linear molecules) directions in which there is no restoring force. The motion which ensues after the system is released is in general a complicated one, with various coordinates vibrating irregularly with a variety of superposed frequencies. However, there exist $(3N - 6)$ [or $(3N - 5)$] orthogonal phase-space vectors $\mathbf{Q}_K$ defining directions such that the ensuing motion after a displacement along any one of them is a simple harmonic oscillation of frequency ω_K. These vectors define the normal modes of the system. Since they span the space of all displacements (when combined with the six translational and rotational vectors), an arbitrary motion can be represented as a sum over normal modes.

These normal coordinates Q_K are defined mathematically as the linear combinations of the ξ_k which bring the kinetic and potential energies, T and V, simultaneously to the diagonal forms

$$T = \tfrac{1}{2}\sum_{K=1}^{3N} (\dot{Q}_K)^2 \qquad V = \tfrac{1}{2}\sum_{K=1}^{3N} \omega_K{}^2 Q_K{}^2 \qquad (7\text{-}53)$$

where $\dot{Q}_K = dQ_K/dt$. (The lack of restoring forces for translation and rotation makes six of the ω_K equal zero.) Introducing the canonical momentum by

$$P_K \equiv \frac{\partial T}{\partial \dot{Q}_K} \qquad (7\text{-}54)$$

which in this case is simply $\dot{Q}_K$, the classical Hamiltonian of the problem consists of a sum of terms, each dealing with only one mode, namely,

$$H = T + V$$

$$= \sum_{K=1}^{3N} H_K = \sum_{K=1}^{3N} \tfrac{1}{2}(P_K{}^2 + \omega_K{}^2 Q_K{}^2) \tag{7-55}$$

Then, using the Hamilton canonical equation

$$\dot{P}_K = -\frac{\partial H}{\partial Q_K} \tag{7-56}$$

we are led to

$$\ddot{Q}_K + \omega_K{}^2 Q_K = 0 \tag{7-57}$$

which has the solution

$$Q_K = Q_K{}^0 \cos(\omega_K t + \varphi) \tag{7-58}$$

where φ is an arbitrary phase angle.

The corresponding quantum-mechanical problem is also trivial, since a simple product solution, with one factor for each term in H, is successful. That is, if we consider

$$H\Psi = \left(\sum_K H_K\right) \prod_K \psi_K(Q_K) = E\Psi$$

and divide through by Ψ, we have

$$\sum_K \frac{H_K \psi_K}{\psi_K} = E$$

Since each term in the sum depends only on one Q_K, all Q_K being independent variables, the sum can equal the constant E only if each term is itself a constant, E_K, say. Thus we have $3N$ equations of the form

$$H_K \psi_K = \tfrac{1}{2}(P_K{}^2 + \omega_K{}^2 Q_K{}^2)\psi_K = E_K \psi_K \tag{7-59}$$

to determine $\Psi = \prod_K \psi_K(Q_K)$ and the equation

$$E = \sum_K E_K$$

to determine the system energy. Using the well-known solution for the harmonic oscillator in quantum theory, we then have

$$E = \sum_K (v_K + \tfrac{1}{2})\hbar\omega_K \tag{7-60}$$

and

$$\Psi = \prod_K N_{v_K} H_{v_K}(\alpha_K Q_K) e^{-(\alpha_k{}^2 Q_k{}^2/2)} \tag{7-61}$$

where the H_{v_K} are Hermite polynomials, $\alpha_K^2 = \omega_K/\hbar$, and the N_{v_K} are normalizing constants. Thus, whether we use classical or quantum mechanics to treat the vibrational problem, once the normal coordinates Q_K which throw H into the form (7-55) are found, the problem is largely solved.

Let us now look more closely at the process of determining the transformation relating the Q_K to the ξ_k. First of all, it is convenient to transform from the ξ_k to "mass-weighted" coordinates

$$q_k = \sqrt{m_k}\xi_k \tag{7-62}$$

where m_k is the mass of the atom having displacement ξ_k. In terms of the q_k, the Hamiltonian is transformed from

$$H = \sum_k \tfrac{1}{2}m_k\dot{\xi}_k^2 + \sum_{k,l} \frac{1}{2}\frac{\partial^2 V}{\partial\xi_k\,\partial\xi_l}\,\xi_k\xi_l$$

to

$$H = \sum_k \tfrac{1}{2}\dot{q}_k^2 + \sum_{k,l} \frac{1}{2}\frac{\partial^2 V}{\partial q_k\,\partial q_l}\,q_k q_l$$

or

$$H = \sum_k \tfrac{1}{2}p_k^2 + \sum_{k,l} \tfrac{1}{2}f_{kl}q_k q_l \tag{7-63}$$

Since the kinetic-energy term is now expressed as a simple sum of squares, any orthogonal transformation of coordinates will preserve this form. To verify this, let us consider new coordinates defined by an orthogonal transformation a_{kK} so that

$$q_k = \sum_k a_{kK}Q_K \tag{7-64a}$$

$$Q_K = \sum_k (a^{-1})_{Kk}q_k = \sum_k a_{kK}q_k \tag{7-64b}$$

Then
$$\sum_k q_k^2 = \sum_{k,K,L} a_{kK}Q_K a_{kL}Q_L = \sum_{K,L}\left(\sum_k a_{kK}a_{kL}\right)Q_K Q_L$$

$$= \sum_{K,L}\left(\sum_k (a^{-1})_{Kk}a_{kL}\right)Q_K Q_L = \sum_{K,L}\delta_{KL}Q_K Q_L = \sum_K Q_K^2$$

Since the transformation a_{kK} is independent of time, the same transformation holds for the $\dot{q}_k$ or for the p_k and this proof shows that $\sum_k p_k^2 = \sum_K P_K^2$, as was desired.

Since the kinetic energy retains its desired simple form under an arbitrary orthogonal transformation, one is free to choose a_{kK} so as to reduce the potential energy to a sum of squared terms with no cross terms. Applying the transformation to the potential-energy term in (7-63), we obtain

$$V = \tfrac{1}{2}\sum_{k,l,K,L} f_{kl}a_{kK}a_{lL}Q_K Q_L$$

$$= \tfrac{1}{2}\sum_{KL} [\mathbf{a}^{-1}\mathbf{f}\mathbf{a}]_{KL}Q_K Q_L$$

We want this to equal a sum of terms of the form $\frac{1}{2}\omega_K^2 Q_K^2$. This will be true if

$$[\mathbf{a}^{-1}\mathbf{fa}]_{KL} = \omega_K^2 \delta_{KL}$$

After some rearrangement, this can be written

$$\sum_l [f_{kl} - \omega_K^2 \delta_{kl}]a_{lK} = 0 \qquad k = 1, 2, \ldots, 3N \qquad (7\text{-}65)$$

For a nontrivial solution to these equations, ω_K must satisfy the secular equation

$$|f_{kl} - \omega_K^2 \delta_{kl}| = 0 \qquad (7\text{-}66)$$

and, for each such ω_K, there is determined a set of coefficients a_{lK} defining a normal mode. When normalized and combined, these sets of a_{lK} determine the entire transformation matrix $\mathbf{a}$. One can show that six (or five for linear molecules) of these ω_K are zero. These correspond to the modes which represent simple translation or rotation of the molecule. Since $\mathbf{a}$ is an orthogonal matrix, all other modes (i.e., those with $\omega_K \neq 0$) must be orthogonal to those of zero frequency.

Although these mass-weighted cartesian coordinates simplify the general theory nicely, they are not usually the most practical for calculation. One can work directly with nonweighted cartesian coordinates with only a slight change in the secular equation. However, the coordinate transformation is no longer orthogonal. More generally, one can use noncartesian coordinates η_k in which there are cross terms in the kinetic energy as well as in the potential energy. Still, a generalized secular equation will determine the eigenfrequencies.

A particularly useful approach is the use of *internal coordinates*. In this method, one eliminates the six redundant coordinates at the very beginning, by using only coordinates internal to the molecule (such as bond lengths and angles). This reduces the size of the secular equation from $3N \times 3N$ to $(3N - 6) \times (3N - 6)$, which will usually greatly simplify its solution. Treating this method in detail would take us too far afield. We refer the reader to the excellent book of Wilson, Decius, and Cross[1] for a thorough treatment, including numerous examples.

7-12   Group Theory and Normal Modes

Let us now consider the implications of molecular symmetry for the nature and degeneracy of the normal modes discussed in a general way in the previous section. For geometric consideration of these modes, it is often convenient to think in terms of unit vectors $\hat{Q}_K$ associated with the normal

[1] E. B. Wilson, Jr., J. C. Decius, and P. C. Cross, "Molecular Vibrations," chap. 4, McGraw-Hill Book Company, Inc., New York, 1955.

coordinates Q_K. As long as the matrix relation between q_k and Q_K is orthogonal, the same matrix $\mathbf{a}$ relates $\hat{q}_k$ to $\hat{Q}_K$. Thus the two viewpoints can be exchanged quite freely. The unit vector $\hat{Q}_K$ when decomposed in terms of the $\hat{q}_k$ indicates the displacements of each atom in the normal mode of vibration. If these displacements are depicted by small arrows on a diagram of the molecular framework, then the effect of symmetry operations can be readily seen by inspection.

Since a molecule is not fundamentally changed by applying a symmetry operator P_R, the mode $P_R\hat{Q}_K$ must have the same frequency as $\hat{Q}_K$ itself. Thus, if Q_K is nondegenerate, then $P_R\hat{Q}_K = \pm\hat{Q}_K$, for all P_R, and $\hat{Q}_K$ forms the basis for a one-dimensional representation of the symmetry group of the molecule. However, if Q_K is degenerate, we can say only that

$$P_R\hat{Q}_K = \sum_{K'} \hat{Q}_{K'}D(R)_{K'K} \tag{7-67}$$

where K' runs over all the modes degenerate with K. This is true because any linear combination of degenerate $\hat{Q}_K$ will oscillate at the same frequency ω_K. Thus, degenerate vibrational modes transform according to irreducible representations of dimensionality greater than 1, whereas nondegenerate modes correspond to one-dimensional representations.

This fact allows us to find the number of distinct vibrational frequencies ω_K and their degeneracy by decomposing the $3N$-dimensional representation of the symmetry group of the molecule which is based on three orthogonal unit vectors on each nucleus. After dropping the representations which account for translation and rotation, the remaining ones must all correspond to vibrational frequencies. Since there can be no coupling between modes belonging to different irreducible representations, we get a factoring of the secular equation which determines the eigenvectors $\hat{Q}_K$ by using the proper symmetry coordinates, which is just as powerful as the factoring of the MO secular equations by use of symmetry orbitals.

Example: H_2O. To illustrate these principles, we consider the water molecule, which belongs to C_{2v}. The character table is:

C_{2v}				E	C_2	σ_v	σ_v'
x^2, y^2, z^2		z	A_1	1	1	1	1
xy	R_z		A_2	1	1	-1	-1
xz	R_y	x	B_1	1	-1	1	-1
yz	R_x	y	B_2	1	-1	-1	1
			Γ	9	-1	3	1

We take σ_v to be reflection in the plane of the molecule and σ_v' to be reflection in the perpendicular plane bisecting the HOH bond angle. The C_2 axis is

z, and the y axis is perpendicular to the plane of the molecule. The $3N =$ nine-dimensional representation has been called Γ, and its character is given at the bottom. This character is worked out by consideration of the transformation of the nine cartesian unit vectors with origins on the three atoms in a way familiar from our consideration of π and σ bonds. Knowing the character, we see that

$$\Gamma = 3A_1 + A_2 + 3B_1 + 2B_2$$

The three translational modes account for the representations of x, y, and z, namely, $A_1 + B_1 + B_2$. The three rotations take out $A_2 + B_1 + B_2$. This leaves only

$$\Gamma_{\text{vib}} = 2A_1 + B_1 \qquad (7\text{-}68)$$

Thus the three normal modes are all nondegenerate.

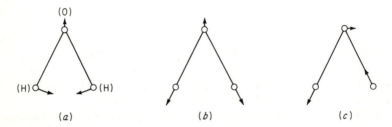

Fig. 7-8. *Normal modes of water molecule. (a)-(b) Symmetry A_1; (c) symmetry B_1.*

The two modes having symmetry A_1 must leave the molecular symmetry undiminished, since A_1 is the totally symmetric representation. In terms of internal coordinates, this restricts the motion to a combination of a change in the bond angle φ and an equal stretch of both OH bond lengths R_1 and R_2. These motions are indicated in Fig. 7-8(a and b). The actual normal coordinates will be two particular linear combinations of the type $Q_K(A_1) = a_K \, \delta\varphi + b_K(\delta R_1 + \delta R_2)$. To get the coefficients a and b, a 2×2 secular equation must be solved. From the character table we see that the B_1 mode must reduce the symmetry in a way which is odd under C_2 and σ_v'. This is satisfied if we choose $Q(B_1) = c(\delta R_1 - \delta R_2)$. Such a mode is indicated in Fig. 7-8c. Actually all these descriptions of the Q's are approximate, since we have treated the oxygen atom as infinitely massive. Since it has a finite mass, it will have to move in such a way as to hold the center of mass stationary and to eliminate any overall rotation.

Example: C_2H_2. As a second illustration we consider the acetylene molecule, C_2H_2. This is a linear molecule belonging to $D_{\infty h}$. Since it is linear, there are only two rotational degrees of freedom. This leaves $3N - 5 = 7$ vibrational degrees of freedom. Analysis of the sort indicated for H_2O shows that the vibrational motions decompose as

$$\Gamma_{\text{vib}} = 2\,\Sigma_g{}^+ + \Sigma_u{}^+ + \Pi_g + \Pi_u \qquad (7\text{-}69)$$

The two $\Sigma_g{}^+$ modes will preserve the full symmetry of the molecule. The $\Sigma_u{}^+$ mode must be odd under inversion or σ_h but still preserve axial symmetry. The Π modes must involve displacements perpendicular to the axis to transform suitably for Π representations. The two degenerate modes in each refer to vibrations in perpendicular planes through the internuclear axis. Appropriate vibrational configurations are sketched in Fig. 7-9.

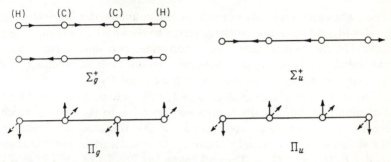

Fig. 7-9. *Normal modes of acetylene molecule. The dashed arrows for the Π modes indicate modes degenerate with those shown by solid arrows.*

Combination and overtone levels. Although primary interest is attached to the fundamental absorption in which one mode becomes singly excited, there is also considerable interest in absorptions involving *combination* and *overtone* levels. A combination level is one in which two or more modes of different frequency are excited out of the ground state, whereas an overtone level is one in which there is more than one unit of excitation of a given frequency but none in modes of other frequencies. We start by considering the symmetry of these types of excited states.

As a first example, let us consider the case of a binary combination level in which $v_L = v_M = 1$, all other v_K being zero. From (7-61) we see that the symmetry of the wavefunction is that of $Q_L Q_M$, since $H_1(Q_K) \sim Q_K$, while $H_0(Q_K) = $ const, and the exponential is invariant under all group operations, since it has the same form as the potential energy. But $Q_L Q_M$ is part of the basis of the direct product of the representations to which Q_L and Q_M, respectively, belong. Thus the combination level has symmetry Γ determined by

$$\Gamma = \Gamma(L) \times \Gamma(M) \tag{7-70}$$

If Q_L and Q_M are both nondegenerate, we can find the symmetry of the resulting nondegenerate combination level by identifying the character $\chi = \chi^{(L)}\chi^{(M)}$ in the table. If Q_L and/or Q_M is degenerate, then the decomposition of the direct product into irreducible representations gives us information about the degeneracy as well as the symmetry of the combination level. For example, if a twofold and a threefold level are singly excited, the total degeneracy is 6.   However, when account is taken of the anharmonic coupling terms, this degeneracy will be broken down as determined by the dimensionality of the irreducible representations found in the direct product. For the specific example of acetylene, treated earlier, if there is single excitation of both the Π_u and Π_g modes, one finds

$$\Pi_u \times \Pi_g = \Sigma_u^+ + \Sigma_u^- + \Delta_u \qquad (7\text{-}71)$$

Thus, when account is taken of anharmonic coupling, this fourfold degenerate level will split into two nondegenerate levels and one doubly degenerate level.   In the general case where more than two modes are excited, this method obviously generalizes to the rule that one finds the resulting symmetry by analyzing the direct product of all the excited Q_K.

Now, if any level is excited to $v_K > 1$, we have to deal with overtone levels. For simplicity, we initially restrict our attention to the case in which only modes of one frequency are excited.   If this frequency is nondegenerate, the situation is very straightforward.   As noted earlier, for nondegenerate modes $P_R Q_K = \pm Q_K$.   Thus all characters consist of $\chi^{(K)}(R) = \pm 1$.   In either case, $P_R Q_K^2 = Q_K^2$, and a similar result holds for all even powers. Also, $P_R Q_K^v = \chi^{(K)}(R) Q_K^v$ for all odd values of v.   Since the Hermite polynomial H_{v_K} consists entirely of even powers of Q_K if v_K is even and entirely of odd powers if v_K is odd, it follows that all even overtone levels belong to the totally symmetric representation, whereas all odd overtone levels belong to the same representation as Q_K, namely, that of the first excited level.

Passing now to the case of overtones of degenerate frequencies, we encounter a more complex situation.   We note first that a simple combinational argument shows that the degeneracy of an overtone level is

$$\binom{v+n-1}{v} = \frac{(v+n-1)!}{v\,(n-1)!}$$

where n is the degeneracy of the mode and v is the total number of quanta of excitation.   For a doubly degenerate frequency, this reduces to simply $(v+1)$, whereas for a triply degenerate frequency it reduces to

$$(v+1)(v+2)/2.$$

As an example, consider a triply excited level of a doubly degenerate mode, The $(v+1) = 4$ possible wavefunctions may be written (apart from invariant factors as)

$$\psi_{30} \sim H_3(Q_{K1})$$
$$\psi_{21} \sim H_2(Q_{K1})H_1(Q_{K2})$$
$$\psi_{12} \sim H_1(Q_{K1})H_2(Q_{K2})$$
$$\psi_{03} \sim H_3(Q_{K2}) \qquad (7\text{-}72)$$

where Q_{K1} and Q_{K2} are the two degenerate normal coordinates with frequency ω_K. Since these four functions form the complete set of degenerate functions, they must form a basis for a representation of the symmetry group. To determine whether this representation is reducible and, if so, how it reduces, we need to find only its character. This process is simplified by noting that, barring accidental degeneracy, Q_{K1} and Q_{K2} must belong to the same representation, since they are degenerate. Next we note that each of the four functions in (7-72) contains a cubic term and a linear term in the Q's. Since the coordinate transformations R of the symmetry group are linear, the cubic terms themselves must form the basis for a four-dimensional representation. (The linear terms form only a two-dimensional one.) Since the character of a representation is invariant under orthogonal linear transformations of the basis functions, we can for convenience choose Q_{K1} and Q_{K2} for each operator R so that the matrix of R is diagonal. In this case, $Q_{K1} \rightarrow R_1 Q_{K1}$ and $Q_{K2} \rightarrow R_2 Q_{K2}$ under R, and $\chi_1(R) = R_1 + R_2$. Accordingly, the character of the representation based on $Q_{K1}{}^3$, $Q_{K1}{}^2 Q_{K2}$, $Q_{K1} Q_{K2}{}^2$, $Q_{K2}{}^3$ will be

$$\chi_3(R) = R_1{}^3 + R_1{}^2 R_2 + R_1 R_2{}^2 + R_2{}^3 \tag{7-73}$$

Similarly, the character of the three-dimensional representation corresponding to $v = 2$ will be

$$\chi_2(R) = R_1{}^2 + R_1 R_2 + R_2{}^2 \tag{7-74}$$

and the character in the original two-dimensional representation of R^v will be

$$\chi_1(R^v) = R_1{}^v + R_2{}^v \tag{7-75}$$

where v is any integer. Combining the relations (7-73) to (7-75), we can obtain a result entirely in terms of characters, which will be independent of basis choice, namely,

$$\chi_3(R) = \tfrac{1}{2}[\chi_1(R)\chi_2(R) + \chi_1(R^3)] \tag{7-76}$$

This equation can be generalized to

$$\chi_v(R) = \tfrac{1}{2}[\chi_1(R)\chi_{v-1}(R) + \chi_1(R^v)] \tag{7-77}$$

which, by iteration, may be used to find the character of any overtone of a doubly degenerate mode. A similar expression can be derived for the case of triply degenerate modes, and explicit (noniterative) expressions can also be derived for both these cases.[1]

To return to the specific example of acetylene, treated earlier, if the Π_g mode is doubly excited, we have

$$\chi_2(R) = \tfrac{1}{2}[\chi_1{}^2(R) + \chi_1(R^2)] \tag{7-78}$$

[1] See, for example, Wilson, Decius, and Cross, *op. cit.*, chap. 7 and app. XIV.

Noting that $C_\varphi{}^2 = C_{2\varphi}$ and that $\sigma_v{}^2 = E$, etc., one can readily calculate $\chi_2(R)$ by this formula. Then, by the standard decomposition formula or by inspection, one finds

$$(\Pi_g)^2 = \Sigma_g{}^+ + \Delta_g \qquad (7\text{-}79)$$

Thus, with anharmonicity allowed for, this overtone level can split into a twofold degenerate Δ_g level and a nondegenerate $\Sigma_g{}^+$ level.

By combining the results of the specialized cases considered in this section, we are now able to work out the symmetry of the most general vibrational level. The general rule is that

$$\Gamma = [\Gamma(1)]^{v_1} \times [\Gamma(2)]^{v_2} \times \cdots \qquad \blacktriangleright \; (7\text{-}80)$$

where the product runs over all distinct vibrational frequencies and $[\Gamma(K)]^{v_K}$ refers to the symmetry of the v_K overtone of the vibration of frequency ω_K, as we have just treated it. In interpreting (7-80), one must of course bear in mind that

$$[\Gamma(K)]^2 \neq \Gamma(K) \times \Gamma(K) \qquad (7\text{-}81)$$

as is illustrated by (7-79), since $\Pi_g \times \Pi_g = \Sigma_g{}^+ + \Sigma_g{}^- + \Delta_g$, which is not the same as $(\Pi_g)^2 = \Sigma_g{}^+ + \Delta_g$.

7-13   Selection Rules for Vibrational Transitions

As long as we confine our attention to electric-dipole transitions (i.e., those which are "infrared active"), the selection rules are determined by the matrix element

$$(v_1' \cdots v_{3N}' \, |\sum_\alpha e_\alpha \mathbf{R}_\alpha| \, v_1 \cdots v_{3N}) \qquad (7\text{-}82)$$

where the e_α are effective charges of the atoms at $\mathbf{R}_\alpha$. Translational and rotational degrees of freedom being neglected, the indicated eigenfunctions are of the form (7-61), namely,

$$|v_1 \cdots v_{3N}) = \exp\left(-\tfrac{1}{2}\sum_K \alpha_K{}^2 Q_K{}^2\right) \prod_K H_{v_K}(\alpha_K Q_K) \qquad (7\text{-}83)$$

As pointed out earlier, the exponential is invariant under any symmetry operation, since it has the same form as the potential energy. Therefore the transformation properties are determined solely by the product of Hermite polynomials. Since we are frequently interested only in the "fundamental absorption" from the ground state (all $v_K = 0$) to an excited state in which one mode has been excited to $v_{K'} = 1$, we first simplify to that case. From the form of the Hermite polynomials, we know that $H_0(x)$ is a constant and $H_1(x)$ is proportional to x. Thus the ground state $|0 \cdots 0)$ transforms as the identical representation Γ_1, and the excited state $|0 \cdots 010 \cdots 0)$ transforms as $Q_{K'}$. Applying our general selection-rule theorem, we conclude that normal mode $Q_{K'}$ can be excited from the ground state by a dipole transition only if $\Gamma(Q_{K'})$ is included in the irreducible representations $\Gamma(\mathbf{r})$ based on x, y, or z.

In molecular spectroscopy one is also interested in *Raman transitions*. In these a quantum of the applied frequency ν is absorbed, and another is emitted with frequency $\nu \pm \nu_{ab}$, where $h\nu_{ab} = E_a - E_b$ is the difference in energy of two states of the system. The frequency of the Raman scattered radiation is $(\nu - \nu_{ab})$ or $(\nu + \nu_{ab})$, depending on whether the incident radiation causes a transition from state b to a, or vice versa. For historical reasons, these two possibilities are called "Stokes" and "anti-Stokes" transitions, respectively. Evidently, Stokes transitions will dominate when $h\nu_{ab} \gg kT$. Detailed quantum-mechanical investigation shows that Raman transitions can occur only if dipole matrix elements exist between both states a and b and some intermediate state i. But if $(a\,|\mathbf{r}|\,i) \times (i\,|\mathbf{r}|\,b) \neq 0$, then under all conditions of interest, $(a\,|\mathbf{rr}|\,b) \neq 0$, where $\mathbf{rr}$ includes all bilinear forms in the coordinates. From this loose argument we conclude that Raman excitation by a Stokes transition is possible for modes such that $\Gamma(Q_K)$ is included in the representations of the bilinear terms x^2, y^2, z^2, xy, yz, zx. The same conclusion is reached by imagining the process to go in two steps: the absorption of the initial photon must take the system to a state i of a representation found in $\Gamma(\mathbf{r})$. The emission of the scattered photon then takes the system from i to a state found in $\Gamma(\mathbf{r}) \times \Gamma(\mathbf{r})$, by the selection-rule theorem. But

$$\Gamma(\mathbf{r}) \times \Gamma(\mathbf{r}) = \Gamma(x^2) + \Gamma(y^2) + \Gamma(z^2) + \Gamma(xy) + \Gamma(yz) + \Gamma(zx).$$

Therefore the overall transition is governed by the transformation properties of these bilinear terms.

In working out selection rules for more general transitions involving combination and overtone levels, the overall symmetry as given by (7-80) must be used. That is, ordinary dipole transitions between states a and b are not forbidden by symmetry if Γ_b is found in the direct product $\Gamma(\mathbf{r}) \times \Gamma_a$. Similarly, Raman transitions are not forbidden if Γ_b is found in $\Gamma(\mathbf{rr}) \times \Gamma_a$.

Note that the Raman-transition operators have even parity, while the direct dipole operator has odd parity. Therefore, in molecules with a center of symmetry, in which therefore parity is a good quantum number for the irreducible representations of the vibrational modes, the same vibrational frequency cannot appear in both the direct and Raman spectra. This important principle is often referred to as the *exclusion rule* of molecular spectra.

We now illustrate these principles by considering the molecular examples treated earlier.

$H_2O(C_{2v})$ ■ Here $\Gamma(z) = \Gamma(x^2) = \Gamma(y^2) = \Gamma(z^2) = A_1$ and $\Gamma(x) = \Gamma(xz) = B_1$. Therefore all the vibrational modes can be excited from the ground state by either direct or Raman transitions.

$C_2H_2(D_{\infty h})$ ■ Here $\Gamma(x^2 + y^2) = \Gamma(z^2) = \Sigma_g{}^+, \Gamma(z) = \Sigma_u{}^+, \Gamma(xz, yz) = \Pi_g$, and $\Gamma(x, y) = \Pi_u$. Therefore only the $\Sigma_u{}^+$ and Π_u modes can be directly excited, whereas only the $\Sigma_g{}^+$ and Π_g modes can be excited in a Raman transition. Actually, the infrared active modes can be picked out by inspection of the sketches in Figs. 7-8 and 7-9 showing the vibrational motions. The infrared active modes are those with an overall oscillating dipole moment, allowing for equivalence of symmetrically related atoms. A less obvious example is the question of whether or not a dipole transition between the overtone level $(\Pi_g)^2$ and the combination level $\Pi_g\Pi_u$ is allowed by symmetry. The decompositions of these two levels are $(\Sigma_g{}^+ + \Delta_g)$ and $(\Sigma_u{}^- + \Sigma_u{}^+ + \Delta_u)$, as found previously. Because of the parity change, only a direct transition is possible. Inspection of the character tables shows that the x and y components can do nothing but that the z component of the dipole moment can couple $\Sigma_g{}^+ \leftrightarrow \Sigma_u{}^+$ and $\Delta_g \leftrightarrow \Delta_u$. Thus, transitions between these pairs of sublevels can in principle occur. They would be expected to be very weak compared with the fundamental absorptions, however.

7-14 Molecular Rotation

We now proceed to the next smaller term in the molecular Hamiltonian. This is the energy associated with the rotational degrees of freedom. The characteristic rotational energies are typically of the order of a few reciprocal centimeters, whereas the vibrational energies were usually of the order of 1,000 cm^{-1}. The rotational degrees of freedom are to a good approximation independent of the electronic and vibrational ones. We may simply average over those relatively rapid internal motions and treat the molecular rotation as that of a rigid body. The errors introduced by this *rigid-rotor* approximation can then be handled by small correction terms for centrifugal stretching and for electron "slip."

The rotational properties of a rigid body are completely specified by its three principal moments of inertia $I_a \leqq I_b \leqq I_c$, which lie on principal axes fixed in the molecule. In terms of these, the Hamiltonian is

$$H = \frac{J_x{}^2}{2I_x} + \frac{J_y{}^2}{2I_y} + \frac{J_z{}^2}{2I_z} \qquad \blacktriangleright \quad (7\text{-}84)$$

where the association between a, b, c and x, y, z is left free for convenience. Thus we need only calculate matrix elements of $J_x{}^2$, $J_y{}^2$, and $J_z{}^2$ with respect to axes gyrating with the molecule. This is readily done after we note that the commutation relations for angular momenta referred to *gyrating* axes have a *reversed* sign, namely,

$$J_x J_y - J_y J_x = -i\hbar J_z \qquad \blacktriangleright \quad (7\text{-}85)$$

and cyclic permutations. This sign change is equivalent to a reversal of the order of the successive operations in the left member, which corresponds to the fact that the successive operations are now to be carried out with respect to rotated axes (see Sec. 5-3). When matrix elements are worked out with the sign change, the major difference is that

$$J^+ = J_x + iJ_y$$

is now a *lowering* operator, rather than a raising operator. To distinguish the two coordinate systems, we introduce the following notation: We write J^+ instead of J_+ to indicate that we are in the gyrating system. Also, we use $g = x, y, z$ for gyrating axes and $F = X, Y, Z$ for space-fixed axes. Finally, we introduce the quantum number K for J_z, retaining M for J_Z.

The basic matrix elements are found to be ($\hbar$ being suppressed throughout)

$$(JK| J_z |JK) = K \tag{7-86a}$$

$$(JK \mp 1| J^\pm |JK) = [(J(J + 1) - K(K \mp 1)]^{\frac{1}{2}} \tag{7-86b}$$

Using these, we can work out the necessary matrix elements by matrix multiplication. We find the diagonal elements to be

$$(JK| J_x^2 |JK) = (JK| J_y^2 |JK) = \frac{J(J + 1) - K^2}{2} \tag{7-87a}$$

$$(JK| J_z^2 |JK) = K^2 \tag{7-87b}$$

so that $J_x^2 + J_y^2 + J_z^2 = J(J + 1)$, as must hold for a diagonal element. The off-diagonal elements are

$$(J, K \pm 2| J_x^2 |JK) = -(J, K \pm 2| J_y^2 |JK)$$

$$= \tfrac{1}{4}[J(J + 1) - K(K \pm 1)]^{\frac{1}{2}}[J(J + 1) - (K \pm 1)(K \pm 2)]^{\frac{1}{2}} \tag{7-88}$$

Therefore the only nonvanishing matrix elements of the Hamiltonian are

$$(JK| H |JK) = \frac{1}{4}\left(\frac{1}{I_x} + \frac{1}{I_y}\right)[J(J + 1) - K^2] + \frac{1}{2I_z} K^2 \tag{7-89a}$$

and $(J, K \pm 2| H |JK) = \dfrac{1}{8}\left(\dfrac{1}{I_x} - \dfrac{1}{I_y}\right)[J(J + 1) - K(K \pm 1)]^{\frac{1}{2}}$

$$\times [J(J + 1) - (K \pm 1)(K \pm 2)]^{\frac{1}{2}} \tag{7-89b}$$

Symmetric top. Before approaching this general Hamiltonian, let us consider the important special case where two of the moments of inertia, I_x and I_y, are equal. Then the $(K \pm 2| H |K)$ elements [(7-89b)] vanish, and the Hamiltonian is exactly diagonal in our representation based on $|JK)$ eigenstates. We can simply read off the energy levels from (7-89a), namely,

$$E(K, J) = \frac{1}{2I_x} [J(J + 1) - K^2] + \frac{1}{2I_z} K^2$$

$$= BJ(J + 1) + (A - B)K^2 \qquad (7\text{-}90)$$

where A and B are conventional symbols for half the reciprocal moments of inertia. The result (7-90) is almost obvious by inspection of the Hamiltonian, since $K^2 = J_z{}^2$ and $J(J + 1) - K^2$ must then be $J_x{}^2 + J_y{}^2$. We note that these levels (except $K = 0$) are doubly degenerate. The values $K = \pm |K|$ have the same energy and correspond simply to rotation in opposite senses about the figure axis.

In the still simpler case of the linear molecule, $I_z = 0$. Therefore the only states with finite energy have $K = 0$, and the energy becomes

$$E(J) = \frac{1}{2I} J(J + 1) = BJ(J + 1) \qquad (7\text{-}91)$$

Selection rules and spectrum. By symmetry the permanent electric-dipole moment of a symmetric top molecule must lie along the figure or z axis. Therefore it transforms as the z component of a vector operator. By the Wigner-Eckart theorem this leads to selection rules for pure rotational transitions requiring that

$$\Delta J = \pm 1, 0 \qquad \Delta K = 0 \qquad (7\text{-}92)$$

Since K does not change in a transition, the energy difference for a transition from J to $J + 1$ is simply

$$h\nu = B[(J + 1)(J + 2) - J(J + 1)] = 2B(J + 1) \qquad (7\text{-}93)$$

Thus, just as in the Landé interval rule for atomic fine structure, the level separation is proportional to the higher of the two angular momenta involved in the transition. Since $J = 0, 1, 2, \ldots$, this formula predicts that successive rotational-transition frequencies should be harmonics of the lowest frequency. This rule is excellently obeyed in practice, as is verified by very precise microwave spectroscopy. The main departures arise from centrifugal distortion, which can be viewed as a change in I_x resulting from the stretching of the molecule by the centrifugal force as one goes to higher rotational levels.

7-15 Effect of Nuclear Statistics on Molecular Rotation

The simple results indicated above need to be modified slightly when the molecule under consideration contains identical nuclei. In that case, the

replacement of the molecule by a classical rigid body is not completely justified. The reason for this is that a rotational motion of the molecule effectively interchanges nuclei. If nuclei are identical, the total wavefunction
must be either symmetric or antisymmetric under such an interchange,
depending on whether the nuclei involved are bosons or fermions. This
requirement may eliminate certain rotational-energy levels or at least change
statistical weights because of the nuclear spin degrees of freedom.

We shall confine our attention to the simple and important case of
homonuclear diatomic molecules.[1] In this case there are just two nuclei to
consider. We can interchange the nuclei, while restoring everything else
to its original position, by the following series of operations:

1. Rotate the entire molecule through an angle π.

2. Invert the electrons back through the center of symmetry.

3. Reflect the electrons in a plane through the molecular axis perpendicular to the axis of rotation.

The rotation 1 introduces a $\pm$ sign depending on whether the rotational
angular momentum J or $\mathscr{K}$ is $\begin{pmatrix} \text{even} \\ \text{odd} \end{pmatrix}$. (The notation $\mathscr{K}$ is used if the
total angular momentum J also contains a spin part.) Operation 2 intro-
duces a $\pm$ sign for $\begin{pmatrix} g \\ u \end{pmatrix}$ states. For $\Sigma^{\pm}$ states, operation 3 introduces a $\pm$.
For Π, Δ, ... states, the twofold degeneracy of $\Lambda = \pm |\Lambda|$ is slightly split
by the presence of rotation, a phenomenon known as Λ doubling. The appro-
priate eigenfunctions[2] in the presence of rotation are $(F_{+}e^{i|\Lambda|\varphi} \pm F_{-}e^{-i|\Lambda|\varphi})$,
where F_{+} and F_{-} are functions independent of φ, variation of φ representing
a rotation of the entire electron cloud about the molecular axis. In this
case, one can show that one of these combinations is odd and the other is
even, so that both signs are available. The overall sign change coming
from the product of the three operations above with that from the nuclear
wavefunction itself must be $\pm$ depending on whether we are dealing with
$\begin{pmatrix} \text{bosons} \\ \text{fermions} \end{pmatrix}$.

Examples. O_2^{16} ■ $^3\Sigma_g^{-}$, $I = 0$. Since there is electronic-spin angular
momentum, we denote the rotational quantum number by $\mathscr{K}$ (so that
$\mathbf{J} = \mathscr{K} + \mathbf{S}$). Then the operations 1, 2, and 3 introduce sign changes as
follows:

$$(-1)^{\mathscr{K}}(+1)(-1) = (-1)^{\mathscr{K}+1}$$

---

[1] For the more general case, see, for example, C. H. Townes and A. L. Schawlow,
"Microwave Spectroscopy," pp. 69 and 102, McGraw-Hill Book Company, Inc., New
York, 1955.

[2] See, for example, R. de L. Kronig, "Band Spectra and Molecular Structure," pp. 63–65,
Cambridge University Press, New York, 1930. Original references are R. de L. Kronig,
Z. Physik, **46**, 814 (1927), **50**, 347 (1928); E. Wigner and E. E. Witmer, *Z. Physik*, **51**, 859
(1928).

The nuclear wavefunction must be symmetric since only one state is available for $I = 0$, namely, $M_I = 0$. Since the nuclear spin is integral, the nuclei are bosons and we must have symmetry (not antisymmetry) under interchange. Thus we require

$$(-1)^{\mathscr{K}+1} = +1$$

or
$$\mathscr{K} = \text{odd}$$

Thus the only rotational levels of O_2 are $\mathscr{K} = 1, 3, 5, \ldots$. These restrictions naturally disappear if the isotopes are different. Thus $O^{16}O^{17}$ has all rotational levels available, including the nonrotating $\mathscr{K} = 0$ state, which is forbidden to O_2^{16}.

H_2 ■ $^1\Sigma_g{}^+$, $I = \frac{1}{2}$. Here the first three operations produce $(-1)^{\mathscr{K}}$, and the final result must be (-1) since we have Fermi statistics. Therefore the nuclear-spin state must be symmetric for odd $\mathscr{K}$ and antisymmetric for even $\mathscr{K}$. The symmetric state is the $F = 1$ state, represented by $\alpha(1)\alpha(2)$, $[\alpha(1)\beta(2) + \beta(1)\alpha(2)]/\sqrt{2}$, and $\beta(1)\beta(2)$. The antisymmetric state is the $F = 0$ state $[\alpha(1)\beta(2) - \beta(1)\alpha(2)]/\sqrt{2}$. Thus the statistical weight of the odd-$\mathscr{K}$ levels is three times that of the even-$\mathscr{K}$ levels because of the nuclear degeneracy. The more populous levels are called *orthohydrogen*, the less populous *parahydrogen*. Because of the weak coupling between the nuclear spins and the rest of the world, the conversion between ortho- and para states is very slow. This can cause anomalous low-temperature specific heats because molecules get "trapped" in the $\mathscr{K} = 1$ state and cannot drop readily into the $\mathscr{K} = 0$ ground state.

The ortho-para ratio is given in general by $(I + 1)/I$, as has been illustrated by the results ∞ for $O_2(I = 0)$ and 3 for $H_2(I = \frac{1}{2})$. To see that this ratio is generally true, consider the sorts of nuclear-spin functions possible for two nuclei having spin I. Since there are $(2I + 1)$ different m_I values for each, there are $(2I + 1)^2$ possible product functions. Of these, $(2I + 1)$ will have $m_I(1) = m_I(2)$, and hence they must obviously be symmetric. The remaining $[(2I + 1)^2 - (2I + 1)]$ functions can be grouped into equal numbers of symmetric and antisymmetric combinations of the sort $[m_I(1)m_I'(2) \pm m_I'(1)m_I(2)]$. From this we see that the number of symmetric combinations is $(2I + 1)(I + 1)$, whereas the number of antisymmetric combinations is $(2I + 1)I$. Taking the ratio, we have the general result

$$\frac{\text{Number of symmetric combinations}}{\text{Number of antisymmetric combinations}} = \frac{I + 1}{I} \tag{7-94}$$

This result may be used to determine a nuclear spin I from the alternating intensities of lines in molecular band spectra. This method was particularly important in early work, and it is still useful because it works independently of the magnitude of any nuclear moments. In fact it works best when

$I = 0$ so that there can be no hyperfine structure at all. This property makes it useful for determining whether the absence of hyperfine structure in a spectrum is actually due to $I = 0$ or merely to a fortuitously small magnetic moment or quadrupole moment.

7-16 Asymmetric Rotor

In an asymmetric rotor, all three moments of inertia are distinct. In this case, (7-89) shows that we have matrix elements which are off-diagonal in $K (= J_z)$ as well as diagonal. However, J is still a rigorous quantum number, as is $M (= J_Z)$, as they must be because of the isotropy of space. Since the energy is independent of M, we can give M any convenient value, such as 0 or J. Thus the overall Hamiltonian matrix for any given M value is composed of $(2J + 1) \times (2J + 1)$ blocks along the diagonal, for $J = 0, 1, 2, \ldots$. There is enough symmetry in the Hamiltonian even here to get some further factoring of the secular equation. First, since the only off-diagonal matrix elements (7-89b) connect states differing by two units in K, each block could be factored into two parts simply by separating the even and odd values of K. Second, inspection of the Hamiltonian matrix elements shows that the matrix is symmetric under reflection across either diagonal—not just across the principal diagonal, as is required for any real Hermitian matrix. This symmetry between $\pm |K|$ arises because of time-reversal symmetry, and it suggests forming the symmetric and antisymmetric combinations $(\psi_K \pm \psi_{-K})/\sqrt{2}$. Since these two types of combinations have distinct symmetries, there are no matrix components of energy between them. The transformation to these symmetrized basis functions is carried out by the Wang symmetrizing transformation.[1]

$$X_J = X_J^{-1} = \frac{1}{\sqrt{2}} \begin{pmatrix} \ddots & & & & & \ddots \\ -1 & 0 & 0 & 0 & 1 \\ 0 & -1 & 0 & 1 & 0 \\ 0 & 0 & \sqrt{2} & 0 & 0 \\ 0 & 1 & 0 & 1 & 0 \\ 1 & 0 & 0 & 0 & 1 \\ \ddots & & & & & \ddots \end{pmatrix} \qquad (7\text{-}95)$$

Thus the $(2J + 1) \times (2J + 1)$ submatrix for a given J, after the transformation

$$H_J' = X_J^{-1} H_J X_J$$

and some reordering of rows and columns to separate even and odd K, is factored into four sub-submatrices, each of dimension approximately $J/2$

[1] S. C. Wang, *Phys. Rev.*, **34**, 243 (1929); J. H. Van Vleck, *Phys. Rev.*, **33**, 467 (1929).

instead of $(2J + 1)$. This decrease in order of the secular equations greatly facilitates solution. Townes and Schawlow[1] tabulate the exact algebraic solutions for low-lying levels. When the secular equations become too large for exact algebraic solution, the continued fraction-expansion method gives a rapidly convergent numerical solution.[2] Extensive numerical tabulations of these rigid-rotor eigenvalues have been compiled by King, Hainer,

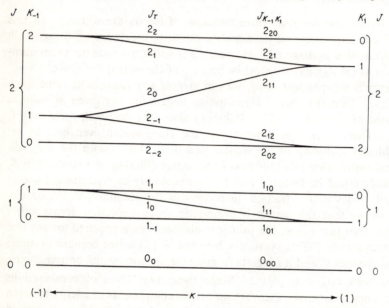

Fig. 7-10. *Qualitative behavior of asymmetric top energy levels showing common notations for labeling of energy levels, as described in the text.*

and Cross.[3] These tables, which are at the heart of much of microwave spectroscopy, are reproduced in the appendixes of the books of Strandberg and of Townes and Schawlow, cited above. They give the energies of all low-lying states for a large number of degrees of asymmetry between the prolate and oblate symmetric top.

The degree of asymmetry is an important parameter, since it determines how far the solutions will deviate from the simple symmetric top results. The most useful asymmetry parameter is that of Ray[4] which is defined by

$$\kappa = \frac{2B - A - C}{A - C} \tag{7-96}$$

[1] Townes and Schawlow, *op. cit.*, p. 90.

[2] See, for example, M. W. P. Strandberg, "Microwave Spectroscopy," p. 11, Methuen & Co., Ltd., London, 1954.

[3] G. W. King, R. M. Hainer, and P. C. Cross, *J. Chem. Phys.*, **11**, 27 (1943); **17**, 826 (1949).

[4] B. S. Ray, *Z. Physik*, **78**, 74 (1932).

where the three rotational constants are ordered so that $A > B > C$. This becomes -1 for a prolate symmetric top ($B = C$) and $+1$ for an oblate symmetric top ($B = A$).

Since K is not a good quantum number in the asymmetric top, an additional label must be introduced to supplement J in denoting a particular level. One system of notation which is used is to give K_{-1} and K_{+1}, the values of K found by letting the asymmetry parameter κ go continuously to -1 and $+1$ from the actual value. Thus the level 4_{32} is one where $J = 4$, $K_{-1} = 3$, and $K_1 = 2$. An alternative notation simply numbers the levels in order of energy with an index $\tau(-J \leqslant \tau \leqslant J)$ so that the J_{-J} level is the lowest and J_J is the highest of the $(2J + 1)$ levels. One can show that $\tau = K_{-1} - K_1$; so there is a simple relation between these nomenclatures. These ideas are illustrated in Fig. 7-10, which shows qualitatively the dependence of energy levels on asymmetry. The energy levels at both $\kappa = -1$ and $+1$ are easily found from (7-90). When $\kappa \neq \pm 1$, the degeneracy of $\pm K$ levels splits, and since there must be a continuous variation with κ, such a connection diagram results.

7-17 Vibration-Rotation Interaction

So far we have discussed what might be called the zero-order Born-Oppenheimer approximation in which the state of the molecule is described as a simple product of an electronic function computed for clamped nuclei, a vibrational eigenfunction, a rotational eigenfunction, a nuclear-spin eigenfunction, and of course in principle a center-of-mass translational eigenfunction. The total energy is expressed as a sum of contributions from each of these degrees of freedom. However, we know from our derivation of the Born-Oppenheimer separation that there are coupling terms in the Hamiltonian which make the separation only approximate. We shall now take up some of the coupling terms to see what their effect is. First we shall consider the interaction between the vibrational and rotational degrees of freedom.

In our treatment of the rotational energy, we treated the nuclear framework as a rigid body. We now take account of the fact that the nuclei are fixed in place only by the finite restoring forces which describe the vibrational motion. As an example, consider the simple diatomic molecule. We take a Hamiltonian describing both rotation and vibration, including an anharmonic potential of the sort considered in (7-47), namely,

$$H = \frac{1}{2\mu R^2}\,\mathbf{J}^2 + a_2 \xi^2 + a_3 \xi^3 + \frac{P^2}{2\mu} \tag{7-97}$$

where, as before, μ is the reduced mass, μR^2 is the moment of inertia, and ξ is a dimensionless vibrational coordinate. Since we no longer treat the molecule as rigid, R cannot be replaced by its equilibrium value R_e in calculating the moment of inertia.   Rather we expand,

$$\frac{1}{R^2} = \frac{1}{R_e^2} [1 - 2\xi + 3\xi^2 + \cdots] \qquad \xi = \frac{R - R_e}{R_e} \qquad (7\text{-}98)$$

Then we rewrite the Hamiltonian as $H = H_0 + H'$, where

$$H_0 = B_e \mathbf{J}^2 + a_2\xi^2 + \frac{P^2}{2\mu} \qquad (7\text{-}99a)$$

and

$$H' = a_3\xi^3 - 2B_e\xi\mathbf{J}^2 + 3B_e\xi^2\mathbf{J}^2 \qquad (7\text{-}99b)$$

and $B_e = (2\mu R_e^2)^{-1}$.   Now, the product of a harmonic-oscillator eigenfunction of ξ and a rotational eigenfunction corresponding to some J forms an eigenfunction for H_0. H' provides the correction terms due to anharmonicity and coupling between rotation and vibration.   The matrix elements of H are readily written down in a $|vJM\rangle$ representation.   Carrying the perturbation theory to second order, and subtracting the vibrational-energy terms which we found earlier in the pure vibrational problem [Eq. (7-52)], we obtain

$$W_{\text{rot}} = [B_e - \alpha_e(v + \tfrac{1}{2})]J(J + 1) - DJ^2(J + 1)^2 \qquad \blacktriangleright \quad (7\text{-}100)$$

where

$$\alpha_e = -B_e\left[\frac{2a_3\lambda^6}{B_e} + 6\lambda^2\right]$$

$$= -B_e\left[\frac{2a_3B_e^2}{(\hbar\omega_e)^3} + \frac{6B_e}{\hbar\omega_e}\right] \qquad (7\text{-}100a)$$

and

$$D = \frac{4B_e^3}{(\hbar\omega_e)^2} \qquad (7\text{-}100b)$$

The term in α_e accounts for the fact that vibrational motion gives a value for $\langle 1/R^2\rangle_{\text{av}}$ different from $1/R_e^2$.   Therefore the moment of inertia depends on the amplitude of the vibration.   The anharmonic coefficient a_3 is negative and large enough to overcome the second term in (7-100a).   Therefore α_e is a positive quantity, and the rotational constant $B_v = B_e - \alpha_e(v + \tfrac{1}{2})$ decreases as one goes to excited vibrational states.   Even in the ground state, the zero-point vibration reduces B_0 below the value B_e which it would have if the nuclei were at the equilibrium separation giving a minimum of potential energy.

The term $DJ^2(J + 1)^2$ arises from centrifugal distortion.   Classically, the centrifugal force is proportional to ω^2 and hence to J^2.   This force then produces a proportional increase in R and in the moment of inertia.   The

D term is the corresponding quantum-mechanical result. Perhaps the energy expression is more easily understood when written as

$$W_{\text{rot}} = B_{vJ}J(J+1) \tag{7-101}$$

where

$$B_{vJ} = B_e\left[1 - \frac{\alpha_e}{B_e}\left(v + \frac{1}{2}\right) - \frac{D}{B_e}J(J+1)\right] \tag{7-101a}$$

directly shows the effect of vibration and centrifugal distortion on the rotational constant. Note that (7-100a) and (7-100b) show that both α_e and D go to zero if the molecule is made rigid ($\hbar\omega_e \to \infty$). This is, of course, as expected from the physical nature of the effects.

We can readily extend these results from the diatomic case to the case of a polyatomic molecule with three principal reciprocal moments of inertia A, B, C. Since each of $(3N - 6)$ normal modes of vibration will enter into changing these moments, we have

$$A_v = A_e - \sum_{i=1}^{3N-6} \alpha_i^A(v_i + \tfrac{1}{2})$$

$$B_v = B_e - \sum_{i=1}^{3N-6} \alpha_i^B(v_i + \tfrac{1}{2}) \tag{7-102}$$

$$C_v = C_e - \sum_{i=1}^{3N-6} \alpha_i^C(v_i + \tfrac{1}{2})$$

where each α_i includes anharmonic as well as harmonic effects. The various coefficients can be evaluated experimentally if data are available on enough excited vibrational states.

It is a bit more difficult to take account of centrifugal-distortion effects in the general asymmetric top. For a symmetric top, however, the generalization is simple. Since the energy can depend only on *squares* of angular momenta, we have

$$W(J, K) = B_vJ(J + 1) + (A_v - B_v)K^2$$
$$- D_JJ^2(J + 1)^2 - D_{JK}J(J + 1)K^2 - D_KK^4 \tag{7-103}$$

The parameters D_J, D_{JK}, D_K are in principle determined by the values of A and B and the various vibrational frequencies. However, in practice they are determined by fitting the experimental data. Since microwave spectroscopic data can be fitted to parts per million by such a function, the moments of inertia can be found to a similar accuracy. From these, the structural parameters (bond lengths and angles) may be determined with very high precision since the atomic masses are known from other experiments.

7-18 Rotation-Electronic Coupling

When one takes cognizance of the fact that a molecule consists of an electronic-charge cloud surrounding a nuclear framework which is rotating, it seems evident that the electronic motion will be perturbed by the rotation, with observable effects.   One simple approach to this problem is to consider the nuclei as an infinitely massive classically rotating set of force centers.[1] Then the Hamiltonian for the electrons has an explicit time dependence, arising from the motion of the nuclei.   However, if one transforms to new coordinates x_i' measured in the rotating system, the problem looks time-independent.   This suggests seeking a solution to the time-dependent Schrödinger equation, which will look stationary in the moving frame. Following this approach, one finds that the electrons must satisfy an effective Hamiltonian given by

$$H_0 - \boldsymbol{\omega} \cdot \mathbf{L} \tag{7-104}$$

where H_0 is the Hamiltonian in absence of rotation, $\mathbf{L}$ is the electronic-orbital angular momentum and $\boldsymbol{\omega}$ is the angular velocity of the molecular framework.   This result is equivalent to the Coriolis force of classical mechanics or the effective magnetic field which enters in Larmor's theorem. Although (7-104) leads to proper results when we replace $\boldsymbol{\omega}$ by $\mathbf{J}/I = 2B\mathbf{J}$ and treat the rotation quantum mechanically, there remain some questions which can be clarified by a unified approach in which a single overall Hamiltonian is used throughout.

This overall Hamiltonian contains the energy of the electrons and that of the rotating nuclei.   If we consider the Hamiltonian of the electrons, referred to a cartesian-coordinate system fixed in the rotating frame, it will retain exactly the same form as before, namely $H_0 = \sum_i \mathbf{p}_i^2/2m + V$.   In this, $p_i = \partial T/\partial \dot{q}_i$, which turns out to be mv_i, where v_i is measured in the *fixed* frame.   Thus the form of the Hamiltonian stays the same, as opposed to the result (7-104), which is based on a different viewpoint.   The rotational energy of the bare nuclear framework will be given by

$$H_{\mathrm{rot}} = \sum_g B_g N_g^2 = \sum_g B_g (J_g - L_g)^2 \tag{7-105}$$

where g runs over the three principal axes in the gyrating system, N_g is the angular momentum of the nuclear frame, L_g is that of the electrons, and J_g is that of the total system.   (For the present, we assume that $S = 0$, restricting attention to singlet states.)   Of these, only J is a good quantum number, because the interaction between $\mathbf{L}$ and $\mathbf{N}$ spoils their individual conservation.   Upon restricting attention to a linear molecule, with $B_x = B_y = B_N$ and $B_z = \infty$, and including the electronic Hamiltonian H_0, (7-105)

[1] G. C. Wick, *Phys. Rev.*, **73**, 51 (1948).

becomes

$$H = H_0 + H_{\text{rot}}$$

$$= H_0 + B_N[(J_x{}^2 + J_y{}^2) - 2(J_xL_x + J_yL_y) + (L_x{}^2 + L_y{}^2)] \quad (7\text{-}106)$$

We note immediately that the cross term exactly reproduces the correction term found in (7-104); so the two approaches are equivalent. Since the last term in (7-106) is small and essentially independent of rotation, it will be dropped in our subsequent discussion. We now briefly consider several observable effects arising from the cross term in (7-106).

Λ **doubling.** After dropping the $(L_x{}^2 + L_x{}^2)$ term, (7-106) becomes

$$H = H_0 + B_N[(J_x{}^2 + J_y{}^2) - 2(J_xL_x + J_yL_y)]$$

Since $\mathbf{J}^2$ and J_z are diagonal operators in a linear molecule, this can be written as

$$H = H_0 + B_N[J(J + 1) - J_z{}^2] - 2B_N(J_xL_x + J_yL_y] \quad (7\text{-}107)$$

Since N_z must equal zero in a linear molecule, $J_z = L_z$, which has the quantized value Λ. Thus the zero-order energy expression is

$$E(n, J, \Lambda) = E_e(n, \Lambda) + B_N[J(J + 1) - \Lambda^2] \quad (7\text{-}108)$$

where n symbolizes any additional quantum numbers required to specify the electronic state of the molecule. This expression displays the usual twofold degeneracy between states with $\Lambda = \pm |\Lambda|$. When the cross term in (7-107) is introduced as a perturbation, this degeneracy is lifted because it has matrix elements with selection rule $\Delta\Lambda = \pm 1$. In the simple case of a Π state, where $\Lambda = \pm 1$, the two degenerate states are coupled via Σ states $(\Lambda = 0)$. By applying perturbation theory, it is clear that the splitting will be of order

$$\frac{B_N{}^2 J(J + 1)}{h\nu_{\Pi\Sigma}}$$

where $h\nu_{\Pi\Sigma}$ is a typical electronic-energy separation from the Π state of interest to nearby Σ states. Since $h\nu_{\Pi\Sigma} \gg B$, this splitting will be small and we get what is called a *doubling* of the levels. For states with $|\Lambda| > 1$, the splitting is still smaller (usually unobservable), since the perturbation theory must be carried to higher order to obtain an effective matrix element between $+ |\Lambda|$ and $- |\Lambda|$. In general, the splitting is of order $B(B/h\nu)^\Lambda$.

A crude physical picture of the origin of this doubling results if we consider the effect of the perturbation on the wavefunctions. For $|\Lambda| = 1$, the unperturbed functions have the dependence $e^{\pm i\varphi}$ on angle about the molecular axis. When mixed by the perturbation, these lead to essentially $\cos \varphi$ and $\sin \varphi$ dependences. The resulting electronic-charge distributions have different moments of inertia about the axis of rotation. This leads to slightly different rotational constants for the two states, and hence to an energy difference proportional to $J(J + 1)$.

Electronic contribution to the moment of inertia. The foregoing discussion shows that it is not correct to calculate the molecular moment of inertia by assuming the entire atomic mass to be concentrated at the nucleus. It is also incorrect to treat the electronic-charge cloud as a rigid body moving with the nuclear framework, because the electrons can "slip" by having circulating currents within the charge cloud as it rotates. In particular, we expect the spherical inner shells of electrons around each nucleus to follow the nuclei, maintaining a fixed orientation in space like the seats on a rotating Ferris wheel. If they do this, slip is complete and these electrons contribute to the moment of inertia exactly as if they were located at the corresponding nucleus. The valence electrons, not being in spherical shells, will tend to rotate rigidly with the framework.

To give a more quantitative treatment which still is simple, let us confine our attention to linear molecules in a $^1\Sigma$ ground state. Then we have

$$H = H_0 + B_N J(J + 1) + H' \tag{7-109}$$

where

$$H' = -2B_N(J_x L_x + J_y L_y) \tag{7-109a}$$

Now, when $J \neq 0$, H' will mix some Π states into the Σ state, since L_x and L_y have $\Delta\Lambda = \pm 1$ selection rules. This produces a circulating electronic current proportional to $BJ/h\nu_{\Pi\Sigma}$. The effect on the energy appears in second-order perturbation theory as

$$\Delta E = 4B_N^2 J(J + 1) \sum_\Pi \frac{|(\Pi |L_x| \Sigma)|^2}{h\nu_{\Pi\Sigma}}$$

where the Σ in the matrix element refers to the ground state and the sum runs over all excited Π states. When combined with (7-109), this result shows that the effective rotational constant including the electrons is

$$B = B_N + B_e$$
$$= B_N - 4B_N^2 \sum_\Pi \frac{|(\Pi |L_x| \Sigma)|^2}{h\nu_{\Pi\Sigma}} \tag{7-110}$$

Since the correction term may be of order $10^{-4}B_N$, it must be evaluated before extremely precise determinations of internuclear distances can be made from observed values of B. As will now be shown, this correction can be determined from the rotational magnetic moment of the molecule.

Rotational magnetic moment. Since the negative charge in the atoms in a molecule is not centered at the nuclei, one would expect a magnetic moment to result from the rotation of even a homonuclear diatomic molecule. Again, however, the circulation produced by the rotation is important. For example, in H_2 the slip is so great that the rotational g factor (defined by $\mu_{rot} = g_r \beta \mathbf{J}$) is $+0.883\ m/M$ compared with the value $1.00\ m/M$ which would result if there were no electrons at all. [Here m/M is the electronic mass in atomic mass units (amu).] This case is unusual, however, and most molecules will have $g_r < 0$ because the electronic contribution slightly outweighs that of the bare nuclei. The electronic contribution is readily shown to be

$$g_r{}^e = -4B_N \sum_\Pi \frac{|(\Pi\ |L_x|\ \Sigma)|^2}{h\nu_{\Pi\Sigma}} = \frac{B_e}{B_N} \tag{7-111}$$

whereas the nuclear contribution is obviously

$$g_r{}^n = \frac{Z}{A}\frac{m}{M} \tag{7-112}$$

Since (7-111) shows that the electronic contribution to B can be expressed in terms of the rotational g factor, we can write

$$\begin{aligned} B &= B_N(1 + g_r{}^e) = B_A(1 + g_r{}^n)(1 + g_r{}^e) \\ &\approx B_A(1 + g_r{}^e + g_r{}^n) = B_A(1 + g_r) \end{aligned} \tag{7-113}$$

where B_A is the rotational constant calculated by using atomic instead of nuclear masses. Since $|g_r|$ is usually much less than $|g_r{}^e|$, the use of atomic masses is a good first approximation. However, to determine internuclear distances to the full accuracy of the data, it is necessary to measure g_r experimentally to evaluate the correction. The rotational g factors for a dozen or so molecules have been measured by observing the small Zeeman energy of the rotational magnetic moment in an external field.

Coupling effects of internal magnetic field. The effective magnetic field produced inside a Σ-state molecule by the rotation will give an energetic coupling to the nuclear spins, and to the electronic spin if the molecule is not in a singlet state. These effects are best visualized by considering the rotation to mix in some Π states, which have orbital angular momentum. These then interact with the nuclear moments via the ordinary h.f.s. coupling and with the electron spin via the spin-orbit coupling $\zeta \mathbf{L} \cdot \mathbf{S}$. The former gives rise to an $\mathbf{I} \cdot \mathcal{K}$ term, while the latter gives rise to a term in the effective Hamiltonian having the form

$$\sim \left[\zeta B \sum_\Pi \frac{|(\Pi\ |L_x|\ \Sigma)|^2}{h\nu_{\Pi\Sigma}} \right] \mathcal{K} \cdot \mathbf{S} \tag{7-114}$$

where $\mathcal{K} = \mathbf{J} - \mathbf{S}$ is the rotational angular momentum, "high-frequency" effects due to the electronic orbital angular momentum $\mathbf{L}$ being neglected.

It should be clear that various other small correction and spin-coupling terms can arise. The fact that many of them depend on similar sums of squared matrix elements allows a certain amount of interrelation of the data to form a coherent picture of the interaction between rotational and electronic degrees of freedom. Examples in which this sort of analysis have been carried particularly far are molecular hydrogen[1] and oxygen.[2]

EXERCISES

7-1 Consider the charge density along the line passing through the nuclei of a hydrogen-molecule ion. Assuming simple LCAO MO's made from ordinary atomic hydrogen $1s$ functions, plot the probability density $\rho = |\psi|^2$ for both the ψ_g and ψ_u functions in Eq. (7-11). Note the difference. Compare these with half the charge density that would result from a classical (i.e., non-wave-mechanical) superposition of the charge densities of two hydrogen atoms. Take the inter-nuclear distance to be $1.06\,A = 2.0$ atomic units, and the overlap integral S to be 0.586. Note the importance of the overlap correction on the normalization.

7-2 As discussed in the text (Sec. 7-3), the molecular orbital $^3\Sigma_g{}^-$ ground state of O_2 is normally taken to be $(\sigma_g 2p)^2 (\pi_u 2p)^4 (\pi_g 2p)^2$ outside of filled $1s$ and $2s$ shells. If we relax this particular assignment among these three types of orbitals, we have eight electrons to assign to 10 spin-orbital functions. (a) Show that this gives rise to 45 possible Slater determinants. (b) Show that there are only three ways of assigning electrons to orbitals (with spin unspecified, but obeying the exclusion principle) so as to produce a Σ_g symmetry. (c) Taking account of the spin degeneracy, show that these three ways lead to three singlets and two triplets, accounting for a total of 9 of the 45 determinants. (d) Considering the triplets, show that both have the symmetry $^3\Sigma_g{}^-$. A simple 2×2 secular equation would determine the optimum admixture of the two configurations.

7-3 Repeat Exercise 7-2, taking account of the possible occupancy of the $(\sigma_u 2p)$ orbital. This leads to 12 states for the eight electrons. Show that there are now 495 possible determinants, 12 ways of assigning electrons for Σ_g symmetry, 15 singlets, 13 triplets, and 3 quintets, of which there are 9 $^3\Sigma_g{}^-$ states. Thus the secular equation for the ground state is only a 9×9 after symmetry has been fully utilized to eliminate extraneous determinants from the 495.[3]

7-4 Consider the dominant configuration in O_2, in which the only unfilled shell is the $(\pi_g 2p)^2$ one. (a) Considering only these two electrons, show explicitly the different correlations of the azimuthal angles φ_1 and φ_2 of the two electrons in the triplet and singlet spin configurations. (b) To allow a rough estimate of the importance of this effect, consider a model in which the electronic wavefunctions are confined to a ring of radius R about the axis of the molecule. Calculate $\langle r_{12} \rangle$ for the singlet and triplet states and for the uncorrelated case of a simple product function (not antisymmetrized). Show that $\langle r_{12} \rangle$ in the singlet and triplet, respectively, are $\frac{14}{15}$ and $\frac{16}{15}$ of $\langle r_{12} \rangle$ for the simple product. Note that, for this model of the wavefunction, $\langle 1/r_{12} \rangle$ would diverge for the singlet and uncorrelated cases, but

[1] N. J. Harrick and N. F. Ramsey, *Phys. Rev.*, **88**, 228 (1952).
[2] M. Tinkham and M. W. P. Strandberg, *Phys. Rev.*, **97**, 937 (1955).
[3] See A. Meckler, *J. Chem. Phys.*, **21**, 1750 (1953).

not for the triplet. This illustrates how the antisymmetry keeps electrons of parallel spin away from each other, with consequent lower electrostatic energy, leading to the Hund rule.

7-5 Consider a hypothetical molecule consisting of three identical atoms arranged at the vertices of an equilateral triangle so that the symmetry of the problem is D_{3h}. (a) Find the symmetries of all symmetry orbitals which can be formed from s and p atomic functions. (b) Find linear combinations of atomic functions forming orthonormal basis functions. (c) Work out expressions similar to (7-37) for the energies of the orbitals based on the s functions.

7-6 Consider the hypothetical triatomic molecule of Exercise 7-5. (a) How many vibrational degrees of freedom does it have? (b) Find the decomposition of the nine-dimensional representation based on cartesian displacements of the three atoms into irreducible representations of D_{3h}, and identify those corresponding to translation, rotation, and vibration. (c) Construct diagrams showing the displacements in each of the vibrational modes. (HINT: Use the basis-function generating machinery, starting with an arbitrary set of displacements having no translational or rotational component.) Which of these modes are infrared-active and which are Raman-active? (d) Work out the symmetries and degeneracies of all combination and overtone levels having a total of two units of excitation.

7-7 Consider again the symmetrical triatomic molecule of Exercises 7-5 and 7-6. Let the internuclear distance be a and the mass of each atom be M. (a) What are the three principal moments of inertia? (b) What are the rotational-energy levels? Make a diagram to scale showing all those for $J \leq 4$, also indicating the degeneracies for each. (c) What transitions, if any, are allowed electric-dipole transitions? Explain.

7-8 Work out general algebraic expressions for all the asymmetric rotor energy levels for $J = 0, 1, 2,$ and 3.

7-9 Consider the water molecule H_2O. Let the OH distances be a and the HOH bond angle be θ. (a) For general values of these parameters, what are the principal axes and the corresponding moments of inertia? Make a drawing to approximate scale, showing these axes for the actual case of $\theta = 105°$. (b) What is the value of Ray's asymmetry parameter κ for a general value θ? (c) Using the results of Exercise 7-8, compute the energy levels of H_2O for $J = 0, 1,$ and 2. (Use $a = 0.957$ A for the bond length and $\theta = 105°$.) Quote your results in units of reciprocal centimeters. Make an energy-level diagram showing these levels to approximate scale.

7-10 Derive expressions for the vibrational and rotational energy of a diatomic molecule analogous to Eqs. (7-52) and (7-100) but retaining a term $a_4\xi^4$ in the vibrational potential in addition to the terms given in Eq. (7-47).

7-11 In O_2, a $^3\Sigma$ molecule, there is a term of the form $2\lambda S_z{}^2$, coupling the spin to the molecular figure axis z, and another term, of the form $\mu\mathcal{K} \cdot \mathbf{S}$, where $\mathcal{K}$ is the rotational angular momentum ($\mathbf{J-S}$), coupling the spin to the rotational motion. [See Exercise 5-10 and Eq. (7-114).] Consider these terms to be added to the usual rotational Hamiltonian $B\mathcal{K}^2$, and solve the resulting Hamiltonian for the energy levels. Note that J and S are good quantum numbers but that $\mathcal{K}$ is not. An exact solution can be obtained for all energy levels, which involves no more than a 2×2 secular equation. Numerical values of the parameters are $B = 1.44$, $\lambda = 1.98$, and $\mu = 0.0084 \text{ cm}^{-1}$. [HINT: It is probably easiest to set up the matrix elements in a Hund-case (a) representation, in which components of the angular momenta along the figure axis are quantized. In this case, the only nonzero component there is $S_z = \Sigma$, since there is no component of the rotational angular momentum of a linear molecule along the axis. Then $S_z^2 = \Sigma^2$, and the matrix elements of $\mathbf{J} \cdot \mathbf{S}$ are also readily written down. The Wang transformation (7-95) applied to the 3×3 matrix for $\Sigma = \pm 1, 0$ will prove useful in reducing the secular equation.[1]]

REFERENCES

COULSON, C. A.: "Valence," Oxford University Press, New York, 1952.

DAUDEL, R., R. LEFEBVRE, and C. MOSER: "Quantum Chemistry," Interscience Publishers, Inc., New York, 1959.

DENNISON, D. M., and K. T. HECHT: Molecular Spectra, chapter in D. R. Bates (ed.), "Quantum Theory," vol. II, Academic Press Inc., New York, 1962.

EYRING, H., J. WALTER, and G. F. KIMBALL: "Quantum Chemistry," John Wiley & Sons, Inc., New York, 1944.

HEINE, V.: "Group Theory," chap. 5, Pergamon Press, New York, 1960.

HERZBERG, G.:" Molecular Spectra and Molecular Structure," vol. I, 2d ed., Spectra of Diatomic Molecules, vol. II, Infrared and Raman Spectra of Polyatomic Molecules, D. Van Nostrand Company, Inc., Princeton, N.J., 1950, 1945.

PAULING, L., and E. B. WILSON, JR.:"Introduction to Quantum Mechanics," McGraw-Hill Book Company, Inc., New York, 1935.

STRANDBERG, M. W. P.: "Microwave Spectroscopy," Methuen & Co., Ltd., London, 1954.

STREITWIESER, A., JR.: "Molecular Orbital Theory for Organic Chemists," John Wiley & Sons, Inc., New York, 1961.

TOWNES, C. H., and A. L. SCHAWLOW: "Microwave Spectroscopy," McGraw-Hill Book Company, Inc., New York, 1955.

WILSON, E. B., JR., J. C. DECIUS, and P. C. CROSS: "Molecular Vibrations," McGraw-Hill Book Company, Inc., New York, 1955.

[1] See R. Schlapp, *Phys. Rev.*, **51**, 342 (1937); Tinkham and Strandberg, *loc. cit.*

8 SOLID-STATE THEORY

To give any sort of proper coverage to the quantum theory of solids would require much more space than is available here. Rather than trying to survey the entire field in a brief chapter, we shall confine our attention to two aspects which are particularly characteristic of the solid state, namely, the electronic-energy-band structure and the symmetry of magnetic crystals. In our treatments of both of these, we shall find group theory to be a powerful aid.

8-1 Symmetry Properties in Solids

We have already noted that the full symmetry group of a crystalline solid, the space group, contains translational-symmetry operators, rotational-symmetry operators from the point group, and, of course, combinations of the two.   The most general space-group element may be represented by the conventional notation $\{\mathbf{R} \mid \mathbf{t}\}$, where $\mathbf{R}$ represents a rotation (proper or improper) and $\mathbf{t}$ represents a translation. The corresponding coordinate transformation is

$$\mathbf{x}' = \mathbf{R}\mathbf{x} + \mathbf{t} \tag{8-1}$$

From this property, it is easy to verify that the multiplication rule for two such operators is

$$\{\mathbf{R} \mid \mathbf{t}\}\{\mathbf{S} \mid \mathbf{t}'\} = \{\mathbf{R}\mathbf{S} \mid \mathbf{R}\mathbf{t}' + \mathbf{t}\} \tag{8-2}$$

Similarly the inverse of $\{\mathbf{R} \mid \mathbf{t}\}$ is given by

$$\{\mathbf{R} \mid \mathbf{t}\}^{-1} = \{\mathbf{R}^{-1} \mid -\mathbf{R}^{-1}\mathbf{t}\} \tag{8-3}$$

From the forms of these results, it is obvious that the rotational parts $\mathbf{R}$ of these operators, obtained by setting $\mathbf{t} = 0$, in themselves form a group. This is the point group of the crystal.   It is also clear that the pure translation operators $\{\mathbf{E} \mid \mathbf{t}\}$ form a subgroup, one which is both invariant and Abelian.   This translational subgroup contains all the translations which can be written in the form

$$\mathbf{t}_n = n_1\mathbf{a}_1 + n_2\mathbf{a}_2 + n_3\mathbf{a}_3 \tag{8-4}$$

where n_1, n_2, and n_3 are integers and $\mathbf{a}_1$, $\mathbf{a}_2$, $\mathbf{a}_3$ are the three primitive translations which define the periodicity of the lattice.   Naturally, these translations are covering operations of the crystal only if it is infinite in extent, or if one neglects surface effects, or if one introduces periodic boundary conditions, so that a finite volume of crystal is imagined to be repeated over and over to fill all space.   For many purposes, we can imagine any one of these to be the case with equivalent results.   It is usually most convenient to use periodic boundary conditions, and they lead naturally to the running (as opposed to standing) waves which are necessary for treating any electronic transport or conduction process.   We shall use them here.

If we ignored all symmetry except the translational symmetry (or if there were no more, as in a triclinic lattice without even inversion symmetry), the entire symmetry classification of an electronic wavefunction in the solid would be given by Bloch's theorem (see Sec. 3-7).   The irreducible representations would be completely labeled by giving the *crystal-momentum* quantum number k, defined by the relation (3-26), which in three dimensions becomes

$$\psi_\mathbf{k}(\mathbf{r} + \mathbf{a}) = e^{i\mathbf{k}\cdot\mathbf{a}}\psi_\mathbf{k}(\mathbf{r}) \tag{8-5}$$

where $\mathbf{a}$ is any of the three primitive translations.   Hence, (8-5) also holds when $\mathbf{a}$ is replaced by any $\mathbf{t}_n$ satisfying (8-4).   Thus

$$\psi_\mathbf{k}(\mathbf{r} + \mathbf{t}_n) = e^{i\mathbf{k}\cdot\mathbf{t}_n}\psi_\mathbf{k}(\mathbf{r}) \qquad \blacktriangleright \tag{8-6}$$

As in one dimension, this property allows us to write the most general wavefunction in the form

$$\psi_k(\mathbf{r}) = u_k(\mathbf{r})e^{i\mathbf{k}\cdot\mathbf{r}} \qquad \blacktriangleright \quad (8\text{-}7)$$

where u_k is periodic so that $u_k(\mathbf{r}) = u_k(\mathbf{r} + \mathbf{t}_n)$ and $\mathbf{k}$ takes on a discrete set of values determined by the repeat distance in the periodic boundary conditions. Wavefunctions in this Bloch form evidently extend over the entire crystal, representing equal probability that the electron is in any unit cell of the crystal, just as with the plane waves of a completely free electron. Such a form is essential if we are to have a convenient theory of electronic conduction in metals.

If, on the other hand, we ignore the translational symmetry and consider only the point-group symmetry, we are led to the considerations we treated in Chap. 4 in conjunction with the crystal-field-splitting problem. This approach would be justified if the atomic wavefunctions located on the various sites in the crystal had so little overlap or interaction that each atom could be treated separately, with no concern for the existence of other identical atoms in adjacent unit cells in the crystal. In that case, the complete symmetry classification of the wavefunctions is provided by the various irreducible representations of the appropriate point group, namely, that giving the local symmetry about the atomic site. This need not be the same as the point group giving the macroscopic crystal symmetry, also referred to as the *crystal class*.

Our present objective is to bring together the two aspects of the problem mentioned above, so that we can treat the complete symmetry classification of the wavefunctions where both the translational and point-group symmetries have nontrivial consequences. The possibilities for complication as one considers various space groups are so great that we shall severely limit the generality of our discussion so as to illustrate the basic ideas in as simple a way as possible. Thus, we shall largely confine our attention to the 73 *simple*, or *symmorphic*, space groups, in which all symmetry operators which are obtained by taking the direct product of the translational subgroup with the point group are in the space group. In this case, the *entire* point group is a subgroup of the space group. This subgroup is isomorphic to the factor group of the space group with respect to the invariant subgroup of translations. In making this restriction, we exclude the 157 nonsimple space groups, containing glide planes and screw axes. These are elements combining a reflection or rotation with a translation which is *not* of the form (8-4) and accordingly is not separately a group element. For example, when there is a screw diad axis in the symmetry of the system, this means that one symmetry operation of the space group is a rotation by π about an axis, coupled with a translation by half the repeat distance along that axis. In this case, neither the twofold rotation nor the half unit of translation is separately a member of the *space* group. Yet the twofold rotation *is* a member of the *point* group, as defined above. Thus, in this case, the entire point group is *not* a subgroup of the space group. Naturally this leads to a somewhat more intricate theory than in the case of the simple space groups. For the requisite generalizations of the treatment given here, the reader is referred to the literature.[1]

8-2   The Reciprocal Lattice and Brillouin Zones

Inspection of (8-6), the relation defining the quantum number, or representation label, $\mathbf{k}$, shows that $\mathbf{k}$ is actually not uniquely specified. This is so because if we add any vector $\mathbf{K}$ to $\mathbf{k}$, such that $\mathbf{K} \cdot \mathbf{t}_n$ is a multiple of 2π for all $\mathbf{t}_n$, then the defining relation (8-6) is as well satisfied by $\mathbf{k} + \mathbf{K}$ as by $\mathbf{k}$ itself. The particular vectors $\mathbf{K}$ are closely related to the vectors $\mathbf{b}$ of the reciprocal lattice, which are defined by the relation

$$\mathbf{b}_1 = \frac{\mathbf{a}_2 \times \mathbf{a}_3}{\mathbf{a}_1 \cdot (\mathbf{a}_2 \times \mathbf{a}_3)} \tag{8-8}$$

and two others obtained by cyclic permutation of indices. These vectors have the dimension of reciprocal length and are said to lie in *reciprocal space*. We note that in the special case where the three primitive translations in direct space, namely, the $\mathbf{a}_i$, are orthogonal, the $\mathbf{b}_i$ are parallel to the $\mathbf{a}_i$ and are of length $1/|a_i|$. In any case, it is evident from (8-8) that

$$\mathbf{a}_i \cdot \mathbf{b}_j = \delta_{ij} \tag{8-9}$$

Thus, if we consider any vector $\mathbf{K}_m$ of the form

$$\mathbf{K}_m = 2\pi(m_1 \mathbf{b}_1 + m_2 \mathbf{b}_2 + m_3 \mathbf{b}_3) \tag{8-10}$$

the scalar product $\mathbf{K}_m \cdot \mathbf{t}_n$ will have the value $2\pi(m_1 n_1 + m_2 n_2 + m_3 n_3)$. Since the m_i and n_i are integers, $\mathbf{K}_m \cdot \mathbf{t}_n = 2\pi \times$ integer, $e^{i\mathbf{K}_m \cdot \mathbf{t}_n} = 1$, and

$$e^{i(\mathbf{K}_m + \mathbf{k}) \cdot \mathbf{t}_n} = e^{i\mathbf{k} \cdot \mathbf{t}_n} \tag{8-11}$$

Thus, as far as the defining relation (8-6) is concerned, it is not possible to distinguish between $\mathbf{k}$ and $\mathbf{k} + \mathbf{K}_m$, where $\mathbf{K}_m$ is given by (8-10), and we must consider these as essentially identical $\mathbf{k}$ values. This shows that we may confine all our attention to $\mathbf{k}$ vectors lying inside a finite zone of this reciprocal space, or $\mathbf{k}$ space, since any $\mathbf{k}$ vector lying outside the zone can be considered identical to one inside the zone, by adding a suitable $\mathbf{K}_m$. It is geometrically evident that the size of this region is determined by taking that part of $\mathbf{k}$

---

[1] See, for example, G. F. Koster, F. Seitz, and D. Turnbull (eds.), "Solid State Physics," vol. V, p. 173, Academic Press Inc., New York, 1957.

space closer to the origin than to any of the other reciprocal-lattice points, specified by reciprocal-lattice vectors $\mathbf{K}_m$.[1] This zone may be constructed by setting up perpendicular bisecting planes on the lines connecting the origin to all reciprocal-lattice points and then taking the volume about the origin enclosed by these intersecting planes. The zone constructed in this way is known as the *first Brillouin zone*, and it plays a central role in all our subsequent discussions because all wavefunctions can be classified as having a wave vector $\mathbf{k}$ lying in this zone. In adding or subtracting a $\mathbf{K}$ vector to obtain an equivalent $\mathbf{k}$ vector in the first Brillouin zone, one obtains the so-called "reduced" wave vector. If all wavefunctions are described in terms of such reduced wave vectors, the *reduced-zone scheme* of description results.

On the other hand, in the extended-zone scheme all $\mathbf{k}$ values are considered, and second, third, and higher Brillouin zones are defined to characterize the outlying regions of $\mathbf{k}$ space. The first and second zones for the square lattice are shown in Fig. 8-1, for example. The higher zones all have the same area as the first, and the various pieces of which they consist can be exactly fitted inside the first zone by translation by reciprocal-lattice vectors. The boundaries of all zones are formed by the set of planes described above. The significance of these planes is that, for any $\mathbf{k}$ on one of them, there is another $\mathbf{k}' = \mathbf{k} + \mathbf{K}$ such that $|\mathbf{k}'| = |\mathbf{k}|$. Hence the second wave is degenerate with the first, and, as we shall see directly, first-order shifts in energy can result from the periodic potential. This gives rise to discontinuities in energy along the zone boundaries in an extended-zone description. The extended-zone scheme is particularly useful for nearly free electrons, where the wavefunction is well approximated by a plane wave of some definite $\mathbf{k}$ value, which need not be in the first zone.

As a simple example of these concepts, let us consider a two-dimensional "solid" with a square lattice with primitive translations of length a along both x and y. The reciprocal lattice vectors $\mathbf{b}_i$ then also form a square lattice, with $|\mathbf{b}_i| = 1/a$. In this case the first Brillouin zone is a square, centered on the origin, and extending to $\pm\pi/a$ along both the k_x and k_y directions as shown in Fig. 8-1. Actually, the same information would be contained within any square of the same size, but it is usually most convenient to take the zone centered on the origin. Now, if the periodic boundary conditions require a repetition of the wavefunctions after a displacement by N unit cells along either the x or the y axis, and thus after a distance $L = Na$, the allowed $\mathbf{k}$ values form a discrete set, with both k_x and k_y being restricted to integral multiples of $2\pi/L$ or $2\pi/Na$ (see Sec. 3-7). Thus there are N discrete allowed values of k_x and of k_y, and as a result N^2 values of $\mathbf{k}$, N^2 being

---

[1] Strictly speaking, $\mathbf{K}_m/2\pi$ is a reciprocal-lattice vector, since it is composed of integral multiples of the vectors $\mathbf{b}_i$. However, we shall follow conventional usage, which allows the same expression to refer to either one, when there is no danger of confusion.

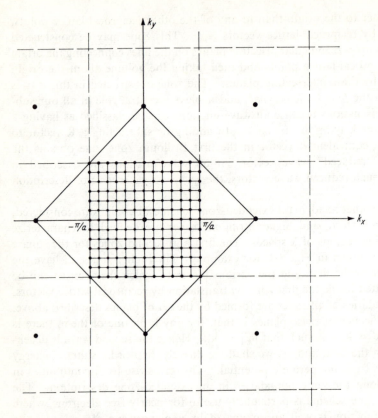

Fig. 8-1. *Schematic diagram of first and second Brillouin zone: The discrete set of allowed k-values is indicated inside the first zone.*

the total number of cells in the coordinate space crystal. This equality is a general result, valid also in three dimensions and independent of the symmetry of the unit cell.

A still simpler example is the one-dimensional case with a period of a along the x axis, the example treated explicitly in Sec. 3-7. If the "crystal" is of length $L = Na$, there are N discrete k values, namely, $2\pi p/L$, with p being an integer lying between $-N/2$ and $+N/2$, so that k runs from $-\pi/a$ to $+\pi/a$. Because we have only a one-dimensional variation of k, we can now plot the energy of the electronic eigenfunction against k in a simple manner. It is instructive to start out by considering the so-called "empty-lattice" case, in which we imagine that the periodic potential associated with the crystal lattice has been "turned off," leaving the electron free. In that case, we know that the eigenfunctions are plane waves $\psi_k \sim e^{ikx}$ and the corresponding energies are $E_k = \hbar^2 k^2/2m$. Although the reciprocal lattice should play no substantive role if there is no interaction between the electron and the crystal, we can still use the reduced-zone scheme to plot all energies in a finite range of k values corresponding to the first Brillouin zone. This is illustrated in Fig. 8-2, where the extended-zone scheme is also indicated in a lighter curve. In obtaining the reduced-zone energies, we have simply translated the energy curves by multiples of the reciprocal-lattice distance $2\pi/a$.

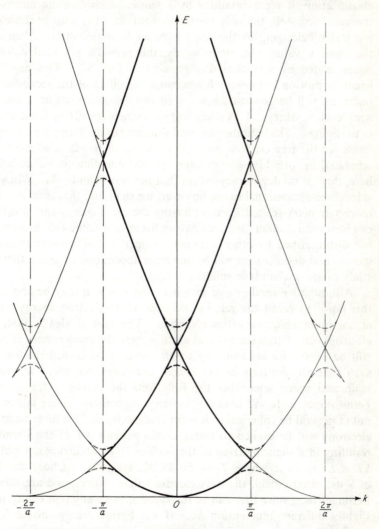

Fig. 8-2. *One-dimensional E(k) curve showing reduced-zone scheme applied to free-electron or empty lattice example. Dotted curves near level crossings represent modifications introduced by finite periodic potential, which results in forbidden bands of energy levels. Recall also that k takes on only a discrete but very finely spaced set of values.*

If we now imagine that the periodic potential of the lattice is turned on but is still weak, then we may consider it as a perturbation on these energy levels. Since by definition the periodic potential $V(x)$ has the periodicity of the lattice, it is invariant under the space group of the crystal, which in this case is simply the cyclic group leading to Bloch's theorem and the classification of representations by k value. Applying the matrix-element theorem (Sec. 4-9), the only matrix elements of $V(x)$ must be those connecting states belonging to the same irreducible representation, hence having the same k value. In other words, the periodic potential couples only states related by a vertical displacement in Fig. 8-2. This results in the usual "repulsion" of energy levels, since the effect of the second-order perturbation will be greatest when the energy denominators are least, namely, near level crossings. This effect is shown schematically by the dotted curves in the figure. The periodic potential thus breaks up the continuum of energy levels for the free electron into a discrete set of bands of allowed energies, separated by forbidden-energy gaps. In the one-dimensional case discussed here, there is no degeneracy except that between k and $-k$. When we consider three-dimensional cases, however, we shall find that there can be many more symmetry-related k values having the same energy and also that there can be several distinct functions having the same k value and the same energy, but distinguished by other symmetry properties. The study of these symmetries and degeneracies will be our main application of group theory in the study of energy bands in solids.

Although we shall not go into it in any detail, it may be appropriate at this point to recall the relation between these electron energy bands and actual observable properties of solids. The central idea is that, because electrons obey Fermi statistics, at absolute zero the energy states of the crystal will be filled with electrons up to an energy level (called the Fermi level) such that all electrons in the system are accounted for. The surface in reciprocal space separating the full from the empty states is called the *Fermi surface*. It will have a symmetry appropriate to the lattice and will not in general be spherical as it is for free electrons. At finite temperatures, electrons will be excited to states within roughly kT of the Fermi energy, resulting in a slight blurring of the surface. (The blurring is slight because $kT \ll E_F = kT_F$, where $T_F \approx 50{,}000°K$, typically.) Under the influence of a d-c electric field, the only electrons that can respond are those at the Fermi surface, since they can be displaced into an adjacent empty state with slightly different momentum $\hbar k$. If the Fermi energy should fall in the energy gap between two bands, where there is no density of states, such a field can produce no current and we have an insulator. However, if the gap is comparable with kT, there will be some electrons excited above the gap to a *conduction band*, where they are able to be accelerated into empty states, and we have a semiconductor. If the Fermi energy lies within an

allowed-energy band, all electrons on the Fermi surface are able to partici-
pate in conduction and we have a metal.   From this brief sketch it is apparent
that a knowledge of the energy-band structure of a solid is fundamental for
determining all its electronic properties.   Because a straightforward solution
is difficult at best, it is worthwhile to draw on group theory for any help
symmetry can provide in reducing the magnitude of the problem.

8-3 Form of Energy-band Wavefunctions

It is instructive to carry on this simple one-dimensional example to consider
the form of the wavefunctions in the crystal for various k values and various
bands.   In one dimension, (8-7) becomes simply

$$\psi_k(x) = u_k(x)e^{ikx} \tag{8-12}$$

where $u_k(x + a) = u_k(x)$.   Evidently, the entire unsolved problem is reduced
to the determination of $u_k(x)$ within one unit cell because the periodicity
then determines it over the rest of the crystal.

In the empty-lattice, or free-electron, approximation, we simply set
$u_k(x) = 1$ for all k.   However, if we are using the reduced-zone scheme,
with $|k| \leq \pi/a$, this choice of $u_k(x)$ can be maintained only for the lowest
band.   When we go to the second band (or more precisely, what becomes the
second band when a nonvanishing periodic potential is introduced), $u_k(x)$
must be chosen to be $e^{i2\pi x/a}$ for $-\pi/a \leq k \leq 0$, and $e^{-i2\pi x/a}$ for $0 \leq k \leq \pi/a$,
in order to reproduce the free-electron wavefunctions of the same energy.
Note that both these have the required periodicity.   For the higher bands,
other periodic functions of the form $e^{\pm i2n\pi x/a}$ are required.   Particular
interest arises at the points of crossing of energy levels on the reduced-zone
scheme, since that is where the periodic potential has the greatest effect.
For example, at $k = \pi/a$, the wavefunctions for the first and second band
are, respectively,

$$\psi_1 = 1 \cdot e^{i\pi x/a} \tag{8-13a}$$

and

$$\psi_2 = e^{-i2\pi x/a} \cdot e^{i\pi x/a} = e^{-i\pi x/a} \tag{8-13b}$$

Since these are degenerate, if there is any nonvanishing matrix element of
$V(x)$ connecting them, the proper eigenfunctions are the sum and difference
of ψ_1 and ψ_2, namely,

$$\psi_+ = \psi_1 + \psi_2 \sim \cos \frac{\pi x}{a} \tag{8-14a}$$

$$\psi_- = \psi_1 - \psi_2 \sim \sin \frac{\pi x}{a} \tag{8-14b}$$

We note in passing that both ψ_+ and ψ_- can still be written in the form
(8-12) as our general theorem requires, but now $u_k(x) = (1 \pm e^{-i2\pi x/a})$.

We also note that ψ_+ and ψ_- differ only in phase, one being shifted by a distance $a/2$ with respect to the other along the x axis. This is precisely the shift required so that one will have the charge density centered at the center of the unit cell, while the other will have a node there, with the charge piled up at the edge of the cell. Obviously, if there is any periodic potential

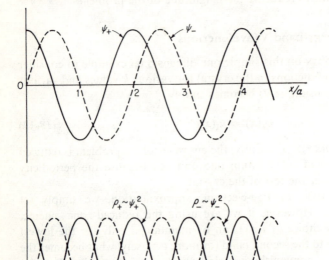

Fig. 8-3. *Band-edge wavefunctions at $k = \pi/a$ in the nearly free-electron approximation.*

at all, these two functions will have different energy, leading to the gap between the two bands indicated by the dotted curve in Fig. 8-2. The width of the gap is just twice the matrix element connecting ψ_1 and ψ_2, as given above. These wavefunctions are illustrated in Fig. 8-3.

If we now imagine that the periodic potential $V(x)$ is quite strong, then the free-electron wavefunctions are no longer suitable approximations to be used in any reasonably rapidly convergent perturbation approximation. In that case, one might consider the opposite limit, in which one considers an electron as tightly bound to one atom along the chain, with the translational symmetry simply superimposed. That is, if $\varphi(x)$ is the wavefunction for the electron bound to one center, one can set up a periodic function $u_k(x)$ by simply superimposing the $\varphi(x)$ from all the centers to form

$$u_k(x) \sim \sum_i \varphi(x - x_i) \qquad (8\text{-}15)$$

When this $u_k(x)$ (which actually has no dependence on k in this approximation) is substituted into (8-12), we obtain the so-called "tight-binding approximation" to the wavefunction for energy-band electrons. From this point of view, the various bands result from various $\varphi(x)$. If we are considering actual energy bands in a three-dimensional solid, these will accordingly be characterized by the atomic state being used for $\varphi(x)$, so that one speaks of a $2s$ band or a $3d$ band, for example. This nomenclature has some meaning even when the binding is not very tight, and it is often used.

As the simplest example of the tight-binding approximation, let us consider the band in one dimension corresponding to the $1s$ band in a solid. In this case, $\varphi(x)$ is a nodeless function, decreasing smoothly from a maximum at the origin. Then the $u_k(x)$ constructed according to (8-15) will be a nodeless function with a maximum in each unit cell at the position of the nucleus, as illustrated in Fig. 8-4a. Now, at $k = 0$, $\psi_k(x) = u_k(x)$, and, instead of being a constant as in the free-electron case, the wavefunction has a peak in each unit cell. If we go to $k \neq 0$, then the e^{ikx} factor modulates the amplitude of the periodic function $u_k(x)$ as we move from one cell to the next in the crystal. This is illustrated in Fig. 8-4b, for the case of a k vector one-third of the way from the origin to the boundary of the first Brillouin zone. We plot only the real part of $\psi(x)$; the imaginary part has a modulating envelope displaced by $3a/2$. As a result the total charge-density distribution is the same in every cell. All that is modulated is really the phase of the wavefunction, just as in a free-electron traveling wave, $\psi \sim e^{ikx}$.

Moving on to the zone boundary, $k = \pi/a$, Fig. 8-4c and d shows the real and imaginary parts of $\psi(x)$. These two functions correspond to the ψ_+ and ψ_- of the free-electron case, illustrated in Fig. 8-3. As in that previous case, one of the functions peaks the charge density at the nuclei, whereas the other puts a node at the nucleus. In other words, Re $\psi(x)$ still has a $1s$-like character, whereas Im $\psi(x)$ has a $2p$-like character. Continuing on to $k = 2\pi/a$, Fig. 8-4e and f shows that the real and imaginary parts of $\psi(x)$ now look qualitatively like $2s$ and $2p$ functions, respectively, added in phase in each cell, corresponding to $k = 0$. Because, in fact $u_k(x)$ cannot be treated as independent of k in a rigorous way, these pictorial considerations must be viewed as highly approximate versions of the actual form of the wavefunctions. Nonetheless, they do show in a simple way that, even in the tight-binding approximation, there is some validity to the picture introduced earlier of the higher bands in the reduced-zone scheme arising from absorbing the extra nodes from the e^{ikx} function into the $u_k(x)$. Thus, Fig. 8-4c approximates the wavefunction from the lowest band at the zone boundary, whereas Fig. 8-4d approximates the second band there. Similarly, Fig. 8-4e approximates a $2s$-like band at the zone center, whereas Fig. 8-4f approximates a $2p$-like band there. We shall later go into more detail on the determination of these functions.

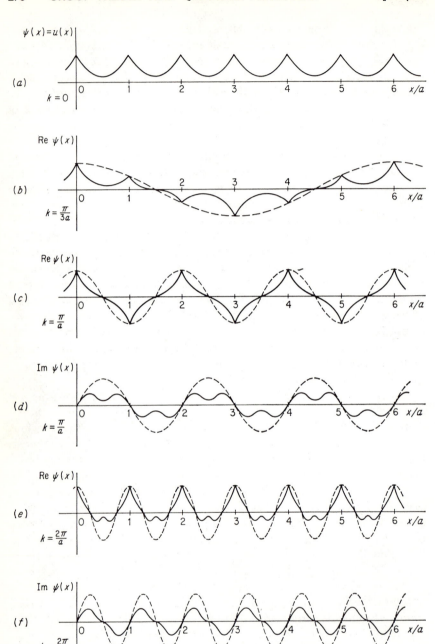

Fig. 8-4. *Representative tight-binding wavefunctions based on* 1 *s function in each unit cell.*

8-4 Crystal Symmetry and the Group of the k Vector

Let us consider the result of applying one of the transformation operators P_R, corresponding to one of the operations R of the point group, to a wavefunction in the Bloch form. That is, we consider

$$P_R \psi_{\mathbf{k}} = P_R u_{\mathbf{k}}(\mathbf{r}) e^{i\mathbf{k}\cdot\mathbf{r}}$$
$$= u_{\mathbf{k}}(\mathbf{R}^{-1}\mathbf{r}) e^{i\mathbf{k}\cdot\mathbf{R}^{-1}\mathbf{r}} \tag{8-16}$$

If we note that applying the same orthogonal transformation to both vectors in a scalar product does not change the value of the product, we have

$$\mathbf{k} \cdot \mathbf{R}^{-1}\mathbf{r} = \mathbf{R}\mathbf{k} \cdot \mathbf{R}\mathbf{R}^{-1}\mathbf{r} = \mathbf{R}\mathbf{k} \cdot \mathbf{r} \tag{8-17}$$

Inserting this result in (8-16) leads to

$$P_R \psi_{\mathbf{k}}(\mathbf{r}) = u_{\mathbf{k}}(\mathbf{R}^{-1}\mathbf{r}) e^{i\mathbf{R}\mathbf{k}\cdot\mathbf{r}} \tag{8-18}$$

Now, so long as we are dealing with a simple space group, all operations from the point group are also symmetry operations from the space group. Thus the function $u_{\mathbf{k}}(\mathbf{R}^{-1}\mathbf{r})$ is periodic if $u_{\mathbf{k}}(\mathbf{r})$ is, because $\mathbf{R}^{-1}\mathbf{a}_i$ must be a primitive translation if $\mathbf{a}_i$ is. Since periodicity is all that is required of $u_{\mathbf{k}}(\mathbf{r})$, we can rename $u_{\mathbf{k}}(\mathbf{R}^{-1}\mathbf{r}) = u'_{\mathbf{R}\mathbf{k}}(\mathbf{r})$ so as to conform to the general form. In that case, (8-18) becomes

$$P_R \psi_{\mathbf{k}}(\mathbf{r}) = u'_{\mathbf{R}\mathbf{k}}(\mathbf{r}) e^{i\mathbf{R}\mathbf{k}\cdot\mathbf{r}} \tag{8-19}$$

where the prime on $u'_{\mathbf{R}\mathbf{k}}$ is to remind us that, if one chooses an $\mathbf{R}$ such that $\mathbf{R}\mathbf{k} = \mathbf{k}$, it need not be true that $u_{\mathbf{k}}(\mathbf{r})$ is also unaffected by the transformation. (In the usual case of no degeneracy, u' can differ from u only by a phase factor; if there is degeneracy, u' can be a linear combination of the degenerate functions u.) The conclusion from this argument, then, is that the effect of operating on $\psi_{\mathbf{k}}$ with P_R is to produce an eigenfunction also in Bloch form, but with the $\mathbf{k}$ vector simply rotated to $\mathbf{R}\mathbf{k}$. This enables us to make a simple geometric inspection to find all the symmetry-related $\mathbf{k}$ vectors, all of which must have associated eigenfunctions of the same energy.

If one carries out successively all the point-group operations R on a given wave vector $\mathbf{k}$, one generates what is known as the *star* of $\mathbf{k}$. A number of typical cases are illustrated in Fig. 8-5, for the case of a two-dimensional square lattice. For example, if one takes a general $\mathbf{k}$, located in no special symmetry direction, then each operation of the group will take it into a new position. In this case, the star will have h "points," one for each element of the group. If the $\mathbf{k}$ vector lies along a symmetry line as in Fig. 8-5b and c, then some of the symmetry operations will leave it unchanged and the star will have fewer points. If the $\mathbf{k}$ vector touches the zone boundary, there will in general be a reduction in the number of distinct $\mathbf{k}$ values in the star because any two which are related by a reciprocal-lattice vector $\mathbf{K}_m$ must be treated as identical and counted only once. This is illustrated for one pair in Fig. 8-5d. If the $\mathbf{k}$ vector lies on a symmetry plane and also on the zone boundary, further reductions in number occur.   Finally, in Fig. 8-5f we reach the point where all the symmetry-related **k** vectors are identical via reciprocal-lattice vector displacements.   In other words, the **k** vector must be considered invariant under all operations of the group, just as if it had been **k** = 0.   Thus the symmetry at the corner of a square lattice is as high as at the center.

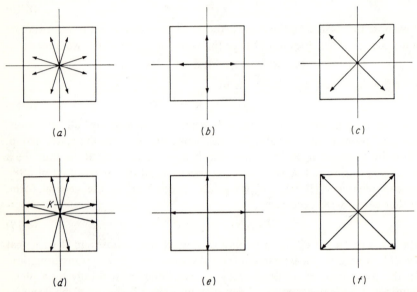

(a)   (b)   (c)

(d)   (e)   (f)

Fig. 8-5. *Typical examples of the star of the wave vector for the case of a two-dimensional square lattice.   (a) General point, eight distinct k values; (b)-(c) symmetry planes, four distinct k values; (d) general point on zone boundary, four distinct k values after allowing for translational equivalence; (e)-(f) special points on zone boundary, two and one distinct k values, respectively.*

It is clear that special interest is attached to symmetry points in **k**-space which are left invariant by some of the group operations, since only there can symmetry-based degeneracy exist among functions with the same **k** value.   Those operators which leave a given **k** invariant form a subgroup of the entire point group, called the *group of the wave vector*.   For example, if **k** lies along the k_x axis, its group will include all rotations about the x axis and all reflections in planes including the x axis.   Now, under an operation of the group **k**, ψ_k may be left unchanged (apart from a phase factor, perhaps), or else it may be transformed into a new function ψ'_k, with the same wave vector.   In the latter case, there will be more than one distinct $u_k(\mathbf{r})$ associated with the same exponential factor $e^{i\mathbf{k}\cdot\mathbf{r}}$.   These various $u_k(\mathbf{r})$ will transform among themselves under the group of **k** according to an irreducible representation of the group which is called the *small representation*.

To summarize, to specify completely the transformation properties of a wavefunction under the operations of the space group, it is necessary to give two bits of information. The first is the **k** vector, which specifies the transformation properties under translation and determines the star of **k**. The second is the small-representation label telling the symmetry properties of $u_k(\mathbf{r})$ under the operations of the group of the wave vector. At a general point in **k** space, however, the group of the wave vector contains only the identity, and the small-representation label is not necessary.

8-5 Pictorial Consideration of Eigenfunctions

Before plunging into the full group-theoretical machinery required for a systematic study of the possible symmetries and degeneracies of energy bands

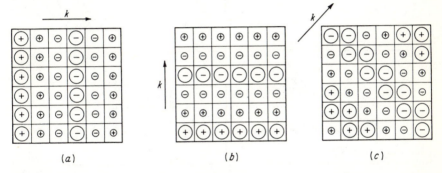

Fig. 8-6. *Schematic illustration of dependence of energy in an s band on the direction of the k vector. (a) and (b) have the same energy, being equivalent by symmetry. (c) need not have this same energy, even if $|k|$ is the same as in (a) and (b), because no symmetry operation connects the wavefunction of (c) with that of (a) or (b). (The real part of ψ is systematically shown in each case.)*

of all sorts, let us try to acquire a firm grasp of the qualitative symmetry considerations which our machinery will be handling. We can do this by inspecting simple sketches illustrating the form of the wavefunctions for successively more complicated situations. Perhaps the simplest case to consider is the case of an s-like band in the tight-binding approximation. In Fig. 8-4, we have already inspected the one-dimensional case. In going even to two dimensions, we can no longer plot the wavefunction in any simple way. Rather, we resort to a schematic diagram, in which the amplitude in each cell is indicated by the size of a pattern representing the symmetry of the cellular wavefunction. Thus in Fig. 8-6 we use a circle in each cell to represent an s-like wavefunction, the size of the circle representing the amplitude of the $e^{i\mathbf{k}\cdot\mathbf{r}}$ function, and the sign is indicated explicitly. In Fig. 8-6*a* and *b*, we show the appearance of the wavefunction for **k** vectors

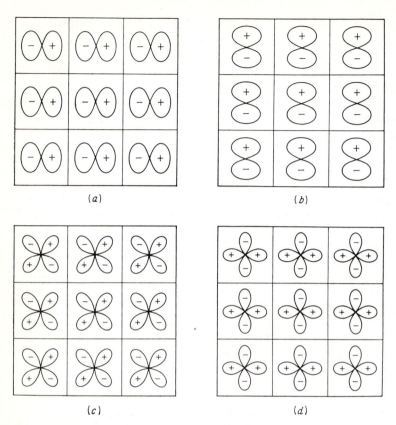

Fig. 8-7. *Symmetry aspects of p-like and d-like bands in cubic lattice at k = 0. (a)-(b) Two p-band wavefunctions related by a fourfold symmetry rotation, hence degenerate. (c)-(d) Representative d-band wavefunctions of symmetry T_{2g} and E_g under the octahedral group O_h; these are not related by any symmetry operation, and hence they will not in general be degenerate.*

of equal magnitude, directed along k_x and k_y, respectively. Evidently the fourfold rotation relating x to y will transform one of the functions into the other, and they must be degenerate. When we consider Fig. 8-6c, however, we see at a glance that there can be no symmetry operation which transforms that wavefunction into either of the previous two. Thus, even if the magnitude of the **k** vector is the same as in the first two cases, the energy need not be the same, because the states are not related by symmetry. In other words, the function $E(\mathbf{k})$ need not be spherical for electrons in bands; it need have only the symmetry of the point group of the crystal.

Another simple case to consider is that of $\mathbf{k} = 0$. Then the $e^{i\mathbf{k}\cdot\mathbf{r}}$ factor in the wavefunction reduces to unity, and the wavefunction must have exactly the same form in each cell. If we consider an s band, there is only one state with $\mathbf{k} = 0$ and no complications arise. However, if we go to a p band, there are three p functions to work with. If the point symmetry is as high as cubic, these are all related by symmetry operations of the group and they will be degenerate. This is illustrated by Fig. 8-7a and b. If the symmetry were lower, they would not be all degenerate (see Sec. 4-6). If we now proceed to a d band, then, even with a cubic point group, all five

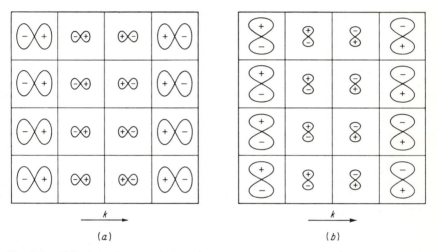

$$(a) \qquad\qquad\qquad (b)$$

Fig. 8-8. *Schematic diagram showing how $k \neq 0$ reduces the symmetry by eliminating the operation connecting p_x and p_y functions.*

functions cannot be related by symmetry operations. Rather they separate into three degenerate functions of T_{2g} symmetry and two degenerate functions of E_g symmetry, as discussed in Chap. 4 in connection with the crystal-field splittings. These are illustrated in Fig. 8-7c and d. All these results are the same as if we had only one site to consider, because, for $\mathbf{k} = 0$, the translational symmetry from cell to cell does not lower the overall symmetry of the wavefunction by introducing a "special" direction.

The interesting novel cases arise when we combine the symmetry-lowering presence of a nonzero $\mathbf{k}$ vector with atomic symmetries of p and higher bands, where degeneracy may exist at $\mathbf{k} = 0$. The simplest example of this is the case of the three p-like functions, degenerate at $\mathbf{k} = 0$ in a cubic lattice, and transforming according to the irreducible representation T_{1u}. As soon as $\mathbf{k} \neq 0$, the three p-band (more precisely, T_{1u}) wavefunctions need not be connected by any symmetry operation. This is illustrated in Fig. 8-8, where the p_x and p_y bands are shown for the case of $\mathbf{k}$ lying along the k_x direction. Inspection immediately shows that these functions are no longer related by any symmetry operation of the point group. Hence they will in general have different energies.   So long as **k** lies along k_x, however, the p_y and p_z bands will remain degenerate.   But if **k** takes on a general direction, the point-group symmetry of the wavefunction is completely destroyed and no degeneracy will remain.   These results are illustrated schematically in Fig. 8-9.   At each stage in this process of lowering the symmetry, the characterization of ψ by irreducible representation labels will of course be done in

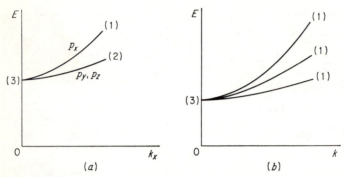

Fig. 8-9.   (a) Decrease of degeneracy in p band as k increases from zero along the k_x axis, leaving p_y and p_z bands still degenerate; (b) complete lifting of degeneracy when k lies in a general direction.

terms of the group of the wave vector for that particular **k**, because that is all the symmetry left in the problem after **k** has been specified.

8-6   Formal Consideration of Degeneracy and Compatibility

Having developed some feeling for the effect of the **k** vector on the symmetry of wavefunctions in crystals, let us now proceed to a more formal mathematical treatment.   Our starting point is the Schrödinger equation for a single electron moving in the periodic crystalline potential $V(\mathbf{r})$, namely,

$$H\psi_{\mathbf{k}} = E\psi_{\mathbf{k}} \tag{8-20}$$

where

$$H = \frac{\mathbf{p}^2}{2m} + V(\mathbf{r}) = -\left(\frac{\hbar^2}{2m}\right)\nabla^2 + V(\mathbf{r})$$

and where $\psi_{\mathbf{k}}$ must be in the Bloch form $u_{\mathbf{k}}(\mathbf{r})e^{i\mathbf{k}\cdot\mathbf{r}}$.   If we substitute this form for $\psi_{\mathbf{k}}$ into (8-20), we are led, after canceling out the exponential factor common to each term, to

$$-\left(\frac{\hbar^2}{2m}\right)[\nabla^2 u + 2i\mathbf{k}\cdot\nabla u - k^2 u] + V(\mathbf{r})u = Eu$$

After a little rearranging, this can be written as

$$-\left(\frac{\hbar^2}{2m}\right)\nabla^2 u + [V(\mathbf{r}) + (\hbar/m)\mathbf{k}\cdot\mathbf{p}]u = E'u \qquad (8\text{-}21)$$

where $E' = E - \hbar^2 k^2/2m$ is the energy measured with respect to the kinetic energy of a free electron with the same $\mathbf{k}$ value. This equation (8-21) might be called a *pseudo-Schrödinger equation* for the periodic part $u(\mathbf{r})$ of the wavefunction. It contains the usual kinetic-energy part in $\nabla^2 u$, which would be the kinetic energy of the $u_{\mathbf{k}}(\mathbf{r})$ by itself. It also contains the periodic potential $V(\mathbf{r})$. However, the effective potential in the brackets contains another term, proportional to $\mathbf{k}\cdot\mathbf{p}$, which acts like a momentum-dependent potential. It is this new term which can be viewed explicitly as lowering the symmetry of the pseudo-Hamiltonian for determining energy levels for a prescribed $\mathbf{k}$ value.

The process of obtaining this pseudo-Hamiltonian may profitably be compared with the similar process followed in solving for the energy levels in the hydrogen atom. There we take advantage of the spherical symmetry to determine the angular form of the wavefunction in terms of spherical harmonics. This reduces the problem to the solution of a one-dimensional pseudo-Schrödinger equation for the radial function, in which the angular motion has been replaced by an effective centrifugal potential in the radial equation. In the present case, because we know from Bloch's theorem how the wavefunction must vary from cell to cell, we are able to reduce the problem to an equivalent one within one cell, but with a peculiar effective potential.

In analyzing the symmetry of the pseudo-Hamiltonian in (8-21), we know that the periodic potential $V(\mathbf{r})$ is invariant under all point-group operations. However, the $\mathbf{k}\cdot\mathbf{p}$ term in the effective potential is invariant only under the group of the wave vector at $\mathbf{k}$. This fact follows from the consideration (8-17) on recalling that $\mathbf{p}$ and $\mathbf{r}$ have the same transformation properties under all proper and improper rotations. Thus the appropriate symmetry group for the analysis of our eigenfunctions is that part of the point group which is also in the group of the wave vector. For example, in a simple cubic structure, the point group is O_h. At $\mathbf{k} = 0$, this is the appropriate group to use, since all point-group operations will leave $\mathbf{k} = 0$. If we then consider a nonzero $\mathbf{k}$ vector along the k_x axis, the symmetry is immediately reduced to the tetragonal group C_{4v}, which contains only the eight elements E, C_2, $2C_4$, $2\sigma_v$, $2\sigma_d$, all of which leave k_x unchanged. Since this group has four one-dimensional and one two-dimensional representation, the highest degeneracy is twofold. Thus the threefold degenerate, p-like T_{1u} states at $\mathbf{k} = 0$ must split for $k_x \neq 0$. Inspection of the character tables shows that T_{1u} must split into an A_1 and an E representation, when the symmetry is lowered from O_h to C_{4v}, in the manner described above. This relationship is described by saying that symmetries A_1 and E of the group C_{4v} for wave-functions with $\mathbf{k}$ on the k_x axis are compatible with the symmetry T_{1u} of the group O_h for a wavefunction at $\mathbf{k} = 0$.

This concept of compatibility (introduced by Bouckaert, Smoluchowski, and Wigner[1]) is the same as the one we considered earlier in conjunction with the successive reduction of degeneracy by crystal fields of lower and lower symmetry; namely, if we consider a small term of the lower symmetry added as a perturbation to the Hamiltonian of higher symmetry, then the original eigenfunctions will still form zero-order eigenfunctions for the new problem and also basis functions for representing the new smaller group. A representation Γ_i of the subgroup is compatible with a representation Γ_j of the larger group if the basis for Γ_i is included in the basis for Γ_j. In terms of the representations themselves, the irreducible representation Γ_i

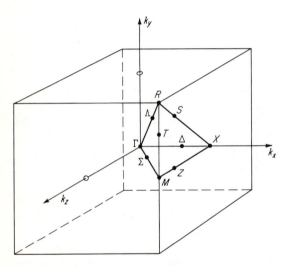

Fig. 8-10. *Simple cubic Brillouin zone with special points labeled.*

of the subgroup is compatible with the irreducible representation Γ_j of the full group if, and only if, Γ_i is present in the decomposition of Γ_j in terms of the irreducible representations of the subgroup. Whether or not this is true may be determined by analysis of only the characters of the represen-tations, as we observed in considering the crystal-field problem.

To carry this discussion forward in concise and explicit terms, let us now introduce some examples of the conventional labeling of special symmetry points and lines in the Brillouin zone. The simplest example in three dimen-sions is the simple cubic case, shown in Fig. 8-10. In terms of the standard notation used in the figure, the case discussed above of a general point along the k_x axis corresponds to the point Δ, where the group of the $\mathbf{k}$ vector is C_{4v}. Let us systematically work out all the compatibility relations between

[1] L. P. Bouckaert, R. Smoluchowski, and E. P. Wigner, *Phys. Rev.*, **50**, 58 (1936).

symmetries at Γ ($\mathbf{k} = 0$) and Δ.   In dealing with this problem in general it is customary to label the irreducible representations by indicating the special point under consideration, with a subscript indicating the particular irreducible representation for the appropriate group of the wave vector, the representations being simply numbered in a conventional order.   Thus the information content of the representation labels commonly used in molecular work is given up in order to give the information on the $\mathbf{k}$ vector which is necessary for characterizing band wavefunctions.   For an unambiguous labeling, it is necessary to specify the conventional order in terms of a character table. To illustrate, the character table at the point Δ may be written as shown in Table 8-1.   We have also given the characters of the representations for the

Table 8-1

Character table at point Δ and characters of representations from Γ

C_{4v}	E	$C_4^{\,2}$	$2C_4$	$2\sigma_v$	$2\sigma_d$
Δ_1	1	1	1	1	1
Δ_2	1	1	-1	1	-1
Δ_3	1	1	-1	-1	1
Δ_4	1	1	1	-1	-1
Δ_5	2	-2	0	0	0
Γ_{1g}	1	1	1	1	1
Γ_{2g}	1	1	-1	1	-1
Γ_{3g}	2	2	0	2	0
Γ_{4g}	3	-1	1	-1	-1
Γ_{5g}	3	-1	-1	-1	1

full group O_h of the wave vector at the point Γ, labeled in the arbitrary number system of Bethe.   In obtaining these characters from the tables for $O_h = O \times i$, we have quoted the even-parity cases and we have used the fact that σ_v is equivalent to $iC_4^{\,2}$, whereas σ_d is equivalent to iC_2.   For the odd-parity cases, such as the p bands of $T_{1u} = \Gamma_{4u}$ symmetry discussed above, the characters for the reflections would be changed in sign because they are improper rotations.   Given the information in this table, one can proceed by inspection of the characters (or by use of the decomposition formula) to find all the compatibility relations between Δ and Γ.   These are given (for the even-parity cases) in Table 8-2.   In the same manner, the compatibility relations between Σ and Γ can be worked out by using the subgroup C_{2v}, with the four one-dimensional representations Σ_1, Σ_2, Σ_3, and Σ_4. Finally, the compatibility between Λ and Γ can be worked out by using the subgroup C_{3v}, which has two one-dimensional representations Λ_1 and Λ_2

and a two-dimensional representation Λ_3. These results are also given in the table.

To proceed with the discussion of compatibility, we now note that, if one lets k increase along one of these symmetry directions toward the Brillouin-zone boundary, the symmetry increases suddenly when the boundary is reached. The reason for this is illustrated in Fig. 8-5, which shows

Table 8-2

Compatibility relations between representations at Γ and those at Δ, Σ, and Λ

Γ_{1g}	Γ_{2g}	Γ_{3g}	Γ_{4g}	Γ_{5g}
Δ_1	Δ_2	Δ_1, Δ_2	Δ_4, Δ_5	Δ_3, Δ_5
Σ_1	Σ_3	Σ_1, Σ_3	$\Sigma_2, \Sigma_3, \Sigma_4$	$\Sigma_1, \Sigma_2, \Sigma_4$
Λ_1	Λ_2	Λ_3	Λ_2, Λ_3	Λ_1, Λ_3

that **k** vectors differing only by reciprocal-lattice vectors are to be considered identical. In our present example of the simple cubic lattice, when the point Δ reaches the zone boundary at the point X, it suddenly becomes equivalent to the **k** vector at $k_x = -\pi/a$. The group of the **k** vector then suddenly is expanded by inclusion of a reflection plane perpendicular to the x direction and other associated symmetry elements, so that the group of the wave vector expands from C_{4v} to D_{4h}. This new group has 16 elements rather than 8, and 10 rather than 5 classes. By the same methods used above, we can now set up a compatibility table relating the representations X_i of D_{4h} at X to the representations Δ_i of the subgroup C_{4v} at Δ.

The sudden increase in symmetry when the point Σ reaches the zone edge at M is even more dramatic, since all four corners of the section of the cube by the $k_x k_z$ plane in Fig. 8-10 are related by reciprocal-lattice vectors. Thus any operation taking M into itself or any of these other three equivalent points is in the group of the wave vector. It is easy to see that the resulting symmetry group is D_{4h}. To find the compatibility relations between the representations M_i of D_{4h} at M and those of the subgroup C_{2v} at Σ, one again proceeds as above.

Finally, if the point Λ is allowed to move all the way out to the zone corner at R, one finds that the symmetry group suddenly expands from C_{3v} to the full octahedral group O_h, because of the identity of all eight corners of the zone. This enables us to use Table 8-2 to specify the compatibility of the various representations R_i at R with those at Λ, since the Γ_i and the R_i will have the same characters and hence the same decompositions.

To complete the discussion of compatibility in this case, we now note that, since the symmetry (C_{4v}) at T is a subgroup of the symmetry groups at both R and M, compatibility relations can be set up between the T_i and the

R_i and M_i. Similarly, relations can be set up between the S_i and the R_i and X_i and between the Z_i and the M_i and X_i. This completes the enumeration of the relations between symmetry points and lines. Evidently there are also entire planes of points having higher symmetry than points just off the plane, and similar arguments can be worked through to treat such cases. Given all these relations, one is finally in a position to determine whether or not a set of 10 symbols of the sort $\Gamma_i \Sigma_j \Delta_k \Lambda_l R_m T_n S_o Z_p M_q X_r$ does in

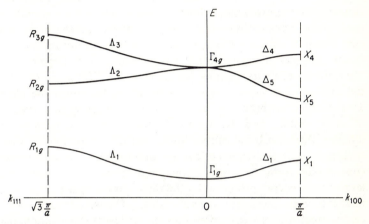

Fig. 8-11. *Schematic diagram of symmetry classifications along energy surfaces in two symmetry directions in k space.*

fact specify a possible set of mutually compatible symmetries of the wavefunction at all symmetry points in the zone. If it does, then it describes the symmetry at all points in the zone of the eigenfunctions belonging to a possible single sheet of the $E(\mathbf{k})$ surface. Figure 8-11 illustrates schematically several sheets of this energy surface along two symmetry directions in **k** space, with compatible classifications according to Table 8-2.

The example of the simple cubic lattice having been worked through in some detail, it should now be clear how other relatively simple cases can be handled. Diagrams showing the special symmetry points in the Brillouin zones of other structures and collections of character tables with various standard conventional labels for the irreducible representations are available in the literature.[1] For more details on more complicated cases, especially on nonsymmorphic space groups, the reader is also referred to the literature.[2]

[1] See, for example, G. F. Koster, in F. Seitz and D. Turnbull (eds.), "Solid State Physics," vol. 5, p. 173, Academic Press Inc., New York, 1957.

[2] The classic paper underlying the present treatment is that of Bouckaert, Smoluchowski, and Wigner, *loc. cit.*; a more recent and pedagogical presentation is given by V. Heine, "Group Theory in Quantum Mechanics," Pergamon Press, New York, 1960.

8-7   Group Theory and the Plane-wave Approximation

Let us now turn from our study of the use of group theory to classify the symmetries of wavefunctions in various parts of the Brillouin zone and consider a related application in which the symmetry is put to use in the simplification of the calculation of the actual wavefunctions and energies. We shall treat the plane-wave approximation, suitable, as discussed above, for the case of very loosely bound electrons. Actually, as pointed out by Herring,[1] the approximation may be made much more rapidly convergent for the real cases of interest by orthogonalizing all the plane waves to the core electrons on each atom, since that automatically introduces the rapid variations on each site which otherwise require a very large number of plane waves in an ordinary plane-wave expansion. This leads to what is called the *method of orthogonalized plane waves,* or for short, the OPW method. This method has been applied by numerous workers in actual calculations of energy bands, and the underlying theory has also been put into many different forms to bring out various aspects.[2] In particular, one can show that there is a large degree of cancellation between the large negative-energy contribution from the strong attractive potential in the core region and the large kinetic energy produced by the rapid variation of ψ required by orthogonalization to the core wavefunctions. The small remaining difference can be treated as a weak "pseudopotential" perturbing simple plane waves. Many aspects of the symmetry properties which we shall discuss here in conjunction with an ordinary plane-wave expansion can also be carried over directly into an OPW calculation; so the method is not without practical importance.

From Bloch's theorem, we know that we can always write the band electron wavefunction in the form $\psi_k(\mathbf{r}) = u_k(\mathbf{r})e^{i\mathbf{k}\cdot\mathbf{r}}$, where $u_k(\mathbf{r})$ is periodic with the periodicity of the crystal lattice. Therefore, $u_k(\mathbf{r})$ can be expanded in a Fourier series (i.e., a series of plane waves $e^{i\mathbf{K}\cdot\mathbf{r}}$, with $\mathbf{K}$ being reciprocal-lattice vectors). That is, we write

$$u_k(\mathbf{r}) = \sum_{\mathbf{K}'} U_k(\mathbf{K}')e^{i\mathbf{K}'\cdot\mathbf{r}} \tag{8-22}$$

This form may now be substituted into (8-21), our pseudo-Schrödinger equation for the periodic part of the wavefunction. This leads to

$$\sum_{\mathbf{K}'} \left[\frac{\hbar^2 K'^2}{2m} + V(\mathbf{r}) + \frac{\hbar^2 \mathbf{k}\cdot\mathbf{K}'}{m} \right] U_k(\mathbf{K}')e^{i\mathbf{K}'\cdot\mathbf{r}} = E_k' \sum_{\mathbf{K}'} U_k(\mathbf{K}')e^{i\mathbf{K}'\cdot\mathbf{r}} \tag{8-23}$$

If we multiply through by $e^{-i\mathbf{K}\cdot\mathbf{r}}$ and integrate, using the orthogonality of the plane waves, we obtain, after rearranging and reintroducing the total

[1] C. Herring, *Phys. Rev.,* **57,** 1169 (1940).

[2] See, for example, J. C. Phillips and L. Kleinman, *Phys. Rev.,* **116,** 287 (1959); and M. H. Cohen and V. Heine, *Phys. Rev.,* **122,** 1821 (1961).

energy $E_k = E'_k + \hbar^2 k^2/2m$,

$$\left[\frac{\hbar^2}{2m}(K^2 + 2\mathbf{K} \cdot \mathbf{k} + k^2) + V_{\mathbf{KK}} - E_k\right]U_k(\mathbf{K}) = -\sum_{\mathbf{K'} \neq \mathbf{K}} V_{\mathbf{KK'}} U_k(\mathbf{K'}) \quad (8\text{-}24)$$

In this,

$$V_{\mathbf{KK'}} = \Omega^{-1} \int V(\mathbf{r}) e^{-i(\mathbf{K}-\mathbf{K'})\cdot\mathbf{r}} \, d\tau \qquad (8\text{-}25)$$

where Ω is the normalization volume. Thus $V_{\mathbf{KK'}}$ is actually a function only of $\mathbf{K}-\mathbf{K'}$. In particular, $V_{\mathbf{KK}}$ is simply the average of the periodic potential over the unit cell, and it can be absorbed into the zero of the total energy E_k. For simplicity, let us assume that this has been done in what follows. Given any periodic potential $V(\mathbf{r})$, the matrix elements $V_{\mathbf{KK'}}$ can be calculated from (8-25) in a straightforward manner. Given them, Eqs. (8-24) form a set of linear algebraic equations to be solved for the coefficients $U_k(\mathbf{K})$ specifying the wave function ψ_k with energy E_k. In general, one would need an infinite set of coefficients, and hence one would have to solve an impossibly large secular equation. However, a reasonable approximation can often be obtained by limiting the expansion to a relatively small number of plane waves. This is particularly true if an OPW expansion is actually being carried out. Thus it is worthwhile to consider the case of a finite set of plane waves to see how they may be most efficiently chosen and handled.

After absorbing the mean potential energy into the total energy, (8-24) can be written as

$$\left[E_k - \frac{\hbar^2}{2m}(\mathbf{K} + \mathbf{k})^2\right]U_k(\mathbf{K}) = \sum_{\mathbf{K'} \neq \mathbf{K}} V_{\mathbf{KK'}} U_k(\mathbf{K'}) \qquad (8\text{-}26)$$

This form is familiar for the case of a perturbation calculation in which the diagonal energies are $\hbar^2(\mathbf{K} + \mathbf{k})^2/2m$ and the off-diagonal elements are the $V_{\mathbf{KK'}}$. As is well known, before perturbation theory can be applied, we must exactly diagonalize the Hamiltonian within any manifold of degenerate functions to get the proper zero-order linear combinations. Group theory is very helpful in carrying this out.

To illustrate this, consider the case of wavefunctions at $\mathbf{k} = 0$. Then the unperturbed energies are simply $\hbar^2 K^2/2m$, so that all plane waves with the same value of $|\mathbf{K}|$ will be degenerate. Barring accidental degeneracy, all such degenerate $\mathbf{K}$ values must be related by symmetry operations of the point group. For example, in the simple cubic case, the lowest nonzero $\mathbf{K}$ value is $(2\pi/a, 0, 0)$, which we abbreviate by introducing the Miller indices (100). Having the same energy as this plane wave will be the five others with indices (010), (001), ($\bar{1}$00), ($0\bar{1}0$), and ($00\bar{1}$), where, for example, ($\bar{1}$00) refers to $(-2\pi/a, 0, 0)$. Evidently, these are all interrelated by the operations of O_h. The next degenerate set, having twice the unperturbed energy, is the set of 12 waves based on permutations of (110), allowing for negative values of the indices.

Now, since the six degenerate- (100) type waves comprise a complete set of degenerate functions transforming into one another under the operations of O_h, they must form the basis for a six-dimensional reducible representation of that group. The character of this representation is readily found by

Table 8-3

Characters of various plane-wave representations

O_h	E	$8C_3$	$6C_4$	$3C_4{}^2$	$6C_2$	i	$8iC_3$	$6iC_4$	$3iC_4{}^2$	$6iC_2$
(000)	1	1	1	1	1	1	1	1	1	1
(100)	6	0	2	2	0	0	0	0	4	2
(110)	12	0	0	0	2	0	0	0	4	2

considering the effect on this set of six **K** vectors of a typical operation from each class in the group, in a manner reminiscent of our analysis of the character of a set of σ bonds in Sec. 4-12. The characters for the first few sets of waves are given in Table 8-3. From these characters, in the usual manner, we find that

$$(000) \to A_{1g} = \Gamma_{1g} \tag{8-27a}$$

$$(100) \to A_{1g} + E_g + T_{1u} = \Gamma_{1g} + \Gamma_{3g} + \Gamma_{4u} \tag{8-27b}$$

$$(110) \to A_{1g} + E_g + T_{1u} + T_{2g} + T_{2u} = \Gamma_{1g} + \Gamma_{3g} + \Gamma_{4u} + \Gamma_{5g} + \Gamma_{5u} \tag{8-27c}$$

The particular linear combinations of the plane waves having these various symmetries can be found by using the basis-function generating machinery of Sec. 3-8 or in many cases simply by inspection. For example, if we consider the six (100) functions, it is evident that the completely symmetric combination schematically indicated by

$$\frac{[(100) + (\bar{1}00) + (010) + (0\bar{1}0) + (001) + (00\bar{1})]}{\sqrt{6}}$$

belongs to Γ_{1g}. Similarly, for Γ_{4u}, by analogy with the three functions (p_x, p_y, p_z) we find

$$\frac{[(100) - (\bar{1}00)]}{\sqrt{2}}, \frac{[(010) - (0\bar{1}0)]}{\sqrt{2}}, \frac{[(001) - (00\bar{1})]}{\sqrt{2}}$$

whereas, for Γ_{3g}, we have

$$\frac{[(100) + (\bar{1}00) - (010) - (0\bar{1}0)]}{2},$$

$$\frac{[(100) + (\bar{1}00) + (010) + (0\bar{1}0) - 2(001) - 2(00\bar{1})]}{\sqrt{12}}$$

The linear combinations of definite symmetry having been found, our general matrix-element selection rule guarantees that the only matrix elements of the Hamiltonian will be between functions of the same symmetry. Thus, at $\mathbf{k} = 0$, even when we consider the 19 plane waves corresponding to (000), (100), and (110), we never have a secular equation larger than 3×3 to solve (for the Γ_{1g} case), and it is only a 2×2 for the Γ_{3g} and Γ_{4u} cases, and a 1×1 for the Γ_{5g} and Γ_{5u} symmetries. This accomplishes an enormous reduction in the algebraic effort, with no loss in accuracy.

If we leave the $\mathbf{k} = 0$ point, the lower resulting symmetry reduces the power of this method, because fewer plane waves will be degenerate. For example, in the simple cubic case at the point Δ, the group of the wave vector is only C_{4v}. In that case, of the (100) waves, only the four (010), (0$\bar{1}$0), (001), and (00$\bar{1}$) are still degenerate. It is easy to show that these form a basis for a representation that decomposes into $\Delta_1 + \Delta_2 + \Delta_5$. The other two, (100) and ($\bar{1}$00), each form bases for one-dimensional representations of symmetry Δ_1. Thus, even within only these six waves, there are three combinations of Δ_1 symmetry, and one no longer has such a large reduction in the size of the maximum secular equations. Obviously, if one went to a point of still lower symmetry, this method would be of still less help, and, at a completely general nonsymmetry point, it would be of no help at all. This emphasizes the fact that making an energy-band calculation of a given accuracy at a point of high symmetry is much easier than at a point of low symmetry because in the former case large numbers of plane waves are taken care of as a single unit by use of the symmetry of the situation. Although we have illustrated this fact only in the plane-wave approximation, it holds true in one form or another in all approximation methods.

8-8 Connection between Tight- and Loose-binding Approximations

Just as it was informative to set up a correlation diagram between the energy levels of molecules in the separated atom and united atom limits, it is also instructive to set up a correlation diagram relating the electronic energies in crystals in the tight- and loose-binding limits. To make such a diagram, one must choose some particular point in $\mathbf{k}$ space, while varying the strength of the crystal potential. For simplicity, we take $\mathbf{k} = 0$, since we have already treated both limits most carefully at that point. In the limit of no crystal potential, the energies are those of the (000), (100), (110), etc., plane waves, as discussed just above. In the opposite limit of extremely tight binding, the energies are just the usual atomic energy levels, as split by the crystalline field (of cubic symmetry for the case we are treating). Thus at either extreme the nature and energetic order of the levels are determined by a relatively simple case. Since the group-theoretical symmetry

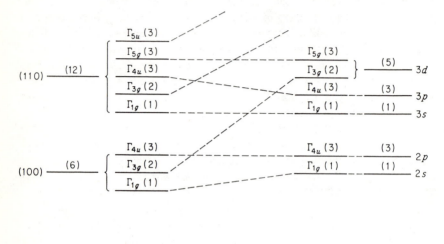

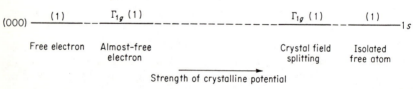

Fig. 8-12. *Correlation diagram between loose-binding and tight-binding energy levels at* $k = 0$ *in simple cubic case.*

classifications are rigorous, regardless of the strength of the potential, they can be followed through from one limit to the other. Noting that levels of the same symmetry cannot cross (because in general there will be an off-diagonal matrix element of the Hamiltonian, causing them to "repel" each other), we can rough in the entire diagram. The actual case would be expected to correspond to the level structure found at some intermediate potential strength in the diagram. Such a diagram is shown in Fig. 8-12 for the case at $k = 0$. Note how strongly the relative positions of some levels shift with the tightness of binding. A similar diagram could be drawn for any other point in the zone, but the power of the symmetry classification would not be as great.

The experimental consequences of this level structure in determining the optical properties of solids, particularly in the ultraviolet region of the spectrum, have allowed quite detailed information to be obtained about the band structure in some cases. The reader is referred to the literature[1] for some specific examples.

8-9 Spin-orbit Coupling in Band Theory

So far, we have been considering the band theory of electrons moving in a periodic potential without taking any account of the electron spin. In this case the Hamiltonian operator is a simple real operator, and the Schrödinger equation reduces to a purely real differential equation whose eigenfunctions and eigenvalues are being sought. In the absence of spin-orbit coupling, these eigenfunctions can be combined with spin eigenfunctions to form solutions of the form

$$\psi_{k+} = u_k(\mathbf{r})e^{(i\mathbf{k}\cdot\mathbf{r})}\alpha \quad \text{and} \quad \psi_{k-} = u_k(\mathbf{r})e^{i\mathbf{k}\cdot\mathbf{r}}\beta \qquad (8\text{-}28)$$

where α and β are the usual spin up and down eigenfunctions, with $S_z = \pm\frac{1}{2}$. This gives the usual twofold spin degeneracy associated with each of the spatial eigenfunctions which we have been treating.

In fact, there will be a spin-orbit coupling term in the Hamiltonian which will make the true situation more complicated than this. This subject was first treated in a thorough manner by Elliott,[2] and we give some of his results here. The form of the spin-orbit coupling term in the Hamiltonian may be found from Eq. (6-47) by the argument used to obtain (6-48), the usual form for the spin-orbit coupling in atoms, namely, $\xi(r)\mathbf{l}\cdot\mathbf{s}$. However, because there is no spherical symmetry in solids, the result must in general be left in terms of the gradient of the potential and the linear momentum operator. Thus one obtains

$$H_{\text{spin-orbit}} = \frac{\hbar}{2m^2c^2}(\nabla V \times \mathbf{p})\cdot\mathbf{s} \qquad (8\text{-}29)$$

as the term to be added to the usual Hamiltonian. There are other relativistic terms, such as those in p^4 and in $\nabla V \cdot \mathbf{p}$, which arise if this result is obtained by systematic reduction of the Dirac equation from four-component to two-component form rather than by the simple argument of Chap. 6. These will be neglected, however, because they do not introduce any qualitatively new effects but simply produce quantitative shifts in the energy eigenvalues. Evidently, the term (8-29) must still have the full symmetry demanded by the space group, and so the Bloch form of the wavefunctions

[1] H. Ehrenreich and H. R. Phillip, *Phys. Rev.*, **128**, 1622 (1962); **129**, 1550 (1963).
[2] R. J. Elliott, *Phys. Rev.*, **96**, 266, 280 (1954).

is maintained. However, because there is a spin-dependent operator in the Hamiltonian, the eigenfunctions will no longer in general be eigenfunctions of S_z. Thus, in general the eigenfunctions will be of the form

$$\psi_{\mathbf{k}} = [u_{\mathbf{k}+}(\mathbf{r})\alpha + u_{\mathbf{k}-}(\mathbf{r})\beta]e^{i\mathbf{k}\cdot\mathbf{r}} \qquad (8\text{-}30)$$

For the case of nondegenerate bands (i.e., those with only the spin degeneracy) the admixture of the reversed spin direction in any given function will be only of the order of the ratio of an atomic spin-orbit energy ζ to the interband energy separation.   At symmetry points, where there is orbital degeneracy, however, there will in general be zero-order mixtures of both spin orientations just as in the atomic case, and this will have important consequences in the determination of symmetries and energy splittings.   For example, when spin is included, we expect the Γ_{4u} functions, which transform like the three p functions, to split into a fourfold Γ_{8u} level corresponding to $j = \frac{3}{2}$ and a twofold Γ_{6u} level corresponding to $j = \frac{1}{2}$.

Considering the effect of including the spin more generally, we are confronted with a situation which is exactly analogous to that treated in Sec. 4-7, when the double groups were introduced to take care of the crystal-field splittings of states with half-integral angular momentum. For the simple space groups, which factor neatly into a group of pure point operations and a group of translations, the rules developed in Sec. 4-7 are adequate to handle the character tables of all the "extra" representations coming from the double-valued spinor functions.   For the nonsimple space groups, complications arise in the enumeration of the group elements, but we shall not go into that here.   Thus we can simply use the machinery of the double groups developed in Chap. 4 to work out the decomposition of the direct product of the representation Γ_i for the spatial part of the wavefunction with $D_{\frac{1}{2}}$ for the spin part.   That is

$$\Gamma_i \times D_{\frac{1}{2}} = \sum_j a_{ij}\Gamma_j \qquad (8\text{-}31)$$

where $D_{\frac{1}{2}}$ stands for the representation obtained from the two spin functions under the group of the wave vector.   For example, at $\mathbf{k} = 0$ in the simple cubic case, where the double group of the wave vector is O'_h, $D_{\frac{1}{2}} = \Gamma_{6g}$. Taking the direct product of Γ_{6g} with the ordinary single-valued representations appropriate for the spatial functions, namely, $\Gamma_{1g,u}$ through $\Gamma_{5g,u}$, one finds the effect of adding spin upon the degeneracies of the energy bands at $\mathbf{k} = 0$.   The results are

$$\Gamma_1 \to \Gamma_6$$
$$\Gamma_2 \to \Gamma_7$$
$$\Gamma_3 \to \Gamma_8$$
$$\Gamma_4 \to \Gamma_6 + \Gamma_8$$
$$\Gamma_5 \to \Gamma_7 + \Gamma_8$$

with the parity being unchanged by the addition of the spin. Thus the only levels which are split by the inclusion of spin are the triply degenerate Γ_4 and Γ_5 levels.

8-10 Time Reversal in Band Theory

In Sec. 5-16 we discussed the nature of the time-reversal symmetry in quantum mechanics and in particular its implications concerning a possible increase in degeneracy because of the increased symmetry entailed. We noted that Wigner had shown that, although time reversal, being antiunitary, cannot be represented within our usual framework, the question of extra degeneracy can be resolved by a consideration of the representations of the usual unitary operators of the group and their complex conjugates. Specifically, he showed that there were three cases a, b, and c, listed in Table 5-4, which determine whether an extra doubling of the degeneracy due to time-reversal symmetry occurs. These considerations may be carried over directly into our considerations here, with the integral spin rules applying for the case of energy bands with neglect of spin and the half-integral spin rules applying for the case of energy bands with spin included.

To determine which of the cases a to c is applicable, the Frobenius-Schur test involving only characters, rather than the complete representations D, D^*, is more convenient. However, this test in its general form requires one to form the sum $\sum_R \chi(R^2)$ over all group elements to see whether it is equal to $\pm h$, or zero. Since a space group contains an infinite number of elements, h, this test as it stands is not practical. Fortunately, Herring[1] has simplified this test for space groups to a form involving only elements from the group of the wave vector. This simplified form is

$$\sum_{Q_0} \chi(Q_0^2) = \begin{cases} n & \text{Case } a \\ 0 & \text{Case } b \\ -n & \text{Case } c \end{cases} \tag{8-32}$$

where the elements Q_0 are those elements of the space group which take $\mathbf{k}$ into $-\mathbf{k}$ and n is the number of such elements. Evidently, Q_0^2 will then take $\mathbf{k}$ into itself, and hence Q_0^2 must be in the group of the wave vector at $\mathbf{k}$. If the group of the wave vector G^k contains the inversion i, then the Q_0 are just the elements of G^k. If G^k does not contain i, then the Q_0 are the elements of iG^k. For example, in the simple cubic case treated earlier in this chapter, at the point Δ, the group of the wave vector is C_{4v}, which does not contain the inversion. Thus the elements Q_0 are those in iC_{4v}. When we form the elements Q_0^2 from these, the inversion drops out but for more general cases we must retain it. Using the characters

[1] C. Herring, *Phys. Rev.*, **52**, 361 (1937).

given in Table 8-1 and the facts that $(C_4^2)^2 = E$ and $\sigma_v^2 = \sigma_d^2 = E$, we see that $\sum_{Q_0} \chi(Q_0^2) = 8 = n$ for all five ordinary representations. This corresponds to case a, and since we are treating the spin-free case, Table 5-4 shows that there is no extra time-reversal degeneracy.

If spin is included, each one-electron function has half-integral spin and the rules from the last column of Table 5-4 apply. If we consider $\mathbf{k}$ at a general point in the zone, where there is no special symmetry, the group of the wave vector contains only the identity. In this case, the time-reversal symmetry doubles the degeneracy, as required by Kramer's theorem, so that every level is doubly degenerate.

Returning for simplicity to the spin-free case, we note that time reversal always introduces inversion symmetry into the $E(\mathbf{k})$ function even if the inversion is lacking from the spatial symmetry of the lattice. This can be seen in an elementary way by noting that in the absence of spin the time-reversed function is found by taking the complex conjugate. Thus, if we start with a function $\psi_\mathbf{k} = u_\mathbf{k}(\mathbf{r})e^{i\mathbf{k}\cdot\mathbf{r}}$, the time-reversed, and necessarily degenerate, function related to this is $\psi_\mathbf{k}^* = u_\mathbf{k}^*(\mathbf{r})e^{-i\mathbf{k}\cdot\mathbf{r}}$. But this latter function can be thought of as $\psi_{-\mathbf{k}} = v_{-\mathbf{k}}(\mathbf{r})e^{i(-\mathbf{k})\cdot\mathbf{r}}$. This $\psi_{-\mathbf{k}}$ has the same energy as $\psi_\mathbf{k}$, but transforms according to $-\mathbf{k}$. Thus the energy surfaces in $\mathbf{k}$ space have inversion symmetry regardless of whether the crystal has this symmetry in coordinate space. A further consequence may be deduced by considering a point at the center of a face of the Brillouin zone where $-\mathbf{k}$ is identical with $\mathbf{k}$ via a reciprocal-lattice translation. Thus $\psi_{-\mathbf{k}} = \psi_\mathbf{k}^* = \psi_\mathbf{k}$, and $\psi_\mathbf{k}$ is real, provided that the symmetry is low enough (as in a triclinic lattice) so that there is no degeneracy at this point. For the case of no degeneracy, one can draw the further conclusion that at such a point the gradient of $E(\mathbf{k})$ is zero. This can be seen analytically by noting that

$$E\left(\frac{\mathbf{K}}{2} - \delta\mathbf{k}\right) = E\left(\frac{-\mathbf{K}}{2} + \delta\mathbf{k}\right) = E\left(\frac{\mathbf{K}}{2} + \delta\mathbf{k}\right) \qquad (8\text{-}33)$$

where we have successively used inversion symmetry and translational symmetry of the $E(\mathbf{k})$ function. Given (8-33), evidently grad $E(\mathbf{K}/2) = 0$ provided only that there is no degeneracy at $\mathbf{K}/2$ and that $E(\mathbf{k})$ has a continuous derivative at that point. This absence of a gradient of energy can also be anticipated by noting that for a real wavefunction, such as we have here at $\mathbf{K}/2$, the perturbation $(\hbar/m)(\mathbf{k}\text{-}\mathbf{K}/2) \cdot \mathbf{p}$ from (8-21), being purely imaginary, can have no first-order effect on the energy. Thus the energy change with $\mathbf{k}$ must start quadratically with components of $(\mathbf{k}\text{-}\mathbf{K}/2)$. In case of degeneracy, this conclusion need not hold, since (8-33) could be satisfied by two $E(\mathbf{k})$ curves crossing at $\mathbf{K}/2$ with finite equal and opposite slopes. Similarly, in the second view, in case of degeneracy, the proof of the reality of $\psi_\mathbf{k}$ fails, and hence the $(\mathbf{k}\text{-}\mathbf{K}/2) \cdot \mathbf{p}$ perturbation can have first-order effects.

8-11 Magnetic Crystal Groups

In the preceding section, we have treated the implications of time-reversal symmetry for the usual case of nonmagnetic crystals. We noted that the time-reversed state, of necessity degenerate, was in some cases distinct, leading to additional degeneracy if the reversed state was not related to the starting state by any other symmetry operation. Even if the reversed state was distinct, however, its charge density was always the same, because time reversal reverses currents but leaves charge density invariant. Thus, so far as energy or charge density is concerned, time reversal is always a symmetry operation. (It is of course understood that any "external" magnetic fields are reversed in the operation of time reversal by the reversal of the currents producing them.) Moreover, because of the equality of energy of the two states, they have equal statistical likelihood of occurrence. This leads to zero average current density in the usual case where thermally induced perturbations cause rapid fluctuation between the two degenerate states. The situation becomes quite different, however, when a (potentially) magnetic crystal is lowered through its Néel temperature T_N. At this temperature long-range magnetic order sets in through a cooperative action of the spins on the various sites, coupled together through the exchange interaction. Because this order extends over a macroscopic system, the fluctuations between time-reversed states can occur only by very unlikely processes. Hence, time-averaged net currents and magnetic moments will exist with a high degree of stability, or, more properly, metastability. The stability of these arrays makes it possible to determine the magnetic structure in such crystals by observing the scattering of neutrons by the magnetic forces arising from the magnetic atoms. Another manifestation of the magnetic ordering is that it introduces a spin-dependent potential in the Hartree-Fock equations, so that the effective one-electron potential is not invariant under time reversal. As a consequence, even Kramer's degeneracies will in general be lifted.

This magnetic ordering sets in as the temperature is *lowered* because the ordering lowers the entropy as well as the energy of the system, and hence the requirement of minimum free energy F $(= U\text{-}TS)$ favors such a transition at low temperatures, where the entropy S is given less heavy negative weight in the free energy. In most cases the transition to the ordered state is of second order. That is, the averaged state of the system changes continuously at the transition point even though the symmetry is lowered discontinuously. This requirement of state continuity at the transition will enable us to draw some conclusions about the compatibility of symmetries in the magnetic and nonmagnetic states of a given crystal.

To provide a systematic basis for discussion of the symmetry properties of magnetic crystals, it is necessary to develop an extension of the ordinary theory of crystallographic space and point groups. These ordinary groups are appropriate for the characterization of the charge density $\rho(\mathbf{r})$ in a crystal, such as is relevant in the analysis of X-ray diffraction data, for example. However, to characterize in addition the symmetries of distributions of currents $\mathbf{j}(\mathbf{r})$ or magnetization $\mathbf{M}(\mathbf{r})$ requires a more general framework. For example, the number of possible space groups increases from 230 to 1,651,[1] and the number of point groups goes from 32 to 122. These enlarged numbers of groups are often referred to as *Shubnikov*,[2] or *color*, groups, because Shubnikov first worked out the theory of symmetry groups in which an operation interchanging black and white colors is considered in addition to the usual geometric operations. This color change is of course analogous to the reversal of spin or current directions under time reversal, although, being a scalar, it can represent only the reversal of a vector of fixed direction. Strictly speaking, since time reversal is an antiunitary operation, the irreducible representations for the magnetic groups are not completely identical with those of the isomorphous groups treated by Shubnikov.

In these magnetic groups, there are two types of elements, those A_i not containing time reversal, and those $M_k = TA_k$, which do contain it. The latter are sometimes referred to as *antioperators*, since they are antiunitary (see Sec. 5-16). Since time reversal commutes with all spatial operators and since T^2 is the identity (as a physical symmetry operator, with no concern about phases of half-integral wavefunctions), we immediately see that

$$A_i A_{i'} = A_{i''} \qquad A_i M_k = M_{k'} \qquad M_k A_i = M_{k''} \qquad M_k M_{k'} = A_i$$

Now, in building up the 122 magnetic point groups, we find three types of groups. The first type is the set of 32 ordinary point groups, with no antiunitary operators. Such a group would be the complete symmetry, for example, in a ferromagnetic crystal with only the identity and the inversion as point operations. Evidently T would reverse the sense of spins and could not be a symmetry operation. The next type of point group is the set of 32 formed from the usual ones by adjoining the time-reversal operator, so that, for all A_i in the group TA_i is also in the group. Since A_i includes E, TA_i includes T itself. But T cannot be a symmetry operation in any magnetic crystal, since it reverses the sign of any magnetic moment. Thus this set of 32 groups would appear to be possible only for diamagnetic or paramagnetic crystals, which have no time-averaged magnetic moments. However, since the point-group operations are those obtained from the

[1] N. V. Belov, N. N. Neronova, and T. S. Smirnova, *Trudy Inst. Krist., Akad. Nauk S.S.S.R.*, **11**, 33 (1955); *Kristallografiya*, **2**, 315 (1957); English translation, *Soviet Phys.-Cryst.*, **2**, 311 (1957).

[2] A. V. Shubnikov, "Symmetry and Antisymmetry of Finite Figures," U.S.S.R. Press, Moscow, 1951.

space group by setting all translations equal to zero, T is a possible point-group operation in an antiferromagnetic crystal if it always appears combined with a translation connecting two oppositely directed spins in identical chemical environments, as illustrated in Fig. 8-13a. Finally, there are the more interesting new cases, the 58 magnetic point groups which contain T only in combination with a spatial rotation or reflection. This restriction eliminates the case just mentioned in which a pure translation connects two states with spins reversed but all charge densities the same. The distinction is illustrated in one dimension in Fig. 8-13. If translation by a is denoted by A, then TA is a symmetry operation for the structure in Fig. 8-13a, but

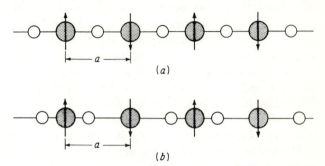

Fig. 8-13. *Two antiferromagnetic structures in one dimension.*

not for that in Fig. 8-13b. Thus (a) would have T in its point group, but (b) would not. Note that ferromagnetic or ferrimagnetic[1] crystals can never have T itself in the point group, even allowing for translations.

Let us now concentrate on the latter set of 58 point groups, since the first two sets of 32 can be handled simply by methods with which we have already dealt. We have noted that these 58 groups do not contain T itself. This can be generalized to the extent of excluding all elements of the form $M_k = TA_k$, where A_k is of odd order n, so that $A_k^n = E$. This is easily proved by noting that, if such an M_k is a group element, so is $M_k^n = (TA_k)^n = T^n(A_k)^n = T$, since n is odd, leading to a contradiction. Among the usual point-group elements, this excludes TC_3 as a possible element of these groups.

Now, designating one of these 58 groups by G, we note that it divides into two subsets of elements, the A_i and the $M_k = TA_k$, so that $G = \{A_i, M_k\}$. Following Tavger and Zaitsev,[2] we note further properties of G which allow us to construct all these groups, as follows:

1. The set of A_k and A_i are distinct; that is, no element A_j appears both with and without T. This is proved by noting that A_j^{-1} must exist

[1] A ferrimagnetic crystal has unequal antiparallel spins, leading to a net moment.
[2] B. A. Tavger and V. M. Zaitsev, *J. Exptl. Theoret. Phys. (U.S.S.R.)*, **30**, 564 (1956); English translation, *Soviet Phys. (JETP)*, **3**, 430 (1956).

in G if A_j does, since G is a group. If TA_j were also to be in the group, then $TA_jA_j^{-1} = T$ would be also, since the group is closed. This would be a contradiction of our requirement that T by itself not be in G.

2. If we replace T by E in the group $G = \{A_i, TA_k\}$, forming the set $G' = \{A_i, A_k\}$, then G' is also a group and in fact is one of the usual 32 point groups. Physically, this must be true, since when we replace T by E, we obtain covering operations of the charge density $\rho(\mathbf{r})$.

3. If we consider the group $G = \{A_i, M_k\}$, then the set $H = \{A_i\}$ forms an invariant unitary subgroup of G which is one of the 32 ordinary point groups. It is obvious that H forms a group, since the product of any two A_i is again an A_i, and the identity and inverses must be included among the A_i since G was a group and since the product of an A_i with any M_k would be an M_k. Since conjugation with any element involves multiplying by T^2 or else not by T at all, either leading to an element of H, H is surely an invariant subgroup. Further, it is also an invariant subgroup of G', since setting $T \to E$ leaves this argument intact. The fact that it is a *point* group follows, since it is a subgroup of a point group.

4. Since H is an invariant subgroup of G (and G'), the number of elements in G (or G') must be an integral multiple of the number of elements in H (see Sec. 2-5). In fact, the number of elements in G must be twice the number in H; in other words, the number of unitary operators A_i must equal the number of antiunitary ones $M_k = TA_k$. This is readily seen, since if we multiply each A_i by some M_k, we produce as many distinct M_k as there are A_i. Hence the number of M_k must be at least as large as the number of A_i. Conversely, if we multiply each M_k by some $M_{k'}$, we produce as many distinct A_i as there are M_k. To reconcile these results, the number of A_i and M_k must be equal. Thus $\{M_k\} = \{TA_k\}$ is the only coset of H in G, and $\{A_k\}$ is the only coset of H in G'. One can further show that, if H is an invariant subgroup of index 2 with respect to *any* point group G', then $G = H + T(G' - H)$ must form a group, which will be one of the type we are considering. In this, $G' - H = C$ is the coset of H with respect to G'. The proof follows immediately upon recalling that the multiplication table of the factor group (see Sec. 2-8) involving only H and C requires that $H \cdot H = H$, $H \cdot C = C \cdot H = C$, and $C \cdot C = H$. Thus

$$G \cdot G = (H + TC) \cdot (H + TC) = (H + TC) = G$$

and G forms a group.

Given the above results, we can set up a simple procedure for finding all the magnetic groups. Start with one of the ordinary point groups G'. Find all its invariant subgroups of index 2, called H_r. Then, for each H_r, set up the magnetic group $G = H_r + T(G' - H_r)$. In the Schoenflies notation, this magnetic group is denoted by $G'(H_r)$, which gives a complete specification. By successively considering all 32 of the G', one finds all 58 of the magnetic point groups G. These are listed in Table 8-4, where we list all H for each G'. (For convenience, we also list the 32 groups not containing T, so that all elements are unitary and $H = G'$.) To facilitate reference to the literature, we also give the corresponding labels in the Shubnikov and the international notations. As explained more fully in Appendix B, in the international notation a number n represents an n-fold axis, $\bar{n}$ represents an n-fold rotation combined with inversion, and the letter m represents a mirror plane. In the magnetic groups, the underscore on an element means that it is to be combined with T. Thus the international notation combines the symbol for G' with underscoring, which gives information equivalent to the specification of the subgroup H. The Shubnikov notation is rather similar to the international notation, except that n refers to S_n rather than iC_n and the single dot refers to parallel elements and the double dot to perpendicular ones. As we shall not be using them, we shall not give further details on these notations here.

To illustrate the general considerations discussed above, let us consider the group C_{2h}, which has the four elements E, C_2, σ_h, i. Since the square of each of these elements is E, we see that there are three ways to choose the invariant subgroup, namely, (E, i), (E, C_2), and (E, σ_h). These correspond to the three successive entries under $G' = C_{2h}$ in Table 8-4 (excepting the group with no antielements). In each case, the other two elements of the group form the coset which is to be multiplied by T. For example, in the magnetic group $C_{2h}(C_2)$, the elements are E, C_2, $T\sigma_h$, Ti. The other magnetic groups $C_{2h}(C_i)$ and $C_{2h}(C_{1h})$ are formed in a similar manner.

8-12 Symmetries of Magnetic Structures

As an illustration of the use of these magnetic groups to characterize structures, consider the hypothetical orthorhombic ferromagnetic unit cell shown in Fig. 8-14. In the disordered state above T_N there would be no averaged magnetic moments on the sites, and the point group would be D_{2h}. When the order sets in as shown, the symmetry changes. The unitary group of symmetry operations decreases to C_{2h}, and the magnetic symmetry group is then $D_{2h}(C_{2h})$. To verify these statements, we note that D_{2h} contains the elements E, C_{2x}, C_{2y}, C_{2z}, σ_x, σ_y, σ_z, and i. (We associate x with the a axis, etc.) Evidently, these are all symmetry operations for the nonmagnetic state. In the magnetic state, however, C_{2x} and C_{2y}, for example, lead to a configuration in which all spins are reversed. Thus they can be in the magnetic group only in conjunction with T. C_{2z} remains in the group, since it leaves the spins unchanged. Recalling that spins or magnetic moments are axial vectors, we see that σ_z leaves the spins unchanged, whereas σ_x and σ_y reverse them. Also, inversion leaves the spin directions unchanged,

Table 8-4

Magnetic point groups

As indicated in the text, each mixed magnetic group can be characterized by giving a point group G' and an invariant unitary subgroup H of index 2. All 58 such groups are listed here, together with the alternative Shubnikov and international notations. In the latter notations, one takes the symbol for G' and underscores those elements to which T has been adjoined. We also list the 32 point groups not containing T at all, in which the unitary subgroup is the entire group G'; hence, in these cases H is of index 1. We do not list the other 32 point groups, which are obtained by adjoining T to all elements of G' to make a group of double size. The final column indicates whether the structure is ferromagnetic (F) or antiferromagnetic (AF). The 32 doubled point groups, which we have omitted, may be either diamagnetic or paramagnetic, or else antiferromagnetic if T is always coupled with a translation.

Crystal system	Group G'	Number of elements	Invariant unitary subgroup H	Shubnikov notation	International notation	Magnetic order
Triclinic	C_1	1	C_1	1	1	F
	$S_2(C_i)$	2	S_2	$\bar{2}$	$\bar{1}$	F
			C_1	$\underline{\bar{2}}$	$\underline{\bar{1}}$	AF
Monoclinic	C_{1h}	2	C_{1h}	m	m	F
			C_1	$\underline{m}$	$\underline{m}$	F
	C_2	2	C_2	2	2	F
			C_1	$\underline{2}$	$\underline{2}$	F
	C_{2h}	4	C_{2h}	$2{:}m$	$2/m$	F
			C_i	$\underline{2}{:}m$	$\underline{2}/m$	F
			C_2	$2{:}\underline{m}$	$2/\underline{m}$	AF
			C_{1h}	$\underline{2}{:}m$	$2/m$	AF
Rhombic	C_{2v}	4	C_{2v}	$2{\cdot}m$	$2mm$	AF
			C_{1h}	$\underline{2}{\cdot}m$	$\underline{2}mm$	F
			C_2	$2{\cdot}\underline{m}$	$2\underline{mm}$	F
	D_2	4	D_2	$2{:}2$	222	AF
			C_2	$2{:}\underline{2}$	$2\underline{22}$	F
	D_{2h}	8	D_{2h}	$m{\cdot}2{:}m$	mmm	AF
			C_{2h}	$\underline{m}{\cdot}2{:}m$	$\underline{m}mm$	F
			C_{2v}	$m{\cdot}2{:}\underline{m}$	$mm\underline{m}$	AF
			D_2	$\underline{m}{\cdot}2{:}\underline{m}$	$\underline{mmm}$	AF
Tetragonal	C_4	4	C_4	4	4	F
			C_2	$\underline{4}$	$\underline{4}$	AF
	S_4	4	S_4	$\bar{4}$	$\bar{4}$	F
			C_2	$\underline{\bar{4}}$	$\underline{\bar{4}}$	AF
	C_{4h}	8	C_{4h}	$4{:}m$	$4/m$	F
			C_{2h}	$\underline{4}{:}m$	$\underline{4}/m$	AF
			C_4	$4{:}\underline{m}$	$4/\underline{m}$	AF
			S_4	$\underline{4}{\cdot}m$	$\underline{4}/\underline{m}$	AF
	D_{2d}	8	D_{2d}	$\bar{4}{\cdot}m$	$\bar{4}2m$	AF
			C_{2v}	$\underline{\bar{4}}{\cdot}m$	$\underline{\bar{4}}2m$	AF
			D_2	$\underline{\bar{4}}{\cdot}m$	$\underline{\bar{4}}2\underline{m}$	AF
			S_4	$\bar{4}{\cdot}\underline{m}$	$\bar{4}2\underline{m}$	F
	C_{4v}	8	C_{4v}	$4{\cdot}m$	$4mm$	AF
			C_{2v}	$\underline{4}{\cdot}m$	$\underline{4}mm$	AF
			C_4	$4{\cdot}\underline{m}$	$4\underline{mm}$	F
	D_4	8	D_4	$4{:}2$	$42(422)$	AF
			D_2	$\underline{4}{:}2$	$\underline{4}2$	AF
			C_4	$4{:}\underline{2}$	$4\underline{2}$	F
	D_{4h}	16	D_{4h}	$m{\cdot}4{:}m$	$4/mmm$	AF
			D_{2h}	$m{\cdot}\underline{4}{:}m$	$\underline{4}/mmm$	AF
			C_{4h}	$\underline{m}{\cdot}4{:}m$	$4/\underline{m}mm$	F
			D_{2d}	$m{\cdot}\underline{4}{:}\underline{m}$	$\underline{4}/mm\underline{m}$	AF
			C_{4v}	$m{\cdot}4{:}\underline{m}$	$4/mm\underline{m}$	AF
			D_4	$\underline{m}{\cdot}4{:}\underline{m}$	$4/\underline{m}m\underline{m}$	AF

Table 8-4—continued

Crystal system	Group G'	Number of elements	Invariant unitary subgroup H	Shubnikov notation	International notation	Magnetic order
Rhombohedral	C_3	3	C_3	3	3	F
	S_6	6	S_6	$\bar{6}$	$\bar{3}$	F
			C_3	<u>$\bar{6}$</u>	<u>$\bar{3}$</u>	AF
	C_{3v}	6	C_{3v}	$3{\cdot}m$	$3m$	AF
			C_3	$3{\cdot}\underline{m}$	$3\underline{m}$	F
	D_3	6	D_3	$3{:}2$	32	AF
			C_3	$3{:}\underline{2}$	$3\underline{2}$	F
	D_{3d}	12	D_{3d}	$\bar{6}{\cdot}m$	$\bar{3}m$	AF
			S_6	$\bar{6}{\cdot}\underline{m}$	$\bar{3}\underline{m}$	F
			C_{3v}	$\underline{\bar{6}}{\cdot}m$	$\underline{\bar{3}}m$	AF
			D_3	$\underline{\bar{6}{\cdot}m}$	$\underline{\bar{3}m}$	AF
Hexagonal	C_{3h}	6	C_{3h}	$3{:}m$	$\bar{6}$	F
			C_3	$3{:}\underline{m}$	$\underline{\bar{6}}$	AF
	C_6	6	C_6	6	6	F
			C_3	$\underline{6}$	$\underline{6}$	AF
	C_{6h}	12	C_{6h}	$6{:}m$	$6/m$	F
			S_6	$\underline{6}{:}m$	$\underline{6}/m$	AF
			C_{3h}	$\underline{6}{:}m$	$\underline{6}/m$	AF
			C_6	$6{:}\underline{m}$	$6/\underline{m}$	AF
	D_{3h}	12	D_{3h}	$m{\cdot}3{:}m$	$\bar{6}m2$	AF
			C_{3v}	$\underline{m}{\cdot}3{:}m$	$\bar{6}m\underline{2}$	AF
			D_3	$\underline{m{\cdot}}3{:}\underline{m}$	$\bar{6}\underline{m}2$	AF
			C_{3h}	$\underline{m}{\cdot}3{:}m$	$\bar{6}\underline{m2}$	F
	C_{6v}	12	C_{6h}	$6{\cdot}m$	$6mm$	AF
			C_{3v}	$\underline{6}{\cdot}m$	$\underline{6}mm$	AF
			C_6	$6{\cdot}\underline{m}$	$6\underline{mm}$	F
	D_6	12	D_6	$6{:}2$	$62(622)$	AF
			D_3	$\underline{6}{:}2$	$\underline{6}2$	AF
			C_6	$6{:}\underline{2}$	$6\underline{2}$	F
	D_{6h}	24	D_{6h}	$m{\cdot}6{:}m$	$6/mmm$	AF
			D_{3d}	$m{\cdot}\underline{6}{:}m$	$6/m\underline{mm}$	AF
			C_{6h}	$\underline{m}{\cdot}6{:}m$	$6/m\underline{mm}$	F
			D_{3h}	$m{\cdot}\underline{6}{:}m$	$6/\underline{m}mm$	AF
			C_{6v}	$m{\cdot}6{:}\underline{m}$	$6/\underline{mmm}$	AF
			D_6	$\underline{m}{\cdot}6{:}\underline{m}$	$6/\underline{mmm}$	AF
Cubic	T	12	T	$3/2$	23	AF
	T_h	24	T_h	$\bar{6}/2$	$m3$	AF
			T	$\underline{\bar{6}}/2$	$\underline{m}3$	AF
	T_d	24	T_d	$3/\bar{4}$	$\bar{4}3m$	AF
			T	$3/\underline{\bar{4}}$	$\underline{\bar{4}}3m$	AF
	O	24	O	$3/4$	$43(432)$	AF
			T	$3/\underline{4}$	$4\underline{3}$	AF
	O_h	48	O_h	$\bar{6}/4$	$m3m$	AF
			T_h	$\bar{6}/4$	$m3\underline{m}$	AF
			T_d	$\underline{\bar{6}}/4$	$\underline{m}3m$	AF
			O	$\bar{6}/\underline{4}$	$m3\underline{m}$	AF

because they are axial vectors. From this enumeration, we see that the magnetic symmetry group contains the elements E, C_{2z}, σ_z, i, TC_{2x}, TC_{2y}, $T\sigma_x$, and $T\sigma_y$, as required for $D_{2h}(C_{2h})$.

As a more realistic example introducing further features, consider the tetragonal antiferromagnetic unit cell shown in Fig. 8-15. This is the

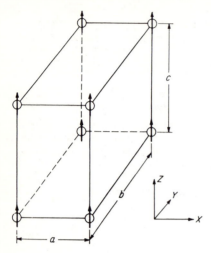

Fig. 8-14.   *Orthorhombic ferromagnetic unit cell of $D_{2h}(C_{2h})$ symmetry.*

structure of the antiferromagnets MnF_2, FeF_2, and CoF_2, for example, which have the rutile structure. The shaded circles represent the metallic ions and the hollow circles the fluorine ions. The translations a_1 and a_2 in the xy plane are of equal length a, and the third orthogonal translation a_3 is of length c in the z direction. Although this diagram shows eight up spins and only one down spin, the structure is antiferromagnetic since each of the eight corner ions is "shared" among eight such cells which touch the same corner. We note that the magnetic lattice has the same primitive

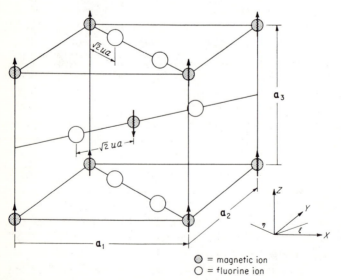

⊚ = magnetic ion
○ = fluorine ion

Fig. 8-15.   *Tetragonal antiferromagnetic unit cell.*   ($|a_1| = |a_2| = a$; $|a_3| = c \neq a$.)

translations as the paramagnetic one, because the different orientation of the fluorine neighbors distinguishes the body-center magnetic ions from the corner ions even in the disordered state. Thus there are two metal ions in the chemical as well as the magnetic unit cell. This would not be the case if the fluorine ions were not present. The presence of two metallic ions per unit cell introduces some symmetry operations involving the nonprimitive translation $\tau_0 = \frac{1}{2}(a_1 + a_2 + a_3)$ connecting the two metallic ions. In the paramagnetic state we have 16 symmetry operations involving either no translation or the nonprimitive translation τ_0. In our space-group-element notation $\{R \mid t\}$, where R represents a rotation or reflection and t a translation, these are:

1. $\{E \mid 0\}$

2. $\{C_2 \mid 0\}$

3. $\{C_{2\xi} \mid 0\}$

4. $\{C_{2\eta} \mid 0\}$

5. $\{C_4 \mid \tau_0\}$

6. $\{C_4^{-1} \mid \tau_0\}$

7. $\{C_{2x} \mid \tau_0\}$

8. $\{C_{2y} \mid \tau_0\}$

9. $\{i \mid 0\}$

10. $\{\sigma_h \mid 0\} = \{C_2 \mid 0\}\{i \mid 0\}$

11. $\{\sigma_{d\xi} \mid 0\} = \{C_{2\xi} \mid 0\}\{i \mid 0\}$

12. $\{\sigma_{d\eta} \mid 0\} = \{C_{2\eta} \mid 0\}\{i \mid 0\}$

13. $\{S_4^{-1} \mid \tau_0\} = \{C_4 \mid \tau_0\}\{i \mid 0\}$

14. $\{S_4 \mid \tau_0\} = \{C_4^{-1} \mid \tau_0\}\{i \mid 0\}$

15. $\{\sigma_{vx} \mid \tau_0\} = \{C_{2x} \mid \tau_0\}\{i \mid 0\}$

16. $\{\sigma_{vy} \mid \tau_0\} = \{C_{2y} \mid \tau_0\}\{i \mid 0\}$

The rotations C_4 and C_2 are about the z axis, whereas $C_{2\xi}$ and $C_{2\eta}$ are rotations about axes parallel to $(a_1 \pm a_2)$. We note that the second eight of these are simply the products of the first eight with the inversion $\{i \mid 0\}$. All space-group operations for this crystal are built up from these 16 operations combined with translations by multiples of the primitive displacements. Now, the rule for finding the point group from this space group is to set all translations, primitive and nonprimitive, equal to zero, since these translations do not affect the macroscopic crystal symmetry under point operations. Thus the point group is just $D_4 \times i = D_{4h}$, even though the operations of the form $\{R \mid 0\}$ actually present in the space group would form only $D_2 \times i = D_{2h}$. This is an example of a nonsimple space group, in which the entire point group is *not* a subgroup of the space group.

Continuing to the ordered state of MnF_2, for example, we have the spin arrangement shown in Fig. 8-15. Considering in turn the 16 operations listed above, we find that the unitary subgroup of symmetry operations contains the following eight elements: $\{E \mid 0\}$, $\{C_2 \mid 0\}$, $\{C_{2x} \mid \tau_0\}$, $\{C_{2y} \mid \tau_0\}$, and the products of these with the inversion. Thus the unitary subgroup is $D_2 \times i = D_{2h}$, and, in the Schoenflies notation, the magnetic group would be $D_{4h}(D_{2h})$. One might note that the spins must lie along the z axis in this crystal to be consistent with this point group. For example, if the spins were to be tipped to lie on the x direction, the magnetic symmetry group would be reduced to $D_{2h}(C_{2h})$, with the C_2 axis being the x axis. Such a reduction in symmetry is obvious, since if the spins point along x, it is no longer equivalent to y and so the point group cannot be tetragonal. Because of magnetostriction, the symmetry of $\rho(\mathbf{r})$ will also be lowered by a tipping of the spins, as is indicated by the change of G' from D_{4h} to D_{2h}.

It is also instructive to consider the symmetry situation which would prevail if the fluorine ions were not present in the structure. In that case, in the paramagnetic state, the unit cell would be only half as large, containing only one magnetic atom, since the translation τ_0 (by itself) would be a symmetry operation of the translation group. In the language of crystallography, the Bravais lattice would be body-centered tetragonal rather than the primitive tetragonal Bravais lattice characteristic of the magnetic structure or of the chemical structure with fluorine atoms present. As a result of the doubling of unit-cell size in going from the disordered to the ordered state, the Brillouin-zone size would be cut in half. This naturally has characteristic effects on the band structure of the crystal. Another thing to notice is that, although the magnetic point group of the actual MnF_2 crystal is $D_{4h}(D_{2h})$, it would be the group $D_{4h} \times T$ (one of the 32 omitted from Table 8-4) if the fluorines were not there. Thus the nonmagnetic ions cannot be ignored in determining questions of symmetry.

Enumeration of ferromagnetic groups. Having gained some familiarity with the significance of these groups, we can now readily work out which of the groups require an antiferromagnetic structure and which can have a net ferromagnetic moment. We have already noted that the 32 groups of the form $G' \times T$ can be either diamagnetic or paramagnetic, or else antiferromagnetic with T coming in only through antitranslations. For the other 90 groups listed in Table 8-4, the structures cannot be diamagnetic or paramagnetic, because then T by itself would be a symmetry operation, and none of these 90 groups has this operation included. Many of them cannot be ferromagnetic (F) because some operation of the group will reverse the sign of an antiaxial vector (such as the magnetization) for all possible orientations. These then must be antiferromagnetic (AF). The results are given in the last column of Table 8-4. To see how these may be obtained, we note that the magnetization vector must in general lie along any two-, three-, four-, or six-fold axes that might be present, and perpendicular to any mirror planes. However, it must also be perpendicular to any twofold antiaxis and parallel to any antimirror plane. With these rules, it is easy to verify the entries in the table and to work out the direction in which the magnetization must point in each ferromagnetic case. There are 31 groups consistent with a ferromagnetic moment in some direction. The others must be antiferromagnetic for all spin orientations. For further details and a discussion of the symmetry restrictions on piezomagnetism, Hall effect, etc., see the review of Le Corre.[1]

One caution should be exercised in using these results: the spatial structure referred to here is that in the magnetic state, which may differ from that in the paramagnetic state. For example, a cubic material cannot be ferromagnetic according to our table. What this means is that, if a moment exists in a crystal which was cubic in the paramagnetic state, it must distort slightly on ordering (the process of magnetostriction) so as to reach a symmetry group compatible with a ferromagnetic moment. For example, if $\mathbf{M}$ lies on a fourfold axis, the group must be one of the tetragonal magnetic groups compatible with $\mathbf{M} \neq 0$ which is an invariant subgroup of the initial group. The index of the subgroup will then give the number of distinct directions in which the magnetization might lie in different domains of the crystal. Thus, if the original group was $O_h \times T$ and the final group is $D_{4h}(C_{4h})$, the number of domain directions would be $(48 \times 2)/16 = 6$. These six directions correspond to the positive and negative x, y, and z directions, which are the directions compatible with the fourfold symmetry.

8-13　The Landau Theory of Second-order Phase Transitions

The above considerations can be put on a more precise basis if one restricts attention to the immediate vicinity of the phase transition between the paramagnetic and the ordered state of the crystal. As stated earlier, this transition will normally be of second order, so that the state of the system on the ordered side of the transition ($T < T_N$) will approach arbitrarily close to the disordered state as $T \rightarrow T_N$. Thus, there is a discontinuity of symmetry at T_N, with no discontinuity of state. We can characterize the change in symmetry most easily by giving the symmetry group G_0 in the paramagnetic state and that G_1 in the ordered state immediately below T_N. One can then apply a general theory due to Landau[2] for treating the symmetry change in such a second-order transition.

We start by defining a generalized density function $\rho(\mathbf{r})$ which expresses the spatial distribution of charge and magnetic moment. In the higher symmetry state with space group G_0, this density will be taken to be $\rho_0(\mathbf{r})$, which must of course be invariant under all the transformations in G_0. Since G_0 contains T, ρ_0 must have no magnetic-moment density. Below T_N, we assume that $\rho(\mathbf{r})$ can be written as $\rho(\mathbf{r}) = \rho_0(\mathbf{r}) + \delta\rho(\mathbf{r})$. Because any function can be decomposed into parts transforming according to the rows of the irreducible representations of any group (Sec. 3-8), we can in

[1] Y. Le Corre, *J. phys. radium*, **19**, 750 (1958).
[2] See, for example, L. D. Landau and E. M. Lifshitz, "Statistical Physics," chap. 14, Pergamon Press, Addison-Wesley Publishing Company, Inc., Reading, Mass., 1958.

particular use the representations of G_0 to write

$$\rho(\mathbf{r}) = \sum_{j,\kappa} c_\kappa^{(j)} \varphi_\kappa^{(j)} \tag{8-34}$$

Since ρ_0 is invariant under G_0, it is represented by only a single term in this expansion, namely, the one from the one-dimensional identical representation. As any contribution to $\delta\rho$ having this symmetry cannot give a magnetic moment, it will not be important in the expansion of $\delta\rho$. Thus the expansion (8-34) actually refers to $\delta\rho$, apart from the term invariant under G_0.

Next we consider the free energy Φ of the crystal. This will be a function of the external parameters pressure and temperature and of the internal-state parameters $c_\kappa^{(j)}$ specifying the state of the system. As the pressure and temperature are varied, the $c_\kappa^{(j)}$ shift to minimize Φ, as is required to maintain the equilibrium state of the system, and of course the $\varphi_\kappa^{(j)}$ will also change, though maintaining their symmetry. Denote the free energy at the transition point by Φ_0. At this point, all $c_\kappa^{(j)} = 0$, except that from the identical representation. Since the change of state is continuous in a second-order phase transition, we know that the other $c_\kappa^{(j)}$ start with vanishingly small values at the transition and increase as T drops below T_N. This suggests that we should be able to expand Φ as a power series in the $c_\kappa^{(j)}$ near the transition point. The form of this series is controlled by the fact that the free energy must be invariant under all symmetry transformations of G_0 because it must be valid through the transition point. Now the $c_\kappa^{(j)}$ must transform under symmetry operations in the same way as the $\varphi_\kappa^{(j)}$ if the product ρ is to have a unique value independent of the choice of co-ordinate system. Thus the same combinations of the $c_\kappa^{(j)}$ as of the $\varphi_\kappa^{(j)}$ form invariants. The transformation properties of products are found from direct products of the irreducible representations. Recalling our matrix-element selection rules, an invariant can be obtained only from the product of a representation with itself. (Throughout this discussion, pairs of complex-conjugate representations are treated as single irreducible representations.) Moreover, by the generalized Unsöld theorem (Sec. 4-9) the sum of squares of the partners in an irreducible representation forms such an invariant. Thus we take the expansion of Φ near the transition to begin with the terms

$$\Phi = \Phi_0 + \sum_j{}' A^{(j)}(P, T) \sum_\kappa [c_\kappa^{(j)}]^2 \tag{8-35}$$

where the prime on the sum denotes the absence of the term from the identical representation.

Since the crystal must be invariant under G_0 at T_N, the values $c_\kappa^{(j)}$ must be all zero there in equilibrium. From a consideration of first and second derivatives, this requires that all the $A^{(j)}$ be nonnegative. If they were all positive, however, all the $c_\kappa^{(j)}$ would remain zero through the transition region and no transition would occur. Thus one of the $A^{(j)}(P, T)$ must vanish at

T_N, as it changes from positive to negative. Denote it by $A^{(j')} = a(T - T_N)$. Then, for $T > T_N$, all the $c_\kappa^{(j)}$ are zero; for $T < T_N$, the $c_\kappa^{(j')}$ for this particular representation j' become nonzero, and we can write

$$\delta\rho = \sum_\kappa c_\kappa^{(j')} \varphi_\kappa^{(j')} \tag{8-36}$$

Thus we have the general result that the magnetic order set up in a second-order phase transition from the paramagnetic state transforms like the basis functions of a single irreducible representation of the space group G_0 of the paramagnetic crystal. The symmetry operators leaving $\rho_0 + \delta\rho$ invariant evidently form a subgroup G_1 of G_0. The determination of the particular linear combination of $\varphi_\kappa^{(j')}$ within a representation is more intricate, since it depends on the higher-order invariants which keep $\delta\rho$ from increasing without bound for $T < T_N$. We shall not go into this here. Upon cooling a finite distance below T_N, one can no longer expect that $\delta\rho$ will consist of basis functions from a single irreducible representation, as in (8-36). Rather, one expects odd "harmonics" from higher-order terms in the expansion to come in. However, the symmetry group G_1 defined by the invariance of (8-36) should remain valid until another transition occurs at still lower temperatures.

These points are illustrated by an example, treated by Dimmock,[1] in which the magnetic-moment density just below T_N has the form

$$\rho \sim \mu_x(T_N - T)^{\frac{1}{2}} \cos \frac{2\pi x}{3a}$$

corresponding to a single irreducible representation of the space group having $k = 2\pi/3a$. This introduces a magnetic superlattice with period $3a$ rather than the chemical-cell period a. At lower temperatures, Dimmock shows that one might expect ρ to include a third harmonic term $\sim \mu_x(T_N - T)^{\frac{3}{2}} \cos(2\pi x/a)$, representing a "sharpening" of the structure. Still higher terms would in general also be expected, but, for a structure with magnetic ions only at $x = 0$, $a/2$, a, $3a/2$, ..., the two terms above represent the most general spatial variation obtainable from higher odd powers of the initial density. The higher terms would affect the temperature dependences, however. We note that although the term in $\cos(2\pi x/a)$ no longer belongs to the same irreducible representation of the paramagnetic space group G_0, as does $\cos(2\pi x/3a)$, it does have the same symmetry group G_1, because any operation leaving $\cos(2\pi x/3a)$ invariant also leaves $\cos(2\pi x/a)$ invariant. This is consistent with our general conclusions stated above.

8-14 Irreducible Representations of Magnetic Groups

For many applications, it is necessary to know the irreducible representations of the groups, not just the structures of the groups themselves. For example,

[1] J. O. Dimmock, *Phys. Rev.*, **130**, 1337 (1963).

the wavefunctions of localized states in a crystal are characterized by irreducible representations of the point group of the symmetry about the local site (which may be considerably smaller than the macroscopic point group of the crystal structure). On the other hand, for discussion of nonlocalized states extending over the periodic structure of the crystal, use of the full space-group representations is essential. The nonlocalized states to be dealt with are of two major sorts: the electronic energy-band states, considered earlier in the chapter for nonmagnetic crystals, and spin wave states. The latter are a type of excitation of the magnetic system in which there are small deviations from alignment by spins on all sites rather than a full quantum unit of spin deviation on a single site. The phases of the deviations at the various sites are related by the same exponential factor $e^{i\mathbf{k}\cdot\mathbf{r}}$ which Bloch's theorem required for the electronic excitations. Within the unit cell, however, the symmetry of the spin wave excitation is that of the lowering operator S_- referred to a quantization axis along the ordered spin direction of the atom concerned, rather than the symmetry of an atomic electronic wavefunction as in the electronic energy bands.

The methods for handling the representations of the magnetic groups have been discussed in some detail by Dimmock and Wheeler[1]. We shall not go into the complications of the full discussion. However, we note several points. The presence of the antiunitary operators associated with time reversal precludes the possibility of ordinary matrix representations, as we noted in Sec. 5-16. Rather, one must settle for corepresentations, which require the insertion of complex-conjugate signs into certain of the product relations. Still, it is possible to reduce questions of degeneracy to a discussion involving only the unitary subgroup of the full magnetic group, in an extension of the work of Herring referred to in Sec. 8-10. In this way, the discussion of compatibility relations between symmetries of the electronic wavefunctions at various symmetry points and lines in the Brillouin zone can be extended to the case of magnetic crystals, in which time reversal is not a symmetry operator. In addition to the compatibility relations among the various parts of the zone, one can also work out compatibilities between the symmetries in the paramagnetic state and in the ordered state below T_N, which will describe how degeneracies are lifted by the reduction of symmetry on going from G_0 to G_1. One difficulty which arises is that if the magnetic unit cell is larger than the chemical one, owing to a magnetic-superlattice structure, then the magnetic Brillouin zone will be smaller than the chemical one. In this case, some $\mathbf{k}$ values which have special symmetry properties because they are on the surface of the magnetic zone will lack these symmetries in the chemical zone. Thus, in these cases, the magnetic symmetry group G_1 is not, in fact, a subgroup of the nonmagnetic group G_0. Although

[1] J. O. Dimmock and R. G. Wheeler, *J. Phys. Chem. Solids*, **23**, 729 (1962); *Phys. Rev.*, **127**, 391 (1962).

these cases require special handling, the discussion of Dimmock and Wheeler shows that no serious difficulty results.

EXERCISES

8-1 For nearly free electrons, it is often useful to use the extended-zone scheme, with the approximation that the wavefunctions are nearly $e^{i\mathbf{k}\cdot\mathbf{r}}$ and the energies are nearly $E(\mathbf{k}) = \hbar^2 k^2/2m$, with $\mathbf{k}$ being allowed to range far beyond the first Brillouin zone, as described in Sec. 8-2. By graphical construction, find the first four zones for the square lattice, the first two being indicated in Fig. 8-1. For the higher zones, show how the pieces fit together within the first zone. Using the free-electron approximation for the energy, sketch energy contours in the reduced-zone scheme for all four of these zones.

8-2 Any one-dimensional periodic potential can be Fourier analyzed into a series $V(x) = \displaystyle\sum_{n=-\infty}^{\infty} V_n\, e^{2\pi i n x/a}$. ($V_{-n} = V_n^*$, since V is real.) Use this potential in a perturbation treatment of the energy levels and wavefunctions of a nearly free electron. Note that special handling of two nearly degenerate functions is required whenever k is nearly equal to an integral multiple of π/a. Show that the two eigenvalues at such k values differ by just twice the matrix element connecting the two zero-order plane-wave states. Show that there is a quadratic variation of energy with k near such points, and relate the coefficient to the gap between the two energies, showing that a small gap produces a rapid variation with k, corresponding to a small *effective mass*. On the other hand, show that, for the lowest state at $k = 0$, the effective mass is always greater than the true mass.

8-3 Generalize Exercise 8-2 to the case of three dimensions with arbitrary reciprocal lattice.

8-4 (*a*) Apply the results of Exercise 8-2 to the case in which the potential consists only of the $|n| = 1$ terms in the expansion, and take V_1 real so that the potential is even about the origin. (*b*) Apply the results of Exercise 8-2 to the case of a *square-wave potential*, in which $V(x)$ is an even function such that $V(x) = -v$ for $|x| < a/4$ and $V(x) = +v$ for $a/4 < |x| < 3a/4$, with a periodicity of a. The results of these perturbation approaches will be valid only for weak potentials. Exact solutions to these problems can be obtained, the first in terms of Mathieu functions[1] and the second by piecing together solutions for constant potentials.[2]

8-5 Consider a two-dimensional square lattice, with symmetry group C_{4v}. The Brillouin zone will also be square, and the various symmetry points and lines are as indicated in Fig. 8-16. Find the group of the wave vector at each of the points Γ, Σ, Δ, X, M, and Z. For each of these groups write down character tables to establish an unambiguous notation. Then work out *all* the compatibility relations between adjacent points, and write out several sets of mutually compatible symmetries for all points.

[1] J. C. Slater, *Phys. Rev.*, **87**, 807 (1952).
[2] R. de L. Kronig and W. G. Penney, *Proc. Roy. Soc. (London)*, **A130**, 499 (1931).

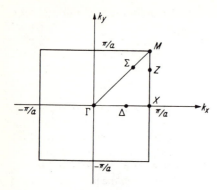

Fig. 8-16. *Special points in square Brillouin zone.*

8-6 Consider a tetragonal lattice whose point group is D_{4h}. The reciprocal lattice is also tetragonal, and the cross section of the first Brillouin zone at the horizontal reflection plane is a square. Within this plane, special points can be labeled as in Fig. 8-16. (a) What is the group of the wave vector at each of the labeled symmetry points? (b) Work out the compatibility relations between the irreducible representations at Γ, Σ, and M.

8-7 Continue the analysis of the plane-wave expansion at $\mathbf{k} = 0$ given in Sec. 8-7, carrying it through the (111), (200), (210), and (211) waves, as well as those treated in the text. Find all the symmetries which can be obtained from each set of degenerate waves. Work out explicit linear combinations for all the symmetries obtainable from the (110)-type waves.

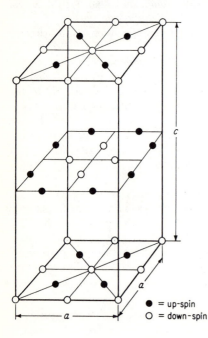

Fig. 8-17. *Tetragonal magnetic structure.*

● = up-spin
○ = down-spin

8-8   Consider the tetragonal magnetic structure shown in Fig. 8-17, where the spin directions up and down are parallel to the fourfold axis.   (*a*) What are the point groups of this structure in the paramagnetic state and in the ordered state shown in the figure?   (*b*) Does symmetry allow a ferromagnetic moment?   Explain in some detail why the apparent cancellation of moments in the figure is not rigorously required by symmetry.   This is an example of what is called *feeble*, or *parasitic*, ferromagnetism.[1]

8-9   In the paramagnetic state, the crystals of MnO, FeO, CoO, NiO, MnS, and MnSe have the face-centered cubic NaCl structure shown in Fig. 8-18.   (Point

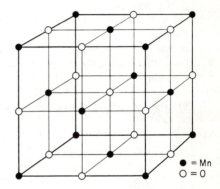

Fig. 8-18.   *Cubic unit cell of* NaCl
structure.

● = Mn
O = O

group is $O_h \times T$.)   In the antiferromagnetic state, all the magnetic ions in a given (111) plane have their spins parallel.   The spin directions of these planes alternate frcm one (111) plane of magnetic ions to the next.   The magnetic symmetry group depends on the axis of orientation of the spins.   Consider the following orientations: (*a*) the [111] direction perpendicular to the planes of parallel spins; (*b*) an [001] direction; (*c*) an [01$\bar{1}$] direction lying in the (111) plane; (*d*) a general direction having no special symmetry.   For each of these orientations, find the magnetic point group, specifying any special directions of group operations with respect to the original cube directions; also find the crystal system of the ordered structure and the number of types of domains which will be formed.[2]

REFERENCES

BELOV, K. P.: "Magnetic Transitions," translated from the Russian by W. H. Furry, Consultants Bureau, New York, 1961.

BOUCKAERT, L. P., R. SMOLUCHOWSKI, and E. P. WIGNER: Theory of Brillouin Zones and Symmetry Properties of Wave Functions in Crystals, *Phys. Rev.*, **50**, 58 (1936).

HEINE, V.: "Group Theory," chap. 6, Pergamon Press, New York, 1960.

[1] See Le Corre, *loc. cit.*
[2] See Le Corre, *loc. cit.*

HERRING, C.: A New Method for Calculating Wave Functions in Crystals, *Phys. Rev.*, **57**, 1169 (1940).

JONES, H.: "The Theory of Brillouin Zones and Electronic States in Crystals," North Holland Publishing Company, Amsterdam, 1960.

KITTEL, C.: "Introduction to Solid State Physics," 2d ed., John Wiley & Sons, Inc., New York, 1956.

KOSTER, G. F.: Space Groups and Their Representations, article in F. Seitz and D. Turnbull (eds.), "Solid State Physics," vol. 5, Academic Press Inc., New York, 1957.

LE CORRE, Y.: Les groupes cristallographiques magnétiques et leurs propriétés, *J. phys. radium*, **19**, 750 (1958).

PEIERLS, R. E.: "Quantum Theory of Solids," Oxford University Press, New York, 1955.

PINCHERLE, L.: Band Structure Calculations in Solids, article in A. C. Stickland (ed.), "Reports on the Progress of Physics," vol. XXIII, Physical Society, London, 1960.

REITZ, J. R.: Methods of the One-electron Theory of Solids, article in F. Seitz and D. Turnbull (eds.), "Solid State Physics," vol. 1, Academic Press Inc., New York, 1955.

SEITZ, F.: "The Modern Theory of Solids," McGraw-Hill Book Company, Inc., New York, 1940.

TAVGER, B. A., and V. M. ZAITSEV: Magnetic Symmetry of Crystals, *J. Exptl. Theoret. Phys.* (*U.S.S.R.*), **30**, 564 (1956); translation, *Soviet Phys.* (*JETP*), **3**, 430 (1956).

WILSON, A. H.: "The Theory of Metals," Cambridge University Press, New York, 1954.

ZIMAN, J. M.: "Electrons and Phonons," Oxford University Press, New York, 1960.

APPENDIX Ⓐ   REVIEW OF VECTORS, VECTOR SPACES, AND MATRICES

Vectors

An n-dimensional vector $\mathbf{v}$ is an ordered n-tuple of numbers, each number being a component of the vector. Two vectors are equal if all components are equal. A vector is a null vector, $\mathbf{v} = \mathbf{0}$, if and only if all components are zero. Vectors obey the rules

$$\mathbf{u} + (\mathbf{v} + \mathbf{w}) = (\mathbf{u} + \mathbf{v}) + \mathbf{w} \qquad \mathbf{u} + \mathbf{v} = \mathbf{v} + \mathbf{u}$$

$$(a + b)\mathbf{v} = a\mathbf{v} + b\mathbf{v} \qquad a(\mathbf{u} + \mathbf{v}) = a\mathbf{u} + a\mathbf{v} \qquad a(b\mathbf{v}) = ab\mathbf{v}$$

where a and b are any complex numbers. There is defined an *inner product* or *Hermitian scalar product* $(\mathbf{v}, \mathbf{w}) = \sum_k v_k^* w_k$. This product has the properties

$$(\mathbf{v}, \mathbf{w})^* = (\mathbf{w}, \mathbf{v}) \tag{A-1a}$$

$$(\mathbf{v}, \mathbf{v}) \geqq 0 \qquad 0 \text{ implies } \mathbf{v} = \mathbf{0} \tag{A-1b}$$

$$(\mathbf{v}, a\mathbf{w}) = a(\mathbf{v}, \mathbf{w}) = (a^*\mathbf{v}, \mathbf{w}) \tag{A-1c}$$

$$(\mathbf{u} + \mathbf{v}, \mathbf{w}) = (\mathbf{u}, \mathbf{w}) + (\mathbf{v}, \mathbf{w}) \tag{A-1d}$$

The length of a vector is given by $|\mathbf{v}| = [(\mathbf{v}, \mathbf{v})]^{\frac{1}{2}}$. If $|\mathbf{v}| = 1$, the vector is said to be *normalized*. We can imagine the cosine of the angle between two vectors in this space to be defined by

$$\cos \theta = \frac{|(\mathbf{v}, \mathbf{w})|}{|\mathbf{v}| \, |\mathbf{w}|}$$

This can be shown to be less than or equal to 1. If it is zero, $\mathbf{v}$ and $\mathbf{w}$ are said to be *orthogonal* vectors. A set of normalized vectors which are mutually orthogonal is often said to be *orthonormal*.

An n-dimensional *vector space* is the totality of vectors which can be formed by allowing the n-tuple components to range over all values. The particular vectors $(1, 0, 0, \ldots)$, $(0, 1, 0, 0, \ldots)$, $\ldots$, $(\ldots, 0, 0, 1)$ are called *basis vectors* for the space. Obviously any vector in the space can be written as a linear combination of these basis vectors. In other words, they are said to *span* the space. Any set of vectors such that an arbitrary vector in the space can be expressed as a linear combination of the set spans the space.

A set of m vectors is said to be *linearly independent* if $c_1\mathbf{v}_1 + c_2\mathbf{v}_2 + \cdots + c_m\mathbf{v}_m = \mathbf{0}$ implies that $c_1 = c_2 = \cdots = c_m = 0$, in other words, if no vector of the set can be expressed in terms of the others of the set alone. In an n-dimensional vector space, only n linearly independent vectors can exist.

317

Matrices and Transformations

A *linear transformation* of coordinates (the components of a vector position) can be expressed as

$$x_1' = \alpha_{11}x_1 + \alpha_{12}x_2 + \cdots + \alpha_{1n}x_n$$

$$x_2' = \alpha_{21}x_1 + \alpha_{22}x_2 + \cdots + \alpha_{2n}x_n$$

$$\cdots\cdots\cdots\cdots\cdots\cdots\cdots\cdots\cdots\cdots\cdots\cdots\cdots \tag{A-2a}$$

$$x_n' = \alpha_{n1}x_1 + \alpha_{n2}x_2 + \cdots + \alpha_{nn}x_n$$

or more compactly as

$$x_i' = \sum_k \alpha_{ik}x_k \tag{A-2b}$$

or still more compactly as

$$\mathbf{r}' = \mathbf{\alpha r} \tag{A-2c}$$

where $\mathbf{r}$ and $\mathbf{r}'$ are vectors and $\mathbf{\alpha}$ is the transformation matrix, a square array of the coefficients α_{ik}. The standard rules for the operation of matrices on column vectors are set up so that this abbreviated vector equation is exactly equivalent to the fully written out set of equations above. Depending on the form of the matrix $\mathbf{\alpha}$, this transformation might induce a rotation of axes, a permutation of axes, or a reflection. For example, the xy plane can be considered to be a two-dimensional vector space, with $x = x_1$ and $y = x_2$. Then the matrix for a counterclockwise rotation of axes by 90° is $\begin{pmatrix} 0 & 1 \\ -1 & 0 \end{pmatrix}$. The transformation equation written out is

$$\begin{pmatrix} x' \\ y' \end{pmatrix} = \begin{pmatrix} 0 & 1 \\ -1 & 0 \end{pmatrix} \begin{pmatrix} x \\ y \end{pmatrix} = \begin{pmatrix} y \\ -x \end{pmatrix}$$

so that $x' = y$ and $y' = -x$, as is correct for the present example. Note that this transformation can be looked at in two ways:

1. As above, the x' are considered coordinates of the same position vector referred to a new orientation of the coordinate axes.

2. Alternatively, the x' can be considered the coordinates of a rotated position vector with respect to the original coordinate axes. In this case, the above matrix represents a rotation of the vector 90° *clockwise*. It is generally true that the sense of an operation is reversed in passing from one of these viewpoints to the other, as is geometrically evident.

Clearly two transformations are equal if and only if all n^2 coefficients in the matrices inducing the two transformations are equal.

The transformation matrix $\mathbf{\alpha}$ can be viewed as a *linear operator*. It is called an operator because it transforms one vector into another. It is linear because

$$\mathbf{\alpha}(b\mathbf{u} + c\mathbf{v}) = b\mathbf{\alpha u} + c\mathbf{\alpha v} \tag{A-3}$$

If we apply a second transformation by a matrix $\mathbf{\beta}$ after applying $\mathbf{\alpha}$, we have

$$x_i'' = \sum_j \beta_{ij}x_j' \qquad x_j' = \sum_k \alpha_{jk}x_k$$

or

$$x_i'' = \sum_{j,k} \beta_{ij}\alpha_{jk}x_k = \sum_k \left(\sum_j \beta_{ij}\alpha_{jk} \right) x_k = \sum_k \gamma_{ik}x_k$$

where γ is a single $n \times n$ matrix which accomplishes the effect of the two transformations applied in succession. In order for this to be true, γ must be defined by

$$\gamma_{ik} = \sum_j \beta_{ij}\alpha_{jk} \tag{A-4}$$

which is the law of *matrix multiplication*.

Matrix Properties

We now enumerate a number of the properties of matrices which we frequently utilize in the text:

1. The *determinant* of the product of two matrices is equal to the product of the determinants of the factors; i.e., if $\gamma = \alpha\beta$, $|\gamma| = |\alpha|\,|\beta|$, where $|\alpha|$ is the determinant of α.

2. Matrix multiplication need *not* be commutative; that is, $\beta\alpha$ need not equal $\alpha\beta$.

3. The *associative* law holds in matrix multiplication; that is, $(\alpha\beta)\gamma = \alpha(\beta\gamma) = \alpha\beta\gamma$.

4. An $n \times n$ *unit matrix* $\mathbf{E}$ exists for any n. Its elements are $E_{ik} = \delta_{ik}$. It has the property that $\mathbf{E}\alpha = \alpha\mathbf{E} = \alpha$, for any α.

5. If $|\alpha| \neq 0$, there is a unique *inverse* to α, denoted by α^{-1}, such that $\alpha^{-1}\alpha = \alpha\alpha^{-1} = \mathbf{E}$. The inverse of a product of matrices is the product of the inverses in inverse order. The proof is obvious since $(\beta^{-1}\alpha^{-1})(\alpha\beta) = \beta^{-1}(\alpha^{-1}\alpha)\beta = \beta^{-1}\beta = \mathbf{E}$.

6. The *null matrix* has all elements equal to zero. Thus $\alpha 0 = 0\alpha = 0$.

7. The *sum* of two matrices is obtained by simply adding corresponding elements. Thus, if $\gamma = \alpha + \beta$, then $\gamma_{ij} = \alpha_{ij} + \beta_{ij}$. Matrix addition is thus *commutative*.

8. When a matrix is multiplied by a number, all elements are multiplied by that number. That is, if $\gamma = c\alpha$, then $\gamma_{ij} = c\alpha_{ij}$.

9. All *diagonal matrices* commute and have diagonal products. For example, if $D_{ik} = D_i\delta_{ik}$, and similarly for D', then $(DD')_{ij} = (D'D)_{ij} = D_i D_i'\delta_{ij}$.

10. The *trace*, or *spur*, of a matrix is the sum of its diagonal elements. That is, $\text{Tr } \alpha = \sum_i \alpha_{ii}$. The trace of a product is independent of the order of the factors; for example, $\text{Tr }\alpha\beta = \sum_i (\alpha\beta)_{ii} = \sum_i \sum_j \alpha_{ij}\beta_{ji} = \sum_j \sum_i \beta_{ji}\alpha_{ij} = \sum_j (\beta\alpha)_{jj} = \text{Tr }\beta\alpha$.

11. A *similarity transformation* applied to a matrix α yields one of the form $\beta^{-1}\alpha\beta$. Such transformations leave the trace invariant, since $\text{Tr }\beta^{-1}(\alpha\beta) = \text{Tr }(\alpha\beta)\beta^{-1} = \text{Tr }\alpha$.

12. If all matrices in a matrix equation are subjected to the same similarity transformation, the matrix equation will hold also for the transformed matrices. That is, if $\alpha' = s^{-1}\alpha s$, etc., and if $\alpha\beta = \gamma$, then it follows that $\alpha'\beta' = \gamma'$. The importance of similarity transformations in our work stems from this property.

Generalizations

As discussed above, we have treated matrices as square arrays, in which the rows and columns have been numbered similarly $1, 2, 3, \cdots, n$. We can generalize to consider *nonsquare* matrices. The simplest case is one where the rows and columns are of equal number but are labeled differently. For example, the rows might be labeled x, y, z and the columns ξ, η, ζ. This generalization causes no difficulty with the properties described above. A further generalization is to the case of *rectangular* matrices, where the number of rows and columns is different. Matrix multiplication is still possible using the standard definition, provided that the number of columns in the first factor is equal to the number of rows in the second. That is, an $m \times n$ matrix times an $n \times p$ matrix yields an $m \times p$ product matrix. An extreme example of this is the multiplication of an $n \times n$ matrix into an $n \times 1$ column vector (matrix) to yield another $n \times 1$ column vector. Another extreme example is the multiplication of a $1 \times n$ row vector into an $n \times 1$ column vector to yield a 1×1 scalar number. Such properties as the associative law still hold, but other concepts, such as the determinant, diagonal sum, and diagonal matrices, become meaningless for rectangular matrices.

It is useful to introduce the notion of the *direct product* of two matrices denoted by $\gamma = \alpha \times \beta$. The elements of the direct-product matrix have a double set of subscripts and are defined by

$$\gamma_{ij,kl} = \alpha_{ik}\beta_{jl} \tag{A-5}$$

Evidently the number of elements in the direct-product matrix is the product of the number of elements in each of the factors, and there is no restriction of equality on the number of rows and columns in the two factor matrices. It is also obvious that the direct product of two diagonal matrices is diagonal. One can readily verify that $(\alpha \times \beta)(\alpha' \times \beta') = \alpha\alpha' \times \beta\beta'$, that is, that taking a direct product commutes with matrix multiplication.

Diagonalization of Matrices

Consider the *secular equation* associated with a matrix α, namely,

$$0 = |\alpha - \lambda E| = \begin{vmatrix} (\alpha_{11} - \lambda) & \alpha_{12} & \cdots \\ \alpha_{21} & (\alpha_{22} - \lambda) & \cdots \\ \cdots & \cdots & (\alpha_{nn} - \lambda) \end{vmatrix} \tag{A-6}$$

Now apply a similarity transformation to α, obtaining the matrix $\beta = s^{-1}\alpha s$, and consider the secular equation of β. This is

$$0 = |\beta - \lambda E| = |s^{-1}\alpha s - \lambda E| = |s^{-1}\alpha s - \lambda s^{-1}Es| = |s^{-1}(\alpha - \lambda E)s|$$
$$= |s^{-1}| \, |\alpha - \lambda E| \, |s| = |\alpha - \lambda E| \, |s^{-1}s| = |\alpha - \lambda E|$$

Thus *all* the roots of the secular equation, which are the *eigenvalues* λ_k of the matrix, are invariant under a similarity transformation. This is a stronger form of our previous results that the *trace*, or sum of eigenvalues, and the *determinant*, or product of eigenvalues, are invariant.

The significance of an eigenvalue λ_k of the matrix α is that for these values there are nontrivial solutions to the matrix equation

$$\alpha v^k = \lambda_k v^k \tag{A-7}$$

The *eigenvectors* $\mathbf{v}^k$ are defined by this equation. They have the property of being unchanged in direction by the operation of the transformation α. That is, they define the *principal axes* of the system. The eigenvalues λ_k give the degree of "stretching" of the vector $\mathbf{v}^k$ produced by α.

Collecting the n eigenvectors (for an $n \times n$ matrix), we can form a matrix $\mathbf{V}$ of the form

$$\mathbf{V} = (\mathbf{v}^1 \mathbf{v}^2 \cdots \mathbf{v}^n) \tag{A-8a}$$

in which

$$V_{ik} = (v^k)_i \tag{A-8b}$$

If we also form a diagonal matrix of the eigenvalues

$$\Lambda_{ik} = \Lambda_{ki} = \lambda_k \delta_{ik} \tag{A-9}$$

we can write a typical example of the eigenvalue equation (A-7) as

$$\sum_j \alpha_{ij}(v^k)_j = \sum_j \alpha_{ij} V_{jk} = \lambda_k(v^k)_i = \lambda_k V_{ik} = \sum_j \lambda_k \delta_{jk} V_{ij} = \sum_j V_{ij} \Lambda_{jk}$$

or

$$\alpha \mathbf{V} = \mathbf{V}\Lambda$$

or

$$\mathbf{V}^{-1}\alpha\mathbf{V} = \Lambda \tag{A-10}$$

Thus we have shown that the (unitary) transformation matrix $\mathbf{V}$ composed of the (normalized) column eigenvectors of α provides a similarity transformation which diagonalizes α, leaving the eigenvalues on the diagonal. In this form, it is clear that the eigenvalues λ_k do, in fact, form the solutions of the secular equation (A-6).

Nomenclature and Properties of Special Matrices

The *transpose* of a matrix α, denoted by $\tilde{\alpha}$, is obtained from α by reflecting across the principal diagonal, so that $\tilde{\alpha}_{ij} = \alpha_{ji}$. Evidently then, $\widetilde{(\alpha\beta)} = \tilde{\beta}\tilde{\alpha}$.

The *complex conjugate* of a matrix α, denoted by α^*, is obtained by taking the complex conjugate of all matrix elements, so that $(\alpha^*)_{ij} = (\alpha_{ij})^*$. Evidently, this leaves products in the same order.

The *adjoint* of a matrix α, denoted by $\alpha^\dagger$, is the transpose of the complex conjugate. That is, $\alpha^\dagger = \tilde{\alpha}^*$, and $(\alpha^\dagger)_{ij} = \alpha_{ji}^*$. The product rule is then $(\alpha\beta)^\dagger = \beta^\dagger \alpha^\dagger$.

A *Hermitian*, or *self-adjoint*, matrix H has the property that $\mathbf{H}^\dagger = \mathbf{H}$, or that $H_{ij} = H_{ji}^*$.

A *unitary* matrix is one whose inverse is the adjoint. That is, $\mathbf{U}^\dagger = \mathbf{U}^{-1}$, so that $(U^{-1})_{ij} = U_{ji}^*$.

A *real orthogonal* matrix can be viewed as a unitary matrix, with the restriction that all elements are real. Thus $\mathbf{R}^{-1} = \mathbf{R}^\dagger = \tilde{\mathbf{R}}$, and $(R^{-1})_{ij} = R_{ji}$. Transformations between orthogonal coordinate systems are carried out by such matrices.

Some properties of these special types of matrices are as follows:

1. The Hermitian scalar product is unchanged if the adjoint of a given operator is transferred to the other factor in the product. For example,

$$(\mathbf{v}, \alpha\mathbf{w}) = (\alpha^\dagger\mathbf{v}, \mathbf{w})$$

2. The rows (or columns) of any unitary matrix form a set of n orthonormal vectors.

3. A unitary transformation applied to both vectors leaves the Hermitian scalar product unchanged. That is,

$$(\mathbf{Uv}, \mathbf{Uw}) = (\mathbf{v}, \mathbf{U}^\dagger\mathbf{Uw}) = (\mathbf{v}, \mathbf{U}^{-1}\mathbf{Uw}) = (\mathbf{v}, \mathbf{w})$$

4. The product of two unitary matrices is unitary, since $(\mathbf{UV})^\dagger = \mathbf{V}^\dagger\mathbf{U}^\dagger = \mathbf{V}^{-1}\mathbf{U}^{-1} = (\mathbf{UV})^{-1}$.

5. Both unitary and Hermitian matrices can always be diagonalized by similarity transformations using unitary matrices as indicated above. (Singular matrices with determinant zero could not be, for example.) The eigenvalues of Hermitian matrices are real. The eigenvalues of unitary matrices are of absolute value unity; if $\mathbf{U}$ is actually real orthogonal, the eigenvalues appear in complex-conjugate pairs.

REFERENCES

HILDEBRAND, F. B.: "Methods of Applied Mathematics," chap. 1, Prentice-Hall, Inc., Englewood Cliffs, N.J., 1952.

MARGENAU, H., and G. M. MURPHY: "The Mathematics of Physics and Chemistry," chap. 10, D. Van Nostrand Company, Inc., Princeton, N. J., 1943.

WIGNER, E. P.: "Group Theory," chaps. 1–3, Academic Press Inc., New York, 1959.

APPENDIX $\mathbb{B}$ CHARACTER TABLES FOR POINT-SYMMETRY GROUPS

In this Appendix we give the character tables for the 32 crystallographic point groups, together with 5 point groups involving fivefold rotations and 2 point groups involving C_∞. These 7 additional groups are not possible in crystals, but they may occur in molecules. When a group is conveniently given as the direct product of a smaller group and the group of the inversion i, this fact is indicated rather than giving the table, since the desired table can be obtained easily from that of the smaller group by use of the method indicated in Sec. 3-9.

Each table is labeled with the group designation in the Schoenflies notation, as developed in the text. To assist in making contact with other references, we also parenthetically quote the group designation in the shortened international notation used in the International Tables for X-ray Crystallography. In this notation, a number n refers to an n-fold rotation axis, i.e., rotation by $2\pi/n$. The notation $\bar{n}$ refers to rotoinversion, obtained by combining an n-fold rotation with the inversion. Thus $\bar{1}$ is just the inversion itself, and $\bar{2}$ is a reflection in the plane perpendicular to the rotation axis. A reflection is usually denoted m, for mirror plane. When the reflection is in the "horizontal" plane, the m is placed in the "denominator" of the symbol, which in the shortened notation is $/m$. For further details on this notation, the reader is referred to the literature.

The irreducible representations are labeled as discussed in the text. Briefly, A and B always refer to one-dimensional representations, E refers to two-dimensional representations, and T refers to three-dimensional representations. Pairs of complex-conjugate representations are bracketed together and labeled as a two-dimensional representation E, because time-reversal symmetry makes them degenerate. The subscripts g and u, for gerade and ungerade, indicate even and odd representations under inversion. In addition to these representation labels, we also indicate the transformation properties of the coordinates x, y, z, of various bilinear combinations of coordinates, and of the infinitesimal rotations R_x, R_y, and R_z.

C_1 (1)	E
A	1

C_2 (2)			E	C_2
x^2, y^2, z^2, xy	R_z, z	A	1	1
xz, yz	$\left.\begin{array}{c} x, y \\ R_x, R_y \end{array}\right\}$	B	1	-1

C_3 (3)			E	C_3	C_3^2
$x^2 + y^2, z^2$	R_z, z	A	1	1	1
$\left.\begin{array}{c}(xz, yz) \\ (x^2 - y^2, xy)\end{array}\right\}$	$\left.\begin{array}{c}(x, y) \\ (R_x, R_y)\end{array}\right\}$	E	$\left\{\begin{array}{c} 1 \\ 1 \end{array}\right.$	$\begin{array}{c}\omega \\ \omega^2\end{array}$	$\begin{array}{c}\omega^2 \\ \omega\end{array}$

$\omega = e^{2\pi i/3}$

C_4 (4)			E	C_2	C_4	C_4^3
$x^2 + y^2, z^2$	R_z, z	A	1	1	1	1
$x^2 - y^2, xy$		B	1	1	-1	-1
(xz, yz)	$\left.\begin{array}{c}(x, y) \\ (R_x, R_y)\end{array}\right\}$	E	$\left\{\begin{array}{c}1 \\ 1\end{array}\right.$	$\begin{array}{c}-1 \\ -1\end{array}$	$\begin{array}{c}i \\ -i\end{array}$	$\begin{array}{c}-i \\ i\end{array}$

C_5 (5)			E	C_5	C_5^2	C_5^3	C_5^4
$x^2 + y^2, z^2$	R_z, z	A	1	1	1	1	1
(xz, yz)	$\left.\begin{array}{c}(x, y) \\ (R_x, R_y)\end{array}\right\}$	E'	$\left\{\begin{array}{c}1 \\ 1\end{array}\right.$	$\begin{array}{c}\omega \\ \omega^4\end{array}$	$\begin{array}{c}\omega^2 \\ \omega^3\end{array}$	$\begin{array}{c}\omega^3 \\ \omega^2\end{array}$	$\begin{array}{c}\omega^4 \\ \omega\end{array}$
$(x^2 - y^2, xy)$		E''	$\left\{\begin{array}{c}1 \\ 1\end{array}\right.$	$\begin{array}{c}\omega^2 \\ \omega^3\end{array}$	$\begin{array}{c}\omega^4 \\ \omega\end{array}$	$\begin{array}{c}\omega \\ \omega^4\end{array}$	$\begin{array}{c}\omega^3 \\ \omega^2\end{array}$

$\omega = e^{2\pi i/5}$

C_6 (6)			E	C_6	C_3	C_2	C_3^2	C_6^5
$x^2 + y^2, z^2$	R_z, z	A	1	1	1	1	1	1
		B	1	-1	1	-1	1	-1
(xz, yz)	$\left.\begin{array}{c}(x, y) \\ (R_x, R_y)\end{array}\right\}$	E'	$\left\{\begin{array}{c}1 \\ 1\end{array}\right.$	$\begin{array}{c}\omega \\ \omega^5\end{array}$	$\begin{array}{c}\omega^2 \\ \omega^4\end{array}$	$\begin{array}{c}\omega^3 \\ \omega^3\end{array}$	$\begin{array}{c}\omega^4 \\ \omega^2\end{array}$	$\begin{array}{c}\omega^5 \\ \omega\end{array}$
$(x^2 - y^2, xy)$		E''	$\left\{\begin{array}{c}1 \\ 1\end{array}\right.$	$\begin{array}{c}\omega^2 \\ \omega^4\end{array}$	$\begin{array}{c}\omega^4 \\ \omega^2\end{array}$	$\begin{array}{c}1 \\ 1\end{array}$	$\begin{array}{c}\omega^2 \\ \omega^4\end{array}$	$\begin{array}{c}\omega^4 \\ \omega^2\end{array}$

$\omega = e^{2\pi i/6}$

C_{2v} (2mm)			E	C_2	σ_v	σ_v'
x^2, y^2, z^2	z	A_1	1	1	1	1
xy	R_z	A_2	1	1	-1	-1
xz	R_y, x	B_1	1	-1	1	-1
yz	R_x, y	B_2	1	-1	-1	1

C_{3v} (3m)			E	$2C_3$	$3\sigma_v$
$x^2 + y^2, z^2$	z	A_1	1	1	1
	R_z	A_2	1	1	-1
$(x^2 - y^2, xy)$ (xz, yz)	(x, y) (R_x, R_y)	E	2	-1	0

C_{4v} (4mm)			E	C_2	$2C_4$	$2\sigma_v$	$2\sigma_d$
$x^2 + y^2, z^2$	z	A_1	1	1	1	1	1
	R_z	A_2	1	1	1	-1	-1
$x^2 - y^2$		B_1	1	1	-1	1	-1
xy		B_2	1	1	-1	-1	1
(xz, yz)	(x, y) (R_x, R_y)	E	2	-2	0	0	0

C_{5v} (5m)			E	$2C_5$	$2C_5^2$	$5\sigma_v$	
$x^2 + y^2, z^2$	z	A_1	1	1	1	1	
	R_z	A_2	1	1	1	-1	$x = \dfrac{2\pi}{5}$
(xz, yz)	(x, y) (R_x, R_y)	E_1	2	$2\cos x$	$2\cos 2x$	0	
$(x^2 - y^2, xy)$		E_2	2	$2\cos 2x$	$2\cos 4x$	0	

C_{6v} (6mm)			E	C_2	$2C_3$	$2C_6$	$3\sigma_d$	$3\sigma_v$
$x^2 + y^2, z^2$	z	A_1	1	1	1	1	1	1
	R_z	A_2	1	1	1	1	-1	-1
		B_1	1	-1	1	-1	-1	1
		B_2	1	-1	1	-1	1	-1
(xz, yz)	(x, y) (R_x, R_y)	E_1	2	-2	-1	1	0	0
$(x^2 - y^2, xy)$		E_2	2	2	-1	-1	0	0

C_{1h} (m)			E	σ_h
x^2, y^2, z^2, xy	R_z, x, y	A'	1	1
xz, yz	R_x, R_y, z	A''	1	-1

C_{2h} $(2/m)$			E	C_2	σ_h	i
x^2, y^2, z^2, xy	R_z	A_g	1	1	1	1
	z	A_u	1	1	-1	-1
xz, yz	R_x, R_y	B_g	1	-1	-1	1
	x, y	B_u	1	-1	1	-1

$C_{3h} = C_3 \times \sigma_h$ $(\bar{6})$			E	C_3	C_3^2	σ_h	S_3	$(\sigma_h C_3^2)$	
$x^2 + y^2, z^2$	R_z	A'	1	1	1	1	1	1	
	z	A''	1	1	1	-1	-1	-1	
$(x^2 - y^2, xy)$	(x, y)	E'	$\begin{cases} 1 \\ 1 \end{cases}$	$\begin{matrix} \omega \\ \omega^2 \end{matrix}$	$\begin{matrix} \omega^2 \\ \omega \end{matrix}$	$\begin{matrix} 1 \\ 1 \end{matrix}$	$\begin{matrix} \omega \\ \omega^2 \end{matrix}$	$\begin{matrix} \omega^2 \\ \omega \end{matrix}$	$\omega = e^{2\pi i/3}$
(xz, yz)	(R_x, R_y)	E''	$\begin{cases} 1 \\ 1 \end{cases}$	$\begin{matrix} \omega \\ \omega^2 \end{matrix}$	$\begin{matrix} \omega^2 \\ \omega \end{matrix}$	$\begin{matrix} -1 \\ -1 \end{matrix}$	$\begin{matrix} -\omega \\ -\omega^2 \end{matrix}$	$\begin{matrix} -\omega^2 \\ -\omega \end{matrix}$	

$$C_{4h} = C_4 \times i \qquad (4/m)$$
$$C_{5h} = C_5 \times \sigma_h \qquad (\overline{10})$$
$$C_{6h} = C_6 \times i \qquad (6/m)$$

S_2 $(\bar{1})$			E	i
$x^2, y^2, z^2, xy, xz, yz$	R_x, R_y, R_z	A_g	1	1
	x, y, z	A_u	1	-1

S_4 $(\bar{4})$			E	C_2	S_4	S_4^3
$x^2 + y^2, z^2$	R_z	A	1	1	1	1
	z	B	1	1	-1	-1
(xz, yz)	(x, y)	E	$\begin{cases} 1 \\ 1 \end{cases}$	$\begin{matrix} -1 \\ -1 \end{matrix}$	$\begin{matrix} i \\ -i \end{matrix}$	$\begin{matrix} -i \\ i \end{matrix}$
$(x^2 - y^2, xy)$	(R_x, R_y)					

$$S_6 = C_3 \times i \qquad (\bar{3})$$

D_2 (222)			E	$C_2{}^z$	$C_2{}^y$	$C_2{}^x$
x^2, y^2, z^2		A_1	1	1	1	1
xy	R_z, z	B_1	1	1	-1	-1
xz	R_y, y	B_2	1	-1	1	-1
yz	R_x, x	B_3	1	-1	-1	1

D_3 (32)			E	$2C_3$	$3C_2'$
$x^2 + y^2, z^2$		A_1	1	1	1
	R_z, z	A_2	1	1	-1
(xz, yz) $(x^2 - y^2, xy)$	(x, y) (R_x, R_y)	E	2	-1	0

D_4 (422)			E	$C_2 = C_4{}^2$	$2C_4$	$2C_2'$	$2C_2''$
$x^2 + y^2, z^2$		A_1	1	1	1	1	1
	R_z, z	A_2	1	1	1	-1	-1
$x^2 - y^2$		B_1	1	1	-1	1	-1
xy		B_2	1	1	-1	-1	1
(xz, yz)	(x, y) (R_x, R_y)	E	2	-2	0	0	0

D_5 (52)			E	$2C_5$	$2C_5{}^2$	$5C_2'$	
$x^2 + y^2, z^2$		A_1	1	1	1	1	
	R_z, z	A_2	1	1	1	-1	
(xz, yz)	(x, y) (R_x, R_y)	E_1	2	$2\cos x$	$2\cos 2x$	0	$x = \dfrac{2\pi}{5}$
$(x^2 - y^2, xy)$		E_2	2	$2\cos 2x$	$2\cos 4x$	0	

D_6 (622)			E	C_2	$2C_3$	$2C_6$	$3C_2'$	$3C_2''$
$x^2 + y^2, z^2$		A_1	1	1	1	1	1	1
	R_z, z	A_2	1	1	1	1	-1	-1
		B_1	1	-1	1	-1	1	-1
		B_2	1	-1	1	-1	-1	1
(xz, yz)	(x, y) (R_x, R_y)	E_1	2	-2	-1	1	0	0
$(x^2 - y^2, xy)$		E_2	2	2	-1	-1	0	0

D_{2d} ($\bar{4}2m$)			E	C_2	$2S_4$	$2C_2'$	$2\sigma_d$
$x^2 + y^2,\ z^2$		A_1	1	1	1	1	1
	R_z	A_2	1	1	1	-1	-1
$x^2 - y^2$		B_1	1	1	-1	1	-1
xy	z	B_2	1	1	-1	-1	1
(xz, yz)	(x, y) (R_x, R_y)	E	2	-2	0	0	0

$$D_{3d} = D_3 \times i \quad (\bar{3}m)$$
$$D_{2h} = D_2 \times i \quad (mmm)$$

$D_{3h} = D_3 \times \sigma_h$ ($\bar{6}m2$)			E	σ_h	$2C_3$	$2S_3$	$3C_2'$	$3\sigma_v$
$x^2 + y^2,\ z^2$		A_1'	1	1	1	1	1	1
	R_z	A_2'	1	1	1	1	-1	-1
		A_1''	1	-1	1	-1	1	-1
	z	A_2''	1	-1	1	-1	-1	1
$(x^2 - y^2, xy)$	(x, y)	E'	2	2	-1	-1	0	0
(xz, yz)	(R_x, R_y)	E''	2	-2	-1	1	0	0

$$D_{4h} = D_4 \times i \quad (4/mmm)$$
$$D_{5h} = D_5 \times \sigma_h \quad (\overline{10}m2)$$
$$D_{6h} = D_6 \times i \quad (6/mmm)$$

T (23)		E	$3C_2$	$4C_3$	$4C_3'$	
	A	1	1	1	1	
	E	1	1	ω	ω^2	
		1	1	ω^2	ω	$\omega = e^{2\pi i/3}$
(R_x, R_y, R_z) (x, y, z)	T	3	-1	0	0	

$$T_h = T \times i \quad (m3)$$

O (432)		E	$8C_3$	$3C_2 = 3C_4{}^2$	$6C_2$	$6C_4$
	A_1	1	1	1	1	1
	A_2	1	1	1	-1	-1
$(x^2 - y^2, 3z^2 - r^2)$	E	2	-1	2	0	0
(R_x, R_y, R_z)						
(x, y, z)	T_1	3	0	-1	-1	1
(xy, yz, zx)	T_2	3	0	-1	1	-1

$$O_h = O \times i \quad (m3m)$$

T_d ($\bar{4}3m$)		E	$8C_3$	$3C_2$	$6\sigma_d$	$6S_4$
	A_1	1	1	1	1	1
	A_2	1	1	1	-1	-1
	E	2	-1	2	0	0
(R_x, R_y, R_z)	T_1	3	0	-1	-1	1
(x, y, z)	T_2	3	0	-1	1	-1

$C_{\infty v}$ (∞m)			E	$2C_\phi$	σ_v
$x^2 + y^2, z^2$	z	$A_1(\Sigma^+)$	1	1	1
	R_z	$A_2(\Sigma^-)$	1	1	-1
(xz, yz)	(x, y)	$E_1(\Pi)$	2	$2\cos\phi$	0
	(R_x, R_y)				
$(x^2 - y^2, xy)$		$E_2(\Delta)$	2	$2\cos 2\phi$	0
		$\cdots$			

$D_{\infty h}$ (∞/mm)			E	$2C_\phi$	C_2'	i	$2iC_\phi$	iC_2'
$x^2 + y^2, z^2$		$A_{1g}(\Sigma_g{}^+)$	1	1	1	1	1	1
		$A_{1u}(\Sigma_u{}^+)$	1	1	1	-1	-1	-1
	R_z	$A_{2g}(\Sigma_g{}^-)$	1	1	-1	1	1	-1
	z	$A_{2u}(\Sigma_u{}^-)$	1	1	-1	-1	-1	1
(xz, yz)	(R_x, R_y)	$E_{1g}(\Pi_g)$	2	$2\cos\phi$	0	2	$2\cos\phi$	0
	(x, y)	$E_{1u}(\Pi_u)$	2	$2\cos\phi$	0	-2	$-2\cos\phi$	0
$(x^2 - y^2, xy)$		$E_{2g}(\Delta_g)$	2	$2\cos 2\phi$	0	2	$2\cos 2\phi$	0
		$E_{2u}(\Delta_u)$	2	$2\cos 2\phi$	0	-2	$-2\cos 2\phi$	0
		$\cdots$						

REFERENCES

EYRING, H., J. WALTER, and G. E. KIMBALL: "Quantum Chemistry," app. VII, John Wiley & Sons, Inc., New York, 1944.

HEINE, V.: "Group Theory," chap. III and apps. J, K, and L, Pergamon Press, New York, 1960.

"International Tables for X-ray Crystallography," Kynoch Press, Birmingham, 1952.

ROSENTHAL, J. E., and G. M. MURPHY: Group Theory and the Vibrations of Polyatomic Molecules, *Revs. Modern Phys.*, **8**, 317 (1936).

APPENDIX C   TABLE OF c^k AND a^k COEFFICIENTS FOR s, p, AND d ELECTRONS

These coefficients are defined in Eqs. (6-37) and (6-40), respectively, of the text. Note that $c^k(lm, l'm') = (-1)^{m-m'} c^k(l'm', lm)$, whereas $a^k(lm, l'm')$ is unchanged under an interchange of arguments. The values of l and l' are indicated by the usual spectroscopic notations s, p, d.

Table C-1

Table of coefficients $c^k(lm, l'm')$

	m	m'	$k = 0$
ss	0	0	1

	m	m'	$k = 1$
sp	0	± 1	$-\sqrt{\frac{1}{3}}$
	0	0	$\sqrt{\frac{1}{3}}$

	m	m'	$k = 2$
sd	0	± 2	$\sqrt{\frac{1}{5}}$
	0	± 1	$-\sqrt{\frac{1}{5}}$
	0	0	$\sqrt{\frac{1}{5}}$

	m	m'	$k = 0$	$k = 2$
pp	± 1	± 1	1	$-\sqrt{\frac{1}{25}}$
	± 1	0	0	$\sqrt{\frac{3}{25}}$
	± 1	∓ 1	0	$-\sqrt{\frac{6}{25}}$
	0	0	1	$\sqrt{\frac{4}{25}}$

Table C-I—*continued*

	m	m'	$k = 1$	$k = 3$
pd	± 1	± 2	$-\sqrt{\frac{6}{15}}$	$\sqrt{\frac{3}{245}}$
	± 1	± 1	$\sqrt{\frac{3}{15}}$	$-\sqrt{\frac{9}{245}}$
	± 1	0	$-\sqrt{\frac{1}{15}}$	$\sqrt{\frac{18}{245}}$
	± 1	∓ 1	0	$-\sqrt{\frac{30}{245}}$
	± 1	∓ 2	0	$\sqrt{\frac{45}{245}}$
	0	± 2	0	$\sqrt{\frac{15}{245}}$
	0	± 1	$-\sqrt{\frac{3}{15}}$	$-\sqrt{\frac{24}{245}}$
	0	0	$\sqrt{\frac{4}{15}}$	$\sqrt{\frac{27}{245}}$

	m	m'	$k = 0$	$k = 2$	$k = 4$
dd	± 2	± 2	1	$-\sqrt{\frac{4}{49}}$	$\sqrt{\frac{1}{441}}$
	± 2	± 1	0	$\sqrt{\frac{6}{49}}$	$-\sqrt{\frac{5}{441}}$
	± 2	0	0	$-\sqrt{\frac{4}{49}}$	$\sqrt{\frac{15}{441}}$
	± 2	∓ 1	0	0	$-\sqrt{\frac{35}{441}}$
	± 2	∓ 2	0	0	$\sqrt{\frac{70}{441}}$
	± 1	± 1	1	$\sqrt{\frac{1}{49}}$	$-\sqrt{\frac{16}{441}}$
	± 1	0	0	$\sqrt{\frac{1}{49}}$	$\sqrt{\frac{30}{441}}$
	± 1	∓ 1	0	$-\sqrt{\frac{6}{49}}$	$-\sqrt{\frac{40}{441}}$
	0	0	1	$\sqrt{\frac{4}{49}}$	$\sqrt{\frac{36}{441}}$

Table C-2

Table of coefficients $a^k(lm, l'm')$

	m	m'	$k = 0$	$k = 2$	$k = 4$
ss	0	0	1	0	0
sp	0	± 1	1	0	0
	0	0	1	0	0
sd	0	± 2	1	0	0
	0	± 1	1	0	0
	0	0	1	0	0
pp	± 1	± 1	1	$\frac{1}{25}$	0
	± 1	0	1	$-\frac{2}{25}$	0
	0	0	1	$\frac{4}{25}$	0
pd	± 1	± 2	1	$\frac{2}{35}$	0
	± 1	± 1	1	$-\frac{1}{35}$	0
	± 1	0	1	$-\frac{2}{35}$	0
	0	± 2	1	$-\frac{4}{35}$	0
	0	± 1	1	$\frac{2}{35}$	0
	0	0	1	$\frac{4}{35}$	0
dd	± 2	± 2	1	$\frac{4}{49}$	$\frac{1}{441}$
	± 2	± 1	1	$-\frac{2}{49}$	$-\frac{4}{441}$
	± 2	0	1	$-\frac{4}{49}$	$\frac{6}{441}$
	± 1	± 1	1	$\frac{1}{49}$	$\frac{16}{441}$
	± 1	0	1	$\frac{2}{49}$	$-\frac{24}{441}$
	0	0	1	$\frac{4}{49}$	$\frac{36}{441}$

INDEX

Abelian groups, 7, 13, 37–39
Abragam, A., 170n., 189n.
Acetylene molecule, 244–250
Accidental degeneracy, 34
Anderson, P. W., 226n.
Angular momentum, commutation
    relations, 97, 250–251
  definition, 95
Angular-momentum operations, 178–181
Antiferromagnetic structures, 301–309
Antilinear operator, 142
Antioperators, definition, 300
Anti-Stokes transitions, 249
Antisymmetric functions, 160–162
Antiunitary operators, 142–147
Asymmetric rotor, 255–257
Asymmetry parameter, Ray's, 256
Axial vector, 83–84, 127–128

Back-Goudsmit effect, 200–201
Ballhausen, C. J., 86n.
Basis functions, 35–43
  combinations of, 112–114
  generation of, 40–42, 228–230
  for irreducible representations, 39–43
Bates, D., 214n.

Belov, N. V., 300n.
Benzene, 228–233
Bethe, H., 67–77, 287
Biedenharn, L. C., 137, 202
Bivnis, R., 122, 137
Blatt, J. M., 137, 202
Bleaney, B., 67, 73
Bloch functions, 232, 268–299
Bloch's theorem, 38, 231, 268, 269, 312
Bonds, double, 89
  π, 89–91
  σ, 89–91
  tetrahedral, 87–88
  trigonal, 88–91
Born-Oppenheimer approximation,
    210–213, 234, 257
Bouckaert, L. P., 69, 286–289
Bravais lattice, 308
Breit, G., 197
Breit-Rabi formula, 200–201
Brillouin zones, 270–273, 286

Calculation of fine structure, 181–187
Cartesian tensor operators, 126–128
Celebrated theorem, 31
Centrifugal distortion, 258–259

Centrifugal potential, 236
Character tables, 27–29, 45
  construction of, 28–29
  for point-symmetry groups, 323–329
Characters, definition, 25
  in direct-product groups, 44–46
  orthogonality of, 26–28
  three-dimensional rotation group, 66
Charge conjugation symmetry, 148–150
Class of elements, 12–13, 15
Class multiplication, 15, 29
Clebsch-Gordan coefficients, 117–124, 131–139, 176, 181
Cohen, M. H., 290n.
Colladay, G. S., 122
Color groups (Shubnikov), 300–309
Combination levels, 245–246
Compatibility of symmetries, 284–289, 312
Condon, E. U., 98, 103, 111, 122, 174n., 175n., 177n., 203
Conduction band, 274
Configuration of atom, 155
Configuration interaction, 156, 221
Conjugate group elements, 12–13, 15
Continuous groups, 98–101
Corepresentations, 144
Coriolis force, 260
Correlation diagram, molecular, 217–219
Cyclic groups, 9, 38, 231
  tight- and loose-binding, 294
Cosets, 9–10, 14–15, 302
Coulson, C. A., 158, 220, 221
Covering operation, 7, 8
Cross, P. C., 242, 247n., 256
Crystal class, 269
Crystal double groups, 75–80
Crystal field, 67–80
Crystal momentum, 268–298
Crystal-symmetry operators, 51–53
Crystal systems, 60–61
Crystallographic point groups, 54–61, 267–270, 323–330

Daudel, R., 233n.
Decius, J. C., 242, 247n.

Degeneracy, accidental, 34
  of band functions, 279–289
  normal, 34
Diagonalization of matrices, 320–321
Dimensionality theorem, 25, 31
Dimmock, J. O., 311, 312
Dipole-dipole interaction, 129–130
Dirac, P. A. M., 166, 189, 226, 295
Direct product of matrices, 320
Direct-product groups, 43–46
Direct-product representations, 46–47
Double bonds, 89
Double groups, 75–80
Duncanson, W. E., 158
Dunham, J. L., 237

Edmonds, A. R., 122
Ehrenreich, H., 295n.
Electric dipole transitions, 83–87
Elliott, R. J., 295
Empty-lattice approximation, 272–276
Equivalent electrons, 155, 171
Eulerian angles, 98, 101–103, 105–106, 110–115
Exchange charge density, 169, 225
Exchange integrals, 166–170, 224–226
Exclusion rule, 249
Eyring, H., 62, 92, 216n., 219n.

Factor group, 14
Falkoff, D. L., 122
Fermi, E., 197–199
Fermi surface, 274
Ferrell, R. A., 198n.
Field-gradient tensor, 128
Fine structure, calculation of, 181–187
Finkelstein, B. N., 215n.
Freeman, A. J., 69n., 226n.
Frobenius-Schur test, 146, 297

Gaunt, J. A., 175–176
Gaunt coefficients, 175–176, 331–333
Goudsmit, S., 197n.

Groups, Abelian, 7, 13, 37–39
  color (Shubnikov), 300–309
  continuous, 98–101
  cyclic, 9, 38, 231
  definition, 6
  direct-product, 43–46
  homomorphic, 14
  isomorphic, 8, 14
  of **k** vector, 279–281, 285–289
  Lie, 98
  magnetic crystal, 299–309
  octahedral, 59–60, 69–80, 282, 285–295
  order of, 7
  permutation, 11–12
  point, 51, 54–80, 267–270, 323–329
  of prime order, 11
  of Schrödinger equation, 33–39
  space, 51, 267–312
Group multiplication, 6, 7
Group-multiplication table, 7, 11

Hainer, R. M., 256
Harrick, N. J., 264*n*.
Hartree, D. R., 159–162
Hartree-Fock functions, 160–162
Hartree-Fock method, 167–170, 233
Hartree functions, 159–160
Haurwitz, E., 158
Heine, V., 76*n*., 122, 170, 222*n*., 289*n*., 290*n*.
Heisenberg, W., 166, 226
Heitler-London method, 87, 220–227
Helium atom, 159
Hermitian matrix, 20, 321
Hermitian scalar product, 35, 317
Herring, C., 226, 290, 297, 312
Horowitz, G. E., 215*n*.
Hückel theory, 228–233
Hund rules, 156, 177, 183, 186–187, 223, 225
Hurwitz integral, 98–100
Hybrid orbitals, 88–92
Hydrogen molecule, 217, 219–227, 254, 264
Hydrogen-molecule ion, 214–216

Hydrogenic functions, 157
Hylleraas, E. A., 159
Hyperfine structure, 193–206
  electric, 201–206
  magnetic, 193–201
    for S-states, 197–199

Improper rotations, 53, 57, 139
Inequivalent electrons, 73, 186
Internal coordinates, 242
International notation for crystallography, 323
Invariant subgroup, 13–15, 302–305
Invariants, construction of, 81, 310
Inversion symmetry, 3–5, 51, 57, 83–87, 139–140, 147–150
Irreducible representations, 19–47
  basis functions for, 39–43
  for crystallographic point groups, 323–329
  for linear molecules, 216–219
  for magnetic groups, 311–313
  of point groups, 62–64
  of three-dimensional rotation group, 65–66
Irreducible tensor operators, 126–130, 137–139
Isomorphic groups, 8, 14

j-j coupling, 156
Jaccarino, V., 194*n*.
James, H. M., 215

k vector group, 279–281, 285–289
Karplus, R., 189
Kimball, G. E., 62, 92, 216*n*., 219*n*.
King, G. W., 256
King, J. G., 194*n*.
Kinoshita, T., 159
Kittel, C., 86
Kleinman, L., 290*n*.
Koopmans' theorem, 207
Kopfermann, H., 197
Koster, G. F., 67*n*., 270*n*., 289*n*.
Kramers' theorem, 78, 143, 299

Kratzer, A., 237
Kroll, N. M., 189

Lambda doubling, 253, 261
Landau, L. D., 309
Landau theory of phase transitions, 309–311
Landé g-factor, 190, 199
Landé interval rule, 184, 195
Le Corre, Y., 309
Lee, T. D., 149
Lefebvre, R., 233n.
Lie groups, 98
Liehr, A. D., 86n.
Lifshitz, E. M., 309n.
Linear molecules, irreducible representations for, 216–219
Lorentz group, 148
Löwdin, P. O., 215n.
L-S coupling, 156
Luttinger, J. M., 86

Magnetic crystal groups, 299–309
Magnetic dipole transitions, 83–87
Magnetic hyperfine structure, 193–201
Magnetic moment, rotational, 262–263
Magnetic point groups, 299–309
    table, 304–305
Matrices, properties of, 318–322
Matrix-element theorems, 80–82, 131–133
Matrix elements between determinantal functions, 162–167
Meckler, A., 264n.
Metropolis, N., 122, 137
Molecular orbitals, 213–223
Molecular rotation, 250–264
Molecular vibration, 234–250
Morita, M., 122
Morse, P. M., 158, 235
Morse potential, 235–238
Moser, C., 233n.
Mulliken, R. S., 87
Multiplet structure, 155

Néel temperature, 299, 309
Neronova, N. N., 300n.
Nesbet, R. K., 226n.
Normal degeneracy, 34
Normal divisor, 13–15, 302–305
Normal modes, 238–250

Octahedral groups, 59–60, 69–80, 282, 285–295
Octupole moment, 194, 206
Opechowski, W., 76
Oppenheimer, R., 210
Orthogonal atomic orbitals, 226–227
Orthogonality, of basis functions, 41–42
    of characters, 26–28
Orthogonality theorem, 20–25
    geometric interpretation of, 25
    statement of, 23
Orthogonalized plane waves, 289–293
Orthohydrogen, 254
Orthonormal vectors, 317
Overlap integral, 214, 224–227
Overtone levels, 245–248
Oxygen molecule, 218, 253–254, 264

Parahydrogen, 254
Parity, 84–87, 140
Partners in basis set, 39–42
Paschen-Back effect, 190, 201
Pauli, W., 149, 193
Pauli matrices, 103, 109
Pauli principle, 74, 155, 172
Pauling, L., 87, 92, 216n.
Pekeris, C. L., 159
Phillip, H. R., 295n.
Phillips, J. C., 69n., 290n.
Poincaré group, 148
Point groups, 54–64, 267–270
    character tables, 323–329
    irreducible representations of, 62–64
Pratt, G. W., Jr., 169n., 170
Projection operators, 40–42
Pseudopotential, 290

Quadrupole moment, 202–206

Rabi, I. I., 200–201
Racah, G., 98, 122, 130, 184, 197
Racah coefficients, 133–139, 202
Raman transitions, 249–250
Ramsey, N. F., 204n., 264n.
Ray, B. S., 256
Ray's asymmetry parameter, 256
Rearrangement theorem, 8–9, 98–100
Reciprocal lattice, 270–274
Reduced matrix element, 125, 132, 139, 189
Reduced zone scheme, 271–274
Representations, 18–47
    character of, 25–28
    dimensionality of, 18, 25, 28
    direct-product, 46–47
    of direct-product groups, 43–46
    equivalent, 19, 36
    faithful, 19
    identical, 19, 28, 37
    irreducible (*see* Irreducible representations)
    matrix, 34–37
    reducible, 19–20, 29–31
        decomposition of, 29–31, 70–80
    regular, 30–31
    unitary, 20–25, 35–36
Roothaan, C. C. J., 233n.
Rose, M. E., 101, 122, 137, 139n., 202
Rotation-electronic coupling, 260–264
Rotation group, 94–139
    homomorphism with unitary group, 103–106
    representations of, 102–103, 109–111
    three-dimensional, 65–67, 94–140
    two-dimensional, 38–39
Rotation-inversion group, 139–140
Rotenberg, M., 122, 123, 137

Saito, R., 122
Sandeman, I., 237
Satten, R. A., 86, 194n., 199
Scalar operators, 124–125

Schawlow, A. L., 204n., 256
Schiff, L. I., 82n., 235n.
Schoenflies notation, 53–61, 323
Schur's lemma, 21–24, 29, 108
Schwinger, J., 123, 189
Schwinger-Pauli-Lüders theorem, 149
Secular equation, definition, 320
Segré, E., 197n.
Seitz, F., 51n.
Selection rules, 81–87, 140, 193, 233–234, 248–250, 252
Sells, R. E., 122
Sherman, A., 92n.
Shortley, G. H., 98, 103, 111, 122, 174n., 175n., 177n., 203
Shubnikov groups (color), 300–309
Simon, A., 122
6 *j*-coefficients, 137
Slater, J. C., 157, 160, 166n., 168–170, 173, 221
Slater determinant, 161–170, 222
Slater sum-rule method, 173, 176–177
Small representation, 280
Smirnova, T. S., 300n.
Smoluchowski, R., 69, 286–289
Sommerfield, C. M., 189
Space groups, 51, 267–313
    definition of, simple, 269
        symmorphic, 269
Spin-orbit coupling, in atoms, 181–188
    in band theory, 295–297
Spinors, 75, 110
Star of wave vector, 279–281
Statistics, nuclear, effect on rotation, 252–255
Statz, H., 67n.
Stereographic projections, 54–55
Sternheimer, R. M., 205
Stevens, K. W. H., 67, 73
Stokes transitions, 249
Strandberg, M. W. P., 256n., 264n.
Streitwieser, A., Jr., 233n.
Stroke, H. H., 194n.
Subgroups, 9–10
Symmetric top, 251–252
Symmetries, of cube, 59
    of magnetic structures, 303–311

Symmetries, of octahedron, 59
   of tetrahedron, 58
Symmetry operators, 3–5, 31–39
   commuting, 4, 34–39, 52–53
   crystalline, 51–60
   of solids, 267–270

Tavger, B. A., 301
Teller, E., 214*n.*
Tensor operators, 124–133
   cartesian, 126–128
   irreducible, 126–130, 137–139
   scalar product of, 129, 137–139
   spherical, 126–130
Tetrahedral bonds, 87
Thomas precession, 181
3-*j* symbol, 122–123
Tight-binding approximation, 276–278
Time-reversal symmetry, 78, 141–150
   in band theory, 297–298
   in magnetic crystals, 299–312
Tinkham, M., 264*n.*
Townes, C. H., 204*n.*, 256
Transformation operators, 3–5, 31–33, 318
   (*See also* Symmetry operators)
Transformation properties, of functions, 62–64
   rotational, 94–140
Transformations, of coordinates, 318–319
   similarity, 19, 319
Trigonal bonds, 88–91
Tubis, A., 158

Unitary group, 103–111
   representations of, 106–111
Unitary matrix, 321
Unrestricted Hartree-Fock method, 169–170
Unsöld theorem, 81, 174
   generalized, 80–81, 231, 310

Valence, directed, 87–92

Van Vleck, J. H., 41, 92*n.*, 166, 189*n.*, 226, 255*n.*
Variational eigenfunctions, 158, 215
Varshni, Y. P., 238
Vector model, 166
   for addition of angular momenta, 115–116
Vector operators, 125–126, 132–133
Vectors, properties of, 317
Vibration, of diatomic molecules, 234–238
   of polyatomic molecules, 238–249
Vibration-rotation interaction, 257–259
Vierergruppe, 11, 57
Virial theorem, 215
Von der Lage, F. C., 69

Walter, J., 62, 92, 216*n.*, 219*n.*
Wang symmetrizing transformation, 255
Wannier functions, 226
Water molecule, 243–244, 249
Watson, R. E., 69*n.*, 226*n.*
Weyl, H., 103
Wheeler, R. G., 312
Wick, G. C., 260*n.*
Wigner, E. P., 20–26, 32–37, 39–44, 69, 70, 74, 120, 122, 144, 149*n.*, 286–289, 297
Wigner coefficients, 117–124, 131–139, 176, 181
   table, for $j_2 = \frac{1}{2}$, 123
      for $j_2 = 1$, 124
Wigner-Eckart theorem, 131–133, 138, 183, 189, 194, 203–205
Wilson, E. B., 216*n.*, 242, 247*n.*
Wood, J. H., 170
Wooten, J. K., Jr., 122, 137

X coefficient, 123

Yang, C. N., 149
Young, L. A., 158

Zaitsev, V. M., 301
Zeeman effect, 188–193, 199–201

Graduate Texts in Mathematics 22

Donald W. Barnes   John M. Mack

An Algebraic Introduction
to
Mathematical Logic

Springer-Verlag   New York Heidelberg Berlin

Donald W. Barnes
John M. Mack

The University of Sydney
Department of Pure Mathematics
Sydney, N.S.W. 2006
Australia

Managing Editors

P. R. Halmos

Indiana University
Department of Mathematics
Swain Hall East
Bloomington, Indiana 47401
USA

C. C. Moore

University of California
at Berkeley
Department of Mathematics
Berkeley, California 94720
USA

AMS Subject Classifications
Primary: 0201
Secondary: 02B05, 02B10, 02F15, 02G05, 02G10, 02G15, 02G20, 02H05, 02H13, 02H15, 02H20, 02H25

Library of Congress Cataloging in Publication Data

Barnes, Donald W
   An algebraic introduction to mathematical logic.
   (Graduate texts in mathematics; v. 22)
   Bibliography: p. 115
   Includes index.
   1. Logic, Symbolic and mathematical. 2. Algebraic
logic. I. Mack, J. M., joint author. II. Title.
III. Series.
QA9.B27   511′.3   74-22241

ISBN 0-387-90109-4 Springer-Verlag New York Heidelberg Berlin
ISBN 3-540-90109-4 Springer-Verlag Berlin Heidelberg New York

Preface

This book is intended for mathematicians. Its origins lie in a course of lectures given by an algebraist to a class which had just completed a substantial course on abstract algebra. Consequently, our treatment of the subject is algebraic. Although we assume a reasonable level of sophistication in algebra, the text requires little more than the basic notions of group, ring, module, etc. A more detailed knowledge of algebra is required for some of the exercises. We also assume a familiarity with the main ideas of set theory, including cardinal numbers and Zorn's Lemma.

In this book, we carry out a mathematical study of the logic used in mathematics. We do this by constructing a mathematical model of logic and applying mathematics to analyse the properties of the model. We therefore regard all our existing knowledge of mathematics as being applicable to the analysis of the model, and in particular we accept set theory as part of the meta-language. We are not attempting to construct a foundation on which all mathematics is to be based—rather, any conclusions to be drawn about the foundations of mathematics come only by analogy with the model, and are to be regarded in much the same way as the conclusions drawn from any scientific theory.

The construction of our model is greatly simplified by our using universal algebra in a way which enables us to dispense with the usual discussion of essentially notational questions about well-formed formulae. All questions and constructions relating to the set of well-formed formulae are handled by our Theorems 2.2 and 4.3 of Chapter I. Our use of universal algebra also provides us with a convenient method for discussing free variables (and avoiding reference to bound variables), and it also permits a simple neat statement of the Substitution Theorem (Theorems 4.11 of Chapter II and 4.3 of Chapter IV).

Chapter I develops the necessary amount of universal algebra. Chapters II and III respectively construct and analyse a model of the Propositional Calculus, introducing in simple form many of the ideas needed for the more complex First-Order Predicate Calculus, which is studied in Chapter IV. In Chapter V, we consider first-order mathematical theories, i.e., theories built on the First-Order Predicate Calculus, thus building models of parts of mathematics. As set theory is usually regarded as the basis on which the rest of mathematics is constructed, we devote Chapter VI to a study of first-order Zermelo-Fraenkel Set Theory. Chapter VII, on Ultraproducts, discusses a technique for constructing new models of a theory from a given collection of models. Chapter VIII, which is an introduction to Non-Standard Analysis, is included as an example of mathematical logic assisting in the study of another branch of mathematics. Decision processes are investigated in Chapter IX, and we prove there the non-existence of decision processes for a number of problems. In Chapter X, we discuss two decision problems from other branches of mathematics and indicate how the results of Chapter IX may be applied.

This book is intended to make mathematical logic available to mathematicians working in other branches of mathematics. We have included what we consider to be the essential basic theory, some useful techniques, and some indications of ways in which the theory might be of use in other branches of mathematics.

We have included a number of exercises. Some of these fill in minor gaps in our exposition of the section in which they appear. Others indicate aspects of the subject which have been ignored in the text. Some are to help in understanding the text by applying ideas and methods to special cases. Occasionally, an exercise asks for the construction of a FORTRAN program. In such cases, the solution should be based on integer arithmetic, and not depend on any special logical properties of FORTRAN or of any other programming language.

The layout of the text is as follows. Each chapter is divided into numbered sections, and definitions, theorems, exercises, etc. are numbered consecutively within each section. For example, the number 2.4 refers to the fourth item in the second section of the current chapter. A reference to an item in some other chapter always includes the chapter number in addition to item and section numbers.

We thank the many mathematical colleagues, particularly Paul Halmos and Peter Hilton, who encouraged and advised us in this project. We are especially indebted to Gordon Monro for suggesting many improvements and for providing many exercises. We thank Mrs. Blakestone and Miss Kicinski for the excellent typescript they produced.

Donald W. Barnes,   John M. Mack

Table of Contents

Preface v

Chapter I   Universal Algebra **1**

§1   Introduction 1
§2   Free Algebras 4
§3   Varieties of Algebras 7
§4   Relatively Free Algebras 8

Chapter II   Propositional Calculus **11**

§1   Introduction 11
§2   Algebras of Propositions 11
§3   Truth in the Propositional Calculus 13
§4   Proof in the Propositional Calculus 14

Chapter III   Properties of the Propositional Calculus . . . **18**

§1   Introduction 18
§2   Soundness and Adequacy of Prop(X) . . . 19
§3   Truth Functions and Decidability for Prop(X) . . . 22

Chapter IV   Predicate Calculus **26**

§1   Algebras of Predicates 26
§2   Interpretations 29
§3   Proof in Pred($V, \mathscr{R}$) 30
§4   Properties of Pred($V, \mathscr{R}$) 32

Chapter V   First-Order Mathematics **38**

§1   Predicate Calculus with Identity 38
§2   First-Order Mathematical Theories . . . 39
§3   Properties of First-Order Theories . . . 43
§4   Reduction of Quantifiers 48

Chapter VI   Zermelo-Fraenkel Set Theory **52**

§1   Introduction 52
§2   The Axioms of ZF 52
§3   First-Order ZF 56
§4   The Peano Axioms 58

Chapter VII   Ultraproducts **62**

§1 Ultraproducts 62
§2 Non-Principal Ultrafilters 64
§3 The Existence of an Algebraic Closure 66
§4 Non-Trivial Ultrapowers 67
§5 Ultrapowers of Number Systems 68
§6 Direct Limits 70

Chapter VIII   Non-Standard Models **74**

§1 Elementary Standard Systems. 74
§2 Reduction of the Order 75
§3 Enlargements 76
§4 Standard Relations 78
§5 Internal Relations. 79
§6 Non-Standard Analysis 80

Chapter IX   Turing Machines and Gödel Numbers . . . **85**

§1 Decision Processes 85
§2 Turing Machines 85
§3 Recursive Functions 89
§4 Gödel Numbers 90
§5 Insoluble Problems in Mathematics. . . . 93
§6 Insoluble Problems in Arithmetic 96
§7 Undecidability of the Predicate Calculus . . . 101

Chapter X   Hilbert's Tenth Problem, Word Problems . . . **105**

§1 Hilbert's Tenth Problem 105
§2 Word Problems 110

References and Further Reading 115

Index of Notations 117

Subject Index 119

An Algebraic Introduction to
Mathematical Logic

Chapter I

Universal Algebra

§1 Introduction

The reader will be familiar with the presentation and study of various algebraic systems (for example, groups, rings, modules) as axiomatic systems consisting of sets with certain operations satisfying certain conditions. The reader will also be aware that ideas and theorems, useful for the study of one type of system, can frequently be adapted to other related systems by making the obvious necessary modifications.

In this book we shall study and use a number of systems whose types are related, but which are possibly unfamiliar to the reader. Hence there is obvious advantage in beginning with the study of a single axiomatic theory which includes as special cases all the systems we shall use. This theory is known as universal algebra, and it deals with systems having arbitrary sets of operations. We shall want to avoid, as far as possible, axioms asserting the existence of elements with special properties (for example, the identity element in group theory), preferring the axioms satisfied by operations to take the form of equations, and we shall be able to achieve this by giving a sufficiently broad definition of "operation". We first recall some elementary facts.

An n-ary relation ρ on the sets $A_1, \ldots, A_n$ is specified by giving those ordered n-tuples $(a_1, \ldots, a_n)$ of elements $a_i \in A_i$ which are in the relation ρ. Thus such a relation is specified by giving those elements $(a_1, \ldots, a_n)$ of the product set $A_1 \times \cdots \times A_n$ which are in ρ, and hence an n-ary relation on $A_1, \ldots, A_n$ is simply a subset of $A_1 \times \cdots \times A_n$. For binary relations, the notation "$a_1 \rho a_2$" is commonly used to express "(a_1, a_2) is in the relation ρ", but we shall usually write this as either "$(a_1, a_2) \in \rho$" or "$\rho(a_1, a_2)$", because each of these notations extends naturally to n-ary relations for any n.

A function $f : A \to B$ is a binary relation on A and B such that, for each $a \in A$, there is exactly one $b \in B$ for which $(a, b) \in f$. It is usual to write this as $f(a) = b$. A function $f(x, y)$ "of two variables" $x \in A, y \in B$, with values in C, is simply a function $f : A \times B \to C$. For each $a \in A$ and $b \in B$, $(a, b) \in A \times B$ and $f((a, b)) \in C$. It is of course usual to omit one set of brackets. There are advantages in retaining the variables x, y in the function notation. Later in this chapter, we will discuss what is meant by variables and give a definition which will justify their use.

Preliminary Definition of Operation. *An n-ary operation on the set A is a function $t : A^n \to A$. The number n is called the arity of t.*

1

Examples

1.1. Multiplication in a group is a binary operation. The $*$-product of two elements a, b is written $a*b$ or simply ab instead of the more systematic $*(a, b)$.

1.2. In a group G, we can define a unary operation $i: G \to G$ by putting $i(a) = a^{-1}$.

1.3. A 0-ary operation on a set A is a function from the set A^0 (whose only element is the empty set $\varnothing$) to the set A, and hence can be regarded as a distinguished element of A. Such an operation arises naturally in group theory, where the 0-ary operation e gives the identity element of the group G.

One often considers several different groups in group theory. If G, H are groups, each has its multiplication operation: $*_G: G \times G \to G$ and $*_H: H \times H \to H$, but one rarely uses distinctive notations for the two multiplications. In practice, the same notation $*$ is used for both, and in fact multiplication is regarded as an operation defined for all groups. The definition of operation given above is clearly not adequate for this usage of the word.

Here is another example demonstrating that our preliminary definition of operation does not match common usage. A ring R is usually defined as a set R with two binary operations $+$, $\times$ satisfying certain axioms. A commonly occurring example of a ring is the zero ring where $R = \{0\}$. In this case, there is only one function $R \times R \to R$, and so $+$, $\times$ are the same function, even though $+$ and $\times$ are still considered distinct operations.

We now give a series of definitions which will overcome the objections raised above.

Definition 1.4. A *type* $\mathscr{T}$ is a set T together with a function $\mathrm{ar}: T \to \mathbf{N}$, from T into the non-negative integers. We shall write $\mathscr{T} = (T, \mathrm{ar})$, or, more simply, abuse notation and denote the type by T. It is also convenient to denote by T_n the set $\{t \in T \,|\, \mathrm{ar}(t) = n\}$.

Definition 1.5. An *algebra A of type T*, or a *T-algebra*, is a set A together with, for each $t \in T$, a function $t_A: A^{\mathrm{ar}(t)} \to A$. The elements $t \in T_n$ are called n-ary *T-algebra operations*.

Observe that each t_A is an operation on the set A in the sense of our preliminary definition of operation. As is usual, we shall write simply $t(a_1, \ldots, a_n)$ for the element $t_A(a_1, \ldots, a_n)$, and we shall denote the algebra by the same symbol A as is used to denote its set of elements.

Examples

1.6. Rings may be considered as algebras of type $T = (\{0, -, +, \cdot\}, \mathrm{ar})$, where $\mathrm{ar}(0) = 0, \mathrm{ar}(-) = 1, \mathrm{ar}(+) = 2, \mathrm{ar}(\cdot) = 2$. We do not claim that such T-algebras are necessarily rings, we simply assert that each ring is an example of a T-algebra for the T given above.

1.7. If R is a given ring, then a module over R may be regarded as a particular example of a T-algebra of type $T = (\{0, -, +\} \cup R, \text{ar})$, where $\text{ar}(0) = 0$, $\text{ar}(-) = 1$, $\text{ar}(+) = 2$, and $\text{ar}(\lambda) = 1$ for each $\lambda \in R$. The first three operations specify the group structure of the module, while the remaining operations correspond to the action of the ring elements.

1.8. Let S be a given ring. Rings R which contain S as subring may be considered as T-algebras, where $T = (\{0, -, +, \cdot\} \cup S, \text{ar})$, $\text{ar}(0) = 0$, $\text{ar}(-) = 1$, $\text{ar}(+) = 2$, $\text{ar}(\cdot) = 2$, and $\text{ar}(s) = 0$ for each $s \in S$. The effect of the S-operations is to distinguish certain elements of R.

Definition 1.9. T-algebras A, B are *equal* if and only if $A = B$ and $t_A = t_B$ for all $t \in T$.

Exercise 1.10. Give an example of unequal T-algebras on the same set A.

Definition 1.11. If A is a T-algebra, a subset B of A is called a T-*subalgebra* of A if it forms a T-algebra with operations the restrictions to B of those on A, i.e., if for all n and for all $t \in T_n$ and $b_1, \ldots, b_n \in B$, we have $t_A(b_1, \ldots, b_n) \in B$.

Any intersection of subalgebras is a subalgebra, and so, given any subset X of A, there is a unique smallest subalgebra containing X—namely, the subalgebra $\cap\{U | U \text{ subalgebra of } A, U \supseteq X\}$. We call this the subalgebra generated by X and denote it by $\langle X \rangle_T$, or if there is no risk of confusion, by $\langle X \rangle$.

Exercises

1.12. A is a T-algebra. Show that $\varnothing$ is a subalgebra if and only if $T_0 = \varnothing$. Show that for all T, every T-algebra has a unique smallest subalgebra.

Many familiar algebraic systems may be regarded as T-algebras for more than one choice of T. However, the subsets which form T-subalgebras may well depend on the choice of T.

1.13. Groups may be regarded as special cases of T-algebras where $T = (\{*\}, \text{ar})$ with $\text{ar}(*) = 2$, or of T'-algebras, where $T' = (\{e, i, *\}, \text{ar})$, $\text{ar}(e) = 0$, $\text{ar}(i) = 1$, $\text{ar}(*) = 2$. Show that every T'-subalgebra of a group is a subgroup, but that not every non-empty T-subalgebra need be a group. Show that if G is a finite group, then every non-empty T-subalgebra of G is itself a group.

Definition 1.14. Let A, B be T-algebras. A *homomorphism* of A into B is a function $\varphi : A \to B$ such that, for all $t \in T$ and all $a_1, \ldots, a_n \in A$ ($n = \text{ar}(t)$), we have

$$\varphi(t_A(a_1, \ldots, a_n)) = t_B(\varphi(a_1), \ldots, \varphi(a_n)).$$

This condition is often expressed as "φ *preserves all the operations of T*".

Clearly, the composition of two homomorphisms is a homomorphism. Further, if $\varphi: A \to B$ is a homomorphism and is invertible, then the inverse function $\varphi^{-1}: B \to A$ is also a homomorphism. In this case we call φ an *isomorphism* and say that A and B are *isomorphic*.

§2   Free Algebras

Definition 2.1.   Let X be any set, let F be a T-algebra and let $\sigma: X \to F$ be a function. We say that F (more strictly (F, σ)) is a *free T-algebra* on the set X of *free generators* if, for every T-algebra A and function $\tau: X \to A$, there exists a unique homomorphism $\varphi: F \to A$ such that $\varphi\sigma = \tau$:

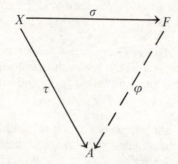

Observe that if (F, σ) is free, then σ is injective. For it is easily seen that there exists a T-algebra with more than one element, and hence if x_1, x_2 are distinct elements of X, then for some A and τ we have $\tau(x_1) \neq \tau(x_2)$, which implies $\sigma(x_1) \neq \sigma(x_2)$.

The next theorem asserts the existence of a free T-algebra on a set X, and the proof is constructive. Informally, one could describe the free T-algebra on X as the collection of all formal expressions that can be formed from X and T by using only finitely many elements of X and T in any one expression. But to say precisely what is meant by a formal expression in the elements of X using the operations of T is tantamount to constructing the free algebra.

Theorem 2.2.   *For any set X and any type T, there exists a free T-algebra on X. This free T-algebra on X is unique up to isomorphism.*

Proof.   (a) *Uniqueness.* We show first that if (F, σ) is free on X, and if $\varphi: F \to F$ is a homomorphism such that $\varphi\sigma = \sigma$, then $\varphi = 1_F$, the identity map on F. To show this, we take $A = F$ and $\tau = \sigma$ in the defining condition. Then $1_F: F \to F$ has the required property for φ, and hence by its uniqueness is the only such map.

Now let (F, σ) and (F', σ') be free on X.

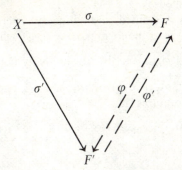

Since (F, σ) is free, there exists a homomorphism $\varphi:F \to F'$ such that $\varphi\sigma = \sigma'$. Since (F', σ') is free, there exists a homomorphism $\varphi':F' \to F$ such that $\varphi'\sigma' = \sigma$. Hence $\varphi'\varphi\sigma = \varphi'\sigma' = \sigma$, and by the result above, $\varphi'\varphi = 1_F$. Similarly, $\varphi\varphi' = 1_{F'}$. Thus φ, φ' are mutually inverse isomorphisms, and so uniqueness is proved.

(b) *Existence.* An algebra F will be constructed as a union of sets F_n ($n \in N$), which are defined inductively as follows.

(i) F_0 is the disjoint union of X and T_0.

(ii) Assume F_r is defined for $0 \leqslant r < n$. Then define

$$F_n = \left\{(t, a_1, \ldots, a_k) \big| t \in T, \, \mathrm{ar}(t) = k, \, a_i \in F_{r_i}, \, \sum_{i=1}^{k} r_i = n - 1\right\}.$$

(iii) Put $F = \bigcup_{n \in N} F_n$.

The set F is now given. To make it into a T-algebra, we must specify the action of the operations $t \in T$.

(iv) If $t \in T_k$ and $a_1, \ldots, a_k \in F$, put $t(a_1, \ldots, a_k) = (t, a_1, \ldots, a_k)$. In particular, if $t \in T_0$, then t_F is the element t of F_0.

This makes F into a T-algebra. To complete the construction, we must give the map $\sigma:X \to F$.

(v) For each $x \in X$, put $\sigma(x) = x \in F_0$.

Finally, we have to prove that F is free on X, i.e., we must show that if A is any T-algebra and $\tau:X \to A$ any map of X into A, then there exists a unique homomorphism $\varphi:F \to A$ such that $\varphi\sigma = \tau$. We do this by constructing inductively the restriction φ_n of φ to F_n and by showing that φ_n is completely determined by τ and the φ_k for $k < n$.

We have $F_0 = T_0 \cup X$. The homomorphism condition requires $\varphi_0(t_F) = t_A$ for $t \in T_0$, while for $x \in X$ we require $\varphi\sigma(x) = \tau(x)$, and so we must have

$\varphi_0(x) = \tau(x)$. Thus $\varphi_0: F_0 \to A$ is defined, and is uniquely determined by the conditions to be satisfied by φ.

Suppose that φ_k is defined and uniquely determined for $k < n$. An element of F_n $(n > 0)$ is of the form $(t, a_1 \ldots, a_k)$, where $t \in T_k$, $a_i \in F_{r_i}$ and $\sum_{i=1}^{k} r_i = n - 1$. Thus $\varphi_{r_i}(a_i)$ is already uniquely defined for $i = 1, \ldots, k$. Furthermore, since $(t, a_1, \ldots, a_k) = t(a_1, \ldots, a_k)$, and since the homomorphism property of φ requires that

$$\varphi(t, a_1, \ldots, a_k) = t(\varphi(a_1), \ldots, \varphi(a_k)),$$

we must define

$$\varphi_n(t, a_1, \ldots, a_k) = t(\varphi_{r_1}(a_1), \ldots, \varphi_{r_k}(a_k)).$$

This determines φ_n uniquely, and as each element of F belongs to exactly one subset F_n, on putting $\varphi(\alpha) = \varphi_n(\alpha)$ for $\alpha \in F_n$ $(n \geq 0)$, we see that φ is a homomorphism from F to A satisfying $\varphi\sigma(x) = \varphi_0(x) = \tau(x)$ for all $x \in X$ as required, and that φ is the only such homomorphism. $\square$

The above inductive construction of the free T-algebra F fits in with its informal description—each F_n is a collection of "T-expressions", increasing in complexity with n. The notion of a T-expression is useful for an arbitrary T-algebra, so we shall formalise it, making use of free T-algebras to do so.

Let A be any T-algebra, and let F be the free T-algebra on the set $X_n = \{x_1, \ldots, x_n\}$. For any (not necessarily distinct) elements $a_1, \ldots, a_n \in A$, there exists a unique homomorphism $\varphi: F \to A$ with $\varphi(x_i) = a_i$ $(i = 1, \ldots, n)$. If $w \in F$, then $\varphi(w)$ is an element of A which is uniquely determined by $a_1, \ldots, a_n$. Hence we may define a function $w_A: A^n \to A$ by putting $w_A(a_1, \ldots, a_n) = \varphi(w)$. We omit the subscript A and write simply $w(a_1, \ldots, a_n)$. If in particular we take $A = F$ and $a_i = x_i$ $(i = 1, \ldots, n)$, then φ is the identity and $w(x_1, \ldots, x_n) = w$.

Definition 2.3. A T-word in the variables $x_1, \ldots, x_n$ is an element of the free T-algebra on the set $X_n = \{x_1, \ldots, x_n\}$ of free generators.

Definition 2.4. A word in the elements $a_1, \ldots, a_n$ of a T-algebra A is an element $w(a_1, \ldots, a_n) \in A$, where w is a T-word in the variables $x_1, \ldots, x_n$.

We have used and even implicitly defined the term "variable" in the above definitions. In normal usage, a variable is "defined" as a symbol for which any element of the appropriate kind may be substituted. We give a formal definition of variable, confirming that our variables have this usual property.

Definition 2.5. A T-algebra variable is an element of the free generating set of a free T-algebra.

Among the words in the variables $x_1, \ldots, x_n$ are the words x_i $(i = 1, \ldots, n)$, having the property that $x_i(a_1, \ldots, a_n) = a_i$. Thus variables may also be regarded as coordinate functions. The concept of a coordinate function certainly provides the most convenient definition of variable for use in analysis. For example, when we speak of a function $f(x, y)$ as a function of two real variables x, y, we have a function f, defined on some subset of $R \times R$, together with coordinate projections $x(a, b) = a$, $y(a, b) = b$ $(a, b \in R)$, and $f(x, y)$ is in fact the composite function $f(a, b) = f(x(a, b), y(a, b))$.

Exercises

2.6. T consists of one unary operation, and F is the free T-algebra on a one-element set X. How many elements are there in F_n? How many elements are there in F?

2.7. If T is empty and X is any set, show that X is the free T-algebra on X.

2.8. T consists of a single binary operation, and F is the free T-algebra on a one-element set X. How many elements are there in F?

2.9. If T consists of one 0-ary operation and one 2-ary operation, and if $X = \varnothing$, then the free T-algebra F on X is countable.

2.10. T is finite or countable, and contains at least one 0-ary operation and at least one operation t with $\mathrm{ar}(t) > 0$. X is finite or countable. Prove that F is countable.

§3 Varieties of Algebras

Let F be the free T-algebra on the countable set $X = \{x_1, x_2, \dots\}$ of variables. Although each element of F is a word in some finite subset $X_n = \{x_1, \dots, x_n\}$, we shall consider sets of words for which there may be no bound to the number of variables in the words.

Definition 3.1. An *identical relation* on T-algebras is a pair (u, v) of elements of F.

There is an n for which u, v are in the free algebra on X_n, and we say that (u, v) is an *n-variable identical relation* for any such n.

Definition 3.2. The T-algebra A *satisfies* the n-variable identical relation (u, v), or (u, v) is a *law* of A, if $u(a_1, \dots, a_n) = v(a_1, \dots, a_n)$ for all $a_1, \dots, a_n \in A$.

Equivalently, (u, v) is a law of A if $\varphi(u) = \varphi(v)$ for every homomorphism $\varphi : F \to A$.

Definition 3.3. Let L be a set of identical relations on T-algebras. The class V of all T-algebras which satisfy all the identical relations in L is called the *variety of T-algebras defined by L*. The *laws of the variety* are all the identical relations satisfied by every algebra of V.

Note that the set of laws of the variety includes L, but may be larger.

Examples

3.4. T consists of a single binary operation $*$, and L has the one element $(x_1*(x_2*x_3), (x_1*x_2)*x_3)$. If A satisfies this identical relation, then $a*(b*c) = (a*b)*c$ for all a, b, $c \in A$. Thus the operation on A is associative and A is a semigroup. The variety defined by L in this case is the class of all semigroups.

3.5. T consists of 0-ary, 1-ary and 2-ary operations e, i, $*$ respectively. L has the three elements

$$(x_1*(x_2*x_3), (x_1*x_2)*x_3),$$

$$(e*x_1, x_1),$$

$$(i(x_1)*x_1, e).$$

The first law ensures that $*$ is an associative operation in every algebra of the variety defined by L. The second shows that the distinguished element e is always a left identity, while the third guarantees that $i(a)$ is a left inverse of the element a. Hence the algebras of the variety are groups.

Exercises

3.6. Show that the class of all abelian groups is a variety.

3.7. R is a ring with 1. Show that the class of unital left R-modules is a variety.

3.8. S is a commutative ring with 1. Show that the class of commutative rings R with $1_R = 1_S$ and which contain S as a subring is a variety.

3.9. Is the class of finite groups a variety?

§4 Relatively Free Algebras

Let V be the variety of T-algebras defined by the set L of laws.

Definition 4.1. A T-algebra R in the variety V is the (*relatively*) *free algebra of V* on the set X of (*relatively*) *free generators* (where a function $\sigma: X \to R$ is given, usually as an inclusion) if, for every algebra A in V and every function $\tau: X \to A$, there exists a unique homomorphism $\varphi: R \to A$ such that $\varphi\sigma = \tau$.

This definition differs from the earlier definition of a free algebra only in that we consider here only algebras in V.

Definition 4.2. An algebra is *relatively free* if it is a free algebra of some variety.

Theorem 4.3. *For any type T, and any set L of laws, let V be the variety of T-algebras defined by L. For any set X, there exists a free T-algebra of V on X.*

Proof: Let (F, ρ) be the free T-algebra on X. A congruence relation on F is defined by putting $u \sim v$ (where $u, v \in F$) if $\varphi(u) = \varphi(v)$ for every homomorphism φ of F into an algebra in V. Clearly $\sim$ is an equivalence relation on F. If now $t \in T_k$ and $u_i \sim v_i$ $(i = 1, \ldots, k)$, then for every such homomorphism φ, $\varphi(u_i) = \varphi(v_i)$, and so

$$\varphi\bigl(t(u_1, \ldots, u_k)\bigr) = t\bigl(\varphi(u_1), \ldots, \varphi(u_k)\bigr) = t\bigl(\varphi(v_1), \ldots, \varphi(v_k)\bigr) = \varphi\bigl(t(v_1, \ldots, v_k)\bigr),$$

verifying that φ is a congruence relation.

We define R to be the set of congruence classes of elements of F with respect to this congruence relation. Denoting the congruence class containing u by $\bar{u}$, we define the action of $t \in T_k$ on R by putting $t(\bar{u}_1, \ldots, \bar{u}_k) = \overline{t(u_1, \ldots, u_k)}$. This definition is independent of the choice of representatives $u_1, \ldots, u_k$ of the classes $\bar{u}_1, \ldots, \bar{u}_k$, and makes R a T-algebra. Also, the map $u \to \bar{u}$ is clearly a homomorphism $\eta : F \to R$. Finally, we define $\sigma : X \to R$ by $\sigma(x) = \overline{\rho(x)}$.

We now prove that (R, σ) is relatively free on X. Let A be any algebra in V, and let $\tau : X \to A$ be any function from X into A. Because (F, ρ) is free, there exists a unique homomorphism $\psi : F \to A$ such that $\psi\rho = \tau$.

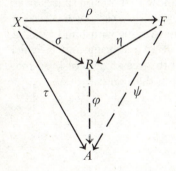

For $\bar{u} \in R$, we define $\varphi(\bar{u}) = \psi(u)$. This is independent of the choice of representative u of the element $\bar{u}$, since if $\bar{u} = \bar{v}$, then $\psi(u) = \psi(v)$. The map $\varphi : R \to A$ is clearly a homomorphism, and $\varphi\sigma = \varphi\eta\rho = \psi\rho = \tau$. If $\varphi' : R \to A$ is another homomorphism such that $\varphi'\sigma = \tau$, then $\varphi'\eta\rho = \tau$ and therefore $\varphi'\eta = \psi$. Consequently for each element $\bar{u} \in R$ we have

$$\varphi'(\bar{u}) = \varphi'\eta(u) = \psi(u) = \varphi(\bar{u}),$$

and hence $\varphi' = \varphi$. $\square$

When considering only the algebras of a given variety V, we may redefine variables and words accordingly. Thus we define a *V-variable* as an element of the free generating set of a free algebra of V, and a *V-word* in the V-variables $x_1, \ldots, x_n$ as an element of the free algebra of V on the free generators $\{x_1, \ldots, x_n\}$.

Examples

4.4. T consists of a single binary operation which we shall write as juxtaposition. Let V be the variety of associative T-algebras. Then all products in the free T-algebra obtained by any bracketing of $x_1, \ldots, x_n$, taken in that order, are congruent under the congruence relation used in our construction of the relatively free algebra, and correspond to the one word $x_1 x_2 \cdots x_n$ of V. We observe that in this example, all elements of the absolutely free algebra F, which map to a given element $x_1 x_2 \cdots x_n$ of the relatively free algebra, come from the same layer F_{n-1} of F.

4.5. T consists of a 0-ary, a 1-ary and a 2-ary operation. V is the variety of abelian groups, defined by the laws given in Example 3.5 together with the law $(x_1 x_2, x_2 x_1)$. In this case, the relatively free algebra on $\{x_1, \ldots, x_n\}$ is the set of all $x_1^{r_1} x_2^{r_2} \cdots x_n^{r_n}$ (or equivalently the set of all n-tuples $(r_1, \ldots, r_n)$) with $r_i \in \mathbb{Z}$. Here the layer property of Example 4.4 does not hold, because, for example, we have the identity $e \in F_0$, $x_1^{-1} \in F_1$, $x_1^{-1} * x_1 \in F_2$ and yet $\bar{e} = \overline{x_1^{-1} * x_1}$.

Exercises

4.6. K is a field. Show that vector spaces over K form a variety V of algebras, and that every vector space over K is a free algebra of V.

4.7. R is a commutative ring with 1 and V is the variety of commutative rings S which contain R as a subring and in which 1_R is a multiplicative identity of S. Show that the free algebra of V on the set X of variables is the polynomial ring over R in the elements of X.

Chapter II

Propositional Calculus

§1 Introduction

Mathematical logic is the study of logic as a mathematical theory. Following the usual procedure of applied mathematics, we construct a mathematical model of the system to be studied, and then conduct what is essentially a pure mathematical investigation of the properties of our model. Since this book is intended for mathematicians, the system we propose to study is not general logic but the logic used in mathematics. By this restriction, we achieve considerable simplification, because we do not have to worry about precise meanings of words—in mathematics, words have precisely defined meanings. Furthermore, we are free of reasoning based on things such as emotive argument, which must be accounted for in any theory of general logic. Finally, the nature of the real world need not concern us, since the world we shall study is the purely conceptual one of pure mathematics.

In any formal study of logic, the language and system of reasoning needed to carry out the investigation is called the meta-language or meta-logic. As we are constructing a mathematical model of logic, our meta-language is mathematics, and so all our existing knowledge of mathematics is available for possible application to our model. We shall make specific use of informal set theory (including cardinal numbers and Zorn's lemma) and of the universal algebra developed in Chapter I.

For the purpose of our study, it suffices to describe mathematics as consisting of assertions that if certain statements are true then so are certain other statements, and of arguments justifying these assertions. Hence a model of mathematical reasoning must include a set of objects which we call statements or propositions, some concept of truth, and some concept of a proof. Once a model is constructed, the main subject of investigation is the relationship between truth and proof. We shall begin by constructing a model of the simpler parts of mathematical reasoning. This model is called the Propositional Calculus. Later, we shall construct a more refined model (known as the First-Order Predicate Calculus), copying more complicated parts of the reasoning used in mathematics.

§2 Algebras of Propositions

The Propositional Calculus considers ways in which simple statements may be combined to form more complex statements, and studies how the truth or falsity of complex statements is related to that of their component

11

statements. Some of the ways in which statements are combined in mathematics are as follows. We often use "and" to combine statements, and we write $p \wedge q$ for the statement "p and q", which is regarded as true if and only if both the statements p, q are true. We frequently assert that (at least) one of two possibilities is true, and we write $p \vee q$ for the statement "p or q", which we consider to be true if at least one of p, q is true and false if both p and q are false. We often assert that some statement is false, and we write $\sim p$ (read "not p") for the statement "p is false", which is regarded as true if and only if p is false. Another common way of linking two statements is through an assertion "if p is true, then so is q". For this we write "$p \Rightarrow q$" (read "p implies q"), which, in mathematical usage, is true unless q is false and p is true.

We want our simple model to imitate the above constructions, so we want our set of propositions to be an algebra with respect to the four operations given above. This could be done by taking the free algebra with these operations, but we know that in ordinary usage, the four operations are not independent. Thus a simpler system is suggested, in which we choose some basic operations which will enable us to define all the above operations. This may be done in many ways, some of which are explored in exercises at the end of Chapter III, where they may be studied more thoroughly. We choose a way which is perhaps not the natural one, but which has advantages in that it simplifies the development of the theory. Our choice rests on the fact that in mathematics, a result is often proved by showing that the denial of the result leads to a contradiction. We introduce into our notation a symbol for a contradiction by specifying that our algebra will have a distinguished element (i.e., a 0-ary operation) F, which we will think of as a contradiction or falsehood.

Definition 2.1.   Let $T = \{F, \Rightarrow\}$, where F is a 0-ary operation and $\Rightarrow$ is a binary operation. Any T-algebra is called a *proposition algebra*.

Definition 2.2.   The *proposition algebra $P(X)$ of the propositional calculus on the set X of propositional variables* is the free T-algebra on X.

Example 2.3.   The algebra $\mathbb{Z}_2$ of integers mod 2 can be made into a proposition algebra by defining $F_{\mathbb{Z}_2} = 0$ and $m \Rightarrow n = 1 + m(1 + n)$.
We shall make frequent use of this example.

In any proposition algebra, we introduce the further operations $\sim$, $\vee$, $\wedge$, $\Leftrightarrow$ by defining

$$\sim p = p \Rightarrow F$$
$$p \vee q = (\sim p) \Rightarrow q$$
$$p \wedge q = \sim(\sim p \vee \sim q)$$
$$p \Leftrightarrow q = (p \Rightarrow q) \wedge (q \Rightarrow p).$$

We point out that the above are not statements in our proposition algebras, because the symbol $=$ is not an operation in our proposition algebras. The first equation says that $\sim p$ is a notation for the element $p \Rightarrow F$ of our algebra. We shall often omit brackets, as we did above in writing $\sim p \vee \sim q$ for $(\sim p) \vee (\sim q)$.

Exercises

2.4. Show that our definitions of $\sim$, $\vee$, $\wedge$, $\Leftrightarrow$ conform to normal usage.
2.5. Express $\sim$, $\vee$, $\wedge$ in $\mathbb{Z}_2$ in terms of multiplication and addition.
2.6. Is $\mathbb{Z}_2$ a free proposition algebra?

§3   Truth in the Propositional Calculus

Having determined the form of our algebra of propositions, we must now find a meaning for the concept of truth applied to our propositions. We are guided here by the observation that in ordinary mathematical usage, the truth or falsity of the compound statement $p \Rightarrow q$ is determined completely once the truth or falsity of each of p, q is specified. Every simple statement is given a value—true or false—and the truth or falsity of any compound statement depends on and is determined by the truth values of its components. This leads us to consider valuations on $P(X)$, i.e., functions which assign to each element $p \in P(X)$ one of two possible values, which for convenience are denoted by 0, 1. We are then considering functions $v : P(X) \to \mathbb{Z}_2$, interpreting $v(p) = 1$ as meaning "p is true", and $v(p) = 0$ as "p is false". In order that a valuation act properly on compound propositions, the functions v must be proposition algebra homomorphisms.

Definition 3.1.   A *valuation* of $P(X)$ is a proposition algebra homomorphism $v : P(X) \to \mathbb{Z}_2$. We say that $p \in P(X)$ is *true with respect to v* if $v(p) = 1$, and that p is *false with respect to v* if $v(p) = 0$.

Since X is a set of free generators of $P(X)$, the values $v(x)$ for $x \in X$ may be assigned arbitrarily. These values, once assigned, determine the homomorphism v uniquely and so determine $v(p)$ for all $p \in P(X)$.

In ordinary usage, the interesting and important notion relating the truth values of statements is that of consequence—a statement q is a consequence of statements $p_1, \ldots, p_n$ if q is true of every mathematical system in which $p_1, \ldots, p_n$ are all true. This idea is incorporated in our model by considering valuations which assign the value 1 to all of $p_1, \ldots, p_n$.

Definition 3.2.   Let $A \subseteq P(X)$ and $q \in P(X)$. We say that q is a *consequence* of the set A of assumptions, or that A *semantically implies q*, if $v(q) = 1$ for every valuation v such that $v(p) = 1$ for all $p \in A$. We shall write this $A \vDash q$, and we shall denote by $\mathrm{Con}(A)$ the set $\{ p \in P(X) | A \vDash p \}$ of all consequences of A.

Definition 3.3. Let $p \in P(X)$. We say that p is *valid*, or is a *tautology*, if $v(p) = 1$ for every valuation v of $P(X)$.

Thus p is a tautology if $\emptyset \vDash p$. We shall write this simply as $\vDash p$. Note that $A \vDash p$ is not a proposition (i.e., not an element of $P(X)$), but simply a statement in the meta-language about our model.

Examples

3.4. $\{q\} \vDash p \Rightarrow q$. For if v is any valuation with $v(q) = 1$, then

$$v(p \Rightarrow q) = v(p) \Rightarrow v(q) = v(p) \Rightarrow 1 = 1 + v(p)(1 + 1) = 1.$$

3.5. $\vDash p \Rightarrow p$. For if v is any valuation, then

$$v(p \Rightarrow p) = v(p) \Rightarrow v(p) = 1 + v(p)(1 + v(p)) = 1,$$

since $x(1 + x) = 0$ for all $x \in \mathbb{Z}_2$.

Exercises

3.6. Show that $\{F\} \vDash p$ for all $p \in P(X)$.
3.7. Show that $\{p, p \Rightarrow q\} \vDash q$ and $\{p, \sim q \Rightarrow \sim p\} \vDash q$ for all $p, q \in P(X)$.
3.8. Show that $p \Rightarrow (q \Rightarrow p)$, $(p \Rightarrow (q \Rightarrow r)) \Rightarrow ((p \Rightarrow q) \Rightarrow (p \Rightarrow r))$ and $\sim \sim p \Rightarrow p$ are tautologies, for all $p, q, r \in P(X)$.

Lemma 3.9. Con *is a closure operation on* $P(X)$, *that is, it has the properties*
   (i) $A \subseteq \text{Con}(A)$,
   (ii) *If* $A_1 \subseteq A_2$, *then* $\text{Con}(A_1) \subseteq \text{Con}(A_2)$,
   (iii) $\text{Con}(\text{Con}(A)) = \text{Con}(A)$.

Proof:
   (i) Trivial.
   (ii) Suppose $q \in \text{Con}(A_1)$. Let v be any valuation such that $v(A_2) \subseteq \{1\}$. Then $v(A_1) \subseteq \{1\}$ and so $v(q) = 1$ since $q \in \text{Con}(A_1)$. Hence $q \in \text{Con}(A_2)$.
   (iii) Suppose $q \in \text{Con}(\text{Con}(A))$, and let v be a valuation such that $v(A) \subseteq \{1\}$. For all $p \in \text{Con}(A)$, we have $v(p) = 1$ by the definition of $\text{Con}(A)$. Thus $v(\text{Con}(A)) \subseteq \{1\}$ and so $v(q) = 1$. Thus $q \in \text{Con}(A)$. $\square$

§4   Proof in the Propositional Calculus

A mathematical system is usually specified by certain statements called assumptions, which describe certain characteristic features of the system. A proof of some other property of the system consists of a succession of statements, ending in a statement of the desired property, in which each statement has been obtained from those before it in some acceptable manner. Apart from the particular assumptions of the system, which are considered accept-able at any step in a proof, we distinguish two methods which permit the addition of a statement to a given acceptable string of statements. There is a specific collection of statements which are considered acceptable additions in any mathematical proof—they can be regarded as underlying assumptions common to every mathematical system—and which we formalise as certain specified propositions which may be introduced at any stage into any proof. Such propositions are called the axioms of our model. The other permissible method consists of rules which specify, in terms of those statements already set down, particular statements which may be adduced. Rules of this kind, when formalised, are called the rules of inference of our model.

For the propositional calculus on the set X, we take as axioms all elements of the subset $\mathscr{A} = \mathscr{A}_1 \cup \mathscr{A}_2 \cup \mathscr{A}_3$ of $P(X)$,

where $\mathscr{A}_1 = \{p \Rightarrow (q \Rightarrow p) | p, q \in P(X)\}$,

$$\mathscr{A}_2 = \{(p \Rightarrow (q \Rightarrow r)) \Rightarrow ((p \Rightarrow q) \Rightarrow (p \Rightarrow r)) | p, q, r \in P(X)\},$$

and

$$\mathscr{A}_3 = \{\sim \sim p \Rightarrow p | p \in P(X)\}.$$

As our one rule of inference, we take the rule known as modus ponens: from p and $p \Rightarrow q$, deduce q. We may now give a formal definition of a proof.

Definition 4.1. Let $q \in P(X)$ and let $A \subseteq P(X)$. In the propositional calculus on the set X, a *proof of q from the assumptions A* is a finite sequence $p_1, p_2, \ldots, p_n$ of elements $p_i \in P(X)$ such that $p_n = q$ and for each i, either $p_i \in \mathscr{A} \cup A$ or for some $j, k < i$, we have $p_k = (p_j \Rightarrow p_i)$.

Definition 4.2. Let $q \in P(X)$ and let $A \subseteq P(X)$. We say that q is a *deduc-tion* from A, or q is *provable* from A, or that A *syntactically implies* q, if there exists a proof of q from A. We shall write this $A \vdash q$, and we shall denote by $\text{Ded}(A)$ the set of all deductions from A.

Definition 4.3. Let $p \in P(X)$. We say that p is a *theorem* of the proposi-tional calculus on X if there exists a proof of p from $\varnothing$.

Thus p is a theorem if $\varnothing \vdash p$, which we write simply as $\vdash p$.

Lemma 4.4. (i) *If $q \in \text{Ded}(A)$, then $q \in \text{Ded}(A')$ for some finite subset A' of A.*
(ii) *Ded is a closure operation on $P(X)$.*

Proof: (i) This holds because a proof of q from A, being a finite sequence of elements of $P(X)$, can contain only finitely many members of A.
(ii) The first two requirements for a closure operation are obviously met by Ded. Suppose now that $q \in \text{Ded}(\text{Ded}(A))$. Then there exists a proof $p_1, \ldots, p_n$ of q from $\text{Ded}(A)$. In this proof, certain (perhaps none) of the p_i,

say $p_{i_1}, \ldots, p_{i_r}$ are in Ded(A). Let $p_{i_j, 1}, p_{i_j, 2}, \ldots, p_{i_j, r_j}$ be a proof of p_{i_j} from A. Replace each of the p_{i_j} in $p_1, \ldots, p_n$ by its proof $p_{i_j, 1}, \ldots, p_{i_j, r_j}$. The resulting sequence is a proof of q from A. $\quad \square$

Examples

4.5. $\vdash p \Rightarrow p$. For any $p \in P(X)$, the following sequence $p_1, \ldots, p_5$ is a proof of $p \Rightarrow p$:

$$p_1 = p \Rightarrow ((p \Rightarrow p) \Rightarrow p), \qquad\qquad (\mathscr{A}_1)$$

$$p_2 = (p \Rightarrow ((p \Rightarrow p) \Rightarrow p)) \Rightarrow ((p \Rightarrow (p \Rightarrow p)) \Rightarrow (p \Rightarrow p)), \qquad (\mathscr{A}_2)$$

$$p_3 = (p \Rightarrow (p \Rightarrow p)) \Rightarrow (p \Rightarrow p), \qquad\qquad (p_2 = p_1 \Rightarrow p_3)$$

$$p_4 = p \Rightarrow (p \Rightarrow p), \qquad\qquad (\mathscr{A}_1)$$

$$p_5 = p \Rightarrow p. \qquad\qquad (p_3 = p_4 \Rightarrow p_5)$$

The proof is the sequence $p_1, \ldots, p_5$. These have been written on successive lines for ease of reading. We have placed notes alongside each step to explain why it can be included at that stage of the proof, but these notes are not part of the proof.

4.6. $\{q\} \vdash p \Rightarrow q$. A proof of this is $q \Rightarrow (p \Rightarrow q), q, p \Rightarrow q$.

4.7. $\vdash F \Rightarrow q$. For any $q \in P(X)$, the following is a proof:

$$p_1 = (\sim \sim q \Rightarrow q) \Rightarrow (F \Rightarrow (\sim \sim q \Rightarrow q)), \qquad\qquad (\mathscr{A}_1)$$

$$p_2 = \sim \sim q \Rightarrow q, \qquad\qquad (\mathscr{A}_3)$$

$$p_3 = F \Rightarrow (\sim \sim q \Rightarrow q), \qquad\qquad (p_1 = p_2 \Rightarrow p_3)$$

$$p_4 = (F \Rightarrow (\sim \sim q \Rightarrow q)) \Rightarrow ((F \Rightarrow \sim \sim q) \Rightarrow (F \Rightarrow q)), \qquad (\mathscr{A}_2)$$

$$p_5 = (F \Rightarrow \sim \sim q) \Rightarrow (F \Rightarrow q), \qquad\qquad (p_4 = p_3 \Rightarrow p_5)$$

$$p_6 = F \Rightarrow (\sim q \Rightarrow F) = F \Rightarrow \sim \sim q, \qquad\qquad (\mathscr{A}_1)$$

$$p_7 = F \Rightarrow q. \qquad\qquad (p_5 = p_6 \Rightarrow p_7)$$

4.8. $\vdash \sim p \Rightarrow (p \Rightarrow q)$. A proof of this is the sequence $p_1, \ldots, p_7$ of Example 4.7, followed by

$$p_8 = (F \Rightarrow q) \Rightarrow (p \Rightarrow (F \Rightarrow q)), \qquad\qquad (\mathscr{A}_1)$$

$$p_9 = p \Rightarrow (F \Rightarrow q), \qquad\qquad (p_8 = p_7 \Rightarrow p_9)$$

$$p_{10} = (p \Rightarrow (F \Rightarrow q)) \Rightarrow ((p \Rightarrow F) \Rightarrow (p \Rightarrow q)), \qquad\qquad (\mathscr{A}_2)$$

$$p_{11} = (p \Rightarrow F) \Rightarrow (p \Rightarrow q) = \sim p \Rightarrow (p \Rightarrow q). \qquad (p_{10} = p_9 \Rightarrow p_{11})$$

The length of the proof needed for such a trivial result as $\sim p \Rightarrow (p \Rightarrow q)$ may well alarm a reader familiar with mathematical theorems and proofs. Ordinary mathematical proofs are very much abbreviated. For example, (allegedly) obvious steps are usually omitted, and previously established results are quoted without proof. Such devices are not available to us, because of the very restrictive nature of our definition of proof in the propositional calculus. We could reduce the lengths of many proofs if we extended our definition to include further rules of inference or abbreviative rules, but by doing so, we would complicate our study of the relationship between truth and proof, which is the principal object of the theory. We remark that in order to show that $\sim p \Rightarrow (p \Rightarrow q)$ is a theorem of the propositional calculus, it suffices to argue as follows: we have $\vdash F \Rightarrow q$, and the sequence $p_7, \ldots, p_{11}$ is a proof of $\sim p \Rightarrow (p \Rightarrow q)$ from the assumption $\{F \Rightarrow q\}$. Thus

$$\sim p \Rightarrow (p \Rightarrow q) \in \text{Ded}(\{F \Rightarrow q\}) \subseteq \text{Ded}(\text{Ded}(\varnothing)) = \text{Ded}(\varnothing),$$

hence $\vdash \sim p \Rightarrow (p \Rightarrow q)$.

This is a mathematical proof of the existence of a proof in the propositional calculus. It is not a proof in the propositional calculus. We shall find other ways of demonstrating the existence of proofs without actually constructing them formally.

Exercises

4.9. Show that $\text{Ded}(A)$ is the smallest subset D of $P(X)$ such that $D \supseteq \mathscr{A} \cup A$ and such that if $p, p \Rightarrow q \in D$, then also $q \in D$.

4.10. Construct a proof in the propositional calculus of $p \Rightarrow r$ from the assumptions $\{p \Rightarrow q, q \Rightarrow r\}$.

We close this chapter with a useful algebraic result.

Theorem 4.11. (The Substitution Theorem). *Let X, Y be any two sets, and let $\varphi : P(X) \to P(Y)$ be a homomorphism of the (free) proposition algebra on X into the (free) proposition algebra on Y. Let $w = w(x_1, \ldots, x_n)$ be any element of $P(X)$ and let A be any subset of $P(X)$. Put $a_i = \varphi(x_i)$.*
  (a) *If $A \vdash w$, then $\varphi(A) \vdash w(a_1, \ldots, a_n)$.*
  (b) *If $A \models w$, then $\varphi(A) \models w(a_1, \ldots, a_n)$.*

Proof: (a) Suppose $p_1, \ldots, p_r$ is a proof of w from A. If $p_i \in A$, then trivially $\varphi(p_i) \in \varphi(A)$. Since φ is a homomorphism, it follows that if p_i is an axiom of the propositional calculus on X, then $\varphi(p_i)$ is an axiom of the propositional calculus on Y. For the same reason, if $p_k = (p_j \Rightarrow p_i)$, then $\varphi(p_k) = \varphi(p_j \Rightarrow p_i) = \varphi(p_j) \Rightarrow \varphi(p_i)$. Thus $\varphi(p_1), \ldots, \varphi(p_r)$ is a proof in the propositional calculus on Y of $\varphi(w)$ from $\varphi(A)$. Since $\varphi(w) = w(a_1, \ldots, a_n)$, the result is proved.

(b) Suppose $A \models w$. Let $v : P(Y) \to \mathbb{Z}_2$ be a valuation of $P(Y)$ such that $v(\varphi(A)) \subseteq \{1\}$. Then the composite map $v\varphi : P(X) \to \mathbb{Z}_2$ is a valuation of $P(X)$, and $v\varphi(A) \subseteq \{1\}$. Since $A \models w$, we have $v\varphi(w) = 1$, i.e. $v(\varphi(w)) = 1$. Thus $\varphi(A) \models \varphi(w)$. $\square$

Chapter III

Properties of the Propositional Calculus

§1 Introduction

The properties of the Propositional Calculus that are of interest are those that arise in studying the relation between truth and proof. These properties are important features in the study of any formal system of reasoning, and we begin with some general definitions.

Definition 1.1. A *logic* $\mathscr{L}$ is a system consisting of a set P of elements (called propositions), a set $\mathscr{V}$ of functions (called valuations) from P into some value set W, and, for each subset A of P, a set of finite sequences of elements of P (called proofs from the assumptions A).

For example, the logic called the Propositional Calculus on the set X, and henceforth denoted by $\mathrm{Prop}(X)$, consists of the set $P = P(X)$ (the free proposition algebra on X), the set $\mathscr{V}$ of all homomorphisms of $P(X)$ into $\mathbb{Z}_2$, and, for each subset A of $P(X)$, the set of proofs as defined in §4 of Chapter II.

The concepts of semantic implication and syntactic implication in $\mathscr{L}$ are defined in terms of valuation and proof respectively, in some manner analogous to that used for the propositional calculus, and the notations $A \models p$, $A \vdash p$ will again be used to denote respectively "p is a consequence of A", "p is a deduction from A". p is a tautology of $\mathscr{L}$ if $\varnothing \models p$ and it is a theorem of $\mathscr{L}$ if $\varnothing \vdash p$. The logic $\mathscr{L}$ for which these assertions are made will always be clear from the context.

Definition 1.2. A logic $\mathscr{L}$ is *sound* if $A \vdash p$ implies $A \models p$.

Definition 1.3. A logic $\mathscr{L}$ is *consistent* if F is not a theorem.

Definition 1.4. A logic $\mathscr{L}$ is *adequate* if $A \models p$ implies $A \vdash p$.

Choosing $A = \varnothing$, we see that a sound logic has the desirable property that theorems are always true, and an adequate logic has the equally desirable property that valid propositions can be proved. While soundness and adequacy each express a connection between truth and proof, consistency is an expression of a purely syntactic property that any logic might be expected to have, namely that one cannot deduce contradictions.

Since the theorems and tautologies of a logic are each of significance, the following decidability properties are also important.

18

Definition 1.5. A logic $\mathscr{L}$ is *decidable for validity* if there exists an algorithm which determines for every proposition p, in a finite number of steps, whether or not p is valid.

Definition 1.6. A logic $\mathscr{L}$ is *decidable for provability* if there exists an algorithm which determines for every proposition p, in a finite number of steps, whether or not p is a theorem.

§2 Soundness and Adequacy of Prop(X)

Theorem 2.1. (The Soundness Theorem) *Let* $A \subseteq P(X)$, $p \in P(X)$. *If* $A \vdash p$, *then* $A \vDash p$.

Proof: Suppose there exists a proof $p_1, \ldots, p_n$ of p from A. We have to show p is a consequence of A.

Let $v : P(X) \to \mathbb{Z}_2$ be a valuation for which $v(A) \subseteq \{1\}$. We shall use induction over the length n of the proof of p from A to show that $v(p) = 1$. Suppose that $n = 1$. Then $p \in A \cup \mathscr{A}$, and since every axiom is a tautology (Exercise 3.8 of Chapter II), we have $v(p) = 1$.

Suppose now $n > 1$, and that $v(q) = 1$ for every q provable from A by a proof of length $< n$. Then $v(p_1) = v(p_2) = \cdots = v(p_{n-1}) = 1$. Either $p_n \in A \cup \mathscr{A}$ and $v(p_n) = 1$, as required, or for some $i, j < n$, we have $p_i = p_j \Rightarrow p_n$. In the latter case, $v(p_j) = v(p_j \Rightarrow p_n) = 1$, and the homomorphism property of v requires $v(p_n) = 1$. $\square$

Corollary 2.2. (The Consistency Theorem) F *is not a theorem of* Prop(X).

Proof: If $\vdash F$, then $\vDash F$ by the Soundness Theorem. Since axioms are tautologies, $v(F) = 1$ for every valuation v, contradicting the definition of valuation. This implies that there are no valuations. But $P(X)$ is free and every map of X into $\mathbb{Z}_2$ can be extended to a valuation. $\square$

Exercise 2.3. Show that Con(A) is closed with respect to modus ponens (i.e., if $p, p \Rightarrow q \in$ Con(A), then $q \in$ Con(A)). Use Exercise 4.9 of Chapter II to prove that Con(A) $\supseteq$ Ded(A). This is another way of stating the Soundness Theorem.

The proof of adequacy for Prop(X) is more difficult, and we first prove a preparatory result of independent interest.

Theorem 2.4. (The Deduction Theorem) *Let* $A \subseteq P(X)$, *and let* p, $q \in P(X)$. *Then* $A \vdash p \Rightarrow q$ *if and only if* $A \cup \{p\} \vdash q$.

Proof: (a) Suppose $A \vdash p \Rightarrow q$. Let $p_1, \ldots, p_n$ be a proof of $p_n = p \Rightarrow q$ from A. Then $p_1, \ldots, p_n, p, q$ is a proof of q from $A \cup \{p\}$.

(b) Suppose $A \cup \{p\} \vdash q$. Then we have a proof $p_1, \ldots, p_n$ of q from $A \cup \{p\}$. We shall use induction over the length n of the proof.

If $n = 1$, then $q \in \mathscr{A} \cup A \cup \{p\}$. If $q \in \mathscr{A} \cup A$, then q, $q \Rightarrow (p \Rightarrow q)$, $p \Rightarrow q$ is a proof of $p \Rightarrow q$ from A. If $q = p$, then $\vdash p \Rightarrow p$ (Example 3.4 of Chapter II), and so $A \vdash p \Rightarrow q$.

Suppose now $n > 1$. By induction, $A \vdash p \Rightarrow p_i$ for $i = 1, 2, \ldots, n - 1$, and we may suppose $q \notin \mathscr{A} \cup A \cup \{p\}$. For some $i, j < n$, we have $p_i = p_j \Rightarrow q$. Thus $A \vdash p \Rightarrow p_j$, $A \vdash p \Rightarrow (p_j \Rightarrow q)$, and there exists a proof $q_1, \ldots, q_k$, q_{k+1} from A with

$$q_k = p \Rightarrow p_j,$$
$$q_{k+1} = p \Rightarrow (p_j \Rightarrow q).$$

We put

$$q_{k+2} = (p \Rightarrow (p_j \Rightarrow q)) \Rightarrow ((p \Rightarrow p_j) \Rightarrow (p \Rightarrow q)), \qquad (\mathscr{A}_2)$$
$$q_{k+3} = (p \Rightarrow p_j) \Rightarrow (p \Rightarrow q), \qquad (q_{k+2} = q_{k+1} \Rightarrow q_{k+3})$$
$$q_{k+4} = p \Rightarrow q. \qquad (q_{k+3} = q_k \Rightarrow q_{k+4})$$

Then $q_1, \ldots, q_{k+4}$ is a proof of $p \Rightarrow q$ from A. □

The Deduction Theorem is useful in establishing a result of the form $A \vdash p \Rightarrow q$, because it is usually much easier to show $A \cup \{p\} \vdash q$. Even if a proof in Prop(X) of $p \Rightarrow q$ from A is required, the method used in proving the Deduction Theorem can be applied to convert a proof of q from $A \cup \{p\}$ into a proof of $p \Rightarrow q$ from A.

Example 2.5.   We show $\{p \Rightarrow q, q \Rightarrow r\} \vdash p \Rightarrow r$. First we show $\{p, p \Rightarrow q, q \Rightarrow r\} \vdash r$, and a proof of this is p, $p \Rightarrow q$, q, $q \Rightarrow r$, r. It follows from the Deduction Theorem that $\{p \Rightarrow q, q \Rightarrow r\} \vdash p \Rightarrow r$.

We now convert the proof of r from $\{p, p \Rightarrow q, q \Rightarrow r\}$ into a proof of $p \Rightarrow r$ from $\{p \Rightarrow q, q \Rightarrow r\}$. We shall write the steps of the original proof in a column on the left. Alongside each, we then write a comment on the nature of the step, and then the corresponding steps of the new proof.

p	Proposition to be deleted from the assumptions	$p \Rightarrow ((p \Rightarrow p) \Rightarrow p)$, $(p \Rightarrow ((p \Rightarrow p) \Rightarrow p)) \Rightarrow ((p \Rightarrow (p \Rightarrow p)) \Rightarrow (p \Rightarrow p))$, $(p \Rightarrow (p \Rightarrow p)) \Rightarrow (p \Rightarrow p)$, $p \Rightarrow (p \Rightarrow p)$, $p \Rightarrow p$.
$p \Rightarrow q$	Retained assumption	$p \Rightarrow q$, $(p \Rightarrow q) \Rightarrow (p \Rightarrow (p \Rightarrow q))$, $p \Rightarrow (p \Rightarrow q)$.
q	Modus ponens	$(p \Rightarrow (p \Rightarrow q)) \Rightarrow ((p \Rightarrow p) \Rightarrow (p \Rightarrow q))$, $(p \Rightarrow p) \Rightarrow (p \Rightarrow q)$, $p \Rightarrow q$.
$q \Rightarrow r$	Retained assumption	$q \Rightarrow r$, $(q \Rightarrow r) \Rightarrow (p \Rightarrow (q \Rightarrow r))$, $p \Rightarrow (q \Rightarrow r)$.
r	Modus ponens	$(p \Rightarrow (q \Rightarrow r)) \Rightarrow ((p \Rightarrow q) \Rightarrow (p \Rightarrow r))$, $(p \Rightarrow q) \Rightarrow (p \Rightarrow r)$, $p \Rightarrow r$.

Of course, the proof we have constructed can be abbreviated, because the first 11 steps serve only to prove the retained assumption $p \Rightarrow q$.

Exercises

2.6. Show that $p \Rightarrow r \in \text{Ded}\{p \Rightarrow q, p \Rightarrow (q \Rightarrow r)\}$. Hence show that if $p \Rightarrow q, p \Rightarrow (q \Rightarrow r) \in \text{Ded}(A)$, then $p \Rightarrow r \in \text{Ded}(A)$, and so prove the Deduction Theorem without giving an explicit construction for a proof in $\text{Prop}(X)$.

2.7. Show that $\vdash p \Rightarrow \sim \sim p$ and construct a proof of $p \Rightarrow \sim \sim p$ in $\text{Prop}(X)$. (Hint: show $\{p, \sim p\} \vdash F$ and use the Deduction Theorem twice.)

2.8. Show that the following are theorems of $\text{Prop}(X)$,

(a) $p \Rightarrow p \vee q$,   (b) $q \Rightarrow p \vee q$,

(c) $(p \vee q) \Rightarrow (q \vee p)$,   (d) $p \wedge q \Rightarrow p$,

(e) $p \wedge q \Rightarrow q$,   (f) $(p \wedge q) \Rightarrow (q \wedge p)$.

Definition 2.9. Let $A \subseteq P(X)$. We say that A is *consistent* if $F \notin \text{Ded}(A)$. A is called a *maximal consistent* subset if A is consistent and if every subset $T \subseteq P(X)$ which properly contains A is inconsistent.

Lemma 2.10. *The subset $A \subseteq P(X)$ is maximal consistent if and only if*

(i) $F \notin A$, *and*

(ii) $A = \text{Ded}(A)$, *and*

(iii) *for all $p \in P(X)$, either $p \in A$ or $\sim p \in A$.*

Proof: (a) Let A be maximal consistent. Since A is consistent, $F \notin \text{Ded}(A)$ and therefore $F \notin A$. Since $\text{Ded}(\text{Ded}(A)) = \text{Ded}(A)$, $\text{Ded}(A)$ is consistent. As $A \subseteq \text{Ded}(A)$, $A = \text{Ded}(A)$ by the maximal consistency of A. Finally, suppose $p \notin A$. Then $F \in \text{Ded}(A \cup \{p\})$, i.e. $A \cup \{p\} \vdash F$. By the Deduction Theorem, $A \vdash p \Rightarrow F$, i.e., $\sim p \in \text{Ded}(A)$.

(b) Suppose A has the properties (i), (ii), (iii). Then $F \notin \text{Ded}(A)$. If T properly contains A, then there exists $p \in T$ such that $p \notin A$. By (iii), $\sim p \in A$, hence $p, \sim p \in T$, and $p, \sim p, F$ is a proof of F from T. Thus A is maximal consistent. $\square$

Lemma 2.11. *Let A be a consistent subset of $P(X)$. Then A is contained in a maximal consistent subset.*

Proof: Let $\Sigma = \{T \subseteq P(X) \mid T \supseteq A, F \notin \text{Ded}(T)\}$. Since $A \in \Sigma$, $\Sigma \neq \varnothing$. Suppose $\{T_\alpha\}$ is a totally ordered family of members of Σ, and put $T = \bigcup_\alpha T_\alpha$. Clearly $T \subseteq P(X)$, $T \supseteq A$. If F is provable from T, F is provable from a finite subset of T, and this subset is contained in some T_α, contrary to $T_\alpha \in \Sigma$. Hence $F \notin \text{Ded}(T)$, and Σ is an inductively ordered set. By Zorn's Lemma, Σ has a maximal member say M. This M is the required maximal consistent subset. $\square$

The next result is the key to the Adequacy Theorem.

Theorem 2.12. (The Satisfiability Theorem) *Let A be a consistent subset of $P(X)$. Then there exists a valuation $v: P(X) \to \mathbb{Z}_2$, such that $v(A) \subseteq \{\mathfrak{t}\}$.*

Proof: Let M be a maximal consistent subset containing A. For $p \in P(X)$, put $v(p) = 1$ if $p \in M$ and $v(p) = 0$ if $p \notin M$. We now prove v is a valuation.

Certainly $v(F) = 0$, because $F \notin M$. It remains to show $v(p \Rightarrow q) = v(p) \Rightarrow v(q)$. If $q \in M$, then $p \Rightarrow q \in M$ because $\{q\} \vdash p \Rightarrow q$, and $v(p \Rightarrow q) = 1 = v(p) \Rightarrow v(q)$. If $p \notin M$, then $p \Rightarrow q \in M$ because $\{\sim p\} \vdash p \Rightarrow q$, and $v(p \Rightarrow q) = 1 = v(p) \Rightarrow v(q)$. If $p \in M$ and $q \notin M$, then $p \Rightarrow q \notin M$, and $v(p \Rightarrow q) = 0 = v(p) \Rightarrow v(q)$. $\square$

Theorem 2.13. (The Adequacy Theorem) *Let $A \subseteq P(X)$, $p \in P(X)$. If $A \vDash p$ in* Prop(X), *then $A \vdash p$ in* Prop(X).

Proof: Suppose $A \vDash p$, so that $v(A) \subseteq \{1\}$ implies $v(p) = 1$ for every valuation v. If $A \cup \{\sim p\}$ is consistent, it follows from the Satisfiability Theorem that there is a valuation v such that $v(A \cup \{\sim p\}) \subseteq \{1\}$, which is not possible. Hence $F \in \text{Ded}(A \cup \{\sim p\})$, i.e., $A \cup \{\sim p\} \vdash F$. By the Deduction Theorem, $A \vdash \sim p \Rightarrow F$. Since $\vdash \sim \sim p \Rightarrow p$, we have $A \vdash p$. $\square$

Exercise 2.14. Show that if $A \vDash p$, then $A_0 \vDash p$ for some finite subset A_0 of A. (This result is known as the Compactness Theorem.)

§3 Truth Functions and Decidability for Prop(X)

Each valuation v of $P(X)$ determines a natural equivalence relation r_v on $P(X)$ given by $p r_v q$ if $v(p) = v(q)$, and which is in fact a congruence relation on $P(X)$. That is, each r_v satisfies the condition that if $p r_v p_1$ and $q r_v q_1$, then $(p \Rightarrow q) r_v (p_1 \Rightarrow q_1)$. The intersection of the relations r_v for all valuations v of $P(X)$ is therefore a congruence relation on $P(X)$, which we call semantic equivalence and denote by $\vDash\!\dashv$. Since $p \vDash\!\dashv q$ if and only if $v(p) = v(q)$ for every valuation v of $P(X)$, we see that $p \vDash\!\dashv q$ if and only if $\{p\} \vDash q$ and $\{q\} \vDash p$.

Definition 3.1. The set of congruence classes of $P(X)$ with respect to $\vDash\!\dashv$ is an $\{F, \Rightarrow\}$—algebra called the *Lindenbaum algebra* on X and denoted by $L(X)$.

Let $X_n = \{x_1, \ldots, x_n\}$. Clearly $L(X_n)$ is a homomorphic image of $P(X_n)$. If $w = w(x_1, \ldots, x_n) \in P(X_n)$ is any word in $x_1, \ldots, x_n$, then its image in $L(X_n)$ is the congruence class $\bar{w} = \bar{w}(x_1, \ldots, x_n)$ say, of all words congruent to w under the relation $\vDash\!\dashv$. Our aim is to show that $\bar{w}$ can be regarded as a function $\bar{w}: \mathbb{Z}_2^n \to \mathbb{Z}_2$.

For any $\bar{w}(x_1, \ldots, x_n) \in L(X_n)$, choose a representative $w(x_1, \ldots, x_n) \in P(X_n)$. If $(z_1, \ldots, z_n) \in \mathbb{Z}_2^n$, then there is a unique valuation $v: P(X_n) \to \mathbb{Z}_2$ such that $v(x_i) = z_i$ for $i = 1, \ldots, n$. We define $\bar{w}(z_1, \ldots, z_n) = v(w(x_1, \ldots, x_n))$, observing that this definition is independent of the choice of representative w of $\bar{w}$, because if w_1 is another representative, then $w \vDash\!\dashv w_1$ and $v(w) = v(w_1)$. In this way we associate with each element $\bar{w}$ of $L(X_n)$ a function $\mathbb{Z}_2^n \to \mathbb{Z}_2$, but, before we identify $\bar{w}$ with this function, we must show that if $\bar{w}$ and $\bar{w}_1$ have the same associated function, then $\bar{w} = \bar{w}_1$.

Suppose that $\bar{w}$ and $\bar{w}_1$ have the same associated function, so that $\bar{w}(z_1, \ldots, z_n) = \bar{w}_1(z_1, \ldots, z_n)$ for all $(z_1, \ldots, z_n) \in \mathbb{Z}_2^n$. Let w, w_1 be representatives of $\bar{w}, \bar{w}_1$ respectively. Then $\bar{w}(z_1, \ldots, z_n) = v(w)$, where v is the valuation for which $v(x_i) = z_i$ $(i = 1, \ldots, n)$, and we have $v(w) = v(w_1)$. The last equation holds for every valuation v, hence $w \models\mid w_1$ and $\bar{w} = \bar{w}_1$. We may therefore identify the elements of $L(X_n)$ with their associated functions.

Definition 3.2. A function $f : \mathbb{Z}_2^n \to \mathbb{Z}_2$ is called a truth function.

Theorem 3.3. $L(X_n)$ is the set of all truth functions $f : \mathbb{Z}_2^n \to \mathbb{Z}_2$.

Proof: The constant functions $0, 1 \in L(X_n)$ since $0 = \bar{F}$ and $1 = (\overline{F \Rightarrow F})$. Thus the result holds for $n = 0$.

If f, g are truth functions $\mathbb{Z}_2^n \to \mathbb{Z}_2$, we define the truth function $f \Rightarrow g$ by $(f \Rightarrow g)(z_1, \ldots, z_n) = f(z_1, \ldots, z_n) \Rightarrow g(z_1, \ldots, z_n)$. For convenience of notation, we denote the ith coordinate function by u_i. We have $u_i = \bar{x}_i \in L(X_n)$.

We now suppose $n > 0$, and shall use induction over n to complete the proof. Let $f = f(u_1, \ldots, u_n)$ be a truth function of n variables. Put

$$g(u_1, \ldots, u_{n-1}) = f(u_1, \ldots, u_{n-1}, 0), \quad h(u_1, \ldots, u_{n-1}) = f(u_1, \ldots, u_{n-1}, 1).$$

Then $g, h \in L(X_{n-1}) \subseteq L(X_n)$. The function $k : \mathbb{Z}_2^n \to \mathbb{Z}_2$, defined by

$$k(u_1, \ldots, u_n) = (\sim u_n \Rightarrow g(u_1, \ldots, u_{n-1})) \wedge (u_n \Rightarrow h(u_1, \ldots, u_{n-1}))$$

is in $L(X_n)$, and

$$
\begin{aligned}
k(u_1, \ldots, u_{n-1}, 0) &= (1 \Rightarrow g(u_1, \ldots, u_{n-1})) \wedge (0 \Rightarrow h(u_1, \ldots, u_{n-1})) \\
&= g(u_1, \ldots, u_{n-1}) \wedge 1 \\
&= g(u_1, \ldots, u_{n-1}) \\
&= f(u_1, \ldots, u_{n-1}, 0).
\end{aligned}
$$

Similarly, one obtains $k(u_1, \ldots, u_{n-1}, 1) = f(u_1, \ldots, u_{n-1}, 1)$. Thus $k = f$ and $f \in L(X_n)$. $\square$

We now apply truth functions to settle the question of decidability for Prop(X).

Lemma 3.4. Let $w = w(x_1, \ldots, x_n) \in P(X)$. Then $\models w$ if and only if its associated truth function $\bar{w} : \mathbb{Z}_2^n \to \mathbb{Z}_2$ is the constant 1.

Proof: Suppose $\bar{w} = 1$. Let $v : P(X) \to \mathbb{Z}_2$ be any valuation of $P(X)$. Put $a_i = v(x_i)$. Then the restriction of v to $P(X_n)$ is a valuation of $P(X_n)$, and $v(w) = \bar{w}(a_1, \ldots, a_n) = 1$. Thus $v(w) = 1$ for every valuation v of $P(X)$, i.e., $\models w$.

Suppose conversely that w is valid. Let $(a_1, \ldots, a_n) \in \mathbb{Z}_2^n$. There exists a valuation v of $P(X)$ with $v(x_i) = a_i$. (We may assign arbitrarily values for elements of $X - \{x_1, \ldots, x_n\}$.) Then the restriction of v to $P(X_n)$ is a valuation of $P(X_n)$, and $\bar{w}(a_1, \ldots, a_n) = v(w) = 1$. Thus $\bar{w} = 1$. $\square$

Theorem 3.5. Prop(X) *is decidable for validity.*

Proof: We give an algorithm for deciding if $w \in P(X)$ is valid. The element w is a word $w(x_1, \ldots, x_n)$ in some finite set $x_1, \ldots, x_n$ of variables. Let $\bar{w} = w(u_1, \ldots, u_n)$ be the associated truth function. For each $(a_1, \ldots, a_n) \in \mathbb{Z}_2^n$, we calculate $\bar{w}(a_1, \ldots, a_n)$. By Lemma 3.4, w is valid if and only if all these values are 1. □

Corollary 3.6. Prop(X) *is decidable for provability.*

Proof. An element $p \in P(X)$ is a theorem if and only if it is valid. □

Exercises

3.7. Show that every truth function $\mathbb{Z}_2^n \to \mathbb{Z}_2$ can be expressed in terms of the coordinate functions and the one operation $|$ defined by $\bar{w}_1 | \bar{w}_2 = \sim(\bar{w}_1 \wedge \bar{w}_2)$.

3.8. A truth function $f(u_1, \ldots, u_n)$ is said to be in disjunctive normal form if it is expressed in one of the forms $f = 0, f = 1$, or $f = v_1 \vee v_2 \vee \cdots \vee v_k$ for $0 < k < 2^n$, where each $v_j = u_{1j} \wedge u_{2j} \wedge \cdots \wedge u_{n_j j}$, and $u_{ij} = u_r$ or $\sim u_r$ for some r.

Show that every truth function is expressible in disjunctive normal form, and specify a procedure for associating with each truth function $\mathbb{Z}_2^n \to \mathbb{Z}_2$ a unique disjunctive normal form.

3.9. (a) Let $p \in P(X)$. Find a $p' \in P(X)$, expressible in a form involving no operations other than $\sim$, $\wedge$ and $\vee$, such that $\vDash p \Leftrightarrow p'$.

(b) Let $p, q \in P(X)$. Find truth functions for $\sim(p \vee q) \Leftrightarrow (\sim p \wedge \sim q)$ and $\sim(p \wedge q) \Leftrightarrow (\sim p \vee \sim q)$.

(c) p and p' are related as in (a). Let p^* be the statement obtained from p' by replacing each $\vee$ by $\wedge$, each $\wedge$ by $\vee$, and each $x \in X$ by $\sim x$. Prove that $\vDash \sim p \Leftrightarrow p^*$.

3.10. A truth function $f(u_1, \ldots, u_n)$ is said to be in conjunctive normal form if it expressed in one of the forms $f = 0, f = 1$, or $f = v_1 \wedge v_2 \wedge \cdots \wedge v_k$ for $0 < k < 2^n$, where each $v_j = u_{1j} \vee u_{2j} \vee \cdots \vee u_{n_j j}$, and $u_{ij} = u_r$ or $\sim u_r$ for some r. Use Exercises 3.8 and 3.9 to specify a procedure for associating with each truth function $\mathbb{Z}_2^n \to \mathbb{Z}_2$ a unique conjunctive normal form.

3.11. Let p, p' and q, q' be related as in Exercise 3.9(a). Let p^d, q^d be the statements obtained from p', q' by replacing each $\vee$ by $\wedge$ and each $\wedge$ by $\vee$. Show that $\vDash p$ if and only if $\vDash \sim p^d$. Deduce that if $\vdash p \Rightarrow q$, then $\vdash q^d \Rightarrow p^d$. (This result expresses a duality principle for Prop(X).)

3.12. Write a FORTRAN program to decide if $w(x_1, x_2, x_3) \in P(X_3)$ is valid.

3.13. Show that Prop(X) is decidable for $\{p_1, \ldots, p_n\} \vDash q$, where $p_1, \ldots, p_n, q \in P(X)$.

3.14. Construct a propositional calculus Prop$_1$(X) with $P_1(X)$ the free $\{\Rightarrow, \sim\}$-algebra. Show that there is a $\{\Rightarrow, \sim\}$-homomorphism φ: $P_1(X) \to P(X)$ which is the identity on X. Is φ a monomorphism? Is φ an epimorphism? Does there exist a $\{\Rightarrow, \sim\}$-homomorphism $\psi : P(X) \to P_1(X)$ which is the identity on X? (Hint: Consider the images of F and of $F \Rightarrow F$ $(= \sim F)$.)

Show that there exists a $\{\Rightarrow, F\}$-homomorphism $\theta : P(X) \to P_1(X)$ which is the identity on X, taking as element F of $P_1(X)$ the element $\sim (x_1 \Rightarrow x_1)$. Show that $w \in P_1(X)$ is valid if and only if $\varphi(w)$ is valid. Show that $p \in P(X)$ is valid if and only if $\theta(p)$ is valid. Establish the Consistency, Adequacy and Decidability theorems for $\text{Prop}_1(X)$.

3.15.   Using the method of 3.14 investigate the following propositional calculi:

(a) $\text{Prop}_2(X)$ with $P_2(X)$ free of type $\{\sim, \vee\}$,

(b) $\text{Prop}'_2(X)$, with $P'_2(X)$ relatively free of type $\{\sim, \vee\}$, with the identical relation $p \vee q = q \vee p$,

(c) $\text{Prop}_3(X)$ with $P_3(X)$ free of type $\{|\}$ (see 3.7),

(d) $\text{Prop}'_3(X)$ with $P'_3(X)$ relatively free of type $\{|\}$, with the identical relation $p|q = q|p$.

Chapter IV

Predicate Calculus

§1   **Algebras of Predicates**

The initial step in our development of the Propositional Calculus was the construction of proposition algebras, which formalise the way in which a given collection of "primitive" statements is enlarged by combining statements. The Propositional Calculus does not analyse the original primitive statements. Our aim now is to construct a more complicated model of mathematical reasoning, which incorporates more of the ordinary features of this reasoning.

Mathematics is usually about something, that is, there is usually some set $\mathscr{U}$ of objects under discussion and investigation. A typical statement in such a discussion would be "u has the property p", where $u \in \mathscr{U}$ and p is some property relevant to elements of $\mathscr{U}$. A convenient notation for this statement is $p(u)$. Such a statement depends on the element u, and may be thought of as a function of u. The phrase "has the property p" is known as a predicate, and p (as used in the notation $p(u)$) is known as a predicate symbol. More generally, if r is an n-ary relation on $\mathscr{U}$, the statement "$(u_1, \ldots, u_n)$ is in the relation r" is denoted by $r(u_1, \ldots, u_n)$, and r is called an n-ary predicate. A 0-ary predicate is a statement which does not depend on any elements of $\mathscr{U}$, and so corresponds to an unanalysed statement.

If p, q are properties, then $p(u) \wedge q(u)$ is true for just those elements u with both properties. Denoting by P the subset of $\mathscr{U}$ consisting of those elements with property p, and by Q the subset of $\mathscr{U}$ of elements with property q, we see that $P \cap Q$ is the subset of those elements u for which $p(u) \wedge q(u)$ is true. Similarly, $P \cup Q$ is the subset of elements u for which $p(u) \vee q(u)$ is true, while the set of elements u satisfying $\sim p(u)$ is the complement of P in $\mathscr{U}$.

Another common form of statement in mathematical discussion is "For all $u \in \mathscr{U}$, $p(u)$". If $\mathscr{U}$ were a finite set, say $\mathscr{U} = \{u_1, \ldots, u_n\}$, then this could be expressed as $p(u_1) \wedge p(u_2) \wedge \cdots \wedge p(u_n)$, but it is not possible to do this if $\mathscr{U}$ is an infinite set. We thus introduce the notation $(\forall u)p(u)$ for the above statement. $(\forall u)$ is called the universal quantifier. Note that the u in $(\forall u)$ is only a dummy—$(\forall u)p(u)$ is in no way dependent on u, and is the same statement about $\mathscr{U}$ as $(\forall v)p(v)$. We do not need additional notations to deal with a limited use of "for all" as in statements such as "For all u such that $p(u)$, we have $q(u)$". This can be expressed as $(\forall u)(p(u) \Rightarrow q(u))$.

Statements of the form "There exists $u \in \mathscr{U}$ with the property p" are also common in mathematics. We write this statement as $(\exists u)p(u)$. The existential

26

quantifier $(\exists u)$ is, however, related to the universal quantifier $(\forall u)$, as follows. When we say "There does not exist u with property p", we are in fact asserting $(\forall u)(\sim p(u))$. Thus $(\exists u)p(u)$ has the same meaning[1] as $\sim((\forall u)(\sim p(u)))$, and we have no need to include the existential quantifier in the construction of our model. We shall define $(\exists u)$ to mean $\sim(\forall u)\sim$.

We now set up an appropriate analogue of a proposition algebra. Proposition algebras are built upon underlying sets of propositional variables. We begin here with an infinite set V whose elements will be called individual variables, and with a set $\mathcal{R}$ (whose elements will be called relation or predicate symbols) together with an arity function $\mathrm{ar}\colon \mathcal{R} \to \mathbb{N}$. The individual variables may be thought of as names to be given to mathematical objects, and the relation symbols as names to be given to relations between these objects. The set of generators we shall use to construct our set P of propositions must clearly contain each element $r(x_1, \ldots, x_n)$ for each $r \in \mathcal{R}$ and $(x_1, \ldots, x_n) \in V^n$, where $n = \mathrm{ar}(r)$. It is also clear that P must be an $\{F, \Rightarrow\}$—algebra, and that for each $x \in V$, we shall need a function $(\forall x)\colon P \to P$.

Let $\tilde{P}(V, \mathcal{R})$ be the free algebra on the set $\{(r, x_1, \ldots, x_n) | r \in \mathcal{R}, x_i \in V, n = \mathrm{ar}(r)\}$ of free generators, of type $\{F, \Rightarrow, (\forall x) | x \in V\}$, where F is a 0-ary operation, $\Rightarrow$ binary, and each $(\forall x)$ unary. We call $\tilde{P} = \tilde{P}(V, \mathcal{R})$ the full first order algebra on $(V, \mathcal{R})$. We use the more usual notation $r(x_1, \ldots, x_n)$ for the generator $(r, x_1, \ldots, x_n)$, and we put $\mathcal{R}_n = \{r \in \mathcal{R} | \mathrm{ar}(r) = n\}$.

We could use this algebra $\tilde{P}$ as our algebra of propositions, but it is more convenient to use a certain factor algebra. If $w \in \tilde{P}$, then w is a word in the free generators of $\tilde{P}$, each of which has the form $r(x_1, \ldots, x_n)$. If $x_1, \ldots, x_m$ are the distinct individual variables occurring in w, then we can think of w as a function $w(x_1, \ldots, x_m)$ of these variables. Now we regard $(\forall x_1)w(x_1, \ldots, x_m)$ as being essentially the same as $(\forall y)w(y, x_2, \ldots, x_m)$, provided only that $y \notin \{x_2, \ldots, x_m\}$. The reason for this has been pointed out before, and is that the x_1 in $(\forall x_1)w(x_1, \ldots, x_m)$ is a dummy, used as an aid in describing the construction of the statement. It serves the same purpose as the variable t does in the definition of the gamma function as $\Gamma(x) = \int_0^\infty e^{-t}t^{x-1}\,dt$.

We shall construct a factor algebra of $\tilde{P}$, in which these elements, considered above as being essentially the same, will be identified. Further identifications are possible. The question of which identifications are made is purely one of convenience. The congruence relation on $\tilde{P}$ which we use needs some care in its construction, and we begin by defining two functions on $\tilde{P}$.

Definition 1.1. Let $w \in \tilde{P}$. The set of *variables involved in* w, denoted by $V(w)$, is defined by

$$V(w) = \cap \{U | U \subseteq V, w \in \tilde{P}(U, \mathcal{R})\}.$$

---

[1] This is very different to the concepts of existence used in other contexts such as "Do flying saucers exist?" or "Does God exist?" or "Do electrons exist?".

Exercise 1.2.   Show that
(i) $V(F) = \varnothing$.
(ii) If $r \in \mathcal{R}$, $\mathrm{ar}(r) = n$, and $x_1, \ldots, x_n \in V$, then $V(r(x_1, \ldots, x_n)) = \{x_1, \ldots, x_n\}$.
(iii) If $w_1, w_2 \in \tilde{P}$, then $V(w_1 \Rightarrow w_2) = V(w_1) \cup V(w_2)$.
(iv) If $x \in V$ and $w \in \tilde{P}$, then $V((\forall x)w) = \{x\} \cup V(w)$.
Show further that (i)–(iv) may be taken as the definition of the function $V(w)$.

Definition 1.3.   Let $w \in \tilde{P}$. The *depth of quantification* of w, denoted by $d(w)$, is defined by
(i) $d(F) = 0$, $d(r(x_1, \ldots, x_n)) = 0$ for every free generator of $\tilde{P}$.
(ii) $d(w_1 \Rightarrow w_2) = \max(d(w_1), d(w_2))$.
(iii) $d((\forall x)w) = 1 + d(w) \ (x \in V)$.

Our desired congruence relation on $\tilde{P}$ may now be defined.

Definition 1.4.   Let $w_1, w_2 \in \tilde{P}$. We define $w_1 \approx w_2$ if
(a) $d(w_1) = d(w_2) = 0$ and $w_1 = w_2$, or
(b) $d(w_1) = d(w_2) > 0$, $w_1 = a_1 \Rightarrow b_1$, $w_2 = a_2 \Rightarrow b_2$, $a_1 \approx a_2$ and $b_1 \approx b_2$, or
(c) $w_1 = (\forall x)a$, $w_2 = (\forall y)b$ and either
(i) $x = y$ and $a \approx b$, or
(ii) there exists $c = c(x)$ such that $c(x) \approx a$, $c(y) \approx b$ and $y \notin V(c)$.

We remark that in part (c) (ii), the notation $c = c(x)$ indicates the way the element concerned is a function of x, and ignores its possible dependence on other variables. We use it so we can represent the effect of substituting y for x throughout. It is therefore unnecessary for us to impose the condition $x \notin V(c(y))$. The notation does not imply $V(c(x)) = \{x\}$, hence we must impose the condition $y \notin V(c(x))$. Thus the condition (c) (ii) is symmetric, and $\approx$ is trivially reflexive. The proof that it is transitive is left as an exercise.

Exercise 1.5.
(i) Given that $z \notin V(w_1) \cup V(w_2)$, show by induction over $d(w_1)$ that the element $c = c(x)$ in (c) (ii) can always be chosen such that $z \notin V(c)$.
(ii) If $u(x) \approx v(x)$ and $y \notin V(u(x)) \cup V(v(x))$, show by induction over $d(u(x))$ that $u(y) \approx v(y)$.
(iii) Prove that $\approx$ is transitive.

Since the relation $\approx$ is an equivalence which is clearly compatible with the operations of the algebra, it is a congruence relation on $\tilde{P}(V, \mathcal{R})$.

Definition 1.6.   The *(reduced) first-order algebra* $P(V, \mathcal{R})$ on $(V, \mathcal{R})$ is the factor algebra of $\tilde{P}(V, \mathcal{R})$ by the congruence relation $\approx$.

The elements of $P = P(V, \mathcal{R})$ are the congruence classes. If $w \in \tilde{P}$ and

$[w]$ is the congruence class of w, then

$$(\forall x)[w] = [(\forall x)w],$$

and

$$[w_1] \Rightarrow [w_2] = [w_1 \Rightarrow w_2],$$

Definition 1.7. Let $w \in P$. We define the set var(w) of (*free*) *variables of* w by putting var$(w) = $ var$(\tilde{w})$, where $\tilde{w} \in \tilde{P}$ is some representative of the congruence class w, and where var$(\tilde{w})$ is defined inductively by

(i) var$(F) = \varnothing$,
(ii) var$(r(x_1, \ldots, x_n)) = \{x_1, \ldots, x_n\}$ for $r \in \mathscr{R}$, $x_1, \ldots, x_n \in V$,
(iii) var$(\tilde{w}_1 \Rightarrow \tilde{w}_2) = $ var$(\tilde{w}_1) \cup $ var$(\tilde{w}_2)$,
(iv) var$((\forall x)\tilde{w}) = $ var$(\tilde{w}) - \{x\}$.

Definition 1.8. Let $A \subseteq P$. Put

$$\text{var}(A) = \bigcup_{p \in A} \text{var}(p).$$

Exercises

1.9. Show that if $\tilde{w}_1 \approx \tilde{w}_2$, then var$(\tilde{w}_1) = $ var$(\tilde{w}_2)$, and conclude that var(w) is defined for $w \in P$.

1.10. Show that for any $w \in P$, there is a representative $\tilde{w}$ of w such that no variable $x \in V$ appears in $\tilde{w}$ more than once in a quantifier $(\forall x)$, and no $x \in $ var(w) appears at all in a quantifier (i.e., $\tilde{w}$ has no repeated dummy variables, and no free variables also appear as dummies).

We assume henceforth that any $w \in P$ is represented by a $\tilde{w} \in \tilde{P}$ having the form described in Exercise 1.10. We shall also usually abuse notation and not distinguish between $p \in \tilde{P}$ and $[p] \in P$.

§2 Interpretations

We want to think of the elements of V as names of objects, and the elements of $\mathscr{R}$ as relations among those objects. If we take a non-empty set U, and a function $\varphi: V \to U$, then we can think of $x \in V$ as a name for the element $\varphi(x) \in U$. Of course, not every element $u \in U$ need have a name, while some elements u may well have more than one name. Next we take a function ψ, from $\mathscr{R}$ into the set of all relations on U, such that if $r \in \mathscr{R}_n$, then $\psi(r)$ is an n-ary relation. It will be convenient to write simply φx for $\varphi(x)$, and ψr for $\psi(r)$. As for valuations, these again should be functions $v: P \to \mathbb{Z}_2$ which will correspond to our intuitive notion of truth. Since our interpretation of the element $r(x_1, \ldots, x_n) \in P$ in terms of U, φ, ψ must obviously be the statement that $(\varphi x_1, \ldots, \varphi x_n) \in \psi r$, we shall require of v that

(a) if $r \in \mathcal{R}_n$ and $x_1, \ldots, x_n \in V$, then $v(r(x_1, \ldots, x_n)) = 1$ if $(\varphi x_1, \ldots, \varphi x_n) \in \psi r$, and is 0 otherwise, while we still require that

(b) v is a homomorphism of $\{F, \Rightarrow\}$-algebras.

It remains for us to define truth for a proposition of the form $(\forall x)p(x)$ in terms of our understanding of it for $p(x)$, and so we use an induction over the depth of quantification. Let $P_k(V, \mathcal{R})$ be the set of all elements p of $P(V, \mathcal{R})$ with $d(p) \leqslant k$. If we take some new variable t, then intuitively, we consider $(\forall x)p(x) \ (=(\forall t)p(t))$ to be true if $p(t)$ is true no matter how we choose to interpret t. This leads to a further requirement for v, namely:

(c_k) Suppose $p = (\forall x)q(x)$ has depth k. Put $V' = V \cup \{t\}$ where $t \notin V$. If for every extension $\varphi' : V' \to U$ of φ and for every $v'_{k-1} : P_{k-1}(V', R) \to \mathbb{Z}_2$, such that $(\varphi', \psi, v'_{k-1})$ satisfy (a), (b) and (c_i) for all $i < k$, we have $v'_{k-1}(q(t)) = 1$, then $v(p) = 1$, otherwise $v(p) = 0$.

Exercise 2.1.   Given U, φ, ψ, prove that there is one and only one function $v : P \to \mathbb{Z}_2$ satisfying (a), (b) and (c_i) for all i.

Briefly, the above exposition of the components of an interpretation of $P(V, \mathcal{R})$ can be expressed as follows.

Definition 2.2.   An *interpretation* of $P = P(V, \mathcal{R})$ in the domain U is a quadruple (U, φ, ψ, v) satisfying the conditions (a), (b) and (c_k) for all k.

As before, we write $A \vDash p$ if $A \subseteq P$, $p \in P$ and $v(p) = 1$ for every inter-pretation of P for which $v(A) \subseteq \{1\}$. We denote by $\text{Con}(A)$ the set of all p such that $A \vDash p$. We write $\vDash p$ for $\varnothing \vDash p$, and any p for which $\vDash p$, is called valid or a tautology.

Exercises

2.3.   Let $w(u_1, \ldots, u_n)$ be any tautology of $\text{Prop}(\{u_1, \ldots, u_n\})$. Let $p_1, \ldots, p_n \in P(V, \mathcal{R})$. Prove that $\vDash w(p_1, \ldots, p_n)$.

2.4. $A \subseteq P(V, \mathcal{R})$ and $p(x) \in A$ for all $x \in V$. Does it follow that $A \vDash (\forall x)p(x)$?

§3   Proof in Pred$(V, \mathcal{R})$

To complete the construction of the logic called the First-Order Predicate Calculus on $(V, \mathcal{R})$, and henceforth denoted by Pred$(V, \mathcal{R})$, we have to define a proof in Pred$(V, \mathcal{R})$.

Definition 3.1.   The *set of axioms* of Pred$(V, \mathcal{R})$ is the set $\mathcal{A} = \mathcal{A}_1 \cup \cdots \cup \mathcal{A}_5$, where

$\mathcal{A}_1 = \{p \Rightarrow (q \Rightarrow p) \,|\, p, q \in P(V, \mathcal{R})\}$,

$\mathcal{A}_2 = \{(p \Rightarrow (q \Rightarrow r)) \Rightarrow ((p \Rightarrow q) \Rightarrow (p \Rightarrow r)) \,|\, p, q, r \in P(V, \mathcal{R})\}$,

$\mathcal{A}_3 = \{\sim \sim p \Rightarrow p \,|\, p \in P(V, \mathcal{R})\}$,

$$\mathscr{A}_4 = \{(\forall x)(p \Rightarrow q) \Rightarrow (p \Rightarrow ((\forall x)q)) \,|\, p, q \in P(V, \mathscr{R}), x \notin \mathrm{var}(p)\},$$
$$\mathscr{A}_5 = \{(\forall x)p(x) \Rightarrow p(y) \,|\, p(x) \in P(V, \mathscr{R}), y \in V\}.$$

We remind the reader that these axioms are stated in terms of elements of the reduced predicate algebra. In $\mathscr{A}_5$, for example, the substitution of y for x in $p(x)$ implies that we have chosen a representative of $[(\forall x)p(x)]$ in which $(\forall y)$ does not appear.

In addition to Modus Ponens, we shall use one further rule of inference, which will enable us to formalise the following commonly occurring argument: we have proved $p(x)$, but x was any element, and therefore $(\forall x)p(x)$. The rule of inference called Generalisation allows us to deduce $(\forall x)p(x)$ from $p(x)$ provided x is general. The restriction on the use of Generalisation needs to be stated carefully.

Definition 3.2 Let $A \subseteq P$, $p \in P$. A *proof of length n* of p from A is a sequence $p_1, \ldots, p_n$ of n elements of P such that $p_n = p$, the sequence $p_1, \ldots, p_{n-1}$ is a proof of length $n - 1$ of p_{n-1} from A, and

(a) $p_n \in \mathscr{A} \cup A$, or

(b) $p_i = p_j \Rightarrow p_n$ for some $i, j < n$, or

(c) $p_n = (\forall x)w(x)$ and some subsequence $p_{k_1}, \ldots, p_{k_r}$ of $p_1, \ldots p_{n-1}$ is a proof (of length $< n$) of $w(x)$ from a subset A_0 of A such that $x \notin \mathrm{var}(A_0)$.

This is an inductive definition of a proof in Pred(V, $\mathscr{R}$). As for Prop(X), we require a proof to be a proof of finite length. The restriction $x \notin \mathrm{var}(A_0)$ in (c) means that no special assumptions about x are used in proving $w(x)$, and is the formal analogue of the restriction on the use of Generalisation in our informal logic.

As before, we write $A \vdash p$ if there exists a proof of p from A. We denote by Ded(A) the set of all p such that $A \vdash p$. We write $\vdash p$ for $\varnothing \vdash p$, and any p for which $\vdash p$ is called a theorem of Pred(V, $\mathscr{R}$).

Example 3.3. We show $\{\sim(\exists x)(\sim p)\} \vdash (\forall x)p$ for any element $p \in P$. (Recall that $(\exists x)$ is an abbreviation for $\sim(\forall x)\sim$.) The following is a proof.

$$p_1 = \sim \sim(\forall x)(\sim \sim p) \Rightarrow (\forall x)(\sim \sim p), \qquad (\mathscr{A}_3)$$

$$p_2 = \sim \sim(\forall x)(\sim \sim p), \qquad \text{(assumption)}$$

$$p_3 = (\forall x)(\sim \sim p), \qquad (p_1 = p_2 \Rightarrow p_3)$$

$$p_4 = (\forall x)(\sim \sim p(x)) \Rightarrow \sim \sim p(y), \qquad (\mathscr{A}_5)$$

Note that by $(\mathscr{A}_5)$, the y in p_4 may be chosen to be any variable. To permit a subsequent use of Generalisation, y must not be in $\mathrm{var}(\sim(\exists x)(\sim p(x)))$. A possible choice for y is the variable x itself.

$$p_5 = \sim \sim p(y), \qquad (p_4 = p_3 \Rightarrow p_5)$$

$$p_6 = \sim \sim p(y) \Rightarrow p(y), \qquad (\mathscr{A}_3)$$

$$p_7 = p(y), \qquad (p_6 = p_5 \Rightarrow p_7)$$

$$p_8 = (\forall y)p(y). \qquad (\text{Generalisation, } y \notin \mathrm{var}(\sim(\exists x)(\sim p(x))))$$

Exercises

3.4. Show that every axiom of Pred$(V, \mathcal{R})$ is valid.

3.5. Construct a proof in Pred$(V, \mathcal{R})$ of $(\forall x)(\forall y)p(x, y)$ from $\{(\forall y)(\forall x)p(x, y)\}$.

§4   Properties of Pred$(V, \mathcal{R})$

We have now constructed the logic Pred$(V, \mathcal{R})$. Its algebra of propositions is the reduced first order algebra $P(V, \mathcal{R})$, its valuations are the valuations associated with the interpretations of $P(V, \mathcal{R})$ defined in §2, and its proofs are as defined in §3.

We can immediately inquire if there is a substitution theorem for this logic, corresponding to Theorem 4.11 of the Propositional Calculus. There, substitution was defined in terms of a homomorphism $\varphi : P_1 \to P_2$ of one algebra of propositions into another. If P_1, P_2 are first order algebras, then as the concept of a homomorphism from P_1 to P_2 requires these algebras to have the same set of operations, it follows that they must have the same set of individual variables. Even in this case, a homomorphism would be too restrictive for our purposes, for we would naturally want to be able to interchange two variables x, y, so mapping elements $p(x)$ of the algebra to $\varphi(p(x)) = p(y)$, but unfortunately such a map is not a homomorphism. For if $p(x) \in P$ is such that $x \in \text{var}(p(x))$, $y \notin \text{var}(p(x))$, then

$$\varphi((\forall x)p(x)) = (\forall y)p(y) = (\forall x)p(x),$$
$$(\forall x)\varphi(p(x)) = (\forall x)p(y).$$

Since $y \in \text{var}((\forall x)p(y))$ but $y \notin \text{var}((\forall y)p(y))$, these elements are distinct and φ is not a homomorphism.

Definition 4.1.   Let $P_1 = P(V_1, \mathcal{R}^{(1)})$ and $P_2 = P(V_2, \mathcal{R}^{(2)})$. A *semi-homomorphism* $(\alpha, \beta) : (P_1, V_1) \to (P_2, V_2)$ is a pair of maps $\alpha : P_1 \to P_2$, $\beta : V_1 \to V_2$ such that
  (a) $\beta(V_1)$ is infinite,
  (b) α is an $\{F, \Rightarrow\}$-homomorphism, and
  (c) $\alpha((\forall x)p) = (\forall x')\alpha(p)$, where $x' = \beta(x)$.

Lemma 4.2.   *Let* $(\alpha, \beta) : (P_1, V_1) \to (P_2, V_2)$ *be a semi-homomorphism. Let* $p \in P_1$ *and suppose* $x \in V_1 - \text{var}(p)$. *Then* $\beta(x) \notin \text{var}(\alpha(p))$.

Proof: We observe first that if $x \neq y$, then $(\forall x)p = (\forall y)p$ if and only if neither x nor y is in var(p).

Since $\beta(V_1)$ is infinite, there is an element $y' \in \beta(V_1)$ such that $y' \neq \beta(x)$ and $y' \notin \beta(\text{var}(p))$. Choosing $y \in V_1$ so that $\beta(y) = y'$, it follows that $(\forall x)p = (\forall y)p$. If $x' = \beta(x)$, then we have

$$(\forall x')\alpha(p) = \alpha((\forall x)p) = \alpha((\forall y)p) = (\forall y')\alpha(p),$$

and it follows again that $x' \notin \text{var}(\alpha(p))$. $\square$

Theorem 4.3. (The Substitution Theorem). *Let $(\alpha, \beta):(P_1, V_1) \to (P_2, V_2)$ be a semi-homomorphism. Let $A \subseteq P$, $p \in P_1$.*

(a) *If $A \vdash p$, then $\alpha(A) \vdash \alpha(p)$.*
(b) *If $A \vDash p$, then $\alpha(A) \vDash \alpha(p)$.*

Proof: (a) Let $p_1, \ldots, p_n$ be a proof of p from A. We use induction over n to show that $\alpha(p_1), \ldots, \alpha(p_n)$ is a proof of $\alpha(p)$ from $\alpha(A)$.

If $a = ((\forall x)(p \Rightarrow q)) \Rightarrow (p \Rightarrow (\forall x)q)$ is an axiom of type A_4, then by Lemma 4.2, the condition $x \notin \mathrm{var}(p)$ is preserved by the semi-homomorphism (α, β), and so $\alpha(a)$ is again an axiom. In all other cases, it is clear that the image of an axiom is an axiom. Thus if $p \in \mathscr{A}^{(1)} \cup A$, then $\alpha(p) \in \mathscr{A}^{(2)} \cup \alpha(A)$, where $\mathscr{A}^{(i)}$ is the set of axioms of Pred(V_i, $\mathscr{R}^{(i)}$). Hence our desired result holds for $n = 1$.

For $n > 1$, we may suppose by induction that $\alpha(p_1), \ldots, \alpha(p_{n-1})$ is a proof of $\alpha(p_{n-1})$ from $\alpha(A)$. If $p_i = p_j \Rightarrow p_n$ for some $i, j < n$, then $\alpha(p_i) = \alpha(p_j) \Rightarrow \alpha(p_n)$, and the result holds. It remains only to consider the case that $p_n = (\forall x)q$, where some subsequence $q_1, \ldots, q_k$ of $p_1, \ldots, p_{n-1}$ is a proof of q from some subset $A_0 \subseteq A$ with $x \notin \mathrm{var}(A_0)$. By induction, $\alpha(q_1), \ldots, \alpha(q_k)$ is a proof of $\alpha(q)$ from $\alpha(A_0)$. For each $w \in A_0$, $x \notin \mathrm{var}(w)$, and by Lemma 4.2, $x' \notin \mathrm{var}(\alpha(w))$, where $x' = \beta(x)$. Thus $x' \notin \mathrm{var}(\alpha(A_0))$, and $\alpha(p_1), \ldots, \alpha(p_{n-1}), (\forall x')\alpha(q)$ is a proof. Since $(\forall x')\alpha(q) = \alpha((\forall x)q) = \alpha(p)$, the result is completely proved.

Part (b) is an easy consequence of (a) once we have proved the Adequacy Theorem, so we omit a proof. We leave as an exercise a direct proof of (b). ☐

Exercises

(The following exercises lead to a direct proof of part (b) of the Substitution Theorem. Throughout, $P_i = P(V_i, \mathscr{R}^{(i)})$ and $(\alpha, \beta):(P_1, V_1) \to (P_2, V_2)$ is a semi-homomorphism.)

4.4. Show that $(\forall x)p(x) = (\forall x)q(x)$ if and only if $p(x) = q(x)$.

4.5. We put $V_i^* = V_i \cup \{y\}$ and $P_i^* = P(V_i^*, \mathscr{R}^{(i)})$, where y is some new variable ($y \notin V_1 \cup V_2$). Show that for each $p(y) \in P_1^* - P_1$, there is a unique $q(y) \in P_2^*$ such that $\alpha((\forall x)p(x)) = (\forall x')q(x')$ for some $x \in V_1$, $x \notin \mathrm{var}(p(y))$ and $x' = \beta(x)$. Hence show that there is a unique semi-homomorphism $(\alpha^*, \beta^*):(P_1^*, V_1^*) \to (P_2^*, V_2^*)$, extending (α, β), such that $\beta^*(y) = y$. Generalise to the addition of n new variables $y_1, \ldots, y_n$.

4.6. Let (U, φ, ψ, v) be an interpretation of P_2. For each $r \in \mathscr{R}_n^{(1)}$, we define an n-ary relation $\psi_1 r$ on U as follows. Take new variables $y_1, \ldots, y_n$, put $V_i^* = V_i \cup \{y_1, \ldots, y_n\}$, and construct the extension (α^*, β^*) of (α, β) as in 4.5. Given $(u_1, \ldots, u_n) \in U^n$, the mapping of y_i to u_i defines a unique extension of (U, φ, ψ, v) to P_2^*, and so assigns a value $v^*(q)$ to each $q \in P_2^*$. We define $(u_1, \ldots, u_n) \in \psi_1 r$ if and only if $v^*(\alpha^*(r(y_1, \ldots, y_n)) = 1$.

Show that $(U, \varphi\beta, \psi_1, v\alpha)$ is an interpretation of P_1. Hence prove part (b) of the Substitution Theorem.

Theorem 4.7. (The Soundness Theorem). *Let $A \subseteq P(V, \mathscr{R})$, $p \in P(V, \mathscr{R})$. If $A \vdash p$, then $A \vDash p$.*

Proof: Let $p_1, \ldots, p_n$ be a proof of p from A. Let (U, φ, ψ, v) be an interpretation of $P(V, \mathscr{R})$ such that $v(A) \subseteq \{1\}$. We have to show that $v(p) = 1$, and we shall use induction on n to prove it. If $n = 1$, $p \in \mathscr{A} \cup A$ and then $v(p) = 1$. Suppose by induction that $n > 1$ and the result holds for proofs of length less than n. If $p_i = p_j \Rightarrow p_n$ for some $i, j < n$, then $v(p_i) = v(p_j) = 1$, and it follows that $v(p) = 1$.

Suppose finally that $p_n = (\forall x)q(x)$ and that $q_1(x), \ldots, q_k(x)$ is a proof of $q(x)$ from the subset A_0 of A with $x \notin \text{var}(A_0)$. We must use condition (c_r) in the definition of interpretation, where r is the depth of p_n. Thus we take a new variable t, we put $V' = V \cup \{t\}$, and we consider extensions $\varphi': V' \to U$ of φ and maps $v'_{r-1}: P_{r-1}(V', \mathscr{R}) \to \mathbb{Z}_2$, as given in condition (c_r). We have to prove that in every case, $v'_{r-1}(q_k(t)) = 1$. But each v'_{r-1} extends uniquely to a valuation $v': P(V', \mathscr{R}) \to \mathbb{Z}_2$ such that (U, φ', ψ, v') is an interpretation of $P(V', \mathscr{R})$. By the Substitution Theorem (Theorem 4.3 (a)), $q_1(t), \ldots, q_k(t)$ is a proof of $q(t)$ from A_0, and so by induction (since $k < n$), $v'(q_k(t)) = 1$. Thus $v((\forall x)q(x)) = 1$ and the theorem is proved. $\square$

Corollary 4.8. (The Consistency Theorem). *F is not a theorem of* $\text{Pred}(V, \mathscr{R})$.

Proof: Let U be any non-empty set, $\varphi: V \to U$ any function, and ψ any function on $\mathscr{R}$ such that if $r \in \mathscr{R}_n$, then $\psi(r)$ is an n-ary relation on U. Then there exists $v: P(V, \mathscr{R}) \to \mathbb{Z}_2$ such that (U, φ, ψ, v) is an interpretation. For every interpretation, and in particular for the one constructed above, $v(F) = 0$. The existence of one interpretation for which $v(F) = 0$ shows that F is not valid. The Soundness Theorem now shows that F is not a theorem. $\square$

Theorem 4.9. (The Deduction Theorem). *Let $A \subseteq P = P(V, \mathscr{R})$ and let $p, q \in P$. Then $A \vdash p \Rightarrow q$ if and only if $A \cup \{p\} \vdash q$.*

Proof: If $A \vdash p \Rightarrow q$, then it follows, as in the case of the Propositional Calculus, that $A \cup \{p\} \vdash q$. Suppose $A \cup \{p\} \vdash q$. We shall again use induction over the length of the proof. The argument used for the case of the Propositional Calculus again applies except in the case where q is obtained by Generalisation. So we suppose $q = (\forall x)r(x)$ and $A_0 \vdash r(x)$, where $A_0 \subseteq A \cup \{p\}$ and $x \notin \text{var}(A_0)$.

(i) $p \notin A_0$. Then $A_0 \subseteq A$ and we have a proof of q from A_0. Follow this proof with the steps $q \Rightarrow (p \Rightarrow q)$, $p \Rightarrow q$ to obtain a proof of $p \Rightarrow q$ from A.

(ii) $p \in A_0$. We have a proof of $r(x)$ from A_0, and so by induction on the proof length, we have $A_1 \vdash p \Rightarrow r(x)$, where $A_1 = A_0 - \{p\}$. By Generalisation, a proof of $p \Rightarrow r(x)$ from A_1 may be followed with $(\forall x)(p \Rightarrow r(x))$. As $p \in A_0$ and $x \notin \text{var}(A_0)$, it follows that $x \notin \text{var}(p)$. We continue the proof with

$$(\forall x)(p \Rightarrow r(x)) \Rightarrow (p \Rightarrow (\forall x)r(x)) \qquad (\mathscr{A}_4)$$

and

$$p \Rightarrow (\forall x)r(x),$$

completing the proof and establishing the theorem. $\square$

Example 4.10.   As we did before, we use the techniques of the proof of the Deduction Theorem to convert the proof $\sim p, (\forall x)(\sim p), \sim(\forall x)(\sim p), F$ of F from $\{(\exists x)p, \sim p\}$ $(x \notin \text{var}(p))$, into a proof of $\sim \sim p$ from $\{(\exists x)p\}$, so proving p from $\{(\exists x)p\}$.

Given proof	Comment	Corresponding Steps of Constructed Proof
$\sim p$	Assumption to be eliminated	$\sim p \Rightarrow ((\sim p \Rightarrow \sim p) \Rightarrow \sim p),$ $(\sim p \Rightarrow ((\sim p \Rightarrow \sim p) \Rightarrow \sim p)) \Rightarrow ((\sim p \Rightarrow (\sim p \Rightarrow \sim p)) \Rightarrow (\sim p \Rightarrow \sim p)),$ $(\sim p \Rightarrow (\sim p \Rightarrow \sim p)) \Rightarrow (\sim p \Rightarrow \sim p), \ \sim p \Rightarrow (\sim p \Rightarrow \sim p), \ \sim p \Rightarrow \sim p.$
$(\forall x)(\sim p)$	Generalisation	$(\forall x)(\sim p \Rightarrow \sim p), ((\forall x)(\sim p \Rightarrow \sim p)) \Rightarrow (\sim p \Rightarrow (\forall x)(\sim p)),$ $\sim p \Rightarrow (\forall x)(\sim p).$
$\sim(\forall x)(\sim p)$	Retained assumption	$\sim(\forall x)(\sim p), (\sim(\forall x)(\sim p)) \Rightarrow (\sim p \Rightarrow \sim(\forall x)(\sim p)),$ $\sim p \Rightarrow (\sim(\forall x)(\sim p)).$
F	Modus ponens	$(\sim p \Rightarrow ((\forall x)(\sim p) \Rightarrow F)) \Rightarrow ((\sim p \Rightarrow (\forall x)(\sim p)) \Rightarrow (\sim p \Rightarrow F)),$ $(\sim p \Rightarrow (\forall x)(\sim p)) \Rightarrow (\sim p \Rightarrow F), \ \sim \sim p.$
	Extension to prove p	$\sim \sim p \Rightarrow p, p.$

Exercises

4.11.   Convert the proof $(\forall x)p(x), ((\forall x)p(x)) \Rightarrow p(x), p(x), (\forall x)(p(x) \Rightarrow q(x)), (\forall x)(p(x) \Rightarrow q(x)) \Rightarrow (p(x) \Rightarrow q(x)), p(x) \Rightarrow q(x), q(x), (\forall x)q(x)$ of $(\forall x)q(x)$ from $\{(\forall x)(p(x) \Rightarrow q(x)), (\forall x)p(x)\}$ into a proof of $(\forall x)p(x) \Rightarrow (\forall x)q(x)$ from $\{(\forall x)(p(x) \Rightarrow q(x))\}$.

4.12.   Prove $\{(\forall x)(p(x) \Rightarrow q(x))\} \vdash (\exists x)p(x) \Rightarrow (\exists x)q(x)$.

We now prove some lemmas which we shall need in establishing the Satisfiability Theorem. As for Prop(X), a subset A is consistent if $F \notin \text{Ded}(A)$.

Lemma 4.13.   *Let A be a consistent subset of $P(V, \mathcal{R})$. Suppose $(\exists x)p(x) \in A$, and $t \notin \text{Var}(A)$. Then $F \notin \text{Ded}(A \cup \{p(t)\})$.*

Proof: Suppose $F \in \text{Ded}(A \cup \{p(t)\})$. Then by the Deduction Theorem, $\sim p(t) \in \text{Ded}(A)$. Since $t \notin \text{Var}(A)$, we may apply Generalisation and obtain $(\forall x)(\sim p(x)) \in \text{Ded}(A)$. But $(\exists x)p(x) = \sim(\forall x)(\sim p(x)) \in A$, and so $F \in \text{Ded}(A)$, contrary to assumption.   □

Lemma 4.14.   *Let A be a consistent subset of $P(V, \mathcal{R})$. Then there exist $V^* \supseteq V$ and $A^* \supseteq A$, where $A^* \subseteq P(V^*, \mathcal{R})$, such that*
   (i) *$F \notin \text{Ded}(A^*)$, and*
   (ii) *for all $p \in P(V^*, \mathcal{R})$, either $p \in A^*$ or $\sim p \in A^*$, and*
   (iii) *if $(\exists x)p(x) \in A^*$, then for some $t \in V^*$, $p(t) \in A^*$.*

Proof: Put $V_0 = V$, $A_0 = A$, $P_0 = P(V, \mathscr{R})$. We construct inductively V_i, $P_i = P(V_i, \mathscr{R})$, A_i' and A_i for $i > 0$. Taking a new variable $t_p^{(i)}$ for each $p \in A_i$ of the form $p = (\exists x)q(x)$, we put

$$V_{i+1} = V_i \cup \{t_p^{(i)} | p \in A_i, p = (\exists x)q(x) \text{ for some } q(x)\},$$
$$A_{i+1}' = A_i \cup \{q(t_p^{(i)}) | p \in A_i, p = (\exists x)q(x), q(x) \in P_i\}.$$

Suppose that $F \notin \mathrm{Ded}(A_i)$. If $F \in \mathrm{Ded}(A_{i+1}')$, then $F \in \mathrm{Ded}(A_i \cup \{q_1(t_{p_1}^{(i)}),$ $\ldots, q_r(t_{p_r}^{(i)})\})$ for some finite set $\{q_1(t_{p_1}^{(i)}), \ldots, q_r(t_{p_r}^{(i)})\}$, which is impossible by Lemma 4.13. Thus $F \notin \mathrm{Ded}(A_{i+1}')$, and by Lemma 2.11 of Chapter II, there exists $A_{i+1} \supseteq A_{i+1}'$ such that A_{i+1} satisfies (i) and (ii). For each $i > 0$, choose[2] such an A_i. Put $V^* = \bigcup_i V_i$, $A^* = \bigcup_i A_i$.

Since any finite subset of A^* is contained in some A_i, it follows that V^* and A^* satisfy (i), (ii) and (iii). $\square$

Theorem 4.15. (The Satisfiability Theorem). *Let A be a consistent subset of $P(V, \mathscr{R})$. Then there exists an interpretation (U, φ, ψ, v) of $P(V, \mathscr{R})$ such that $v(A) \subseteq \{1\}$.*

Proof: If $V^* \supseteq V$ and $P(V^*, \mathscr{R}) \supseteq A^* \supseteq A$, then any interpretation of $P(V^*, \mathscr{R})$ for which $v(A^*) \subseteq \{1\}$ clearly restricts to an interpretation of $P(V, \mathscr{R})$ with $v(A) \subseteq \{1\}$. We may therefore suppose, without any loss of generality, that V, A satisfy the conditions (i), (ii) and (iii) of Lemma 4.14. To construct our interpretation, we take $U = V$, and $\varphi : V \to U$ the identity map. For each $r \in \mathscr{R}_n$, we put $\psi r = \{(x_1, \ldots, x_n) \in V^n | r(x_1, \ldots, x_n) \in A\}$. For each $p \in P(V, \mathscr{R})$, we put $v(p) = 1$ if $p \in A$ and $v(p) = 0$ otherwise. It is easily checked that (U, φ, ψ, v) satisfies the conditions (a), (b) of the definition of an interpretation, and we are left with showing that the condition (c_k) is satisfied for all k.

Let t be some new variable, and let $p = (\forall x)q(x)$ have depth $k + 1$. Suppose first that $p \in A$. Let φ' be any extension of φ to $V' = V \cup \{t\}$, and let $v_k' : P_k(V', \mathscr{R}) \to \mathbb{Z}_2$ be as required for condition (c_{k+1}). Put $y = \varphi'(t)$. Since, by induction, v satisfies (c_i) for $i \leqslant k$, it follows that for all $w(x) \in P_k$, $v'(w(t)) = v(w(y))$. Now $(\forall x)q(x) \in A$, therefore $q(y) \in \mathrm{Ded}(A) = A$, since A is a maximal consistent subset, and this holds for all $y \in V$. Thus $v'(q(t)) = v(q(y)) = 1$ and condition (c_{k+1}) is satisfied in this case.

Suppose that $p = (\forall x)q(x) \notin A$. As $\{\sim(\exists x)(\sim q(x))\} \vdash (\forall x)q(x)$, it follows that $\sim(\exists x)(\sim q(x)) \notin A$. Hence $(\exists x)(\sim q(x)) \in A$, and so for some $y \in V$, $\sim q(y) \in A$. Consider the extension φ' of φ to V' with $\varphi'(t) = y$, and the corresponding $v_k' : P(V', \mathscr{R}) \to \mathbb{Z}_2$. Then $v'(q(t)) = v(q(y)) = 0$. As $v(p) = 0$, we see again that condition (c_{k+1}) is satisfied. $\square$

Theorem 4.16. (The Adequacy Theorem). *Let $A \subseteq P(V, \mathscr{R})$, $p \in P(V, \mathscr{R})$. If $A \vDash p$, then $A \vdash p$.*

---

[2] The proof of Lemma 2.11 involved an application of Zorn's Lemma. We also use the (countable) axiom of choice here to select the A_i.

Proof: If $F \notin \mathrm{Ded}(A \cup \{\sim p\})$, then by the Satisfiability Theorem, there exists an interpretation (U, φ, ψ, v) of $P(V, \mathscr{R})$ such that $v(A \cup \{\sim p\}) \subseteq \{1\}$, which contradicts the hypothesis $A \vDash p$. Therefore $A \cup \{\sim p\} \vdash F$. Hence, by the Deduction Theorem, $A \vdash \sim \sim p$, and the result follows.   □

Corollary 4.17.   (The Compactness Theorem). *If $A \vDash p$, then $A_0 \vDash p$ for some finite subset A_0 of A.*

The Soundness Theorem and the Adequacy Theorem together show that if $A \subseteq P(V, \mathscr{R})$ and $p \in P(V, \mathscr{R})$, then $A \vDash p$ if and only if $A \vdash p$. This result is usually called Gödel's (or the Gödel-Henkin) Completeness Theorem. It was first proved by Gödel in 1930. The method of proof we have used, depending on the Satisfiability Theorem, is due to Henkin.

We have now established for Pred($V, \mathscr{R}$) all the properties previously established for Prop(X), with the exception of decidability. We have good reason for not attempting to prove Pred($V, \mathscr{R}$) is decidable. If $\mathscr{R}$ contains at least one relation symbol of arity greater than 1, then Pred($V, \mathscr{R}$) is undecidable. The precise meaning of this statement, and its proof (which is due to Church and Kalmar), are given in Chapter IX.

Exercise 4.18.   An element $p \in P(V, \mathscr{R})$ is said to be expressed in prenex normal form when it is expressed in the form $p = Q_1 Q_2 \cdots Q_k q$, where Q_i is either $(\forall x_i)$ or $(\exists x_i)$, $x_1, \ldots, x_k$ are distinct, and q is a quantifier-free element of $P(V, \mathscr{R})$. Give an algorithm which constructs from any $p \in P(V, \mathscr{R})$, an element p' in prenex normal form such that $\vdash (p \Rightarrow p') \wedge (p' \Rightarrow p)$.

Chapter V

First-Order Mathematics

§1 Predicate Calculus with Identity

In this chapter, we shall reconstruct some parts of ordinary mathematics within the logical system constructed in Chapter IV. A piece of mathematics constructed within the first-order predicate calculus will be called a first-order theory. By comparing a first-order theory with the informal theory on which it is modelled, we may gain insight into the influence of our logical system on our mathematics.

One feature common to all mathematical theories is the concept of equality or identity. A statement of the form $a = b$ always means that a and b denote the same mathematical object. A consequence of $a = b$ is that, in any statement involving a, we may replace any of the occurrences of a by b without altering the truth or falsity of the statement. We therefore begin by investigating how to formalise in $\text{Pred}(V, \mathcal{R})$ the concept of identity. We clearly require a binary relation symbol $\mathscr{I} \in \mathcal{R}_2$. As the axioms of identity, we take the set $I \subseteq P(V, \mathcal{R})$ consisting of $(\forall x).\mathscr{I}(x, x)$ and the elements $(\forall x_1) \cdots (\forall x_n)(\forall y)(\mathscr{I}(x_j, y) \Rightarrow (r(x_1, \ldots, x_n) \Rightarrow r(x_1, \ldots, x_{j-1}, y, x_{j+1}, \ldots, x_n)))$, for all $r \in \mathcal{R}_n$, all n, and all $j \leqslant n$.

Exercises

1.1. Prove $I \vdash \mathscr{I}(x, y) \Rightarrow \mathscr{I}(y, x)$.

1.2. Prove $I \vdash \mathscr{I}(x, y) \Rightarrow (\mathscr{I}(y, z) \Rightarrow \mathscr{I}(x, z))$.

1.3. Let $w(x, z)$ be any element of P, possibly involving other variables besides x, z. Show that $I \vdash \mathscr{I}(x, y) \Rightarrow (w(x, x) \Rightarrow w(y, x))$. (Hint: use induction over the number of steps in the construction of $w(x, y)$ from V and $\mathcal{R}$.)

1.4. Let (U, φ, ψ, v) be an interpretation of $P(V, \mathcal{R})$ such that $\psi\mathscr{I}$ is the identity relation on U. Let U' be any set containing U, and let $\pi: U' \to U$ be any function such that $\pi(u) = u$ for all $u \in U$. Let $\varphi': V \to U'$ be the composition of φ with the inclusion map $U \to U'$. For each $r \in \mathcal{R}_n$, define the n-ary relation $\psi'r$ on U' by $(u'_1, \ldots, u'_n) \in \psi'r$ if and only if $(\pi(u'_1), \ldots, \pi(u'_n)) \in \psi r$. Show that this defines an interpretation $(U', \varphi', \psi', v')$ of $P(V, \mathcal{R})$, and that for $p \in P(V, \mathcal{R})$, we have $v'(p) = v(p)$. Show that $\psi'\mathscr{I}$ is an equivalence relation on U', but that, no matter what the interpretation (U, φ, ψ, v), U' and π can be constructed such that $\psi'\mathscr{I}$ is not the relation of identity in U'.   .

According to Exercise 1.4, no matter what subset $I' \supseteq I$ of $P(V, \mathcal{R})$ we choose as our axioms of identity, we cannot thereby force $\psi\mathscr{I}$ to be the relation of identity in every interpretation of $P(V, \mathcal{R})$ such that $v(I') = \{1\}$, unless of course we have $F \in \text{Ded}(I')$ and so exclude the existence of such interpretations. We overcome this by constructing a modified form of the first-order predicate calculus, in which the only interpretations allowed will be those for which $\psi \mathscr{I}$ is the identity relation.

Definition 1.5. Suppose $\mathscr{I} \in \mathscr{R}_2$. A *proper interpretation* of $P(V, \mathscr{R})$ is an interpretation (U, φ, ψ, v) such that $\psi \mathscr{I}$ is the relation of identity on U.

Definition 1.6. $\text{Pred}_\mathscr{I}(V, \mathscr{R})$ is the logic with algebra of propositions $P(V, \mathscr{R} \cup \{\mathscr{I}\})$, valuations those arising from proper interpretations, and with proof of p from A in $\text{Pred}_\mathscr{I}(V, \mathscr{R})$ defined as a proof of p from $I \cup A$ in $\text{Pred}(V, \mathscr{R} \cup \{\mathscr{I}\})$.

We shall always assume $\mathscr{I} \in \mathscr{R}$, and so have $P(V, \mathscr{R})$ as the algebra of propositions. We write $A \vdash_\mathscr{I} p$ and $p \in \text{Ded}_\mathscr{I}(A)$ to indicate that p is provable from A in $\text{Pred}_\mathscr{I}(V, \mathscr{R})$, i.e., that $A \cup I \vdash p$ or equivalently $p \in \text{Ded}(A \cup I)$. We say that p is a *proper consequence* of A, written $A \vDash_\mathscr{I} p$ or $p \in \text{Con}_\mathscr{I}(A)$, if $v(p) = 1$ for every proper interpretation of $P(V, \mathscr{R})$ with $v(A) \subseteq \{1\}$. Because of the restriction on the interpretations considered, $A \vDash_\mathscr{I} p$ would appear to be weaker than $A \cup I \vDash p$. We shall see shortly that they are in fact equivalent.

Theorem 1.7. (The Satisfiability Theorem) *Suppose $F \notin \text{Ded}_\mathscr{I}(A)$. Then there exists a proper interpretation of $P(V, \mathscr{R})$ with $v(A) \subseteq \{1\}$.*

Proof: Since $F \notin \text{Ded}(A \cup I)$, there exists an interpretation (U, φ, ψ, v) of $P = P(V, \mathscr{R})$ such that $v(A \cup I) = \{1\}$. The relation $\psi \mathscr{I}$ is an equivalence relation on U. For $u \in U$, denote by $\bar{u}$ the equivalence class $\{u' \in U | (u, u') \in \psi \mathscr{I}\}$, and let $\bar{U}$ be the set of all these equivalence classes. Define $\bar{\varphi} : V \to \bar{U}$ by $\bar{\varphi}(x) = \overline{\varphi(x)}$ for all $x \in V$. For each $r \in \mathscr{R}_n$, ψr has the property that if $(u_i, u_i') \in \psi \mathscr{I}$, then $(u_1, \ldots, u_n) \in \psi r$ if and only if $(u_1', \ldots, u_n') \in \psi r$. Hence we can define a relation $\bar{\psi} r$ on $\bar{U}$ by putting $(\bar{u}_1, \ldots, \bar{u}_n) \in \bar{\psi} r$ if and only if $(u_1, \ldots, u_n) \in \psi r$. This defines a function $\bar{\psi}$ from $\mathscr{R}$ into the relations on $\bar{U}$, and it is easily checked that $(\bar{U}, \bar{\varphi}, \bar{\psi}, v)$ is a proper interpretation of $P(V, \mathscr{R})$. The valuation v is unchanged, consequently we have a proper interpretation with $v(A) \subseteq \{1\}$. $\square$

Corollary 1.8.
(i) $\text{Con}_\mathscr{I}(A) = \text{Con}(A \cup I)$
(ii) *If $A \vDash_\mathscr{I} p$, then $A \vdash_\mathscr{I} p$.*
The soundness and consistency of $\text{Pred}_\mathscr{I}(V, \mathscr{R})$ both follow immediately from the corresponding properties of $\text{Pred}(V, \mathscr{R})$.

§2   First-Order Mathematical Theories

A branch of mathematics is defined by listing the properties and relationships to be studied and by listing the assumptions (usually known as the axioms of the branch of mathematics) made about them. For example, in plane projective geometry, the only properties considered are those of being called a point or line (the actual nature of the objects is irrelevant, only the way they are divided into the two classes matters) and we are concerned with the one relationship of a point lying on a line. (It is taken for granted that we also use the relationship of identity.) The axioms of plane projective geometry are that through any two distinct points there is one and only one line, that any two distinct lines have one and only one common point, and the non-triviality axiom that there exist four points such that no three of them are collinear.

We shall define a mathematical theory in terms of lists of relations and axioms. It is convenient also to include a list of any special objects named in the axioms.

Definition 2.1. A *first-order mathematical theory* is a triple $\mathcal{T} = (\mathcal{R}, A, C)$ where $\mathcal{I} \in \mathcal{R}$, $A \subseteq P(V, \mathcal{R})$ for some $V \supset C$ such that $V - C$ is infinite, and $\text{var}(A) = C$. The set A is called the set of (*mathematical*) *axioms* of $\mathcal{T}$, the set C is called the set of (*individual*) *constants* of $\mathcal{T}$, while the language[1] of $\mathcal{T}$ is the subset $\mathcal{L}(\mathcal{T}) = \{p \in P(V, \mathcal{R}) | \text{var}(p) \subseteq C\}$ of $P(V, \mathcal{R})$. A *theorem* of $\mathcal{T}$ is an element $p \in \mathcal{L}(\mathcal{T})$ such that $A \vdash_{\mathcal{T}} p$.

We point out that the set V is not specified in $\mathcal{T}$, and that any suitable set V may be taken. The set $\mathcal{L}(\mathcal{T})$ is independent of the choice of V. Later, we shall occasionally need a standardised set V of variables, such that $V - C$ is countably infinite. We select as standard variable set the set $V_0 = C \cup \{x_i | i \in \mathbf{N}\}$, where the x_i are disjoint from C.

Definition 2.2. The *algebra* of $\mathcal{T}$ is the set $P(\mathcal{T}) = P(V_0, \mathcal{R})$, where V_0 is the standard variable set. An element $p \in P(\mathcal{T})$, such that $\text{var}(p) \subseteq \{x_1, \ldots, x_n\} \cup C$, is called an *n-variable formula* of $\mathcal{T}$.

The following notations will be used in discussing first-order theories $\mathcal{T}$. If $U \subseteq P(V, \mathcal{R})$ and $p \in P(V, \mathcal{R})$, then we write $U \vdash_{\mathcal{T}} p$ for $A \cup U \vdash_{\mathcal{T}} p$, $\mathcal{T} \vdash p$(or $\vdash_{\mathcal{T}} p$) for $A \vdash_{\mathcal{T}} p$, and $U \vDash_{\mathcal{T}} p$, $\mathcal{T} \vDash_{\mathcal{T}} p$(or $\vDash_{\mathcal{T}} p$), for $A \cup U \vdash_{\mathcal{T}} p$ and $A \vDash_{\mathcal{T}} p$ respectively.

Examples

2.3. (First-order plane projective geometry) We take two unary predicate symbols p, ℓ, interpreting $p(x)$ as "x is a point", and $\ell(x)$ as "x is a line". We take a binary predicate symbol $\in$, and interpret $\in(x, y)$ as "x lies on y". These express the basic concepts of plane projective geometry, so we take $\mathcal{R} = \{p, \ell, \in, \mathcal{I}\}$. Our axiom set is the set $A = \{a_1, \ldots, a_6\}$, where

$$a_1 = (\forall x)((p(x) \vee \ell(x)) \wedge \sim (p(x) \wedge \ell(x))),$$

---

[1] The reader is warned that most authors use this term for $P(V, \mathcal{R})$.

$a_2 = (\forall x)((\exists y) \in (x, y) \Rightarrow p(x)),$

$a_3 = (\forall x)((\exists y) \in (y, x) \Rightarrow \ell(x)),$

$a_4 = (\forall x)(\forall y)(p(x) \wedge p(y) \wedge \sim \mathscr{I}(x, y) \Rightarrow (\exists z)(\in(x, z) \wedge \in(y, z)$

$\qquad \wedge (\forall t)(\in(x, t) \wedge \in(y, t) \Rightarrow \mathscr{I}(z, t)))),$

$a_5 = (\forall x)(\forall y)(\ell(x) \wedge \ell(y) \wedge \sim \mathscr{I}(x, y) \Rightarrow (\exists z)(\in(z, x) \wedge \in(z, y)$

$\qquad \wedge (\forall t)(\in(t, x) \wedge \in(t, y) \Rightarrow \mathscr{I}(z, t)))),$

$a_6 = (\exists x_1)(\exists x_2)(\exists x_3)(\exists x_4)(p(x_1) \wedge p(x_2) \wedge p(x_3) \wedge p(x_4) \wedge \sim \mathscr{I}(x_1, x_2) \wedge$

$\qquad \sim \mathscr{I}(x_1, x_3) \wedge \sim \mathscr{I}(x_1, x_4) \wedge \sim \mathscr{I}(x_2, x_3) \wedge \sim \mathscr{I}(x_2, x_4) \wedge \sim \mathscr{I}(x_3, x_4)$

$\qquad \wedge \sim c(x_1, x_2, x_3) \wedge \sim c(x_1, x_2, x_4) \wedge \sim c(x_1, x_3, x_4) \wedge \sim c(x_2, x_3, x_4))$

where in the non-triviality axiom a_6, $c(x_1, x_2, x_3)$ denotes $(\exists z)(\in(x_1, z)$ $\wedge \in(x_2, z) \wedge \in(x_3, z))$. The axiom a_1 says that each object is either a point or a line, but not both. Axioms a_2 and a_3 say that $\in$ is a relation between a point and a line, while axioms a_4 and a_5 are the usual incidence axioms. For this theory, the set $C = \varnothing$.

There is a very useful notation which abbreviates axioms such as a_4 and a_5. We write $(\exists!x)w(x)$ for $(\exists x)(w(x) \wedge (\forall y)(w(y) \Rightarrow \mathscr{I}(x, y)))$, where $w(x)$ is any element of $P(V, \mathscr{R})$. $(\exists!x)w(x)$ may be read "There exists a unique x such that $w(x)$". In this notation, we have

$$a_4 = (\forall x)(\forall y)(p(x) \wedge p(y) \wedge \sim \mathscr{I}(x, y) \Rightarrow (\exists!z)(\in(x, z) \wedge \in(y, z))).$$

2.4. (Elementary group theory) We take $\mathscr{R} = \{\mathscr{I}, m\}$, where m is a ternary relation symbol. We interpret $m(x, y, z)$ as "$xy = z$". For axioms, we take $A = \{a_1, \ldots, a_4\}$, where

$a_1 = (\forall x)(\forall y)(\exists!z)m(x, y, z),$

$a_2 = (\forall x)(\forall y)(\forall z)(\forall a)(\forall b)(\forall c)(\forall d)(m(x, y, a) \wedge m(a, z, b)$

$\qquad \wedge m(y, z, c) \wedge m(x, c, d) \Rightarrow \mathscr{I}(b, d)),$

$a_3 = (\forall x)m(e, x, x),$

$a_4 = (\forall x)(\exists y)m(y, x, e).$

Axiom a_1 asserts that m defines a function, a_2 is the associative law, a_3 asserts that e is a left identity and a_4 asserts the existence of left inverses. We have $C = \{e\}$. We could reformulate the theory without individual constants by replacing a_3 and a_4 by $(\exists e)(a_3 \wedge a_4)$.

This theory is too restrictive for the study of group theory. Within it, we can prove results such as that e is a right identity or that the identity is unique. But we have no way of expressing properties of subsets, so we cannot discuss subgroups. Nor can we discuss relationships between groups. We called this theory *elementary* group theory because it is restricted to the relationships between elements of a group (as distinct from the relationships between subsets of a group). We shall use the word "elementary" with this meaning in relation to other theories.

Exercises

2.5. Show that $(\forall x)m(x, e, x)$, $(\forall e')((\forall x)m(e', x, x) \Rightarrow \mathcal{I}(e, e'))$, $(\forall x)(\forall y)(m(y, x, e) \Rightarrow m(x, y, e))$ are theorems of the elementary group theory of Example 2.4.

2.6. Show that the formal analogue of the statement "There exist four distinct lines, no three of which are concurrent" is a theorem of the first-order plane projective geometry of Example 2.3.

2.7. $\mathcal{T} = (\mathcal{R}, A, C)$ is a first-order theory and $(\alpha, \beta):(P(V, \mathcal{R}), V) \to (P(V, \mathcal{R}), V)$ is a semi-homomorphism such that $\alpha(A) \subseteq \text{Ded}_{\mathcal{J}}(A)$. If $\mathcal{T} \vdash p$, prove that $\mathcal{T} \vdash \alpha(p)$.

2.8. $\mathcal{T}$ is the first-order plane projective geometry of Example 2.3. The dual $\bar{w}$ of an element $w \in P(V, \mathcal{R})$ is the element obtained from w by interchanging p and ℓ and replacing $\in(x, y)$ by $\in(y, x)$ (all $x, y \in V$) throughout Show that if α is the map $\alpha(w) = \bar{w}$, and if β is the identity map, then (α, β) is a semi-homomorphism (in fact an automorphism) of $P(V, \mathcal{R})$, satisfying the condition of Exercise 2.7. Hence prove that the dual of a theorem of $\mathcal{T}$ is a theorem of $\mathcal{T}$.

The examples given above show how particular mathematical systems may be used to construct first-order theories. We regard the concept of a first-order theory as fundamental to our study of the relationship between reasoning and mathematics, and our direction is set firmly by the next definition. We denote by rel(M) the set of all relations on a set M.

Definition 2.9. A *model* of the first-order theory $\mathcal{T} = (\mathcal{R}, A, C)$ is a set M together with functions $v:C \to M$, $\psi:\mathcal{R} \to \text{rel}(M)$, such that for some set V of variables ($V \supset C$, $V - C$ infinite), there exists a proper interpretation (M, φ, ψ, v) of $P(V, \mathcal{R})$ for which $\varphi|_C = v$ and $v(A) \subseteq \{1\}$.

We think of a model (M, v, ψ) of the theory $\mathcal{T}$ as the essential part of a proper interpretation of $\mathcal{T}$ for which the axioms of $\mathcal{T}$ are true (i.e., for which $v(A) \subseteq \{1\}$). Although the valuation v of $P(V, \mathcal{R})$ is determined by φ and ψ, the restriction $v|_{\mathcal{L}(\mathcal{T})}$ is completely determined by v and ψ, and is independent of the choice of V and the interpretation. Hence there is a well-determined valuation v of $\mathcal{L}(\mathcal{T})$ corresponding to each model (M, v, ψ) of $\mathcal{T}$, and we say that $p \in \mathcal{L}(\mathcal{T})$ is *true for the model* (M, v, ψ) of $\mathcal{T}$ if $v(p) = 1$. We shall refer to the model (M, v, ψ) of $\mathcal{T}$ as the model M of $\mathcal{T}$, whenever this abuse of notation does not lead to confusion.

Example 2.10. Let G be a group with multiplication written as juxtaposition and with identity element 1. We put $v(e) = 1$, and we put $\psi m = \{(x, y, z) \in G^3 | xy = z\}$. $\psi \mathcal{I}$ will of course be the identity relation. Then $G = (G, v, \psi)$ is a model of the elementary group theory of Example 2.4. A model of the elementary group theory is essentially a group.

Given a model (M, v, ψ) of the theory $\mathcal{T}$, some relations on the set M are derived naturally from $\mathcal{T}$, in the following manner. Let $p(x_1, \ldots, x_n)$ be an

n-variable formula of $\mathcal{T}$. For any $m_1, \ldots, m_n \in M$, there is an interpretation (M, φ, ψ, v) of $P(\mathcal{T})$ such that $\varphi|_C = v$ and $\varphi(x_i) = m_i$ for $i = 1, \ldots, n$. These conditions on φ determine $v(p)$, and if $v(p) = 1$, we say that $(m_1, \ldots, m_n)$ satisfies $p(x_1, \ldots, x_n)$, or (by abuse of language) that $p(m_1, \ldots, m_n)$ is true in M. Hence $p(x_1, \ldots, x_n)$ defines an n-ary relation on M, which (by abuse of notation) we denote by $\psi(p)$:

$$\psi(p) = \{(m_1, \ldots, m_n) \in M^n | p(m_1, \ldots, m_n) \text{ is true in } M\}.$$

This leads to the following definition.

Definition 2.11. The n-ary relation ρ on the model M of $\mathcal{T}$ is said to be *definable in $\mathcal{T}$* if $\rho = \psi(p)$ for some n-variable formula p of $\mathcal{T}$. The function $f : M^n \to M$ is called a *definable function* if there is an $(n + 1)$-variable formula p of $\mathcal{T}$ such that

(i) for all $a_1, \ldots, a_n, b \in M$, $f(a_1, \ldots, a_n) = b$ if and only if $p(a_1, \ldots, a_n, b)$ is true, and

(ii) $\mathcal{T} \vdash (\forall x_1) \cdots (\forall x_n)(\exists! y) p(x_1, \ldots, x_n, y)$.

Example 2.12. Conjugacy is a definable relation in elementary group theory. It is defined by the formula

$$p(x_1, x_2) = (\exists x_3)(\exists x_4)(\exists x_5)(m(x_3, x_4, e) \wedge m(x_3, x_1, x_5) \wedge m(x_5, x_4, x_2)).$$

Inverse is a definable function, defined by the formula

$$q(x_1, x_2) = m(x_2, x_1, e).$$

§3 Properties of First-Order Theories

Definition 3.1. The first-order theory $\mathcal{T}$ is called *consistent* if F is not a theorem of $\mathcal{T}$.

The Soundness and Satisfiability Theorems for $\mathrm{Pred}_f(V, \mathcal{R})$ immediately give the following result.

Theorem 3.2. *The theory $\mathcal{T}$ is consistent if and only if there exists a model of $\mathcal{T}$.*

Definition 3.3. The theory $\mathcal{T}$ is called *complete* if, for every $p \in \mathcal{L}(\mathcal{T})$, either $\mathcal{T} \vdash p$ or $\mathcal{T} \vdash \sim p$.

Elementary group theory is not complete, for consider $p = (\forall x).\mathcal{I}(x, e) \in \mathcal{L}(\mathcal{T})$. If p were a theorem of the theory, it would be true for every model. But p is true for the group G if and only if the order of G is 1. As there are groups of order greater than 1, p cannot be a theorem. As there are groups of order 1, $\sim p$ cannot be a theorem.

Theorem 3.4. *The first-order theory $\mathcal{T}$ is complete if and only if every $p \in \mathcal{L}(\mathcal{T})$ which is true in one model of $\mathcal{T}$ is true in every model of $\mathcal{T}$.*

Proof: The result is trivial if $\mathcal{T}$ is inconsistent, so we suppose $\mathcal{T}$ consistent. Suppose that $\mathcal{T}$ is complete, and that $p \in \mathcal{L}(\mathcal{T})$ is true in the model M. Since $\sim p$ is false for M, it is not a theorem of $\mathcal{T}$, and, since $\mathcal{T}$ is complete, it follows that $\mathcal{T} \vdash p$. Therefore p is true in every model of $\mathcal{T}$.

Suppose conversely that for all $p \in \mathcal{L}(\mathcal{T})$, p true in one model implies p true in every model. Take some model M of $\mathcal{T}$, and let $p \in \mathcal{L}(\mathcal{T})$. If p is true in M, then p is true in every model, i.e., $\mathcal{T} \vDash p$, and hence by the Adequacy Theorem $\mathcal{T} \vdash p$. If p is false in M, then $\sim p$ is true in M and so $\mathcal{T} \vdash \sim p$. Thus $\mathcal{T}$ is complete. $\square$

Examples of complete theories are easily produced, as we now show.

Theorem 3.5. *Let* $\mathcal{T} = (\mathcal{R}, A, C)$ *be a consistent theory. Then there exists* $A' \subseteq \mathcal{L}(\mathcal{T})$, *with* $A' \supseteq A$ *and such that* $(\mathcal{R}, A', C)$ *is consistent and complete.*

Proof: Since $\mathcal{T}$ is consistent, it has a model M say. Put $A' = \{p \in \mathcal{L}(\mathcal{T}) | p$ true in $M\}$. Then A' has the required properties. $\square$

Definition 3.6. Let (M_1, v_1, ψ_1) and (M_2, v_2, ψ_2) be models of the theory $\mathcal{T}$. We say that M_1 is *isomorphic* to M_2 if there exists a bijective map $\alpha : M_1 \to M_2$ such that $\alpha v_1 = v_2$ and $(m_1, \ldots, m_n) \in \psi_1 r$ if and only if $(\alpha(m_1), \ldots, \alpha(m_n)) \in \psi_2 r$ for all $r \in \mathcal{R}_n$, all $m_1, \ldots, m_n \in M_1$, and all $n \in \mathbf{N}$.

Definition 3.7. The theory $\mathcal{T}$ is called *categorical* if all models of $\mathcal{T}$ are isomorphic.

Examples

3.8. Two models G_1, G_2 of elementary group theory are isomorphic as models if and only if they are isomorphic in the group theoretic sense. Since there exist groups G_1, G_2 which are not isomorphic, elementary group theory is not categorical.

3.9. We form trivial group theory by adding the further axiom $(\forall x) \mathcal{I}(x, e)$ to elementary group theory. A model of trivial group theory is a group of order 1. Any two such groups are isomorphic, thus trivial group theory is categorical.

Observe that if M_1, M_2 are isomorphic models of the theory $\mathcal{T}$ and if $p \in \mathcal{L}(\mathcal{T})$ is true for M_1, then it is true for M_2. The definition of categoricity, together with Theorem 3.4, immediately yields the next theorem.

Theorem 3.10. *If the theory* $\mathcal{T}$ *is categorical, then* $\mathcal{T}$ *is complete.*

We now generalise the concept of categoricity. We shall denote the cardinal of any set X by $|X|$.

Definition 3.11. The *cardinal* of a model $M = (M, v, \psi)$ is the cardinal $|M|$ of the set M, and will be denoted by $|M|$.

Note that isomorphic models have the same cardinal.

Definition 3.12. Let χ be a cardinal number. The theory $\mathscr{T}$ is called χ-*categorical* or *categorical in cardinal* χ if all models of $\mathscr{T}$ which have cardinal χ are isomorphic.

Example 3.13. Elementary group theory is categorical in cardinal 1. It is not categorical in cardinal 4, because there are two distinct isomorphism classes of groups of order 4.

Provided that χ is a finite cardinal, there is in the language $\mathscr{L}(\mathscr{T})$ of any theory $\mathscr{T}$ an element which specifies χ as the cardinal of a model of $\mathscr{T}$. For if we denote by al(n) the proposition

$$(\exists a_1) \cdots (\exists a_n)(\sim \mathscr{I}(a_1, a_2) \wedge \sim \mathscr{I}(a_1, a_3) \wedge \cdots \wedge \sim \mathscr{I}(a_1, a_n)$$
$$\wedge \sim \mathscr{I}(a_2, a_3) \wedge \cdots \wedge \sim \mathscr{I}(a_{n-1}, a_n)),$$

then any model of $\mathscr{T}$ in which al(n) is true has at least n elements. Any model in which al$(n) \wedge \sim$al$(n + 1)$ is true has exactly n elements.

Theorem 3.14. *Suppose the theory $\mathscr{T}$ has models of arbitrarily large finite cardinal. Then $\mathscr{T}$ has an infinite model.*

Proof: Let $\mathscr{T} = (\mathscr{R}, A, C)$, and put $\mathscr{T}' = (\mathscr{R}, A', C)$ where $A' = A \cup \{\text{al}(n) | n \in \mathbf{N}^+\}$. We show that $\mathscr{T}'$ is consistent. If $A' \vdash_\mathscr{I} F$, then $A \cup N \vdash_\mathscr{I} F$ for some finite subset N of $\{\text{al}(n) | n \in \mathbf{N}^+\}$. Let $n_0 = \max \{n | \text{al}(n) \in N\}$. By hypothesis, there exists a model M of $\mathscr{T}$ with $|M| \geqslant n_0$. This M is a model of $(\mathscr{R}, A \cup N, C)$, which contradicts the hypothesis $A \cup N \vdash_\mathscr{I} F$. Hence $\mathscr{T}'$ is consistent, and so it has a model. Any model M of $\mathscr{T}'$ must satisfy $|M| \geqslant n$ for all $n \in \mathbf{N}^+$, hence $|M|$ is infinite. $\square$

Exercises

3.15. R is a ring with 1. Construct an elementary theory (i.e., one concerned with elements and not with subsets or maps) Mod$_R$, of unital (left) R-modules, such that the models of the theory are precisely all unital R-modules. (Hint: take each $r \in R$ as a binary relation symbol, interpreting $r(m_1, m_2)$ as $rm_1 = m_2$.)

3.16. Construct an elementary theory of fields with constants 0, 1. In the language $\mathscr{L}(\mathscr{F})$ of this theory $\mathscr{F}$, construct a proposition char(n), which asserts that the characteristic divides n ($n \in \mathbf{N}^+$). Hence construct a theory $\mathscr{F}_0$ of fields of characteristic 0 such that $\mathscr{L}(\mathscr{F}_0) = \mathscr{L}(\mathscr{F})$ and the set A_0 of axioms of $\mathscr{F}_0$ includes the set A of axioms of $\mathscr{F}$. Show that for each theorem p of $\mathscr{F}_0$, there is a number $n_p \in \mathbf{N}$ such that p is true for all fields of characteristic greater than n_p. Show that no set A_1 of axioms, such that $\mathscr{L}(\mathscr{F}) \supseteq A_1 \supseteq A$ and A_1 contains only finitely many elements of $\mathscr{L}(\mathscr{F}) - A$, can axiomatise fields of characteristic 0.

Definition 3.17. The *cardinal* $|\mathscr{T}|$ of the theory $\mathscr{T} = (\mathscr{R}, A, C)$ is $|\mathscr{R} \cup A|$. $\mathscr{T}$ is called *finite* if $\mathscr{R} \cup A$ is finite. $\mathscr{T}$ is called *finitely axiomatised* if A is finite.

Since each element of C is a variable of some axiom, and since each axiom involves only finitely many variables, we have either that C and A are both finite, or that $|C| \leqslant |A|$. If $\mathcal{R} \cup A$ is infinite, then $|\mathcal{L}(\mathcal{T})| = |\mathcal{R} \cup A|$, while $\mathcal{L}(\mathcal{T})$ is countable if $\mathcal{R} \cup A$ is finite. We remark that a relation symbol not occurring in any axiom would be of little interest, as it could be interpreted as any relation and so could occur in a theorem of $\mathcal{T}$ only in an essentially trivial way. It would not be a serious restriction to require every relation symbol to appear in some axiom, in which case we would have either $\mathcal{R}$ and A finite or $|\mathcal{R}| \leqslant |A| = |\mathcal{R} \cup A|$. When A is finite, the actual value of $|A|$ is of no real interest, because an axiom set $A = \{a_1, \dots, a_n\}$ can always be replaced by $A' = \{a_1 \wedge \cdots \wedge a_n\}$ without making any essential change in the theory.

The following theorem is the main result of the present chapter, and is in fact the fundamental theorem of model theory.

Theorem 3.18. (Löwenheim-Skolem Theorem). *Let $\mathcal{T}$ be a first-order theory of cardinal χ, and let $\aleph$ be any infinite cardinal such that $\aleph \geqslant \chi$. Suppose $\mathcal{T}$ has an infinite model. Then $\mathcal{T}$ has a model of cardinal $\aleph$.*

Proof: Suppose $\mathcal{T} = (\mathcal{R}, A, C)$. Choose some set $V_0 \supset C$ such that $|V_0 - C| = \aleph$. Then $|P(V_0, \mathcal{R})| = \aleph$. Put

$$A_0' = I \cup A \cup \{\sim \mathcal{I}(x, y) | x, y \in V_0 - C, x \neq y\}.$$

This gives a theory $\mathcal{T}' = (\mathcal{R}, A_0', V_0)$ which we prove consistent. If $\mathcal{T}'$ is inconsistent, then F is provable from A and some finite subset of $\{\sim \mathcal{I}(x, y) | x, y \in V_0 - C, x \neq y\}$, which contradicts the hypothesis that $\mathcal{T}$ has an infinite model. Therefore $\mathcal{T}'$ is consistent.

We follow the method used to prove the Satisfiability Theorem (cf Lemma 4.14 of Chapter IV), and construct inductively sets V_n, A_n' and A_n. We put

$$V_{n+1} = V_n \cup \{t_q^{(n)} | q(x) \in P(V_n, \mathcal{R}), (\exists x) q(x) \in A_n\},$$
$$A_{n+1}' = A_n \cup \{q(t_q^{(n)}) | q(x) \in P(V_n, \mathcal{R}), (\exists x) q(x) \in A_n\},$$

and take for A_{n+1} a maximal consistent subset of $P(V_{n+1}, \mathcal{R})$ containing A_{n+1}'. We put $V^* = \bigcup_n V_n$, $A^* = \bigcup_n A_n$, $P^* = P(V^*, \mathcal{R}) = \bigcup_n P(V_n, \mathcal{R})$.

Since $A_0 \supseteq A_0'$ is a maximal consistent subset of $P(V_0, \mathcal{R})$, and since $|P(V_0, \mathcal{R})| = \aleph$, we have $|A_0| = \aleph$. Then, from $|P(V_n, \mathcal{R})| = |A_n| = \aleph$, it follows that $|V_{n+1}| = \aleph$, and so that $|P(V_{n+1}, \mathcal{R})| = |A_{n+1}| = \aleph$. By induction, $|P(V_n, \mathcal{R})| = \aleph$ for all n, and therefore $|P^*| = \aleph$.

As in the proof of the Satisfiability Theorem, we construct an interpretation (P^*, φ, ψ, v) for which $v(A^*) = \{1\}$. In this interpretation, $\psi \mathcal{I}$ is an equivalence relation, and by replacing elements of P^* by their equivalence classes, we obtain a proper interpretation $(P^*/\psi \mathcal{I}, \bar{\varphi}, \bar{\psi}, v)$. Restricting $\bar{\varphi}$ to $\mathcal{L}(\mathcal{T}')$ gives a model M of $\mathcal{T}'$. Since $|P^*| = \aleph$, $|M| = |P^*/\psi \mathcal{I}| \leqslant \aleph$. But the construction of A_0' ensures that any model of $\mathcal{T}'$ has cardinal at least $\aleph$. Therefore $|M| = \aleph$. Restricting to $\mathcal{L}(\mathcal{T})$ converts M into a model of $\mathcal{T}$. $\square$

Corollary 3.19. *If the first-order theory $\mathscr{T}$ has an infinite model, then $\mathscr{T}$ is not categorical.* (Proof obvious.)

Corollary 3.20. *Suppose the theory $\mathscr{T}$ has cardinal χ, and has only infinite models. Suppose also that $\mathscr{T}$ is categorical in some infinite cardinal $\aleph \geqslant \chi$. Then $\mathscr{T}$ is complete.*

Proof: Let $p \in \mathscr{L}(\mathscr{T})$, and suppose that neither p nor $\sim p$ is a theorem of $\mathscr{T} = (\mathscr{R}, A, C)$. Since $\sim p$ is not a theorem, $\mathscr{T}$ has a model (infinite) in which p is true, and so the theory $\mathscr{T}' = (\mathscr{R}, A \cup \{p\}, C)$ has an infinite model. Since $|\mathscr{T}'| \leqslant \aleph$, $\mathscr{T}'$ has a model M' of cardinal $\aleph$. Similarly $\mathscr{T}'' = (\mathscr{R}, A \cup \{\sim p\}, C)$ has a model M'' of cardinal $\aleph$. But M' and M'' are each models of $\mathscr{T}$ of cardinal $\aleph$, and hence are isomorphic, contrary to p being true in M' and false in M''. $\quad\square$

Exercises

3.21. A dense linearly ordered set is a non-empty set with a binary relation $<$ such that
(a) for all x, y, exactly one of $x < y$, $x = y$, $y < x$ holds,
(b) if $x < y$ and $y < z$, then $x < z$,
(c) if $x < y$, then there exists z such that $x < z < y$,
(d) for each x, there exist y, z such that $y < x < z$.

Using a binary relation symbol ℓ, with $\ell(x, y)$ to be thought of as $x < y$, and also $\mathscr{I}$, construct a finite theory $\mathscr{D}$ whose models are precisely the dense linearly ordered sets. Show that every model of $\mathscr{D}$ is infinite. Prove that $\mathscr{D}$ is categorical in cardinal $\aleph_0$. (Hint: given two countable models of $\mathscr{D}$, enumerate each domain, and then define inductively a mapping, preserving $<$. Show that this map is onto by proving that there can be no first element in any omitted subset of the range space.) Deduce that $\mathscr{D}$ is a complete theory.

3.22. The theory $\mathscr{T} = (\mathscr{R}, A, C)$ has a finite model of cardinal χ. Show that there exists $p \in \mathscr{L}(\mathscr{T})$ such that the models of $\mathscr{T}' = (R, A \cup \{p\}, C)$ are precisely those models of $\mathscr{T}$ which have cardinal χ.

3.23. The theory $\mathscr{T} = (\mathscr{R}, A, C)$ has a model of infinite cardinal χ. Show that there is no subset T of $\mathscr{L}(\mathscr{T})$ such that $\mathscr{T}' = (\mathscr{R}, A \cup T, C)$ is a consistent theory, all of whose models have cardinal χ.

3.24. K is a field. Construct an elementary theory $\mathscr{V}_K$ of vector spaces over K, and, by adding extra axioms, an elementary theory $\mathscr{V}_K^\infty$ of infinite vector spaces over K. Show that $\mathscr{V}_K$ is categorical in any infinite cardinal greater than $|K|$. Hence show that $\mathscr{V}_K^\infty$ is complete. Show that $\mathscr{V}_K$ is not complete.

3.25. $\mathscr{T} = (\mathscr{R}, A, C)$ is a complete theory, which has a finite model $M = (M, \varphi, \psi)$ of cardinal n.
(i) Prove that every model of $\mathscr{T}$ has cardinal n.
Let $M' = (M', \varphi', \psi')$ be another model of $\mathscr{T}$, and let $\alpha : M \to M'$ be bijective. We say α preserves constants if $\alpha\varphi(c) = \varphi'(c)$ for all $c \in C$. We say

that α preserves the relation $r \in \mathcal{R}_t$ if, for all $(m_1, \ldots, m_t) \in M^t$, we have $(m_1, \ldots, m_t) \in \psi r$ if and only if $(\alpha m_1, \ldots, \alpha m_t) \in \psi' r$. We say that α preserves the subset $\mathcal{S}$ of $\mathcal{R}$ if it preserves every $r \in \mathcal{S}$.

(ii) Show that a bijective map $\alpha : M \to M'$ is an isomorphism of models if and only if it preserves constants and preserves $\mathcal{R}$.

Let $a_1, \ldots, a_n$ be the elements of M, so numbered that $\varphi(C) = \{a_1, \ldots, a_k\}$, and let $c_1, \ldots, c_k \in C$ be such that $\varphi(c_i) = a_i$.

(iii) Show that a bijective map α preserves constants if and only if $\alpha(a_i) = \varphi'(c_i)$ for $i = 1, \ldots, k$.

For $r \in \mathcal{R}_t$ and $(i_1, \ldots, i_t) \in \mathbb{Z}_m^t$, we put

$$q(i_1, \ldots, i_t) = \begin{cases} r(x_{i_1}, \ldots, x_{i_t}) \text{ if } (a_{i_1}, \ldots, a_{i_t}) \in \psi r, \\ \sim r(x_{i_1}, \ldots, x_{i_t}) \text{ if } (a_{i_1}, \ldots, a_{i_t}) \notin \psi r, \end{cases}$$

and

$$r^*(x_1, \ldots, x_n) = \bigwedge_{(i_1, \ldots, i_t)} q(i_1, \ldots, i_t).$$

Write $\mathrm{dist}(x_1, \ldots, x_n)$ for $\sim \mathcal{I}(x_1, x_2) \wedge \sim \mathcal{I}(x_1, x_3) \wedge \cdots \wedge \sim \mathcal{I}(x_1, x_n)$ $\wedge \cdots \wedge \sim \mathcal{I}(x_{n-1}, x_n)$.

(iv) Show that

$$(\exists x_{k+1}) \cdots (\exists x_n)(\mathrm{dist}(c_1, \ldots, c_k, x_{k+1}, \ldots, x_n) \wedge$$
$$r^*(c_1, \ldots, c_k, x_{k+1}, \ldots, x_n))$$

is a theorem of $\mathcal{T}$, and hence show that there exists $\alpha : M \to M'$ which is bijective, preserves constants and preserves r. Extend this argument to show for any finite subset $\mathcal{S}$ of $\mathcal{R}$, that there is a bijective map $\alpha : M \to M'$ preserving constants and $\mathcal{S}$.

(v) Using the fact that there are only finitely many bijective maps $\alpha : M \to M'$, and observing that if some given α preserving constants is not an isomorphism, then there is some $r \in \mathcal{R}$ not preserved by α, prove that M and M' are isomorphic. Hence prove that $\mathcal{T}$ is categorical.

§4   Reduction of Quantifiers

In any study of the decidability properties of a theory $\mathcal{T} = (\mathcal{R}, A, C)$, one expects those elements $q \in P(\mathcal{T})$ which involve no quantifiers to pose the least difficulty. If $q \in P(\mathcal{T})$ is quantifier-free, it is a propositional combination of primitive propositions $r(v_1, \ldots, v_n)$ $(r \in \mathcal{R}_n, v \in V)$, whose truth or falsity for any given interpretation of P is easily determined. Truth functions then decide the truth or falsity of q. There are theories $\mathcal{T}$ having the property that, for any $p \in P(\mathcal{T})$, a quantifier-free element $q \in P(\mathcal{T})$ can be found such that $\mathcal{T} \vdash p \Leftrightarrow q$ and $\mathrm{var}(q) \subseteq \mathrm{var}(p)$. Such a process of quantifier elimination could be useful in investigating the completeness or decidability of $\mathcal{T}$. In practice, it is rarely possible to eliminate quantifiers completely, and one must be content with a quantifier-reduction procedure. The resulting element q is then a propositional combination of relatively simple propositions which possibly involve quantifiers. We shall call these "simple" elements of $P(\mathcal{T})$ *primary propositions*.

Definition 4.1. Let $\Pi \subseteq P(\mathcal{T})$. We say that $\mathcal{T}$ admits Π-*reduction of quantifiers* if there is a process which assigns to each $p \in P(\mathcal{T})$ an element $q \in P(\mathcal{T})$ such that

(i) q is a propositional combination of elements of Π,

(ii) $\mathrm{var}(q) \subseteq \mathrm{var}(p)$,

and

(iii) $\mathcal{T} \vdash p \Leftrightarrow q$.

The utility of such a reduction procedure for any investigation depends on the relative simplicity of the elements of Π compared to the elements of $P(\mathcal{T})$. Every theory admits the useless reduction given by $\Pi = P(\mathcal{T})$ and $q = p$. We give a more helpful example.

Example 4.2. Let $\mathscr{E} = (\{\mathscr{I}\}, \varnothing, \varnothing)$ be the theory of equality, with $V = V_0 = \{x_n | n \in \mathbf{N}\}$. We write $\mathscr{I}(x, y)$ as $x = y$, and $\sim\mathscr{I}(x, y)$ as $x \neq y$. As the set of primary propositions, we take

$$\Pi = \{\mathrm{al}(n) | n \in \mathbf{N}\} \cup \{x_i = x_j | i, j \in \mathbf{N}\}.$$

We introduce an abbreviation which we shall use in describing the reduction process. For $1 \leqslant r \leqslant s$, put

$$\mathrm{dist}_r(x_1, \ldots, x_s) = \bigvee_\alpha \left(\bigwedge_{i < j \leqslant r} (x_{\alpha_i} \neq x_{\alpha_j}) \wedge \bigwedge_{i=r+1}^{s} \left(\bigvee_{j=1}^{r} (x_{\alpha_i} = x_{\alpha_j}) \right) \right),$$

where α ranges over the permutations $(\alpha_1, \ldots, \alpha_s)$ of $(1, \ldots, s)$. Observe that $\mathrm{dist}_r(x_1, \ldots, x_s)$ is true in an interpretation if and only if the interpretation of $x_1, \ldots, x_s$ gives exactly r distinct elements of the model. Now put

$$\mathrm{only}(x_1, \ldots, x_s) = \bigvee_{r=1}^{s} (\mathrm{dist}_r(x_1, \ldots, x_s) \wedge \sim\mathrm{al}(r + 1)),$$

and observe that this is true for an interpretation if and only if the interpretations of $x_1, \ldots, x_s$ are all the elements of the model. (Thus, $\mathrm{only}(x_1, \ldots, x_s)$ is true if and only if $(\forall x)((x = x_1) \vee (x = x_2) \vee \cdots \vee (x = x_s))$ is true.) It is clear that $\mathrm{only}(x_1, \ldots, x_s)$ is a propositional combination of elements of Π.

The following set of instructions constitutes the reduction process:

Step 0. If p is quantifier-free, put $q = p$ and stop. Otherwise, express p in prenex normal form (see Exercise 4.18 of Chapter IV) $Q_1 Q_2 \cdots Q_r p_1$, where the Q_i are quantifiers and p_1 is quantifier-free (and hence a propositional combination of elements of the form $x_i = x_j$).

Step 1. We have $p = Q_1 Q_2 \cdots Q_r p_1$, where the Q_i are quantifiers and p_1 is a propositional combination of elements of Π. If $r = 0$, put $q = p$ and stop. If Q_r is a universal quantifier ($\forall x$), proceed to Step 2. If Q_r is an existential quantifier ($\exists x$), then replace Q_r by $\sim(\forall x)\sim$.

Step 2. We have $p = Q(\forall x)p_1$, where Q consists of a (possibly empty) string of quantifiers and possibly a negation, and p_1 is a propositional combination of elements of Π. If $x \notin \text{var}(p_1)$, replace p by Qp_1 and begin again at Step 1. If $x \in \text{var}(p_1)$, express p_1 in conjunctive normal form (see Exercise 3.10 of Chapter III):

$$p_1 = a_1 \wedge a_2 \wedge \cdots \wedge a_k,$$

where $a_j = d_{1j} \vee d_{2j} \vee \cdots \vee d_{n_j,j}$, with each d_{ij} either a primary proposition or the negation of a primary proposition.

Step 3. We have $p = Q(\forall x)p_1$, with $p_1 = \bigwedge\limits_{j=1}^{k} \bigvee\limits_{i=1}^{n_j} d_{ij}$, where each d_{ij} is primary or the negation of a primary proposition. For each $j = 1, 2, \ldots, k$, delete a_j from p_1 if there is an i such that $d_{ij} = (v = v)$ for some $v \in V$, unless this holds for all j, in which case replace $(\forall x)p_1$ by $F \Rightarrow F$ and begin again at Step 1.

Step 4. We have $p = Q(\forall x)p_1$, with $p_1 = \bigwedge\limits_{j=1}^{k} \bigvee\limits_{i=1}^{n_j} d_{ij}$, where d_{ij} is primary or the negation of a primary proposition. For each $j = 1, 2, \ldots, k$, delete from a_j every d_{ij} of the form $(v \neq v)$ with $v \in V$, unless for some j every d_{ij} has this form, in which case replace $(\forall x)p_1$ by F and begin again at Step 1.

Step 5. We have $p = Q(\forall x)p_1$, with $p_1 = \bigwedge\limits_{j=1}^{k} \bigvee\limits_{i=1}^{n_j} d_{ij}$, where d_{ij} is primary or the negation of a primary proposition, and where no d_{ij} has the form $(v = v)$ or the form $(v \neq v)$. Put $a'_j = \bigvee \{d_{ij} | x \in \text{var}(d_{ij})\}$, and $a''_j = \bigvee\{d_{ij} | x \notin \text{var}(d_{ij})\}$, so that $a_j = a'_j \vee a''_j$. Since the only elements π of Π for which $x \in \text{var}(\pi)$ are the elements $x = v$ for $v \in V$, it follows that each nonempty a'_j has the form

$$a'_j = (x = v_1) \vee (x = v_2) \vee \cdots \vee (x = v_s) \vee (x \neq w_1) \vee \cdots \vee (x \neq w_t)$$

for elements $v_1, \ldots, v_s, w_1, \ldots, w_t$ of $V - \{x\}$. (Terms $x = x$ or $x \neq x$ are excluded by Steps 3, 4.) For each j such that a'_j is non-empty, then

(a) if $t = 0$, replace a'_j by only $(v_1, \ldots, v_s)$.
(b) if $t = 1$ and $s = 0$, replace a'_j by F.
(c) if $t > 0$ and $(s, t) \neq (0, 1)$, replace a'_j by
$(w_1 \neq w_2) \vee (w_1 \neq w_3) \vee \cdots \vee (w_1 \neq w_t) \vee (v_1 = w_1) \vee \cdots \vee (v_s = w_1)$.

Finally, delete $(\forall x)$. Now return to Step 1.

In the above procedure, each step replaces the given proposition by one equivalent to it (i.e., true for precisely the same interpretations). In Step 5, for example, we note that $(\forall x)(a_1 \wedge \cdots \wedge a_k)$ is equivalent to $(\forall x)a_1 \wedge (\forall x)a_2 \wedge \cdots \wedge (\forall x)a_k$, and consider each $(\forall x)a_j$ separately. At each return to Step 1, the number of quantifiers in the prefix has been reduced, so the process must stop.

We illustrate the use of quantifier reduction by proving that $\mathscr{E}$ is decidable.

Theorem 4.3. *The theory $\mathscr{E} = (\{\mathscr{I}\},\ \varnothing,\ \varnothing)$ is decidable.*

Proof. Let $p \in \mathscr{L}(\mathscr{E})$. The reduction process described above gives an element q, equivalent to p, which is a propositional combination of elements of Π such that $\mathrm{var}(q) \subseteq \mathrm{var}(p)$. Since $\mathrm{var}(p) = \varnothing$, q is a propositional combination of elements of the form $p_n = \mathrm{al}(n + 1)$ $(n \geqslant 1)$. Hence q is a propositional combination of $p_1, p_2, \ldots, p_k$ for some k. Let $f : Z_2^k \to Z_2$ be the corresponding truth function. Then $\mathscr{E} \vdash q$ if and only if $f(x_1, \ldots, x_k) = 1$ for all $(x_1, \ldots, x_k) \in Z_2^k$ such that, for some n $(0 \leqslant n \leqslant k)$, $x_1 = x_2 = \cdots = x_n = 1$ and $x_{n+1} = x_{n+2} = \cdots = x_k = 0$. This is so because these are the only possible combinations of truth values for $p_1, \ldots, p_k$ in models of $\mathscr{E}$. $\quad\square$

We note that there is no need for a formal definition of decidability of a first-order theory when one is proving constructively that a particular theory is decidable—the proof is self-sufficient. Formality is required if one is to show the nonexistence of a decision process, as we shall do in Chapter IX. We also remark that the above result, on the decidability of the theory of equality, is not in conflict with the theorem of Kalmar mentioned in Chapter IV and proved in Chapter IX. Although the theory of equality involves a binary predicate symbol, it also includes the axioms of identity.

Exercise 4.4. Show that the theory $\mathscr{D}$ (Exercise 3.21) of dense linear order admits Π-reduction of quantifiers with $\Pi = \{(x_i = x_j),\ (x_i < x_j) \mid i, j \in \mathbf{N}\}$. Hence show that $\mathscr{D}$ is decidable and complete.

Chapter VI

Zermelo-Fraenkel Set Theory

§1   Introduction

All the ordinary mathematical systems are constructed in terms of sets. If we wish to study the reasoning used in mathematics, our model of mathematics must include some form of set theory, for otherwise our study must be restrictive. For example, Elementary Group Theory formalises almost nothing of group theory. The pervasive role of set theory in mathematics implies that any reasonable model of set theory will in effect contain a model of all of mathematics (including the mathematics of this book).

The informal way in which properties of sets are used in mathematics often means that one is aware of some of the more useful axioms of set theory without necessarily having seen or studied sets as an axiomatic theory. In those parts of mathematics where a careful account of set theory is needed, the axiomatisation usually chosen is the one known as Zermelo-Fraenkel Set Theory. We shall set out the axioms of this theory (which we denote by ZF) with some brief comments on the significance of the various axioms. We shall then see how this theory ZF may be formalised within $\text{Pred}_A(V, \mathcal{R})$. Finally, we shall consider the significance of some of the results of Chapter V for our formalised set theory. The reader interested in a more detailed account of ZF is referred to [4].

§2   The Axioms of ZF

ZF is the study of a single type of object. Objects of this type will be called sets. We shall admit another type of object, called a property of a set, but the objects which make up any set will themselves be sets. Since one customarily forms sets whose members are mathematical or physical objects of diverse types, the requirement that members of sets must themselves be sets may therefore seem restrictive. Experience has shown that with some exceptions (which can be accommodated by an extension of the theory), all the objects used in mathematics can be constructed as sets, while we can avoid the need to form sets of physical objects by assigning mathematical names to the objects and using the set of names.

In ZF, we study a single relationship[1] between sets. This relationship is called membership and will be denoted by $\in$. Thus $x \in y$ is read "(the set) x

---

[1] We cannot formalise this relationship as a set of pairs, for we are after all just beginning to define our set theory. Later, when we have constructed ZF, we shall see that the collection of pairs involved cannot be a set within ZF.

is a member of (the set) y", or "x belongs to y". We also study property relationships, which are of the form "the set x has the property π".

In the list of axioms of ZF which follows, some are described as axioms, others as axiom schemas. The distinction will be explained when we construct First-Order ZF.

(ZF1) Axiom of Extension. *If a and b are sets, and if for all sets x we have $x \in a$ if and only if $x \in b$, then $a = b$.*

Thus two sets are equal if and only if they have the same members. We shall write $a \subseteq b$ if $x \in a$ implies $x \in b$.

(ZF2) Axiom Schema of Subsets. *For any set a and any property π, there is a set b such that $x \in b$ if and only if $x \in a$ and has the property π.*

By (ZF1), this set is unique. We denote it by $\{x \in a | x \text{ has } \pi\}$. Assuming that at least one set a exists, we can form the set $\varnothing = \{x \in a | x \neq x\}$. Then for all x we have $x \notin \varnothing$. This set $\varnothing$, which is called the empty set, is independent of the choice of the set a used in its construction. By (ZF1), $\{x \in a_1 | x \neq x\} = \{x \in a_2 | x \neq x\}$. It is clear that for all sets b, $\varnothing \subseteq b$.

(ZF2) restricts the way in which a property may be used to form a set, and thereby, the Russell paradox is avoided. It used to be assumed that, for any property π, one could form the set of all objects with that property. Russell considered the property of not being a member of itself. If b is the set of all sets which are not members of themselves, then consideration of whether or not b is a member of itself leads at once to a contradiction. Using (ZF2), one can only form $b = \{x \in a | x \notin x\}$ starting from some given set a. We then find that $b \in b$ is impossible, hence $b \notin b$ and so $b \notin a$. The argument does not lead to a contradiction, but instead proves that for any a, there is a b such that $b \notin a$. Thus there is no set of all sets.

(ZF3) Axiom of Pairing. *If a and b are sets, then there exists a set c such that $a \in c$ and $b \in c$.*

Using (ZF2) with this set c, we can form the set $\{x \in c | x = a \text{ or } x = b\}$. This is independent of the particular set c having a and b as members, and we call $\{x \in c | x = a \text{ or } x = b\}$ the unordered pair whose members are a and b, and denote it by $\{a, b\}$. In the special case where $a = b$, (ZF2) asserts the existence of a set having a as a member. The unordered pair $\{a, a\}$ has only the one member a, and we denote it by $\{a\}$. The ordered pair (a, b) is now defined to be $\{\{a\}, \{a, b\}\}$.

Exercise 2.1. If $(a, b) = (c, d)$, prove $a = c$ and $b = d$. Make sure that your proof allows for the possibility that $a = b$.

For any two sets a, b, we can form $a \cap b = \{x \in a | x \in b\}$. For any non-empty set c, we can form $\cap c = \{x \in b | x \in a \text{ for all } a \in c\}$, where b is some member of c. $\cap c$ is, of course, independent of the choice of b.

Exercise 2.2. Prove that $a \cap b = b \cap a = \cap \{a, b\}$.

Although the axioms already given allow the formation of intersections, the formation of unions requires a further axiom.

(ZF4) Axiom of Union.   *For every set c, there exists a set a such that, if $x \in b$ and $b \in c$, then $x \in a$.*

We can now form $\cup c = \{x \in a | x \in b \text{ for some } b \in c\}$ where a is as in (ZF4). $\cup c$ is again independent of the particular a used, so we write simply $\cup c = \{x | x \in b \text{ for some } b \in c\}$. For any sets a and b, we can form $a \cup b = \cup\{a, b\}$.

Exercise 2.3.   Show that the ordered pairs (a, b) for which $a \in b$ do not form a set. (Assume that there is a set $e = \{(a, b) | a \in b\}$ and show that $\cup(\cup e)$ is the set of all sets.)

The formation of ordered pairs is permitted by the axioms so far given, but not the formation of the set of all ordered pairs of members of given sets. The next axiom remedies this deficiency.

(ZF5) Axiom of the Power Set.   *For each set a, there exists a set b such that, if $x \subseteq a$, then $x \in b$.*

Using (ZF2), we obtain the existence of the power set of a: $\text{Pow}(a) = \{x \in b | x \subseteq a\} = \{x | x \subseteq a\}$, which is clearly independent of the choice of b.

(ZF5) allows formation of the cartesian product $a \times b = \{(x, y) | x \in a \text{ and } y \in b\}$. To show this, we need only produce a set c whose members include all the required ordered pairs (x, y). But $(x, y) = \{\{x\}, \{x, y\}\}$, $\{x\} \subseteq a \cup b$, $\{x, y\} \subseteq a \cup b$, and so both $\{x\}$ and $\{x, y\}$ are members of $\text{Pow}(a \cup b)$. Thus $\{\{x\}, \{x, y\}\} \subseteq \text{Pow}(a \cup b)$, and consequently $(x, y) \in \text{Pow}(\text{Pow}(a \cup b))$ for all $x \in a$ and $y \in b$.

With the cartesian product available, we can now define a relation between two sets a, b as a subset of $a \times b$, and then a function $f : a \rightarrow b$ as a special type of relation. The set of all functions from a to b can be constructed as a subset of $\text{Pow}(\text{Pow}(a \times b))$. For a set c, we define the cartesian product (of the members) of c by $\prod c = \{f : c \rightarrow \cup c | f(x) \in x \text{ for all } x \in c\}$.

Exercises

2.4.   What is $\prod \emptyset$?

2.5.   For any set a, prove that there is no surjective function $f : a \rightarrow \text{Pow}(a)$. (Consider $b = \{x \in a | x \notin f(x)\}$.)

Definition 2.6.   The *successor* of the set x is the set $x^+ = x \cup \{x\}$. The set a is called a *successor set* if $\emptyset \in a$ and $x^+ \in a$ for all $x \in a$.

(ZF6) Axiom of Infinity.   *There exists a successor set.*

This is the first axiom asserting unconditionally that sets exist. In particular, it asserts the existence of $\emptyset$ as this is used in the definition of a successor set. We can now define the set ω of natural numbers:

$$\omega = \{x | x \in a \text{ for every successor set } a\},$$

using (ZF2) and some successor set. We use the usual symbols

$$0 = \varnothing,$$
$$1 = 0^+ = \varnothing \cup \{\varnothing\} = \{\varnothing\} = \{0\},$$
$$2 = 1^+ = 1 \cup \{1\} = \{0, 1\},$$
$$3 = 2^+ = \{0, 1\} \cup \{2\} = \{0, 1, 2\},$$

and so on. The set ω together with the usual operations of addition and multiplication will be denoted by N.

Exercises

2.7. The set n is called transitive if $x \in y$ and $y \in n$ imply that $x \in n$. Show that if n is transitive, then so is n^+.

2.8. If s is a successor set, show that $\{n \in s | n \text{ is transitive}\}$ is a successor set. For all $n \in \omega$, prove that n is transitive.

2.9. Given that $n = \{x \in \omega | x \subset n\}$ and that $n \in \omega$, show that $n^+ = \{x \in \omega | x \subset n^+\}$. Hence prove for all $n \in \omega$ that

(a) $n = \{x \in \omega | x \subset n\}$,
(b) $n \notin n$,
(c) for all $x \in n, n \not\subseteq x$.

$(a \subset b$ means $a \subseteq b$ and $a \neq b$.)

2.10. Show that $0, 1, 2, \ldots$ are all different.

(ZF7) Axiom of Choice. *For each set a, there exists a function $f : \{x \in \text{Pow}(a) | x \neq \varnothing\} \to a$, such that for every non-empty subset x of a, $f(x) \in x$.*

The function f, called a *choice function*, selects from each non-empty subset of a, a member of that subset.

(ZF8) Axiom Schema of Replacement. *If π is a property of pairs of sets such that for all $x \in a$, (x, y) and (x, z) both having π implies that $y = z$, then there exists a set b such that $y \in b$ if and only if there is an $x \in a$ such that (x, y) has π.*

Intuitively, the property π defines a function on some subset of a, and b is the set of images under this function. But a function $f : a \to b$ is a subset of $a \times b$, and this requires b to be a set. The point of this axiom is that although we are not given a function in the formal sense, the type of correspondence it considers does in fact define a function.

(ZF9) Axiom Schema of Restriction. *If π is any property of sets and if there exists a set with π, then there exists a set a with π such that for all $x \in a$, x does not have π.*

(ZF9) excludes the possibility of an infinite sequence $a_1, a_2, \ldots$ of sets such that $a_{i+1} \in a_i$ for all i. To see this, simply take π to be the property of being the first member of some such sequence. By (ZF9), if there exists a set with this property π, then there exists a set a with π such that no member of a

has π. But then a is the first member a_1 of some such sequence $a_1 \ni a_2 \ni \cdots$, and clearly $a_2 \in a$ and has π. Thus there can be no sets with this property.

Exercises

2.11. Show that (ZF9) implies the Axiom of Regularity: For any set $a \neq \varnothing$, there exists $b \in a$ such that $b \cap a = \varnothing$.

2.12. From the Axiom of Regularity, prove that for every set a, $a \notin a$.

2.13. Prove that if $a \subseteq a \times a$, then $a = \varnothing$.

§3   First-Order ZF

We formalise ZF as a first-order theory, which we shall denote by $\mathscr{S}$. We take as relation symbols just $\mathscr{I}, \in$, both binary. We shall use no individual constants in our construction. Where axioms are obvious formalisations of the corresponding informal axioms, we set them down without comment. For ease of understanding, we shall write $x \in y$ and $x = y$ rather than the formally correct $\in(x, y)$ and $\mathscr{I}(x, y)$, and the negations of these statements will be written $x \notin y$ and $x \neq y$.

(ZF1) $(\forall a)(\forall b)(((\forall x)(x \in a \Leftrightarrow x \in b)) \Rightarrow a = b)$.

In the informal version of (ZF2), we used a property π of sets. The informal statement "x has property π." becomes for us the predicate $\pi(x)$, where π is an element of $P(V, \mathscr{R})$, the notation $\pi(x)$ simply describing the dependence of π upon x. (The notation $\pi(x)$ does not imply that $\operatorname{var}(\pi(x)) = \{x\}$.) For given $\pi(x)$, (ZF2) becomes

$$(\forall a)(\exists b)(\forall x)(x \in b \Leftrightarrow (x \in a \wedge \pi(x))),$$

but we must clearly restrict this by requiring that $b \notin \operatorname{var}(\pi(x))$. Moreover, the theory $\mathscr{S}$ is to be without constants, and $\operatorname{var}(\pi(x))$ could have members other than a and x. Thus, if $x_1, \ldots, x_r$ are these other variables, we take as our axiom

$$(\forall x_1) \cdots (\forall x_r)(\forall a)(\exists b)(\forall x)(x \in b \Leftrightarrow (x \in a \wedge \pi(x))).$$

To simplify the notation, we introduce the convention that if $p \in P(V, \mathscr{R})$ and $\operatorname{var}(p) = \{x_1, \ldots, x_n\}$, then $(\forall)p$ denotes $(\forall x_1) \cdots (\forall x_n)p$. The order in which $x_1, \ldots, x_n$ are taken will not matter in any use we make of this notation. Using this convention, the axiom schema becomes

(ZF2) $(\forall)(\forall a)(\exists b)(\forall x)(x \in b \Leftrightarrow (x \in a \wedge \pi(x)))$ *for all* $\pi(x) \in P(V, \mathscr{R})$ *such that* $b \notin \operatorname{var}(\pi(x))$.

Unlike (ZF1), which was a single element of $P(V, \mathscr{R})$, (ZF2) is an infinite collection of axioms, one for each $\pi(x) \in P(V, \mathscr{R})$ satisfying $b \notin \operatorname{var}(\pi(x))$. This is the reason for calling (ZF2) an axiom schema.

Exercise 3.1. Some later axioms of $\mathscr{S}$ will have the form $(\forall)(\exists a)(\forall x)$ $(p(x) \Rightarrow x \in a)$ for certain elements $p(x) \in P(V, \mathscr{R})$. Show that $\{(\forall)(\exists a)(\forall x)$ $(p(x) \Rightarrow x \in a)\} \vdash {}_{\mathscr{S}}(\forall)(\exists a)(\forall x)(p(x) \Leftrightarrow x \in a)$.

We introduce further useful abbreviations. We write $a \subseteq b$ for $(\forall x)(x \in a \Rightarrow x \in b)$, $a = \{x|p(x)\}$ for $(\forall x)(x \in a \Leftrightarrow p(x))$, $a = \{a_1, \ldots, a_n\}$ for $(\forall x)(x \in a \Leftrightarrow (x = a_1 \vee \cdots \vee x = a_n))$, and $c = (a, b)$ for $c = \{\{a\}, \{a, b\}\}$, which itself is an abbreviation whose meaning has been explained. In particular, $a = \varnothing$ is an abbreviation for $(\forall x)(x \notin a)$. We may now write down relatively concise formal versions of four more axioms.

(ZF3) $(\forall a)(\forall b)(\exists c)(a \in c \wedge b \in c)$.
(ZF4) $(\forall c)(\exists a)(\forall x)(((\exists b)(x \in b \wedge b \in c)) \Rightarrow x \in a)$.
(ZF5) $(\forall a)(\exists b)(\forall x)(x \subseteq a \Rightarrow x \in b)$.
(ZF6) $(\exists a)(((\exists b)(b = \varnothing \wedge b \in a)) \wedge (\forall x)(x \in a \Rightarrow$
        $(\exists y)(y = x \cup \{x\} \wedge y \in a)))$.

Exercises

3.2.   Prove $\mathscr{S} \vdash (\forall a)(\forall b)(\exists c)(c = \{a, b\})$.
3.3.   Formalise and prove the formal result that if $(a, b) = (c, d)$, then $a = c$ and $b = d$.
3.4.   Prove $\mathscr{S} \vdash c \neq \varnothing \Rightarrow (\exists d)(\forall x)(x \in d \Leftrightarrow (\forall y)(y \in c \Rightarrow x \in y))$.

In (ZF6), $y = x \cup \{x\}$ is of course an abbreviation for $(\forall z)(z \in y \Leftrightarrow (z = x \vee z \in x))$. We further preserve our informal notations for certain sets by writing $b = \text{Pow}(a)$ for $(\forall x)(x \in b \Leftrightarrow x \subseteq a)$ and $c = a \times b$ for $(\forall x)(x \in c \Leftrightarrow (\exists y)(\exists z)(x = (y, z) \wedge y \in a \wedge z \in b))$.

To make possible a formal version of (ZF7) of reasonable length, we introduce three more abbreviations. We shall write $(\exists!x)p(x)$ for $(\exists x)(p(x) \wedge (\forall y)(p(y) \Leftrightarrow y = x))$ (as in Chapter V), $f: a \to b$ for

$((\exists c)((c = a \times b) \wedge (f \subseteq c)))$
$\wedge (\forall x)(x \in a \Rightarrow (\exists!y)((y \in b) \wedge (\exists z)(z = (x, y) \wedge z \in f)))$,

and $y = f(x)$ for $(\exists z)(z = (x, y) \wedge z \in f)$.

(ZF7) $(\forall a)(\forall b)((b = \{x|(x \subseteq a) \wedge (x \neq \varnothing)\}) \Rightarrow (\exists f)((f: b \to a) \wedge$ $(\forall y)(\forall z)(z = f(y) \Rightarrow z \in y)))$.
(ZF8)   *For every* $p(x, y) \in P(V, \mathscr{R})$, $(\forall)(((\forall x)(\forall y)(\forall z)((x \in a \wedge p(x, y) \wedge$ $p(x, z)) \Rightarrow y = z)) \Rightarrow (\exists b)(\forall y)(y \in b \Leftrightarrow (\exists x)(x \in a \wedge p(x, y))))$.
(ZF9)   *For every* $p(x) \in P(V, \mathscr{R})$,

$$(\forall)((\exists x)p(x) \Rightarrow (\exists a)(p(a) \wedge (\forall x)(x \in a \Rightarrow \sim p(x))))$$

This completes the formalisation of the axioms of our informal set theory, and so completes the list of mathematical axioms of our first-order theory $\mathscr{S}$. By its construction, $\mathscr{S}$ is clearly a consistent theory if our informal set theory is consistent, because any proof of F in $\mathscr{S}$ has an informal equivalent. Since this book (and also much of mathematics) is written in the context of the informal set theory of ordinary mathematics, and since all of this is destroyed if that set theory is inconsistent, we assume the consistency of informal set theory. With this assumption, $\mathscr{S}$ is a consistent theory.

We now observe that the language $\mathscr{L}(\mathscr{S})$ is in fact independent of the choice of the infinite set V of variables used, for if V_0 is a countable subset of V, and if $p \in P(V, \mathscr{R})$ has $\mathrm{var}(p) = \varnothing$, then $p \in P(V_0, \mathscr{R})$, and the result follows on recalling that there are no individual constants in our construction of $\mathscr{S}$. We may therefore suppose that V is countable. Since $\mathscr{R} = \{\mathscr{I}, \in\}$, it follows that $P(V, \mathscr{R})$ is countable and hence that $\mathscr{S}$ is countable. By the Löwenheim-Skolem Theorem, $\mathscr{S}$ has a countable model.

A theorem of ordinary set theory asserts the existence of uncountable sets, and this theorem (with its proof) can be formalised in the theory $\mathscr{S}$. Hence there exists a countable model of a theory which has as a theorem the existence of uncountable sets! The paradox is resolved when we realise that it arises by using the word "set" in two ways. Let us distinguish words used in their ordinary sense from the same words used in the sense of the model by using the adjectives real or model respectively. "$\mathscr{S}$ has a countable model" then becomes "$\mathscr{S}$ has a real countable model", i.e., there is a real function from the real set of natural numbers onto the underlying set of the model. For this model, every model set is at most real countable. But a model set is model countable only if there is a model function from the model set of natural numbers onto it, and the real function which counts it need not be a model function.

§4   The Peano Axioms

We have seen how the natural numbers may be constructed in terms of set theory. We now give an axiomisation of the natural numbers, and study the relationship between this axiomatic system and Zermelo-Fraenkel set theory.

Since addition and multiplication can be defined in terms of the successor function[2], it is sufficient to axiomatise this function. We denote the successor of x by $s(x)$. The Peano axioms for the natural number system N are:

P_1: *0 is a natural number.*

P_2: *If x is a natural number, then $s(x)$ is a uniquely determined natural number* (i.e., s is a function $s: \mathrm{N} \to \mathrm{N}$).

P_3: *If x, y are natural numbers and if $s(x) = s(y)$, then $x = y$.*

P_4: *For each natural number x, $s(x) \neq 0$.*

P_5: *If π is any property such that 0 has π, and such that if x has π then $s(x)$ has π, then every natural number has π.*

---

[2] However, addition and multiplication are not definable within the theory $\mathscr{P}$ we are about to construct. To be able to formalise their definitions, we have to add to $\mathscr{R}$ relation symbols for addition and multiplication. See Exercises 4.2–4.10.

It is a well-known theorem that these axioms determine the system N to isomorphism, i.e., if sets A, A', with functions s, s' respectively, each satisfy the axioms, then there exists a bijective function $f : A \rightarrow A'$ such that $f(0) = 0'$ and $f(s(a)) = s'(f(a))$ for all $a \in A$. An informal proof of this runs as follows. We define $f(0) = 0'$, and, if $f(a) = a'$, define $f(s(a)) = s'(a')$. Taking $\pi(a)$ to be the property that $f(a)$ is uniquely defined by this rule, P_5 then gives the result that f is a function from A to A'. Similarly, we obtain a function $g : A' \rightarrow A$. Taking now $\pi(a)$ to be $g(f(a)) = a$, P_5 gives the result that gf is the identity. Similarly fg is the identity, and so f is the required isomorphism.

The Peano axioms are easily formalised as a first-order theory $\mathscr{P}$. We take one unary relation symbol θ, with $\theta(x)$ to mean $x = 0$, and one binary relation symbol s, with $s(x, y)$ to mean x is the successor of y. The axioms then become

P_1: $(\exists! x)\theta(x)$.
P_2: $(\forall x)(\exists! y)s(y, x)$.
P_3: $(\forall x)(\forall y)(\forall z)((s(z, x) \wedge s(z, y)) \Rightarrow x = y)$.
P_4: $(\forall x)(\forall y)(s(x, y) \Rightarrow \sim \theta(x))$.
P_5: $(\forall)(((\exists x)(\theta(x) \wedge \pi(x)) \wedge (\forall y)(\forall z)(\pi(z) \wedge s(y, z) \Rightarrow \pi(y))) \Rightarrow (\forall y)\pi(y))$,
for all $\pi(x) \in P(V, \mathscr{R})$ such that $y, z \notin \mathrm{var}(\pi(x))$.

$\mathscr{P}$ is clearly a countable theory, and has N as a model. By the Löwenheim-Skolem Theorem, $\mathscr{P}$ is not categorical. This result appears to contradict the theorem that the Peano axioms determine N to isomorphism. But in formalising P_5, we have restricted the application of the axiom to those properties π which are expressible in terms of s and θ, and the properties π used in the uniqueness proof are certainly not of this form. This argument is however only part of the whole story.

Within $\mathscr{P}$, we cannot hope to formalise a proof of the uniqueness theorem. We cannot even state the theorem in $\mathscr{L}(\mathscr{P})$. We need set theory for this, so let us reformulate the Peano axioms within our formal set theory $\mathscr{S}$, as a set of assumptions on a triple $(N, s, 0)$ of sets. We shall take $\mathscr{P}(N, s, 0)$ to be the subset of the first-order algebra of $\mathscr{S}$ consisting of the elements

P_1: $0 \in N$,
P_2: $s : N \rightarrow N$,
P_3: $(\forall x)(\forall y)(\forall z)((z = s(x) \wedge z = s(y)) \Rightarrow x = y)$,
P_4: $(\forall x) \sim (0 = s(x))$,
and all elements of the form
P_5: $(\forall)((\pi(0) \wedge (\forall x)(\forall y)((y = s(x) \wedge \pi(x)) \Rightarrow \pi(y))) \Rightarrow (\forall z)\pi(z))$,
where $y, z \notin \mathrm{var}(\pi(x))$.
We write $(N, s, 0) \simeq (N', s', 0')$ as an abbreviation for

$$(\exists f)((f : N \rightarrow N') \wedge (\forall x)(\forall y)(\forall z)((z = f(x) \wedge z = f(y)) \Rightarrow x = y)$$
$$\wedge (\forall x)(x \in N' \Rightarrow (\exists y)(x = f(y))) \wedge (\forall x)(\forall y)(\forall z)(\forall t)((y = s(x)$$
$$\wedge z = f(x) \wedge t = f(y)) \Rightarrow t = s(z))).$$

It can be shown that $\mathscr{P}(N, s, 0) \cup \mathscr{P}(N', s', 0') \vdash {}_{\mathscr{S}}(N, s, 0) \simeq (N', s', 0')$.

Hence within $\mathscr{S}$, the Peano axioms as now formulated in fact determine N to isomorphism. The axioms of $\mathscr{S}$, together with the assumptions $\mathscr{P}(N, s, 0)$, still do not determine N to isomorphism in the sense of our metalogic. There are non-isomorphic models of $\mathscr{S}$, and the systems of natural numbers within these models may well be non-isomorphic. Our theorem asserts that models of the natural numbers within a given model of $\mathscr{S}$ are isomorphic. Our informal proof worked because we were working within an assumed set theory.

Exercises

4.1. Rephrase our very informal proof of the uniqueness of the natural numbers more carefully in terms of informal axiomatic set theory. (This may be found in [12].) Note that the function f to be constructed is a subset of $N \times N'$ and must be constructed in a way permitted by the axioms. (The inductive construction of f needs justification.) Set out the steps of the argument in sufficient detail for it to become clear that it can be formalised to give a proof that $\mathscr{P}(N, s, 0) \cup \mathscr{P}(N', s', 0') \vdash {}_{\mathscr{A}}(N, s, 0) \simeq (N', s', 0')$.

4.2. Addition is usually defined in terms of the successor function by

(i) $x + 0 = x$, and
(ii) $x + s(y) = s(x + y)$.

Assuming the informal Peano axioms, show that (i) and (ii) define a function $+ : N \times N \to N$, and that

(a) $0 + x = x$,
(b) $s(x) + y = s(x + y)$,
(c) $x + y = y + x$,
(d) $(x + y) + z = x + (y + z)$,
(e) $x + y = x + z$ implies $y = z$.

Give a similar definition of multiplication in terms of addition and the successor function, and establish its basic properties.

In the following exercises, $x_i = n$ (where $n \in N$) is used as an abbreviation for

$$(\exists y_0)(\exists y_1) \cdots (\exists y_{n-1})(\theta(y_0) \wedge s(y_1, y_0) \wedge \cdots \wedge s(x_i, y_{n-1})$$

if $n > 0$, and means $\theta(x_i)$ if $n = 0$. The expression $x_i = x_j + n$ means $x_i = x_j$ if $n = 0$, $s(x_i, x_j)$ if $n = 1$, and

$$(\exists y_1)(\exists y_2) \cdots (\exists y_{n-1})(s(y_1, x_j) \wedge s(y_2, y_1) \wedge \cdots \wedge s(x_i, y_{n-1}))$$

if $n > 1$.

4.3. $\mathscr{P}^*$ is the theory formed from $\mathscr{P}$ by replacing the induction axiom scheme P_5 by

$$P_{5,0}^* : (\forall x)(((\forall y) \sim s(x, y)) \Rightarrow \theta(x)),$$
$$P_{5,n}^* : (\forall x)(x \neq x + n) \qquad (n > 0).$$

$M_I = N \cup (Z \times I)$, where I is some index set. For $m \in M_I$, $\theta(m)$ is interpreted as true if and only if $m = 0 \in N$, and $s(m_1, m_2)$ is true if and only if either $m_2 \in N$ and $m_1 = m_2 + 1$, or $m_2 = (z, i)$, where $z \in Z$ and $i \in I$, and $m_1 = (z + 1, i)$. Show that M_I is a model of $\mathscr{P}^*$, and that every model of $\mathscr{P}^*$ is isomorphic to M_I for some I.

4.4. Prove that every theorem of $\mathscr{P}^*$ is a theorem of $\mathscr{P}$. Hence show that every model of $\mathscr{P}$ is isomorphic to M_I for some I.

4.5. Show that $\mathscr{P}^*$ admits Π-reduction of quantifiers, where

$$\Pi = \{x_i = x_j + n, \, x_i = n \,|\, i, j, n \in N\}.$$

Hence prove that $\mathscr{P}^*$ is decidable and complete.

4.6. Let $\pi(x_0, x_1, \ldots, x_n) \in P(\mathscr{P}) = P(\mathscr{P}^*)$, and let $a_1, \ldots, a_n \in M_I$. Put

$$X = \{m \in M_I \,|\, \pi(m, a_1, \ldots, a_n) \text{ is true in } M_I\}.$$

Using the Π-reduction of quantifiers, show that X is either finite or has finite complement in M_I. Hence prove that M_I satisfies the induction axiom scheme P_5, and so is a model of $\mathscr{P}$.

4.7. From the completeness of $\mathscr{P}^*$ and the fact that every theorem of $\mathscr{P}^*$ is a theorem of $\mathscr{P}$, deduce that every theorem of $\mathscr{P}$ is a theorem of $\mathscr{P}^*$. Hence prove that every M_I is a model of $\mathscr{P}$.

4.8. (Proof that M_I is a model of $\mathscr{P}$ not using reduction of quantifiers.) Show that $\mathscr{P}$ and $\mathscr{P}^*$ are α-categorical for every uncountable cardinal α, and so are complete. As in 4.7, deduce that every M_I is a model of $\mathscr{P}$.

4.9. The theory $\mathscr{A}$ consists of $\mathscr{P}$ together with a ternary relation symbol a and the additional axioms

$$(\forall x)(\forall y)(\exists! z)a(x, y, z),$$
$$(\forall x)(\forall y)(\theta(y) \Rightarrow a(x, y, x)),$$
$$(\forall x)(\forall y)(\forall z)(\forall t)(\forall u)(s(z, y) \wedge a(x, y, t) \wedge a(x, z, u) \Rightarrow s(u, t)).$$

Show that there is no relation on $M_{\{0\}}$ which, taken as ψa, makes $M_{\{0\}}$ a model of $\mathscr{A}$. Hence show that addition is not definable in $\mathscr{P}$.

4.10. Show that not every model of $\mathscr{P}$ is embeddable in a model of $\mathscr{S}$(ZF set theory).

4.11. Taking $x \leqslant y$ as an abbreviation for $(\exists z)a(x, y, z)$, show that the axioms of a total order are theorems of $\mathscr{A}$.

Chapter VII

Ultraproducts

§1   Ultraproducts

In many branches of mathematics, where one is studying a system of some particular type, it is of interest to find out ways of forming new systems of the given type from known examples. One useful method that can often be applied is based on the cartesian product construction. In this section we investigate this construction in the case where the underlying system is a first-order theory $\mathcal{T} = (\mathcal{R}, A, C)$, and (M_i, v_i, ψ_i) for $i \in I$ is a family of models of $\mathcal{T}$. We therefore investigate the possibility of making $M = \prod_{i \in I} M_i$ into a model of $\mathcal{T}$, independently of the particular nature of $\mathcal{T}$.

An element of $\prod_{i \in I} M_i$ is a function $a: I \to \bigcup_{i \in I} M_i$ such that $a(i) \in M_i$. We shall when convenient denote $a(i)$ by a_i, and call it the i-component of a. There is now an obvious way to proceed. We define $v: C \to M$ by putting $v(c)_i = v_i(c)$, and we define ψr, for $r \in \mathcal{R}_n$, by putting $(a^{(1)}, \ldots, a^{(n)}) \in \psi r$ if $(a_i^{(1)}, \ldots, a_i^{(n)}) \in \psi_i r$ for all $i \in I$.

This construction gives a model M of $\mathcal{T}$ in some cases. For example, since a cartesian product of groups is a group, the method works for the case of elementary group theory. However, the method does not work in the case of elementary field theory, because a cartesian product of fields is a commutative ring with 1 having non-zero noninvertible elements. (This is easily seen, because all operations are defined componentwise, and hence $a \in M$ has an inverse if and only if each a_i is invertible. Take an a in which some but not all a_i are invertible.) Hence the above construction must be modified if it is to work for all theories $\mathcal{T}$. We shall have to define $\psi: \mathcal{R} \to \text{rel}(M)$ in such a way that for *every* $p(x_1, \ldots, x_n) \in P(V, \mathcal{R})$, the relation ψp given by p on M corresponds to the relations $\psi_i p$ given on the M_i in precisely the way that the ψr for $r \in \mathcal{R}$ correspond to the $\psi_i r$.

We simplify notation and work only with one variable formulae $p(x)$. (The n-variable case is covered by regarding x as an n-tuple $(x_1, \ldots, x_n)$.) We shall modify the definition of ψ by taking $a \in M$ to be in $\psi p(x)$ if $a_i \in \psi_i p(x)$ for all i in some "suitable" subset of I, where we have yet to decide which subsets of I are to be considered suitable. Since the definition is to apply to all $p \in P$, it applies to $\mathscr{I}(x, y)$. This means that if any subset other than I itself is allowed, $\psi \mathscr{I}$ will not be the identity relation on $\prod_{i \in I} M_i$, but merely an equivalence relation. Therefore we must reduce modulo $\psi \mathscr{I}$ in order to obtain a model of $\mathcal{T}$—the equivalence classes will be the elements of the model.

We now investigate the conditions a family of "suitable" subsets of I must satisfy. Denote such a family by $\mathcal{F}$. Let $p(x), q(x) \in P$, $a \in \prod_{i \in I} M_i$, and let $A = \{i \in I \,|\, a_i \text{ satisfies } p(x)\}$, $B = \{i \in I \,|\, a_i \text{ satisfies } q(x)\}$. Since a formula

62

should hold for some i if it is to hold at all, we have

(i) $\varnothing \notin \mathcal{F}$.

If $A \in \mathcal{F}$, then a satisfies $p(x)$ and so must satisfy $p(x) \vee q(x)$, whatever $q(x)$ may be. Thus for any $B \subseteq I$, $A \cup B \in \mathcal{F}$ if $A \in \mathcal{F}$. Hence

(ii) Every subset of I which contains a set of $\mathcal{F}$ belongs to $\mathcal{F}$.

If $A \in \mathcal{F}$ and $B \in \mathcal{F}$, then a satisfies $p(x)$ and $q(x)$ and so must satisfy $p(x) \wedge q(x)$. Thus $A \cap B \in \mathcal{F}$ if $A, B \in \mathcal{F}$. Generalising to finite subfamilies of $\mathcal{F}$, we have

(iii) Every finite intersection of sets of $\mathcal{F}$ belongs to $\mathcal{F}$.

Finally, since a must satisfy exactly one of $p(x)$ or $\sim p(x)$, we have

(iv) For each $A \subseteq I$, exactly one of A and $I - A$ belongs to $\mathcal{F}$.

Definition 1.1. A set $\mathcal{F}$ of subsets of I satisfying the conditions (i), (ii) and (iii) above is called a *filter* on I. A filter which satisfies (iv) is called an *ultrafilter*.

The filters on I, being subsets of $\mathrm{Pow}(I)$, are partially ordered by inclusion. The ultrafilters are the maximal elements of the set of filters.

Examples

1.2. If $I \neq \varnothing$, $\{I\}$ is a filter on I.

1.3. If k is a fixed element of I, $F = \{J \subseteq I \mid k \in J\}$ is an ultrafilter on I. (Ultrafilters constructed in this way are called *principal* ultrafilters.)

1.4. If I is infinite, the complements of the finite subsets of I form a filter. (When $I = \mathbf{N}$, this filter is called the Fréchet filter.)

Exercise 1.5. $\mathcal{F}$ is an ultrafilter on I and $J \in \mathcal{F}$. Prove that $\mathcal{F}_J = \{A \cap J \mid A \in \mathcal{F}\}$ is an ultrafilter on J, and that for $A \subseteq I$, $A \in \mathcal{F}$ if and only if $A \cap J \in \mathcal{F}_J$. ($\mathcal{F}_J$ is called the *restriction* of $\mathcal{F}$ to J.)

Let $a, b \in \prod_{i \in I} M_i$ and let $\mathcal{F}$ be an ultrafilter on I. We write $a \equiv b \bmod \mathcal{F}$ if $\{i \in I \mid a_i = b_i\} \in \mathcal{F}$, and denote the congruence class containing a by $a\mathcal{F}$. The set of all congruence classes is denoted by $\prod_{i \in I} M_i / \mathcal{F}$. For each $r \in \mathcal{R}$, we define the relation ψr on $\prod_{i \in I} M_i / \mathcal{F}$ by $a\mathcal{F} \in \psi r$ if $\{i \in I \mid a_i \in \psi_i r\} \in \mathcal{F}$. (Here, a is an n-tuple if $r \in \mathcal{R}_n$.) This definition is clearly independent of the choice of representative of the congruence class. To complete the construction, we define $v(c)$ for $c \in C$ to be the congruence class of the function $I \to \bigcup_{i \in I} M_i$ whose i-component is $v_i(c)$.

Theorem 1.6. $\prod_{i \in I} M_i / \mathcal{F}$ *is a model of* $\mathcal{T} = (\mathcal{R}, A, C)$. *An element* $a\mathcal{F}$ *of* $\prod_{i \in I} M_i / \mathcal{F}$ *satisfies* $p(x) \in P$ *(where* a, x *may be* n-*tuples) if and only if* $\{i \in I \mid a_i \text{ satisfies } p(x)\} \in \mathcal{F}$.

Proof: $\prod_{i \in I} M_i / \mathcal{F}$ is clearly a model of $\mathcal{T}' = (\mathcal{R}, \varnothing, C)$. To show that it is a model of $\mathcal{T}$, we have to show that $v(p) = 1$ for all $p \in A$. Since for $p \in A$, $\{i \in I \mid p \text{ is true in } M_i\} = I \in \mathcal{F}$, this will be an immediate consequence of the second assertion of the theorem. We shall prove this latter assertion by induction over the length of p.

If $p = r(x)$, where $r \in \mathscr{R}$, then the result holds by the definition of ψr. If $p = q_1 \Rightarrow q_2$, then $v(p) = 0$ if and only if we have $v(q_1) = 1$ and $v(q_2) = 0$. By induction, this holds precisely when $J_1 = \{i \in I | v_i(q_1) = 1\}$ and $J_2 = \{i \in I | v_i(q_2) = 0\}$ are both in $\mathscr{F}$. Put $J_3 = J_1 \cap J_2$. If $J_3 \in \mathscr{F}$, then $J_1 \in \mathscr{F}$ and $J_2 \in \mathscr{F}$ by condition (ii), while $J_1 \in \mathscr{F}$ and $J_2 \in \mathscr{F}$ imply $J_3 \in \mathscr{F}$ by condition (iii). Thus $v(p) = 0$ if and only if $J_3 = \{i \in I | v_i(p) = 0\} \in \mathscr{F}$. By condition (iv), $v(p) = 1$ if and only if $I - J_3 = \{i \in I | v_i(p) = 1\} \in \mathscr{F}$.

If $p(x) = (\forall y)q(x, y)$, then $a\mathscr{F}$ satisfies $p(x)$ if and only if for every $b\mathscr{F} \in \prod_{i \in I} M_i/\mathscr{F}$, $(a\mathscr{F}, b\mathscr{F})$ satisfies $q(x, y)$. By induction, the latter holds if and only if for all $b\mathscr{F}$, $\{i \in I | (a_i, b_i) \text{ satisfies } q(x, y)\} \in \mathscr{F}$. Let $J = \{i \in I | a_i \text{ satisfies } p(x)\}$. Suppose $J \in \mathscr{F}$. Then for all $i \in J$ and all $b\mathscr{F}$, we have (a_i, b_i) satisfies $q(x, y)$ since a_i satisfies $(\forall y)q(x, y)$. Thus $a\mathscr{F}$ satisfies $p(x)$. Suppose $J \notin \mathscr{F}$. Then for each $i \in K = I - J$, there exists an element $b_i \in M_i$ such that (a_i, b_i) does not satisfy $q(x, y)$. Thus there exists $b \in \prod_{i \in I} M_i$ such that, for all $i \in K$, (a_i, b_i) does not satisfy $q(x, y)$. Since $K \in \mathscr{F}$, $(a\mathscr{F}, b\mathscr{F})$ does not satisfy $q(x, y)$ and $a\mathscr{F}$ does not satisfy $p(x)$. $\square$

Definition 1.7. The model $\prod_{i \in I} M_i/\mathscr{F}$ of $\mathscr{T}$ is called the *ultraproduct* of the models M_i with respect to the ultrafilter $\mathscr{F}$.

Exercises

1.8. Let p_i be the ith prime and let F_i be a field of characteristic p_i. Let $\mathscr{F}$ be an ultrafilter on the set I of positive integers, such that no member of $\mathscr{F}$ is a singleton. Prove that $\prod_{i \in I} F_i/\mathscr{F}$ is a field of characteristic zero.

1.9. $\mathscr{F}$ is an ultrafilter on I, M_i ($i \in I$) are models of the theory $\mathscr{T}$, and $J \in \mathscr{F}$. Prove

$$\prod_{i \in I} M_i/\mathscr{F} \simeq \prod_{j \in J} M_j/\mathscr{F}_J,$$

where $\mathscr{F}_J$ is the restriction of $\mathscr{F}$ to J.

§2   Non-Principal Ultrafilters

Principal ultrafilters on I, as constructed in Exercise 1.3, are of no use for the construction of new models, because an ultraproduct with respect to a principal ultrafilter is always isomorphic to one of the factors.

Exercises

2.1. If $k \in I$ and $\mathscr{F} = \{J \subseteq I | k \in J\}$, prove that

$$\prod_{i \in I} M_i/\mathscr{F} \simeq M_k.$$

2.2. $\mathscr{F}$ is an ultrafilter on I and $A \in \mathscr{F}$ is a finite subset of I. Prove that $\mathscr{F}$ is principal.

We now investigate conditions on a set S of subsets of I for the existence of an ultrafilter $\mathscr{F} \supseteq S$. By an appropriate choice of S, we shall be able to ensure that every such ultrafilter is non-principal.

Definition 2.3. The set S of subsets of I is said to have the *finite intersection property* if every finite subset of S has non-empty intersection.

Lemma 2.4. *Let S be a set of I. There exists a filter on I containing S if and only if S has the finite intersection property.*

Proof: The necessity of the condition is immediate, so we prove its sufficiency. Suppose S has the finite intersection property, and put

$$T = \{U \subseteq I \,|\, U = J_1 \cap \cdots \cap J_n \text{ for some } n \text{ and some } J_1, \ldots, J_n \in S\}.$$

Let

$$\mathscr{F} = \{F \subseteq I \,|\, F \supseteq U \text{ for some } U \in T\}.$$

We prove that $\mathscr{F}$, which clearly contains S, is a filter. By the finite intersection property of S, $\varnothing \notin T$ and so $\varnothing \notin \mathscr{F}$. Also, condition (ii) for a filter is clearly satisfied by $\mathscr{F}$. Finally, if $F_1, \ldots, F_n \in \mathscr{F}$, then for $i = 1, \ldots, n, F_i \supseteq \bigcap_{j=1}^{m_i} J_{ij}$ for some m_i and $J_{i1}, \ldots, J_{im_i} \in S$. Hence.

$$\bigcap_{i=1}^{n} F_i \supseteq \bigcap_{i=1}^{n} \bigcap_{j=1}^{m_i} J_{ij},$$

and so belongs to $\mathscr{F}$. Thus condition (iii) is satisfied and $\mathscr{F}$ is a filter. $\square$

Lemma 2.5. *Let $\mathscr{F}$ be a filter on I. Then there exists an ultrafilter $\mathscr{F}^* \supseteq \mathscr{F}$ on I.*

Proof: The set of filters containing $\mathscr{F}$ is an inductive set. By Zorn's Lemma, it has a maximal member $\mathscr{F}^*$. $\square$

Exercises

2.6. Let $\alpha = |I|$ and suppose $\alpha \geq \beta \geq \aleph_0$. Put $S = \{J \subseteq I \,|\, |I - J| < \beta\}$. Prove that S is a filter and that if $\mathscr{F}$ is an ultrafilter containing S, then no member of $\mathscr{F}$ has cardinal less than β.

2.7. An ultrafilter $\mathscr{F}$ on I is called uniform if $|J| = |I|$ for all $J \in \mathscr{F}$. If $\mathscr{F}$ is a non-principal ultrafilter, show that there exists $J \in \mathscr{F}$ such that $\mathscr{F}_J$ is uniform.

2.8. Let I be a countable set, and $\mathscr{F}$ an untrafilter on I. If $\sigma : I \to I$ is a permutation, show that $\sigma\mathscr{F}$ is also an ultrafilter on I. The collection $\{\sigma\mathscr{F} \,|\, \sigma \text{ a permutation of } I\}$ may be called the orbit of $\mathscr{F}$. Show that if $\mathscr{F}$ is non-principal, its orbit contains exactly $2^{\aleph_0}$ distinct ultrafilters.

2.9. A family $\mathscr{A}$ of infinite subsets of an infinite set X is called almost disjoint (AD) if distinct members of $\mathscr{A}$ have finite intersection. $\mathscr{A}$ is called maximal almost disjoint (MAD) if it is maximal among the AD families. Prove or disprove each of the following:

(a) Given any MAD family $\mathscr{A}$, there is a non-principal ultrafilter $\mathscr{F}$ such that $\mathscr{A}$ and $\mathscr{F}$ are disjoint.

(b) Given any non-principal ultrafilter $\mathscr{F}$ there is a MAD family $\mathscr{A}$ such that $\mathscr{A}$ and $\mathscr{F}$ are disjoint.

§3   The Existence of an Algebraic Closure

We can now apply the theory of ultraproducts to prove a theorem of considerable importance in algebra.

Theorem 3.1.   *Let F be a field. Then there exists an algebraic closure of F.*

Proof: Let $\mathscr{T}$ be elementary field theory augmented by the addition of the elements of F to the set of constants, and of all the relations $a_1 + a_2 = a_3$, $b_1 b_2 = b_3$ holding in F to the set of axioms. The models of $\mathscr{T}$ are the extension fields of F. Put $R = F[x]$, the ring of polynomials over F. For each $r \in R$, let F_r be a splitting field of r. Put

$$J_r = \{s \in R | r \text{ splits over } F_s\}.$$

Since $r_1 r_2 \cdots r_n \in J_{r_1} \cap J_{r_2} \cap \cdots \cap J_{r_n}$, the set $\mathscr{J} = \{J_r | r \in R\}$ has the finite intersection property. By Lemmas 2.4 and 2.5, there exists an ultrafilter $\mathscr{F}$ on R containing $\mathscr{J}$. Put $F^* = \prod_{r \in R} F_r / \mathscr{F}$. Then F^* is a model of $\mathscr{T}$ and so is an extension field of $\mathscr{F}$.

Let $r = x^n + r_1 x^{n-1} + \cdots + r_n$ be a monic polynomial over F. We prove that r splits over F^*. We put

$$p = (\exists a_1) \cdots (\exists a_n)((a_1 + \cdots + a_n = -r_1) \wedge (a_1 a_2 + a_1 a_3 + \cdots +$$
$$a_{n-1} a_n = r_2) \wedge \cdots \wedge (a_1 a_2 \cdots a_n = (-1)^n r_n)).$$

Then p is true for precisely those models of $\mathscr{T}$ over which r splits. But $\{s \in R | p \text{ is true in } F_s\} = J_r \in \mathscr{F}$. By Theorem 1.6, p is true in F^* and so r splits over F^*.

The proof of Theorem 3.1 is completed by the following purely algebraic lemma.

Lemma 3.2.   *Let F^* be an extension of the field F such that every monic polynomial over F splits over F^*. Let $\bar{F}$ be the set of all elements of F^* which are algebraic over F. Then $\bar{F}$ is an algebraic closure of F.*

Proof: Let $f(x)$ be a monic polynomial over $\bar{F}$. Then $f(x) = (x - a_1) \cdots (x - a_n)$ for some $a_1, \ldots, a_n$ in the splitting field of $f(x)$ considered as a polynomial over F^*. But the a_i, being algebraic over $\bar{F}$, are algebraic over F. Let $m_i(x)$ be the minimum polynomial of a_i over F. Since $m_i(x)$ splits over F^*, its roots lie in F^* and, being algebraic over F, are therefore in $\bar{F}$. Thus $a_1, \ldots, a_n \in \bar{F}$ and $f(x)$ splits over $\bar{F}$. Hence $\bar{F}$ is algebraically closed.   □

Exercises

3.3. If F is not algebraically closed, prove that the ultrafilter used in the proof of Theorem 3.1 is not principal.

3.4. In the notation of the proof of Theorem 3.1, show that if F is finite, then $\{s \in R | F_s = F_r\} \notin \mathscr{F}$.

3.5. If F is not algebraically closed, prove that F^* (constructed as above) is not algebraic over F. (If F is infinite, show that elements $a_r \in F_r$ can be chosen such that a_r and a_s have the same minimum polynomial only for $r = s$. If F is finite, show that the elements $a_r \in F_r$ can be chosen such that a_r and a_s have the same minimum polynomial only for those r, s for which $F_r = F_s$.)

3.6. F is a field. For all $i \in I$, take $F_i = F$ and form the ultraproduct $K = \prod_{i \in I} F_i / \mathscr{F}$ with respect to the ultrafilter $\mathscr{F}$. Prove that K is a pure transcendental extension of F.

§4   Non-trivial Ultrapowers

An ultraproduct $\prod_{i \in I} M_i / \mathscr{F}$ in which $M_i = M$ for all $i \in I$ is called an *ultrapower* of M and denoted by $M^I / \mathscr{F}$. There is a natural embedding $\theta : M \rightarrow M^I / \mathscr{F}$ of M in $M^I / \mathscr{F}$ given by $\theta(m) = f_m \mathscr{F}$, where $f_m : I \rightarrow M$ is the constant function $f_m(i) = m$ for all $i \in I$. By identifying m with $\theta(m)$, we may regard M as a subset of $M^I / \mathscr{F}$. (Alternatively, we may replace the theory $\mathscr{T}$ by the theory $\mathscr{T}'$ formed from $\mathscr{T}$ by replacing C by $C \cup M$. By Theorem 1.6, $M^I / \mathscr{F}$ is a model of $\mathscr{T}'$. Since each element $m \in M$ is a constant of $\mathscr{T}'$, this also gives a map $v' : M \rightarrow M^I / \mathscr{F}$.)

Exercise 4.1.   Prove that the maps θ, $v' : M \rightarrow M^I / \mathscr{F}$ coincide.

We shall always make this identification of M with $\theta(M)$, and we omit specific mention of the map θ. The ultrapower $M^I / \mathscr{F}$ is regarded as trivial if $M^I / \mathscr{F} = M$, so we shall look for conditions which ensure non-triviality.

Exercise 4.2.   If M is finite, prove that $M^I / \mathscr{F} = M$.

Definition 4.3.   Let α be a cardinal. The ultrafilter $\mathscr{F}$ on I is called α-*complete* if, for every subset $\mathscr{G} \subseteq \mathscr{F}$ of cardinal α, we have $\cap \mathscr{G} \in \mathscr{F}$. Otherwise, $\mathscr{F}$ is called α-*incomplete*. (It is usual in this context to denote $|\mathbb{N}|$ by ω.)

Lemma 4.4.   *Let α be an infinite cardinal and let $\mathscr{F}$ be an α-incomplete ultrafilter on I. Then there exists a partition of I into α disjoint subsets, none of which is in $\mathscr{F}$.*

Proof: The cardinal α is an ordinal, $\alpha = \{\beta | \beta \text{ ordinal}, \beta < \alpha\}$. Since $\mathscr{F}$ is α-incomplete, there exists $\mathscr{G} \subseteq \mathscr{F}$ such that $|\mathscr{G}| = \alpha$ and $\cap \mathscr{G} \notin \mathscr{F}$. We index the members of $\mathscr{G}$ with the ordinals less than α, so that $\mathscr{G} = \{G_\beta | \beta < \alpha\}$. For each ordinal $\beta \leqslant \alpha$, put $X_\beta = \cap \{G_\gamma | \gamma < \beta\}$ (interpreting this for $\beta = 0$ to mean $X_0 = I$), and put $Y_\beta = X_\beta - X_{\beta+1}$ for $\beta < \alpha$. For $\beta = \alpha$, put

$Y_\alpha = X_\alpha$. Then $\{Y_\beta|\beta \leqslant \alpha\}$ is a partition of I into α disjoint subsets. Since $Y_\alpha = \cap \mathscr{G}$, we have $Y_\alpha \notin \mathscr{F}$. Suppose $Y_\beta \in \mathscr{F}$ for some $\beta < \alpha$. Then $X_\beta - X_{\beta+1} \in \mathscr{F}$. Since also $G_\beta \in \mathscr{F}$, we have $(X_\beta - X_{\beta+1}) \cap G_\beta \in \mathscr{F}$. But $(X_\beta - X_{\beta+1}) \cap G_\beta = \varnothing \notin \mathscr{F}$. $\square$

Lemma 4.5. *Let $\mathscr{F}$ be an α-complete ultrafilter on I. Then for every partition of I into a set $\mathscr{G}$ of α disjoint subsets, some member of $\mathscr{G}$ is in $\mathscr{F}$.*

Exercise 4.6. Prove Lemma 4.5.

Theorem 4.7. *Let $\mathscr{F}$ be an ultrafilter on I and let $\alpha = |M|$. Then $M = M^I/\mathscr{F}$ if and only if $\mathscr{F}$ is α-complete.*

Proof: Suppose $\mathscr{F}$ is α-complete. An element of $M^I/\mathscr{F}$ is $f\mathscr{F}$ for some $f:I \to M$. For each $m \in M$, put $J_m = \{i \in I|f(i) = m\}$. Then $\{J_m|m \in M\}$ is a partition of I into α disjoint subsets. By Lemma 4.6, $J_m \in \mathscr{F}$ for some $m \in M$. This implies $f\mathscr{F} = m$.

Suppose now that $\mathscr{F}$ is α-incomplete. Then α must be infinite, and so, by Lemma 4.4, there is a partition of I into α disjoint subsets, none of which is in $\mathscr{F}$. We may index these subsets with the elements of M. Let $\{J_m|m \in M\}$ be such a partition of I, and let $f(i)$ be the unique $m \in M$ such that $i \in J_m$. This defines a function $f:I \to M$ such that $f\mathscr{F} \notin M$. $\square$

Exercise 4.8. $\mathscr{F}$ is a non-principal ultrafilter on I, and β is the smallest cardinal for which $\mathscr{F}$ is β-incomplete. $|M| = \alpha \geqslant \beta$. Prove that $|M^I/\mathscr{F}| \geqslant \alpha^\beta$.

It can be proved (cf [1], p. 112, Theorem 1.11) that if $|I|$ is less than the first strongly inaccessible cardinal, then every non-principal ultrafilter on I is ω-incomplete. This means that if $\mathscr{F}$ is non-principal, then $M^I/\mathscr{F} \neq M$ except when I is very large or when M is finite.

Exercise 4.9. Let α be an infinite cardinal and let A be a set of cardinal α. Put $M = A \cup \mathrm{Pow}(A)$, and $\mathscr{R} = \{\mathscr{I}, \in, e, s\}$ where e, s are unary and $\mathscr{I}, \in$ are binary. Interpreting $e(x)$ as x is an element of A, $s(x)$ as x is a subset of A, and $\in(x, y)$ as x is a member of y (for x in A, y in $\mathrm{Pow}(A)$), form the theory $\mathscr{T} = (\mathscr{R}, A, C)$ with $\mathscr{R}$ as above, $C = M$ and $A = \{p \in \mathscr{L}(\mathscr{T})|p \text{ true in } M\}$. For any model N of $\mathscr{T}$, put $B = \{n \in N|e(n) \text{ is true}\}$ and $D = \{n \in N|s(n) \text{ is true}\}$. Show that each $d \in D$ is determined by the set $\{b \in B|\in(b, d) \text{ is true}\}$ and hence identify D with a subset of $\mathrm{Pow}(B)$. In the special case where $\mathscr{F}$ is an α-complete ultrafilter on a set I of cardinal 2^α and $N = M^I/\mathscr{F}$, prove that $B = A$ and $N = M$. Hence prove that an α-complete ultrafilter on a set of cardinal 2^α is principal.

§5   Ultrapowers of Number Systems

We have seen that the theory of N (i.e., the theory $\mathscr{T} = (\mathscr{R}, A, C)$ where $\mathscr{R} = \{\mathscr{I}, +, \times, <\}$, $C = \mathrm{N}$ and $A = \{p \in \mathscr{L}(\mathscr{T})|p \text{ true in } \mathrm{N}\}$ cannot be categorical, and the same is true for the theories of the other standard systems Z, Q, R and C. We use ultrapowers to produce models of these theories which are not isomorphic to their standard models. We take the set N as index set I. Let $\mathscr{F}$ be a non-principal ultrafilter on I. Since $\mathscr{F}$ contains no finite sets, every subset of I with finite complement is in $\mathscr{F}$ (i.e., $\mathscr{F}$ is an extension of the Fréchet filter on I). Trivially, $\mathscr{F}$ is ω-incomplete. By Theorem 4.7, if M is any of N, Z, Q, R or C, then $M^I/\mathscr{F} \neq M$.

An element of N^I is just a sequence of natural numbers. When we form $N^I/\mathscr{F}$, we are, among other things, identifying sequences which are the same from some point onwards. Consider the element $u\mathscr{F}$ of $N^I/\mathscr{F}$ given by the function $u: N \to N$ defined by $u(i) = i$. This element $u\mathscr{F}$ is infinite, in the sense that $u\mathscr{F} > n$ for all $n \in N$. To see this, let $k_n: N \to N$ be the constant function $k_n(i) = n$ for all $i \in N$. Then $n = k_n\mathscr{F}$ and $\{i \in N | u(i) > k_n(i)\} = \{i \in N | i > n\} \in \mathscr{F}$. Hence by Theorem 1.6, $u\mathscr{F} > k_n\mathscr{F}$. Similarly, we can show that $Q^I/\mathscr{F}$ has infinitesimal elements. For if $v: N \to Q$ is defined by $v(i) = 1/i$ for $i > 0, v(0) = 1$, then $k_0\mathscr{F} < v\mathscr{F} < k_r\mathscr{F}$ for all $r \in Q$ such that $r > 0$. We clearly have natural inclusions $N^I/\mathscr{F} \subseteq Z^I/\mathscr{F} \subseteq Q^I/\mathscr{F} \subseteq R^I/\mathscr{F} \subseteq C^I/\mathscr{F}$.

Exercise 5.1. Let N be a model of the theory of N which properly contains N. Show that N has infinite elements. Show also that if Q is a model of the theory of Q which properly contains Q, then Q has non-zero infinitesimal elements.

Let $\mathscr{T}$ be the theory of N. Form the theory $\mathscr{T}'$ by adding a new constant u and the new axioms $u > n$ for all $n \in N$. This theory $\mathscr{T}'$ is consistent, indeed $N^I/\mathscr{F}$, with u interpreted as $u\mathscr{F}$, is a model of $\mathscr{T}'$. As $\mathscr{T}'$ is a countable theory, the Löwenheim-Skolem Theorem (Theorem 3.18 of Chapter V) shows that it has a countable model. The model $N^I/\mathscr{F}$ is uncountable, and it is natural to try to modify the ultrapower construction so as to obtain a countable model. We shall take a subset S of the set N^I of all functions from I into N and reduce this set S modulo an ultrafilter $\mathscr{F}$.

Let $\mathscr{T} = (\mathscr{R}, A, C)$ be any theory, $\{M_i | i \in I\}$ a family of models of $\mathscr{T}$, and $\mathscr{F}$ an ultrafilter on I. If S is any non-empty subset of $\prod_{i \in I} M_i$, which includes all the functions $k_c: I \to \prod_{i \in I} M_i$ defined by $k_c(i) = v_i(c)$ for $c \in C$, $i \in I$, then $S/\mathscr{F}$ is a model of $(\mathscr{R}, \varnothing, C)$.

Definition 5.2. $S/\mathscr{F}$ is called a *subultraproduct* of $\{M_i | i \in I\}$ with respect to the ultrafilter $\mathscr{F}$.

A subultraproduct $S/\mathscr{F}$ of models M_i of $\mathscr{T} = (\mathscr{R}, A, C)$ is a model of $(\mathscr{R}, \varnothing, C)$, but unless further conditions are imposed on S or on A, it need not be a model of $\mathscr{T}$. If we examine the proof of Theorem 1.6, we see that it applies unaltered, except for the section which shows that $a\mathscr{F}$ satisfies $p(x) = (\forall y)q(x, y)$ only if the set $J = \{i \in I | a_i \text{ satisfies } p(x)\}$ is in $\mathscr{F}$. For each $i \in I - J$, there exists an element $b_i \in M_i$ such that (a_i, b_i) does not satisfy $q(x, y)$, but we can no longer conclude from this that there exists a function $b \in S$ such that for all $i \in I - J$, (a_i, b_i) does not satisfy $q(x, y)$. If S is chosen so that these functions always exist, then the assertions of Theorem 1.6 will continue to hold for $S/\mathscr{F}$. The next theorem shows how to achieve this. The reader is asked to recall Definition 2.11 of Chapter V.

Theorem 5.3. *Let $\mathscr{T}$ be the theory of* N. *Let S be the set of all functions* $s: \mathrm{N} \to \mathrm{N}$ *which are definable in $\mathscr{T}$. Let $\mathscr{F}$ be an ultrafilter on* N. *Then the subultraproduct $S/\mathscr{F}$ is a countable model of $\mathscr{T}$. If $\mathscr{F}$ is not principal, then* $S/\mathscr{F} \neq \mathrm{N}$.

Proof: To show that $S/\mathscr{F}$ is a model of $\mathscr{T}$, it remains for us to show that if $a \in S$ and $J = \{i \in \mathrm{N} | a_i \text{ satisfies } (\forall y)q(x, y)\} \notin \mathscr{F}$, then there exists $b \in S$ such that, for all $i \in \mathrm{N} - J$, we have that (a_i, b_i) does not satisfy $q(x, y)$. We put

$$b_i = 0 \text{ if } a_i \text{ satisfies } (\forall y)q(x, y),$$

$$b_i = \text{ the least } n \text{ for which } (a_i, n) \text{ does not satisfy } q(x, y)$$
$$\text{if } a_i \text{ does not satisfy } (\forall y)q(x, y).$$

Since a is a definable function, so is b, and the assertion follows. Since P is countable, so is S and hence so is $S/\mathscr{F}$. The function $u: \mathrm{N} \to \mathrm{N}$ given by $u(i) = i$ is clearly definable, as are the constant functions k_n. If $\mathscr{F}$ is non-principal, then $\{i \in \mathrm{N} | u(i) = k_n(i)\} = \{n\} \notin \mathscr{F}$ and so $S/\mathscr{F} \neq \mathrm{N}$. $\square$

§6   Direct Limits

There is a connection between the idea of a subultraproduct, introduced in §5, and the idea of a direct limit, which arises in a number of algebraic contexts, and which we now explain. A directed set is a partially ordered set $(I, \leqslant)$ such that for any $i, j \in I$, there is a $k \in I$ such that $i \leqslant k$ and $j \leqslant k$. A direct family[1] in a category $\mathscr{C}$ is a set $\{A_i | i \in I\}$ of objects A_i of $\mathscr{C}$ indexed by a directed set I, together with a morphism $f_j^i: A_i \to A_j$ for each pair $i \leqslant j$ in I, such that

(i) $f_i^i = 1_{A_i}$ for all $i \in I$,
(ii) $f_k^j f_j^i = f_k^i$ for all $i \leqslant j \leqslant k$ in I.

Definition 6.1. A *direct limit* in $\mathscr{C}$, or more precisely a *limit of the direct family* $\{A_i, f_j^i | i, j \in I\}$ in $\mathscr{C}$, is an object L of $\mathscr{C}$ together with morphisms $\varphi^i: A_i \to L$ such that

(i) $\varphi^j f_j^i = \varphi^i$ for all $i \leqslant j$ in I, and
(ii) for any object M of $\mathscr{C}$ and family of morphisms $\psi^i: A_i \to M$ satisfying $\psi^j f_j^i = \psi^i$ for all $i \leqslant j$ in I, there exists a unique morphism $\theta: L \to M$ such that $\theta \varphi^i = \psi^i$ for all $i \in I$.

_____

[1] More neatly but less intuitively defined as follows: regard the directed set as a category with objects the elements of I and morphisms from i to j the pairs (i, j) with $i \leqslant j$. A direct family in $\mathscr{C}$ is then a functor from this category into $\mathscr{C}$.

Let $\mathcal{T} = (\mathcal{R}, A, C)$ be a theory. We associate with $\mathcal{T}$ the category $\text{Mod}(\mathcal{T})$, whose objects are the models of $\mathcal{T}$ and whose morphisms are the maps $f: M \to M'$ between models (M, v, ψ) and (M', v', ψ'), such that $fv(c) = v'(c)$ for all $c \in C$, and such that $(f(m_1), \ldots, f(m_n)) \in \psi' r$ for all $r \in \mathcal{R}_n$ and $(m_1, \ldots, m_n) \in \psi r$. As an aid to the study of $\text{Mod}(\mathcal{T})$, we use the theory $\mathcal{T}_\varnothing = (\mathcal{R}, \varnothing, C)$ and its associated category $\text{Mod}(\mathcal{T}_\varnothing)$. Every object of $\text{Mod}(\mathcal{T})$ is an object of $\text{Mod}(\mathcal{T}_\varnothing)$, and the morphisms in $\text{Mod}(\mathcal{T})$ between any two of its objects are precisely the morphisms between them in $\text{Mod}(\mathcal{T}_\varnothing)$.

Exercise 6.2. R is a ring and $\mathcal{T}$ is the elementary theory of R-modules. Show that the category of R-modules is precisely the category $\text{Mod}(\mathcal{T})$.

Lemma 6.3. *For any theory $\mathcal{T}$, direct limits exist in $\text{Mod}(\mathcal{T}_\varnothing)$.*

Proof: Let $\{M_i, f^i_j | i, j \in I, i \leqslant j\}$ be a direct family in $\text{Mod}(\mathcal{T}_\varnothing)$. We construct a limit as a subultraproduct. Put $F_i = \{j \in I | j \geqslant i\}$, and let $\mathcal{F}$ be any ultrafilter containing all the F_i. (Actually, we do not need $\mathcal{F}$ to be an ultrafilter—any filter containing the F_i will do.) Put

$$S = \{s: I \to \bigcup_{i \in I} M_i | s_i \in M_i \text{ for all } i \in I; \text{ for some } F \in \mathcal{F}, f^i_j s_i = s_j$$
$$\text{for all } i, j \in F \text{ with } i \leqslant j\}.$$

Then put $L = S/\mathcal{F}$ and define $\varphi^i: M_i \to L$ by $\varphi^i(m) = s\mathcal{F}$, where s is given by $s_j = f^i_j m$ for all $j \in F_i$, and s_j is chosen arbitrarily for $j \notin F_i$. The element $s\mathcal{F}$ of L is clearly independent of the choice of s_j for $j \notin F_i$. L, being a subultraproduct of models M_i of $\mathcal{T}$, is a model of $\mathcal{T}_\varnothing$. The φ^i are clearly morphisms of $\text{Mod}(\mathcal{T}_\varnothing)$ satisfying (i) of Definition 6.1.

Now let N be a model of $\mathcal{T}_\varnothing$, and let $\psi^i: M_i \to N$ be morphisms in $\text{Mod}(\mathcal{T}_\varnothing)$ such that $\psi^j f^i_j = \psi^i$ for all $i \leqslant j$ in I. Let $s \in S$ satisfy $f^i_j s_i = s_j$ for all $i, j \in F$ such that $i \leqslant j$, where F is a member of $\mathcal{F}$. If $i \in F$ and if $m = s_i \in M_i$, then $s\mathcal{F} = \varphi^i(m)$. The condition on the map $\theta: L \to N$ to be constructed requires that $\theta(s\mathcal{F}) = \psi^i(m)$. Hence θ, if it exists, is unique. We define θ by putting $\theta(s\mathcal{F}) = \psi^i(s_i)$ for some $i \in F$, and we must show that this definition is independent of the choice of i.

Suppose that $j \in F$. Then there exists $k \in I$ such that $i \leqslant k$ and $j \leqslant k$. Since $F \cap F_k \in \mathcal{F}$, $F \cap F_k \neq \varnothing$, and so there exists an $r \in F \cap F_k$. We have $s_r = f^i_r s_i = f^j_r s_j$, and so

$$\psi^i s_i = \psi^r f^i_r s_i = \psi^r s_r = \psi^j s_j.$$

Thus θ is well-defined. Clearly, θ is a morphism satisfying the requirements of condition (ii) of the definition. $\square$

We are interested in direct limits in $\text{Mod}(\mathcal{T})$. The next lemma reduces this problem to an investigation of the subultraproduct constructed above.

Lemma 6.4. *The direct family $\{M_i, f^i_j | i, j \in I, i \leqslant j\}$ has a limit in $\text{Mod}(\mathcal{T})$ if and only if the subultraproduct $S/\mathcal{F}$ constructed above is a model of $\mathcal{T}$.*

Proof: If $S/\mathscr{F}$ is a model of $\mathscr{T}$, then it is clearly a limit in $\mathrm{Mod}(\mathscr{T})$ of the given family. If L is a limit in $\mathrm{Mod}(\mathscr{T})$ of the family, then L is also a limit in $\mathrm{Mod}(\mathscr{T}_\varnothing)$, and so is isomorphic to $S/\mathscr{F}$. Thus $S/\mathscr{F}$ is a model of $\mathscr{T}$. $\square$

We now investigate conditions on $\mathscr{T}$ for $S/\mathscr{F}$ to be a model of $\mathscr{T}$.

Definition 6.5. An element $p \in P(V, \mathscr{R})$ is called *universal* if it has the form $(\forall y_1) \cdots (\forall y_s) q(x_1, \ldots, x_r, y_1, \ldots, y_s)$, where $q(x_1, \ldots, x_r, y_1, \ldots, y_s)$ contains no quantifiers.

The argument used in proving Theorem 1.6 shows that the element $s\mathscr{F}$ will satisfy $p(x) = (\forall y)q(x, y)$, where $q(x, y)$ contains no quantifiers, if $\{i \in I \mid s_i \text{ satisfies } p(x)\} \in \mathscr{F}$. (This includes the case where x, y are n-tuples.) Thus every axiom of $\mathscr{T}$ which is universal is satisfied in $S/\mathscr{F}$.

Definition 6.6. A theory $\mathscr{T}$ is called *algebraic* if every axiom of $\mathscr{T}$ is either universal or has the form

$$(\forall x_1) \cdots (\forall x_r)(\exists y_1) \cdots (\exists y_s)p(x_1, \ldots, x_r, y_1, \ldots, y_s, c_1, \ldots, c_t),$$

where $c_i \in C$ and p is constructed from primitive elements of $P(V, \mathscr{R})$ by using $\vee$, $\wedge$ only. The category $\mathrm{Mod}(\mathscr{T})$ is called *algebraic* if $\mathscr{T}$ is algebraic.

The reason for the name is that if the relation symbol $r \in \mathscr{R}_{n+1}$ is to correspond to an n-ary operation of an algebra, then we require the axioms $(\forall x_1) \cdots (\forall x_n)(\exists y)r(x_1, \ldots, x_n, y)$ and $(\forall x_1) \cdots (\forall x_n)(\forall y)(\forall z)(r(x_1, \ldots, x_n, y) \wedge r(x_1, \ldots, x_n, z) \Rightarrow y = z)$. Note that the second of these axioms is universal, and the first is admissible for an algebraic theory.

Theorem 6.7. *Let $\mathscr{T}$ be an algebraic theory. Then direct limits exist in $\mathrm{Mod}(\mathscr{T})$.*

Proof: Let $q = (\forall x)(\exists y)p(x, y, c)$ be an axiom of $\mathscr{T}$, where $p(x, y, c)$ is constructed from primitive elements of $P(V, \mathscr{R})$ using only $\vee$, $\wedge$. (x, y, c may denote n-tuples.) Let $a\mathscr{F} \in S/\mathscr{F}$ and let $\sigma_i = v_i(c)$. We have, for some $F \in \mathscr{F}$, $f_j^i a_i = a_j$ for all i, $j \in F$ with $i \leqslant j$. Take an $i \in F$, and put $F' = F \cap F_i$. Since q is satisfied in M_i, there exists $m_i \in M_i$ such that (a_i, m_i, σ_i) satisfies $p(x, y, c)$. For $j \in F'$, put $m_j = f_j^i m_i$. Since $a_j = f_j^i a_i$ and $\sigma_j = f_j^i \sigma_i$, it follows from the nature of p and the fact that f_j^i is a morphism that (a_j, m_j, σ_j) satisfies $p(x, y, c)$. Choose m_j arbitrarily for $j \notin F'$. Then $m \in S$ and $(a\mathscr{F}, m\mathscr{F}, \sigma\mathscr{F})$ satisfies $p(x, y, c)$. Hence q is satisfied in $S/\mathscr{F}$. $\square$

Corollary 6.8. *Direct limits exist in any variety of universal algebras.*

Exercises

6.9. Show that direct limits exist in the category whose objects are fields and whose morphisms are ring homomorphisms, but that not even finite direct sums exist in this category. Show that the algebraic closure of a field is obtainable as a direct limit.

6.10. Show that the conditions we have imposed on the existential axioms cannot be weakened either by (a) allowing the negation of a primitive relation, or (b) allowing an existential quantifier to precede a universal quantifier. (Take the direct family, indexed by $\mathbf{N}$, of the sets $\{0, 1, \ldots, n\}$ and inclusion maps, (a) with property $p(x)$ true in $\{0, 1, \ldots, n\}$ for $0, 1, \ldots, n - 1$, (b) with relation $\leqslant$ and axiom $(\exists x)(\forall y)(y \leqslant x)$.)

6.11. Show that compact topological spaces and continuous maps do not form an algebraic category.

Chapter VIII

Non-Standard Models

§1   Elementary Standard Systems

Much of mathematics is concerned with the study of "standard" mathematical systems such as the natural numbers, the rationals, the real numbers and the complex numbers, each of which is regarded as a unique system. When we attempt to study one of these systems by axiomatising it within the first-order predicate calculus, we find that our axiomatisation cannot be categorical, and that there exist models of our axiomatic theory not isomorphic to the system we wish to study. Such models have been constructed as ultrapowers in Chapter VII. In this chapter, we investigate ways of exploiting such models in the study of a standard system. We begin by considering elementary systems, i.e., systems in which relations between elements, but not properties of subsets, can be studied.

Definition 1.1.   An *elementary standard system* S is a set S together with a subset $\mathscr{R}$ of the set of relations on S such that $\mathscr{I} \in \mathscr{R}$.

$\mathscr{R}$ is the set of relations on S considered to be of interest. It is usual to denote the underlying set S of S by the same symbol S, and we shall do so.

Example 1.2.   The elementary real number system R consists of the set R of real numbers together with the set $\mathscr{R} = \{\mathscr{I}, +, \times, <\}$ of relations on R. Here, $+$ is the ternary relation $(a, b, c) \in +$ if and only if $a + b = c$, and $\times$ is defined similarly.

Let $S = (S, \mathscr{R})$ be an elementary standard system. We take a set $V \supset S$, such that $V - S$ is countably infinite, and form the first-order algebra $P(S) = P(V, \mathscr{R})$. In this algebra, we think of elements of S and $\mathscr{R}$ as names for themselves[1].

Definition 1.3.   The *language* of S is the subset $\mathscr{L}(S) = \{p \in P(V, \mathscr{R})|\ \mathrm{var}(p) \subseteq S\}$ of $P(V, \mathscr{R})$.

Interpreting each element of S and $\mathscr{R}$ as itself assigns a truth value $v(p)$ to each $p \in \mathscr{L}(S)$.

Definition 1.4.   The *theory* of S is the theory $\mathscr{T}(S) = (\mathscr{R}, A, S)$ where $A = \{p \in \mathscr{L}(S)|v(p) = 1\}$.

---

[1] If we wish to distinguish between the objects and their names, we take for each element $a \in S$ and $\rho \in \mathscr{R}$ elements a', ρ', and use these in the construction of P.

74

$\mathcal{T}(S)$ is a complete theory with S as a model. The theorems of this theory are its axioms, and consist of all elements of $\mathcal{L}(S)$ which are true in S or in any other model of $\mathcal{T}(S)$. If the axiom set A were fully known, then $\mathcal{T}(S)$ could give us no new information about S. However, our knowledge of S is usually incomplete, and any method which extends our knowledge of A in fact extends our knowledge of S. If we can choose a model $*S$ of $\mathcal{T}(S)$ such that the truth or falsity of certain statements $p \in \mathcal{L}(S)$ is more easily determined (by argument in the meta-language) for $*S$ than for S, then we have a method of utilising $\mathcal{T}(S)$ to discover properties of S. Our aim is to construct some useful models $*S$.

Exercise 1.5.   It is assumed above that the theory with relation symbols $\mathcal{R}$ and axioms A has S as its set of constants. Prove this.

Definition 1.6.   Let $*S$ be any model of $\mathcal{T}(S)$. We say that $*S$ is a *standard model* of $\mathcal{T}(S)$ if $*S$ is isomorphic to S, and otherwise $*S$ is called a *non-standard model* of $\mathcal{T}(S)$.

Let $*S = (*S, \varphi, \psi)$ be any model of $\mathcal{T}(S)$. Then $\varphi: S \rightarrow *S$ embeds S in $*S$, since if a, b are distinct elements of S, then $(a \neq b) \in A$ and so is true in $*S$, i.e., $\varphi(a) \neq \varphi(b)$. Similarly, for any n-ary relation $\rho \in \mathcal{R}$, the restriction to $\varphi(S)$ of the relation $\psi(\rho)$ is precisely the relation on $\varphi(S)$ which corresponds under φ to the relation ρ on S. We shall therefore always identify S with its image under φ in $*S$, and so regard the model $*S$ as containing the standard model S.

§2   Reduction of the Order

First-order logic does not permit us to study properties of relations, or to discuss statements such as "For all n-ary relations, ...". This restriction excludes from consideration most of the material in a subject such as real analysis, where functions of various types occupy a dominant place. The general consideration of properties of relations requires a higher-order logic. Fortunately, there is a trick which enables us to bring within the scope of our first-order predicate calculus all these higher-order concepts for any one mathematical system. For any set S, let rel(S) denote the set of all relations on S.

Definition 2.1.   Let S be any set. We define the set $\mathcal{O}^k(S)$, of *kth-order objects* on S, by $\mathcal{O}^0(S) = S$, and $\mathcal{O}^{k+1}(S) = \mathcal{O}^k(S) \cup \text{rel}(\mathcal{O}^k(S))$. Further, we put $\mathcal{O}^\infty(S) = \bigcup_{k \geqslant 0} \mathcal{O}^k(S)$.

For each n, we introduce an $(n + 1)$-ary relation symbol $\in^n$. If ρ is an n-ary relation on S, and if $a_1, \ldots, a_n \in S$, we can now formalise the statement that $(a_1, \ldots, a_n)$ is in ρ as $\in^n(\rho, a_1, \ldots, a_n)$, as well as by $\rho(a_1, \ldots, a_n)$. We have made ρ into an individual constant of a larger theory, and we may if we wish omit ρ from the set of relation symbols. Among the unary relations on S, there is S itself, and those elements of the extended system $\mathcal{O}^{\infty}(S)$ which belong to S are distinguished as those for which the formal statement $\in^1(S, a)$ is true. The statement that $\rho \in \mathcal{O}^1(S)$ is an n-ary relation on S can be formalised as

$$\tau_n(\rho) = (\exists x_1) \cdots (\exists x_n)(\in^1(S, x_1) \wedge \cdots \wedge \in^1(S, x_n) \wedge \in^n(\rho, x_1, \ldots, x_n)),$$

while the statement that ρ is an n-ary relation on the subsets $S_1, \ldots, S_n$ of S can be formalised as

$$\tau_n(\rho, S_1, \ldots, S_n) = \tau_n(\rho) \wedge (\forall x_1) \cdots (\forall x_n)(\in^n(\rho, x_1, \ldots, x_n)$$
$$\Rightarrow \in^1(S_1, x_1) \wedge \cdots \wedge \in^1(S_n, x_n)).$$

We can now handle second-order concepts on a standard system S by forming $\mathcal{T}(\mathcal{O}^1(S))$, where the set of relation symbols includes the symbols $\in^n$ and those required for the properties of relations we wish to study. The statement that all n-ary relations have the property π can then be formalised as $(\forall x)(\tau_n(x) \Rightarrow \pi(x))$.

This process may be applied to still higher-order concepts. $(k + 1)$th-order objects can be studied in $\mathcal{T}(\mathcal{O}^k(S))$, which we call the $(k + 1)$th-order theory of S. We call $\mathcal{O}^k(S)$, together with an appropriate set of relation symbols, a $(k + 1)$th-order standard (mathematical) system, and $\mathcal{O}^{\infty}(S)$ an infinite order standard system. We point out that $\mathcal{T}(\mathcal{O}^k(S))$ (including the case $k = \infty$) is still a first-order theory of an elementary standard system, namely the system $\mathcal{O}^k(S)$. Theorems proved about elementary standard systems thus become applicable to higher-order standard systems.

§3   Enlargements

We recall the definition of a definable n-ary relation ρ on a standard system S. We say that ρ is definable in $\mathcal{T}(S)$ if there is an element $p(x_1, \ldots, x_n) \in P(S)$, where $x_1, \ldots, x_n \in V - S$ and $\text{var}(p(x_1, \ldots, x_n)) \subseteq \{x_1, \ldots, x_n\} \cup S$, such that

$$\rho = \{(a_1, \ldots, a_n) \in S^n | p(a_1, \ldots, a_n) \text{ is true in } S\}.$$

Any such $p(x_1, \ldots, x_n)$ is called a description of ρ. In our work, it will not matter which description of a definable relation ρ we use. We write $\rho(x_1, \ldots, x_n)$ for some description of ρ. In the special case where ρ is a definable subset of S, we use $(x \in \rho)$ to denote an arbitrary description of ρ.

Let S be a standard system, and let $\rho(x, y)$ define a binary relation in $\mathcal{T}(S)$. We define the domain of ρ to be the set D_ρ, where

$$D_\rho = \{a \in S | \rho(a, b) \text{ is true in } S \text{ for some } b \in S\}.$$

Definition 3.1.   A concurrent relation of S is a definable binary relation $\rho = \rho(x, y)$ in $\mathcal{T}(S)$ such that $D_\rho \neq \varnothing$ and, for every finite subset $\{a_1, \ldots, a_n\}$ of D_ρ, there is a $b \in S$ such that $\rho(a_i, b)$ is true for $i = 1, \ldots, n$.

Example 3.2. We consider the (elementary) real number system $\mathbf{R}$. In $\mathbf{R}$, $<$ is a concurrent relation with domain $D_< = \mathbf{R}$. For any finite set $\{x_1, \ldots, x_n\}$ of real numbers, $y = 1 + \max_i(x_i)$ satisfies $x_i < y$ for $i = 1, \ldots, n$.

Now let σ be any set of concurrent relations of the standard system S. For each $\rho \in \sigma$, we take a new variable $c_\rho \notin V$ and form $V^\sigma = V \cup \{c_\rho | \rho \in \sigma\}$, $P^\sigma = P(V^\sigma, \mathscr{R})$. We put

$$A_\sigma = \{\rho(x, c_\rho) | \rho \in \sigma \text{ and } x \in D_\rho\}.$$

Definition 3.3. The *enlargement* of $\mathscr{T} = (\mathscr{R}, A, \mathsf{S})$ with respect to σ is the theory $\mathscr{T}^\sigma = (\mathscr{R}, A \cup A_\sigma, \mathsf{S} \cup \{c_\rho | \rho \in \sigma\})$. When σ is the set of all concurrent relations of S, we call $\mathscr{T}^\sigma$ the *full enlargement* of $\mathscr{T}$, and denote it by $*\mathscr{T}$.

Theorem 3.4. *Let $\mathscr{T} = \mathscr{T}(\mathsf{S})$ and let σ be any set of concurrent relations of S. Then $\mathscr{T}^\sigma$ is consistent.*

Proof. Suppose $\mathscr{T}^\sigma \vdash F$. Then $A_0 \vdash_{\mathscr{T}} F$ for some finite subset A_0 of A_σ. Let $A_0 = \{\rho_j(x_{ij}, c_{\rho_j}) | i = 1, \ldots, r_j; \; j = 1, \ldots, n\}$, where $x_{ij} \in D_{\rho_j}$. Since $\{x_{1j}, \ldots, x_{r_j j}\}$ is a finite subset of D_{ρ_j}, there exists $b_j \in \mathsf{S}$ such that $\rho_j(x_{ij}, b_j)$ is true in S for $i = 1, \ldots, r_j$. Mapping c_{ρ_j} to b_j for $j = 1, \ldots, n$ makes S a model of the theory $\mathscr{T}' = (\mathscr{R}, A \cup A_0, \mathsf{S} \cup \{c_{\rho_j} | j = 1, \ldots, n\})$. Hence $\mathscr{T}'$ is consistent, which contradicts the assumption that $A \cup A_0 \vdash F$. Thus $\mathscr{T}^\sigma$ is consistent. $\square$

Since $\mathscr{T}^\sigma$ is consistent, it has a model. Let S^σ be any model of $\mathscr{T}^\sigma$. As we have already indicated, S^σ has the standard model S of $\mathscr{T}$ embedded in it. We call S^σ a σ-enlargement of S. A model $*\mathsf{S}$ of $*\mathscr{T}$ is called a full enlargement of S.

Suppose that $\mathsf{S}^\sigma = \mathsf{S}$. Then for each $\rho \in \sigma$, the constant c_ρ of $\mathscr{T}^\sigma$ is interpreted as some $b_\rho \in \mathsf{S}$ which satisfies $\rho(x, b_\rho)$ for all $x \in D_\rho$. Thus all we have achieved is the introduction of a new name c_ρ for the element b_ρ. The new axioms $\rho(x, c_\rho)$ reduce to axioms of $\mathscr{T}$ if we replace c_ρ by b_ρ, and so if we add to $\mathscr{T}^\sigma$ the further axioms $c_\rho = b_\rho$ for all $\rho \in \sigma$, the resulting theory is equivalent to $\mathscr{T}$ in the sense that the two theories have the same models. Such an enlargement $\mathscr{T}^\sigma$ is of little use in studying $\mathscr{T}$ and is called a trivial enlargement.

Exercise 3.5. Use the ultrapower construction studied in §5 of Chapter VII to give an alternative proof of Theorem 3.4.

One standard system may be contained in another, as in the case of the integers $\mathbf{Z}$ and the reals $\mathbf{R}$. We shall now obtain a useful result on enlargements of systems related in this way.

Definition 3.6. Let $\mathsf{S} = (S, \mathscr{R})$ and $\mathsf{S}_1 = (S_1, \mathscr{R}_1)$ be standard systems. We say that S is a *subsystem* of S_1, and write $\mathsf{S} \leqslant \mathsf{S}_1$, if $\mathscr{R} \subseteq \mathscr{R}_1$ and if S is a definable subset of S_1.

Examples

3.7.   For any S, $S \leqslant \mathcal{O}^1(S) \leqslant \mathcal{O}^2(S) \leqslant \cdots \leqslant \mathcal{O}^\infty(S)$.

3.8.   Take $\mathbb{R} = (R, \{\mathcal{I}, +, \times, <, i\})$, where $i(x)$ is interpreted as "x is an integer". Then $\mathbb{Z} = (Z, \{\mathcal{I}, +, \times, <\})$ is a subsystem of $\mathbb{R}$.

3.9. $\mathcal{O}^k(\mathbb{Z}) \leqslant \mathcal{O}^{k+1}(\mathbb{R})$ for all $k \geqslant 0$.

Let $S \leqslant S_1$, and let $\rho(x, y)$ be a concurrent relation of S. Since S is definable, the relation ρ on S can be defined in $P(S_1)$ by

$$\rho_1(x, y) = (x \in S) \wedge (y \in S) \wedge \rho(x, y).$$

This ρ_1 is a concurrent relation on S_1, consisting of precisely those pairs of elements which are in ρ. In general, if σ, σ_1 are sets of concurrent relations of S, S_1 respectively, we say $\sigma \leqslant \sigma_1$ if, for every $\rho \in \sigma$, we have that the corresponding $\rho_1 \in \sigma_1$.

Theorem 3.10.   *Let $S \leqslant S_1$, and let σ, σ_1 be sets of concurrent relations of S, S_1 such that $\sigma \leqslant \sigma_1$. Let $S_1^{\sigma_1}$ be an enlargement of S_1 with respect to σ_1. Then $S_1^{\sigma_1}$ contains an enlargement S^σ of S with respect to σ. In particular, any full enlargement of S_1 contains a full enlargement of S.*

Proof.   Put $S^\sigma = \{a \in S_1^{\sigma_1} | p(a)$ is true in $S_1^{\sigma_1}\}$, where $p(x) \in P(S_1)$ is such that $S = \{a \in S_1 | p(a)$ is true in $S_1\}$. S^σ is clearly a model of $\mathcal{T}(S)$, and we must show that for each $\rho \in \sigma$, there is a $b_\rho \in S^\sigma$ such that $\rho(a, b_\rho)$ is true in S^σ for all $a \in D_\rho$. Now ρ_1 is a concurrent relation of S_1, hence there is a $b_{\rho_1} \in S_1^{\sigma_1}$ such that $\rho_1(a, b_{\rho_1})$ is true in $S_1^{\sigma_1}$ for all $a \in D_{\rho_1}$. But $D_{\rho_1} = D_\rho$, and by the construction of ρ_1, $\rho_1(a, b_{\rho_1})$ true implies that $b_{\rho_1} \in S^\sigma$. Thus we can take $b_\rho = b_{\rho_1}$. $\quad\square$

§4   Standard Relations

Definition 4.1.   Let S^σ be an enlargement of S. A *standard n-ary relation* on S^σ is a relation

$$\rho^\sigma = \{(a_1, \ldots, a_n) \in (S^\sigma)^n | \rho(a_1, \ldots, a_n) \text{ is true in } S^\sigma\}$$

for some definable n-ary relation ρ on S.

We also define a *standard element* of S^σ to be an element of S.

Exercises

4.2.   Show that the standard relation ρ^σ is independent of the choice of description of ρ.

4.3.   Show that the one-element subset $\{a\}$ of S^σ is standard if and only if a is standard.

Theorem 4.4.   *Let $*S$ be a full enlargement of S and let u be a definable subset of S. Then $*u = u$ if and only if u is finite.*

Proof. Suppose that $u = \{u_1, \ldots, u_n\}$ is finite. Then

$$u(x) = (x = u_1) \vee (x = u_2) \vee \cdots \vee (x = u_n)$$

is a description of u, and

$$*u = \{a \in *S \mid (a = u_1) \vee (a = u_2) \vee \cdots \vee (a = u_n) \text{ is true in } *S\} = u.$$

Suppose that u is infinite. Let ρ be the binary relation on S defined by $\rho(x, y) = (x \in u) \wedge (y \in u) \wedge (x \neq y)$. Then $D_\rho = u$ and, since u is infinite, for any $u_1, \ldots, u_n \in D_\rho$, there exists $y \in u$ distinct from $u_1, \ldots, u_n$ and thus satisfying $\rho(u_i, y)$ for all i. Therefore ρ is a concurrent relation, and so there is a $b_\rho \in *S$ such that $\rho(x, b_\rho)$ is true for all $x \in D_\rho$. This says that $b_\rho \in *u$ and $b_\rho \notin u$. $\square$

Corollary 4.5. *Suppose the enlargement $\mathcal{T}^\sigma$ of $\mathcal{T}(S)$ is both full and trivial. Then S is finite.*

Proof. S is a definable subset with description $p(x) = \sim F$. $\square$

Corollary 4.6. *Let ρ be a definable n-ary relation on S. Then $*\rho = \rho$ if and only if ρ is finite.*

Proof. If ρ is finite, we can give a description which lists its members and it follows that $*\rho = \rho$. If ρ is infinite, put

$$u_i(x) = (\exists x_1) \cdots (\exists x_{i-1})(\exists x_{i+1}) \cdots (\exists x_n)\rho(x_1, \ldots, x_{i-1}, x, x_{i+1}, \ldots, x_n),$$

and let u_i be the subset of S defined by $u_i(x)$. Then for some i, u_i is infinite, and the theorem implies that $*u_i \neq u_i$, i.e., that $*\rho \neq \rho$. $\square$

Theorem 4.7. *Let ρ be a definable relation on S which defines a function $f : D \to S$ on some definable[2] subset D of S. Then ρ^σ defines a function $f^\sigma : D^\sigma \to S^\sigma$ on the subset D^σ of S^σ.*

Proof. We have $\mathcal{T} \vdash (\forall x)((x \in D) \Rightarrow (\exists! y)\rho(x, y))$. Interpreting this in S^σ gives the result. $\square$

The same argument applies to show that if $f : U \to V$ is a definable function, where U is a definable subset of S^r and V is a definable subset of S^s, then f^σ is a function from U^σ to V^σ.

§5 Internal Relations

Let S be any standard system. For each n, let $\mathcal{R}_n^{(1)}$ be the set of all first-order n-ary relations on S. Then each element of $\mathcal{R}_n^{(1)}$, and also the set $\mathcal{R}_n^{(1)}$ itself are all definable in $\mathcal{T}(\mathcal{O}^1(S))$. If $(\mathcal{O}^1(S))^\sigma$ is an enlargement of $\mathcal{O}^1(S)$ with respect to some set σ of concurrent binary relations of $\mathcal{O}^1(S)$, then every element of $(\mathcal{R}_n^{(1)})^\sigma$ is an n-ary relation on S^σ.

---

[2] Note that $\{a \in S \mid (\exists! y)\rho(a, y) \text{ is true}\}$ is a possible choice for D, but it is not always the most convenient choice.

Definition 5.1.   An *internal* first-order n-ary relation on S^σ is an element of $(\mathscr{R}_n^{(1)})^\sigma$. A relation on S^σ which is not internal is called *external*.

Higher-order internal n-ary relations may be defined similarly, by using the set $\mathscr{R}_n^{(k)}$ of n-ary relations on $\mathscr{O}^{k-1}(S)$.

If $\rho \in \mathscr{R}_n^{(1)}$ is in fact a definable relation on S, we have that

$$\mathscr{T}(\mathscr{O}^1(S)) \vdash (\forall x_1) \cdots (\forall x_n)(\in^n(\rho, x_1, \ldots, x_n) \Rightarrow \rho(x_1, \ldots, x_n)),$$

while each $u \in (\mathscr{R}_n^{(1)})^\sigma$ is the relation

$$u = \{(a_1, \ldots, a_n) \in (\mathscr{O}^1(S))^\sigma \mid \in^n(u, a_1, \ldots, a_n) \text{ is true in } (\mathscr{O}^1(S))^\sigma\}.$$

It follows that the definable relation ρ, considered as an element of $(\mathscr{R}_n^{(1)})^\sigma$, is the relation ρ^σ on S^σ. Hence every standard relation is internal. The converse is not true, for if S is infinite, then $\mathscr{R}_n^{(1)}$ is infinite, and by Theorem 4.4, $*(\mathscr{R}_n^{(1)}) \neq \mathscr{R}_n^{(1)}$.

Lemma 5.2.   *Let u, v be internal n-ary relations on S^σ. Then $u \cap v, u \cup v$, and the complement $u^\sim$ of u are also internal.*

Proof.   We have that

$$\mathscr{T}(\mathscr{O}^1(S)) \vdash (\forall x)(\forall y)((x \in \mathscr{R}_n^{(1)}) \wedge (y \in \mathscr{R}_n^{(1)}) \Rightarrow$$
$$(\exists z)((z \in \mathscr{R}_n^{(1)}) \wedge (\forall t)(t \in z \Leftrightarrow (t \in x) \wedge (t \in y)))).$$

It follows that $u \cap v \in (\mathscr{R}_n^{(1)})^\sigma$ for all $u, v \in (\mathscr{R}_n^{(1)})^\sigma$. Similar proofs apply for $u \cup v$ and for $u^\sim$.   □

§6   Non-Standard Analysis

Let R be the set of real numbers, with relation symbols $\mathscr{R} = \{\mathscr{I}, \times, +, <\}$. We form a full enlargement $*(\mathscr{O}^k(R))$ for some $k \geqslant 1$. Within this, we have standard subsets $*R > *Q > *Z > *N$, which are full enlargements of the reals, rationals, integers and natural numbers respectively. The relations on $*R$ defined by $\times, +, <$ shall be denoted by the same symbols, instead of by the correct but more cumbersome $* \times$, etc. The function $|\,| : R \to R$, defined by $|x| = x$ if $x \geqslant 0$, $|x| = -x$ if $x < 0$, yields the standard function $*R \to *R$ defined in the same way and which we shall denote by the same symbol $|\,|$. We shall call the elements of $*R$ real numbers, distinguishing those in R by calling them standard real numbers.

Theorem 6.1.   *$*R$ is a non-archimedean ordered field.*

Proof.   The axioms of ordered fields are theorems of $\mathscr{T}(R)$ and so hold for $*R$, showing that $*R$ is an ordered field. The relation $x < y$ is a concurrent relation of R with domain R, and consequently there is an element $a \in *R$ such that $r < a$ for all $r \in R$. This implies that for any $r \in R$, and for all $n \in N$, $\sum\limits_{i=1}^{n} r < a$. Hence the ordering on $*R$ is non-archimedean.   □

The archimedean axiom can indeed be expressed in the language of $\mathcal{O}^k(\mathrm{R})$ as

$$(\forall x)(\forall y)((x \in \mathrm{R}) \wedge (y \in \mathrm{R}) \wedge (x > 0) \Rightarrow (\exists n)((n \in \mathrm{N}) \wedge (nx > y))),$$

where $nx > y$ is an abbreviation for $(\forall z)(\times(n, x, z) \Rightarrow (z > y))$. This is a theorem of $\mathcal{T}(\mathcal{O}^k(\mathrm{R}))$ and so holds in *R. It does not assert the archimedean property for *R, as it asserts that if $x, y \in$ *R and if $x > 0$, then there is $n \in$ *N such that $nx > y$. For *R to be archimedean, we need to have $n \in \mathrm{N}$.

Definition 6.2. An element $a \in$ *R is called *finite* if there exists a standard real number b such that $|a| < b$. Otherwise, a is called *infinite*. A (finite) element a is called *infinitesimal* if $|a| < b$ for all standard real numbers $b > 0$.

0 is infinitesimal, and since $0 < a < b$ holds if and only if $0 < 1/b < 1/a$, it follows that if $a \neq 0$, then a is infinitesimal if and only if $1/a$ is infinite.

The proof of Theorem 6.1 contains a proof of the existence of infinite real numbers, and it follows that infinite natural numbers also exist.

Lemma 6.3. *There is no smallest infinite natural number. The set of infinite natural numbers is an external set.*

Proof. If n is a natural number and $n \neq 0$, then $n = m + 1$ for some natural number m, since this result is a theorem of $\mathcal{T}(\mathrm{N})$. If n is the smallest infinite natural number, then $m = n - 1$ is also infinite, and $m < n$, giving a contradiction.

It is a theorem of $\mathcal{T}(\mathcal{O}^1(\mathrm{N}))$ that every non-empty subset of N has a least member. Hence every non-empty internal subset of *N has a least member, and the set of infinite natural numbers cannot be internal. ☐

Lemma 6.4. *Suppose $n \in$ *N. Then n is finite if and only if $n \in \mathrm{N}$.*

Proof. If $n \in \mathrm{N}$, n is clearly finite. Suppose that n is finite. Then $n < b$ for some standard real number b, and $b < m$ for some standard natural number m. Put $u = \{x \in \mathrm{N} | x < m\}$. Then $n \in$ *$u = \{x \in$ *N$|x < m\}$, and *$u = u$ since u is finite. Thus $n \in \mathrm{N}$. ☐

Theorem 6.5. *Each of N, R, the set of infinite real numbers, and the set of infinitesimal real numbers is an external set.*

Proof.
(a) By Lemma 6.3, the set of infinite natural numbers is an external set, and by Lemma 6.4, N is its complement in the internal set *N. Hence N is external by Lemma 5.2.

(b) If R is internal, then so is $\mathrm{N} = \mathrm{R} \cap$ *N, contradicting (a). Similarly, Z and Q are also external.

(c) Let R_∞ be the set of infinite real numbers. If it is internal, then so is $R_\infty \cap$ *N, contradicting Lemma 6.3.

(d) If R_1 is the set of infinitesimal real numbers, then R_1 is bounded above and has no greatest member. It is a theorem of $\mathcal{T}(\mathcal{O}^1(R))$ that if u is a non-empty subset of R which is bounded above and has no greatest element, then $\{x \in R \mid x > y \text{ for all } y \in u\}$ has a least element. If R_1 is internal, then $v = \{x \in {}^*R \mid x > r \text{ for all } r \in R_1\}$ has a least element. But if $x \in v$, then $\frac{1}{2}x \in v$ and $\frac{1}{2}x < x$. Hence v has no least element, and so R_1 is external. $\square$

Let $a, b \in {}^*R$. We write $a \simeq b$ if $a - b$ is infinitesimal. $\simeq$ is clearly an equivalence relation on ${}^*R$.

Exercise 6.6. Show that if $r \in {}^*R$, then there exists $q \in {}^*Q$ such that $q \simeq r$.

Definition 6.7. The *monad* of the finite real number a is the set $\mu(a) = \{r \in {}^*R \mid r \simeq a\}$.

Theorem 6.8. *If a is a finite real number, then $\mu(a)$ contains exactly one standard real number. If R_0 is the set of finite real numbers and R_1 the set of infinitesimal real numbers, then R_0 is a ring, R_1 is an ideal of R_0 and R_0/R_1 is isomorphic to R.*

Proof. If $r, s \in \mu(a)$ and r, s are standard, then $|r - s|$ is an infinitesimal standard real number. Thus $|r - s| = 0$ and $r = s$. We have to show that there is a standard real number in $\mu(a)$. This is so if a is standard, so we suppose a is not standard. Put $L = \{x \in R \mid x < a\}$ and $U = \{x \in R \mid x > a\}$. Since a is finite, there exists $b \in R$ such that $|a| < b$, i.e., $-b < a < b$, showing that L and U are both non-empty. L is bounded above by b and so has a least upper bound α say, which is also the greatest lower bound of U. If $\alpha \in L$, then $U = \{x \in R \mid x > \alpha\}$, and $\alpha \leqslant a < \alpha + r$ for all standard real numbers $r > 0$. Thus $|a - \alpha| = a - \alpha < r$ for all standard $r > 0$, and so $a - \alpha$ is infinitesimal. Similarly, if $\alpha \in U$, we obtain $|a - \alpha| = \alpha - a$ is infinitesimal. Hence $\alpha \in \mu(a)$.

Trivially, R_0 is a ring and R_1 is an ideal of R_0. The map sending a to $\mu(a)$ is the natural homomorphism $R_0 \to R_0/R_1$. Mapping $\mu(a)$ to the standard real number in $\mu(a)$ is an isomorphism. $\square$

Finally, as an introduction to the use of enlargements in the study of analysis, we shall show how a few of the familiar results on limits can be proved with the aid of infinitesimal and infinite elements in an enlargement. We begin with the concept of a limit of a sequence. A sequence is a function $s: N \to R$, and corresponding to any sequence, we have the standard function ${}^*s: {}^*N \to {}^*R$.

Theorem 6.9. *Let $r \in R$ and let $s: N \to R$ be a sequence. Then $\text{Lim}_{n \to \infty} s(n) = r$ if and only if ${}^*s(n) \in \mu(r)$ for all infinite natural numbers n.*

Proof. Suppose that $\text{Lim}_{n \to \infty} s(n) = r$. Then for every standard real number $\varepsilon > 0$, there exists $n_0 \in N$ such that $|s(n) - r| < \varepsilon$ for all $n > n_0$. For this ε and n_0, $(\forall n)((n \in N) \wedge (n > n_0) \Rightarrow |s(n) - r| < \varepsilon)$ is a theorem of

$\mathcal{T}(\mathcal{O}^1(\mathbf{R}))$. Therefore $(\forall n)((n \in {}^*\mathbf{N}) \wedge (n > n_0) \Rightarrow |{}^*s(n) - r| < \varepsilon)$ holds in ${}^*(\mathcal{O}^1(\mathbf{R}))$. If n is an infinite natural number, then $|{}^*s(n) - r| < \varepsilon$, and this is true for all standard real numbers $\varepsilon > 0$. Hence ${}^*s(n) \in \mu(r)$ for all infinite numbers n.

Suppose conversely that ${}^*s(n) \in \mu(r)$ for all infinite natural numbers n. If n_0 is an infinite natural number, then for every standard real number $\varepsilon > 0$, we have $|{}^*s(n) - r| < \varepsilon$ for all $n > n_0$. Thus $(\exists n_0)((n_0 \in {}^*\mathbf{N}) \wedge (\forall n)((n \in {}^*\mathbf{N}) \wedge (n > n_0) \Rightarrow |{}^*s(n) - r| < \varepsilon))$, being true in ${}^*\mathcal{O}^1(\mathbf{R})$, is a theorem of $\mathcal{T}(\mathcal{O}^1(\mathbf{R}))$. Hence there exists $n_0 \in \mathbf{N}$ such that $|s(n) - r| < \varepsilon$ for all $n > n_0$. ☐

By a similar argument, one obtains the following result.

Theorem 6.10.   *Let U be a subset of $\mathbf{R}$, and suppose U contains a neighbourhood of $a \in \mathbf{R}$. Let $f : U \to \mathbf{R}$ be a function defined on U. If $\ell \in \mathbf{R}$, then $Lim_{x \to a} f(x) = \ell$ if and only if ${}^*f(x) \in \mu(\ell)$ for all $x \neq a$ in $\mu(a)$.*

Corollary 6.11.   *The function f is continuous at a if and only if ${}^*f(x) \simeq {}^*f(a)$ for all $x \simeq a$.*

Exercise 6.12.   Prove Theorem 6.10.

For a real function f defined on an arbitrary subset U of $\mathbf{R}$, the necessary and sufficient condition that f be continuous on U is that for each $a \in U$, if $x \in {}^*U$ and $x \simeq a$, then ${}^*f(x) \simeq {}^*f(a)$. The meaning of this condition is altered if we write it formally using $(\forall a)$, as we now show. For then the statement becomes the following: for all $a, x \in {}^*U$, if $x \simeq a$ then ${}^*f(x) \simeq {}^*f(a)$. If this new statement holds, then for an infinitesimal positive real number δ, and for any standard real number $\varepsilon > 0$,

$$(\forall a)(\forall x)((a \in {}^*U) \wedge (x \in {}^*U) \wedge (|x - a| < \delta) \Rightarrow |{}^*f(x) - {}^*f(a)| < \varepsilon)$$

is true in ${}^*(\mathcal{O}^1(\mathbf{R}))$, and so

$$(\exists \delta)((\delta > 0) \wedge (\forall a)(\forall x)((a \in U) \wedge (x \in U)$$
$$\wedge (|x - a| < \delta) \Rightarrow |f(x) - f(a)| < \varepsilon))$$

holds in $\mathcal{O}^1(\mathbf{R})$. But this is precisely the condition that the function f be *uniformly* continuous on U. We have proved the following theorem.

Theorem 6.13.   *Let f be a real-valued function defined on the subset U of $\mathbf{R}$. Then f is uniformly continuous on U if and only if for all $x, y \in {}^*U, x \simeq y$ implies that ${}^*f(x) \simeq {}^*f(y)$.*

It is now a simple matter to prove the following well-known result.

Theorem 6.14.   *Let U be a closed bounded interval $[p, q]$. If the real-valued function f is continuous on U, then it is uniformly continuous on U.*

Proof.   Take any $x \in {}^*U$. x is a finite real number, hence there is a unique $r \in \mathbf{R}$ such that $r \simeq x$. If $r < p$, then $x = r + (x - r) < r + (p - r) = p$, since $x - r$ is infinitesimal and $p - r$ is a standard positive real number. As $x \geqslant p$, we have a contradiction, and so $p \leqslant r$. Similarly, $r \leqslant q$, and $r \in U$. If $y \in {}^*U$ and $y \simeq x$, then $y \simeq r$ and ${}^*f(y) \simeq {}^*f(r)$. In particular, ${}^*f(x) \simeq {}^*f(r)$.

Consequently $*f(x) \simeq *f(r) \simeq *f(y)$, and we have the condition for uniform continuity on U. ∎

Exercise 6.15.   Where does the above method of proof fail if U is taken as the open interval $\{x : p < x < q\}$?

Our final application is to the study of sequences of real-valued functions $s_n(x)$ defined on a subset U of R. The usual necessary and sufficient condition that $s_n(x) \to r(x)$ on U as $n \to \infty$, when expressed in terms of our non-standard analysis, is that for each $x \in U$ and for all infinite n, $*s_n(x) \simeq *r(x)$. Again, the meaning of the condition is altered if we express it in terms of $(\forall x)$, as the next result indicates.

Theorem 6.16.   *The sequence of functions $s_n(x)$ converges uniformly on U to $r(x)$ if and only if for all $x \in *U$ and for all infinite n, $*s_n(x) \simeq *r(x)$.*

Exercise 6.17.   Prove Theorem 6.16 by suitably modifying the argument leading to Theorem 6.13.

Theorem 6.18.   *Suppose the functions $s_n(x)$ are continuous on U, and converge uniformly on U to $r(x)$. Then $r(x)$ is continuous on U.*

Proof.   Let $a \in U$. If $x \in *U$ and $x \simeq a$, and if n is infinite, then by Corollary 6.11 and Theorem 6.14,

$$*r(s) \simeq *s_n(x) \simeq *s_n(a) \simeq *r(a),$$

showing that r is continuous at a. ∎

Exercises

6.19.   Given that $f(x) \to r$ and $g(x) \to s$ $(\neq 0)$ as $x \to a$, prove that $f(x)/g(x) \to r/s$ as $x \to a$.

6.20. $f(x)$ is defined in a neighbourhood of a. Show that $f'(a) = c$ if and only if $\dfrac{*f(x) - *f(a)}{x - a} \simeq c$ for all $x \simeq a, x \neq a$.

6.21.   Prove that if $f(x)$ is differentiable at $x = a$, then $f(x)$ is continuous at $x = a$.

6.22.   R is complete, i.e., every Cauchy sequence in R has a limit in R. Formalise this and interpret it for *R. Is *R complete?

We refer the reader to [8] for further reading and references on the subject of non-standard analysis.

Chapter IX

Turing Machines and Gödel Numbers

§1 Decision Processes

In §3 of Chapter III, we gave a procedure for determining whether or not an element p of $P(X)$ is a theorem of $\mathrm{Prop}(X)$. In §4 of Chapter IV, we asserted that no such procedure exists for $\mathrm{Pred}(V, \mathcal{R})$. Before attempting to prove this non-existence theorem, we must say more precisely what we mean by "procedure". The procedures we shall discuss are called decision processes, and informally we think of a decision process as a list of instructions which can be applied in a routine fashion to give one of a finite number of specified answers. A decision process for $\mathrm{Pred}(V, \mathcal{R})$ is then a finite list of instructions such that for any element $p \in P(V, \mathcal{R})$, there corresponds a unique finite sequence of instructions from the list. The sequence terminates with an instruction to announce a decision of some prescribed kind (e.g., "p is a theorem of $\mathrm{Pred}(V, \mathcal{R})$."). Thus at each step of the process, exactly one instruction of the list is applicable, producing a result to which exactly one instruction is applicable, until after a finite (but not necessarily bounded) number of steps, the process stops and a decision is announced.

The mechanical nature of the process just described suggests that we could think of it as a computer program, carried out on a suitable computer. We shall formalise our ideas by considering processes which could be performed by an idealised computer known as a Turing machine.

§2 Turing Machines

A Turing machine is imagined as consisting of two parts—the machine proper, being a device with a finite set $\mathfrak{Q} = \{q_0, q_1, \ldots, q_m\}$ of possible internal states, and a tape (at least potentially infinite) on which suitably coded instructions to the machine may be printed, and on which the machine can print its response. The tape is divided lengthwise into squares which can be indexed by the integers $\mathbb{Z}$. On each square of the tape is printed one symbol selected from a fixed finite set $\mathfrak{S} = \{s_0, s_1, \ldots, s_k\}$, called the alphabet of the machine. Since we think in terms of finite lists of instructions, we must allow squares to be blank, and hence the alphabet $\mathfrak{S}$ must contain a symbol corresponding to 'blank'. This symbol will always be denoted by s_0. Only finitely many squares of the tape have printed on them a symbol other than s_0. The tape is fed into the machine so that at any time, the machine is scanning exactly one square of the tape.

85

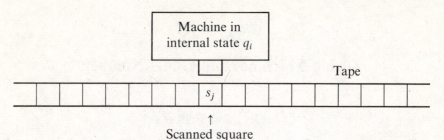

Scanned square

We assume that the machine must be in internal state q_0 to commence operating. The machine operates in discrete steps, its action at any stage being determined by its internal state q_i together with the symbol s_j printed on the square being scanned. The possible actions of the machine are of the following kinds:

(i) The machine replaces the symbol s_j by a symbol s_ℓ and changes its internal state to q_r.

(ii) The machine moves the tape so as to scan the square immediately on the right of the one being scanned, and changes its internal state to q_r.

(iii) The machine moves the tape so as to scan the square immediately on the left of the one being scanned, and changes its internal state to q_r.

(iv) The machine stops.

Since the machine must have no choice of action, exactly one of the above actions will occur at each step.

A Turing machine is specified by giving its set $\mathfrak{Q}$ of internal states, its alphabet $\mathfrak{S}$, and its response to each pair (q_i, s_j) consisting of an internal state and a scanned symbol. Since there are only finitely many pairs (q_i, s_j), the machine response is specified by a finite list. A response of the type (i) can be indicated by quadruples (q_i, s_j, s_ℓ, q_r). Responses (ii) and (iii) can be indicated by the quadruples (q_i, s_j, R, q_r) and (q_i, s_j, L, q_r) respectively, where we have assumed that neither R nor L is in $\mathfrak{S}$. Response (iv) can be specified by having no quadruple beginning with the pair q_i, s_j. Our requirement that the machine be deterministic means that the list of responses has at most one quadruple beginning with each pair q_i, s_j.

We can now expect the following formal definition of a Turing machine to make sense.

Definition 2.1. A *Turing machine* with (finite) alphabet $\mathfrak{S}$ and (finite) set $\mathfrak{Q}$ of internal states is a subset M of $\mathfrak{Q} \times \mathfrak{S} \times (\mathfrak{S} \cup \{L, \mathscr{R}\}) \times \mathfrak{Q}$ $(L, R \notin \mathfrak{S})$, such that if (a, b, c, d) and $(a, b, c', d') \in M$, then $c = c'$ and $d = d'$.

To discuss the operation of a Turing machine M, we need a convenient way of describing its state at each stage of a computation. The state of M at any stage is determined by the contents of the tape, the number of the tape square being scanned, and the internal state of the machine[1]. Denote the

---

[1] Thus the state of M is a description of the total machine configuration, including the internal state of M.

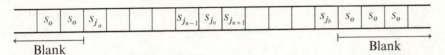

| S_0 | S_0 | S_{j_a} | | | | $S_{j_{n-1}}$ | S_{j_n} | $S_{j_{n+1}}$ | | | | S_{j_b} | S_0 | S_0 | S_0 |

Blank   Blank

symbol printed on square number n by s_{j_n}. Since there are always only finitely many non-blank squares, there exist integers a, b (not unique) such that $j_n = 0$ for all $n < a$ and all $n > b$. a and b can always be selected so that the square currently being scanned, say square number n, lies between square number a and square number b, so that $a \leqslant n \leqslant b$. (We note that only the ordering of the tape squares is important—it is customary to shift the origin each time the machine shifts the tape, so that the square being scanned becomes the origin.) The contents of the tape are completely specified by the finite string $s_{j_a} s_{j_{a+1}} \cdots s_{j_n} \cdots s_{j_b}$, and we shall indicate that the machine is in internal state q_i, scanning square n of a tape with these symbols on it, by writing the string

$$s_{j_a} s_{j_{a+1}} \cdots s_{j_{n-1}} q_i s_{j_n} \cdots s_{j_b}.$$

Definition 2.2. An *instantaneous description* of a Turing machine M with alphabet $\mathfrak{S}$ and set $\mathfrak{Q}$ of internal states is a finite string $s_{\alpha_1} s_{\alpha_2} \cdots s_{\alpha_r} q s_{\beta_1} s_{\beta_2} \cdots s_{\beta_t}$, where $s_{\alpha_i}, s_{\beta_j} \in \mathfrak{S}$ and $q \in \mathfrak{Q}$.

The strings $s_{\alpha_1} \cdots s_{\alpha_r}$ and $s_{\beta_1} \cdots s_{\beta_t}$ are often denoted by single symbols such as σ, τ. An instantaneous description $d = s_{\alpha_1} s_{\alpha_2} \cdots s_{\alpha_r} q s_{\beta_1} s_{\beta_2} \cdots s_{\beta_t}$ is then written simply as $d = \sigma q \tau$. Each of σ, τ may be the empty string.

Since we are interested in the state of M, rather than in descriptions of the state of M, we need to know when two descriptions determine the same state. The previous discussion shows that the only freedom in the definition of description is in the choice of a and b. Thus two descriptions $d = \sigma q \tau$ and $d' = \sigma' q' \tau'$ describe the same state if and only if $q = q'$, σ' is obtainable from σ by adding or deleting a number of symbols s_0 on the left, and τ' is obtainable from τ by adding or deleting a number of symbols s_0 on the right. Descriptions related in this way are called equivalent, and the equivalence class containing the description d is denoted by $[d]$ and called the *state* described by d. For each state $[d]$, there is a unique description $d = \sigma q \tau$ such that the first symbol (if any) of σ, and the last symbol (if any) of τ are distinct from s_0. This description is called the *shortest description* of $[d]$.

Definition 2.3. The Turing machine M takes the state $[d]$ into the state $[d']$, written $[d] \xrightarrow{M} [d']$, if for some representatives $d = \sigma q \tau$ and $d' = \sigma' q' \tau'$, where $\tau = s_\alpha \tau_1$, either

   (i) $(q, s_\alpha, s_{\alpha'}, q') \in M$ and $\sigma' = \sigma$, $\tau' = s_{\alpha'} \tau_1$, or
   (ii) $(q, s_\alpha, R, q') \in M$ and $\sigma' = \sigma s_\alpha$, $\tau' = \tau_1$, or
   (iii) $(q, s_\alpha, L, q') \in M$ and $\sigma = \sigma' s_\beta$, $\tau' = s_\beta \tau$ for some $s_\beta \in S$.

Exercise 2.4. Prove that there is at most one state $[d']$ such that $[d] \xrightarrow{M} [d']$. When $[d']$ exists, show that to each $d \in [d]$, there corresponds a

$d' \in [d']$ so that d and d' are related as in (i), (ii) or (iii) of the definition (appropriately modified if σ or τ is empty).

Definition 2.5.   A state $[\sigma q \tau]$ is called *initial* if $q = q_0$. A state $[\sigma q s_\alpha \tau_1]$ is called *terminal* if there is no quadruple (q, s_α, c, d) in M.

Exercise 2.6.   Show that $[d]$ is terminal if and only if there does not exist a state $[d']$ such that $[d] \overset{M}{\to} [d']$.

Definition 2.7.   A *computation* by the machine M is a finite sequence $[d_0], [d_1], \ldots, [d_p]$ of states such that $[d_0]$ is initial, $[d_p]$ is terminal and $[d_i] \overset{M}{\to} [d_{i+1}]$ for $i = 0, 1, \ldots, p - 1$.

Computations are by definition finite. Given M and $[d]$, there is no guarantee that M, started in state $[d]$ and allowed to operate, will ever stop (i.e., will execute a computation).

Definition 2.8.   We say that M *fails* for the input $[d_0]$ if there is no computation by M beginning with the state $[d_0]$.

For each state $[d_i]$, there is a unique $[d_{i+1}]$ such that $[d_i] \overset{M}{\to} [d_{i+1}]$. Hence failure of M for the input $[d_0]$ means that the sequence of states taken by M and beginning with $[d_0]$ is infinite—i.e., the machine never stops.
Henceforth, the state $[d]$ will be denoted simply by some description d. The context will make clear the sense in which symbols such as d, d_i are being used.

Exercises

2.9.   A stereo-Turing machine M has its tape divided into two parallel tracks. The symbols on a pair of squares (one above the other) are read simultaneously. Show that there is a (mono-)Turing machine M' which will perform essentially the same computations as M.

2.10.   The operator of the Turing machine M has been asked to record the output of M (i.e., the symbols printed on the tape) at the end of each computation by M. Does the operator have any problems? Show that a machine M' can be designed so as to perform essentially the same computations as M, and which in addition will place marker symbols (not in the alphabet of M) either at the furthest out points of the tape used in each computation, or alternatively at the nearest points such that the stopping position of M', and all non-blank symbols, lie between them.

2.11.   A dual-Turing machine M with alphabet $\mathfrak{S}$ has two tapes which can move independently. Show that there is a Turing machine with alphabet $\mathfrak{S} \times \mathfrak{S}$ which will, when given an initial state corresponding to the pair of initial states of a computation by M, perform a computation whose terminal state corresponds to the pair of terminal states of M.

2.12. M_1 and M_2 are Turing machines with the same alphabet $\mathfrak{S}$. A computation by M_1 and M_2 consists of a computation by each of M_1 and

M_2 such that, if $\sigma q_i \tau$ is the output of M_1, then $\sigma q_0 \tau$ is the input for M_2. Show that there is a Turing machine M, whose alphabet contains $\mathfrak{S}$, such that if M is started in an initial state of a computation by M_1 and M_2 with terminal state $\sigma q_j \tau$, then M executes a computation with terminal state $\sigma q_k \tau$ for some q_k, while M fails if started in any other initial state.

2.13. $M_1, \ldots, M_n$ are Turing machines with the same alphabet. An algorithm requires that at each step, exactly one of $M_1, \ldots, M_n$ be applied to the result of the previous step. The Turing machine M, applied to the output of any step, determines which of $M_1, \ldots, M_n$ is to be applied for the next step. Show that there is a single Turing machine which can execute the algorithm and give the same ultimate output.

2.14. Most digital computers can read and write on magnetic tape. The tapes are finite, but the operator can replace them if they run out. Show that such computers can be regarded as Turing machines. In fact, the most sophisticated computers can be regarded as Turing machines. (This is not a mathematical exercise. The reader is asked to review his experience of computers and to see that the definitions given so far are broad enough to embrace the computational features of the computers he has used.)

§3   Recursive Functions

Let M be a Turing machine with alphabet $\mathfrak{S}$. We show how to use M to associate with each pair (k, ℓ) of natural numbers a subset $U_M^{(k, \ell)}$ of $\mathbf{N}^k$ and a function $\Psi_M^{(k, \ell)} : U_M^{(k, \ell)} \to \mathbf{N}^\ell$. For $(n_1, \ldots, n_k) \in \mathbf{N}^k$, put

$$\text{code}(n_1, \ldots, n_k) = s_1^{n_1} s_0 s_1^{n_2} s_0 \cdots s_1^{n_{k-1}} s_0 s_1^{n_k},$$

where the notation s^n denotes a string of n consecutive symbols s. There may or may not be a computation by M whose initial state is the state $d_0 = q_0 \, \text{code}\,(n_1, \ldots, n_k)$. If there is, let $d_t = \sigma q \tau$ be its (uniquely determined) terminal state. Choose a description d_t of this terminal state which has at least ℓ occurrences of s_0 in τ, and determine $(a_1, \ldots, a_\ell) \in \mathbf{N}^\ell$ by defining a_1 to be the number of times s_1 occurs in τ before the first occurrence of s_0, and a_i (for $2 \leqslant i \leqslant \ell$) to be the number of times s_1 occurs in τ between the $(i-1)$th and the ith occurrences of s_0. Let $U_M^{(k, \ell)}$ be the subset of $\mathbf{N}^k$ consisting of all $(n_1, \ldots, n_k) \in \mathbf{N}^k$ for which there exists a computation by M with initial state $q_0 \, \text{code}(n_1, \ldots, n_k)$, and so for which an element $(a_1, \ldots, a_\ell) \in \mathbf{N}^\ell$ is defined. The function $\Psi_M^{(k, \ell)}$, with domain $U_M^{(k, \ell)}$, is defined by the rule

$$\Psi_M^{(k, \ell)}(n_1, \ldots, n_k) = (a_1, \ldots, a_\ell).$$

Definition 3.1. A function $\Psi_M^{(k, \ell)}$ defined as above in terms of a Turing machine M is called a *partial recursive function*[2]. The function $\Psi_M^{(k, \ell)}$ is called a *(total) recursive function* if $U_M^{(k, \ell)} = \mathbf{N}^k$.

---

[2] These functions are usually called Turing computable functions, with a different definition being given for recursive functions. The equivalence of the two definitions is a significant result, but the proof is tedious. The reader is referred to §1 of Chapter X for further information, and to [10], pp. 120–121, 207–237 for full details.

Exercises

3.2. $f: U \to \mathbf{N}^\ell$ is a partial recursive function with domain $U \subseteq \mathbf{N}^k$. Show that there is a Turing machine M such that $\Psi_M^{(k,\ell)} = f$ and such that, for each $(n_1, \ldots, n_k) \in U$, the computation $d_0, d_1, \ldots, d_t$ by M which begins with $d_0 = q_0 \operatorname{code}(n_1, \ldots, n_k)$ ends with $d_t = q \operatorname{code} f(n_1, \ldots, n_k)$ for some internal state q of M.

3.3.   Prove that the composition of (partial) recursive functions is (partial) recursive.

3.4.   The real number r has decimal expansion $t = r_0 \cdot r_1 r_2 r_3 \cdots$. Given that the function $f: \mathbf{N} \to \mathbf{N}$ defined by $f(n) = r_n$ is not recursive, prove that r is transcendental.

3.5.   A subset U of $\mathbf{N}$ is called *recursively enumerable* if it is the range of a recursive function $f: \mathbf{N} \to \mathbf{N}$, or else is empty. Show that $U \subseteq \mathbf{N}$ is recursively enumerable if and only if it is the domain of a partial recursive function.

3.6.   A subset U of $\mathbf{N}$ is called *recursive* if its characteristic function is recursive. Prove that $U \subseteq \mathbf{N}$ is recursive if and only if both U and $\mathbf{N} - U$ are recursively enumerable.

3.7.   Write a FORTRAN program for calculating the greatest common divisor of two integers of unlimited size (possibly beyond the storage capacity of the machine), assuming the availability of unlimited magnetic tape.

§4   Gödel Numbers

We are interested in delimiting the scope of computations performable by Turing machines, and we are also interested in using Turing machines to formalise the notion of decidability for a logical or mathematical system. To do these things, we need some way of listing all the essentially different Turing machines. From the definition of a Turing machine, it is clear that if two machines M, M' differ only in the labels given their internal states and their alphabets (i.e., if there are bijective maps $\mathfrak{Q} \to \mathfrak{Q}'$, $\mathfrak{S} \to \mathfrak{S}'$ which extend naturally to a bijection $M \to M'$), then M and M' perform essentially the same computations (i.e., the bijection $M \to M'$ extends to a bijection between the sets of computations of M and M'). We may therefore suppose that all Turing machines have alphabets chosen from the universal alphabet $\mathfrak{S}^* = \{s_i \mid i \in \mathbf{N}\}$ (with s_0 corresponding to "blank"), and also that they have lists of internal states chosen from the universal list $\mathfrak{Q}^* = \{q_i \mid i \in \mathbf{N}\}$ (with q_0 corresponding to "initial internal state"). Each machine uses a finite subset of $\mathfrak{Q}^*$, containing q_0, and a finite subset of $\mathfrak{S}^*$, containing s_0. Hence we may think of a Turing machine M as a finite subset of $\mathfrak{Q}^* \times \mathfrak{S}^* \times (\mathfrak{S}^* \cup \{L, R\}) \times \mathfrak{Q}^*$. Further, any tape written in the alphabet $\mathfrak{S}^*$ may be used on an arbitrary Turing machine, for a machine will stop if it scans some symbol not in its alphabet.

We now attach to each Turing machine M a number, called the *Gödel number* of M. Denote an element $(a, b, c, d) \in M$ by the string $abcd$. The strings of M have a natural lexicographic order, and by taking all the strings of M in this order we associate with M a unique finite sequence of strings of symbols. We shall define the Gödel number $G(M)$ of M by defining in turn Gödel numbers for every symbol, for every string of symbols and lastly for every finite sequence of strings of symbols.

Define a function $G: \{L, R\} \cup \mathfrak{S}^* \cup \mathfrak{Q}^* \to \mathbb{N}$ by

$$G(L) = 1, \ G(R) = 3, \ G(s_i) = 4i + 5, \ G(q_j) = 4j + 7 (i, j \in \mathbb{N}).$$

If now the symbols a_i have Gödel numbers $G(a_i) = n_i$ $(i = 1, \ldots, r)$, then we define the Gödel number of the string $a_1 \cdots a_r$ by

$$G(a_1 \cdots a_r) = p_1^{n_1} \cdots p_r^{n_r},$$

where p_k denotes the kth prime (so that $p_1 = 2$, $p_2 = 3, \ldots$). The empty string has no Gödel number attached to it.

If $\sigma_1, \ldots, \sigma_s$ are strings of symbols, then we define the Gödel number of the sequence $\sigma_1, \ldots, \sigma_s$ by

$$G(\sigma_1, \ldots, \sigma_s) = p_1^{G(\sigma_1)} p_2^{G(\sigma_2)} \cdots p_s^{G(\sigma_s)}.$$

Finally, the Gödel number of the Turing machine M is defined to be the Gödel number of the unique finite sequence of strings associated with M in the way described before.

Exercises

(In many subsequent exercises, the reader is required to construct a Turing machine. The reader is asked to interpret this as follows: he should convince himself that the required machine can be constructed (perhaps by using previously constructed machines or the results of previous exercises), rather than formally construct the machine as a set of quadruples.)

4.1. Show that, provided each symbol a_j is distinguished from the one element string a_j, and each string σ is distinguished from the sequence σ of length one, then G as defined above is an injective function whose range is a proper subset of $\mathbb{N}$.

4.2. Given a non-empty string σ not containing the symbol s_0, construct a Turing machine which computes $G(\sigma)$ from the initial state $d_0 = q_0\sigma$.

4.3. $f: \mathbb{N} \to \mathbb{N}$ is defined by

$f(n) = 0$ if n is not a Gödel number,
$f(n) = 1$ if n is the Gödel number of a symbol,
$f(n) = 2$ if n is the Gödel number of a string,
$f(n) = 3$ if n is the Gödel number of a finite sequence of strings.

Show that f is recursive.

4.4. The function $f: \mathbb{N} \to \mathbb{N}$ is defined by

$f(n) = 0$ if n is not the Gödel number of a Turing machine,
$f(n) = 1$ if n is the Gödel number of a Turing machine.

Show that f is recursive.

4.5. Turing machines can be ordered by the size of their Gödel numbers. Let $f(n)$ be the Gödel number of the $(n + 1)$-th Turing machine. Show that $f : N \rightarrow N$ is recursive.

4.6. Use cardinality considerations to prove that there exists a non-recursive function $f : N \rightarrow N$.

4.7. Show that there is a Turing machine U with the property that, for each Turing machine M and shortest description d of an initial state, U started in the state q code$(G(M), G(d))$

(i) fails if M fails in the state d,

(ii) computes $G(d_t)$ if d_t is the shortest description of the terminal state reached by M starting from d.

A machine such as U is called a universal Turing machine.

In order to apply Turing machines to questions of decidability of mathematical theories, we must be able to encode elements of the appropriate algebras of propositions. We do this by again constructing a universal alphabet and then defining more Gödel numbers. As we can only hope to code countable theories, we confine our attention to them. For each $i \in N$, the subset $\mathscr{R}_i$ of the set $\mathscr{R}$ of relations of any countable theory is at most countable, so we take a universal set $\mathscr{R}^* = \{r_{ij} | i, j \in N\}$ of relation symbols, where, for each j, $r_{ij} \in \mathscr{R}_i^*$. Likewise, we take a set $C^* = \{c_j | j \in N\}$ of constants, and a set $X^* = \{x_j | j \in N\}$ of variables, and put $V^* = C^* \cup X^*$. For operation symbols we take F, $\Rightarrow$ and $\{(\forall x_j) | j \in N\}$. We now have a universal alphabet in which every countable theory can be written. Each element of the algebra $P(V^*, \mathscr{R}^*)$ of such a theory has a representative which can be written as a finite string of symbols of this alphabet, for we can replace any $(\forall c_j)$ which occurs, and brackets are unnecessary—we write $a \Rightarrow b$ as $\Rightarrow ab$, $r_{2j}(x_1, x_2)$ as $r_{2j}x_1x_2$, etc. Each string of symbols then has at most one meaning as an element of $\tilde{P}(V^*, \mathscr{R}^*)$.

Exercise 4.8. Prove that each string of symbols has at most one meaning as an element of $\tilde{P}(V^*, \mathscr{R}^*)$.

Gödel numbers are now assigned to our universal alphabet as follows:

$$G(F) = 2, \ G(\Rightarrow) = 3, \ G(r_{ij}) = 5^{i+1}7^{j+1}, \ G(c_j) = 11^{j+1},$$
$$G(x_j) = 13^{j+1}, \ G((\forall x_j)) = 17^{j+1}.$$

For a string $a_1 a_2 \cdots a_n$ of symbols, we put $G(a_1 a_2 \cdots a_n) = p_1^{G(a_1)} p_2^{G(a_2)} \cdots p_n^{G(a_n)}$, with p_i denoting the ith prime, as before. For sequences of strings, we also use the method given before. Finally, we define the Gödel number of an element $p \in P(V^*, \mathscr{R}^*)$ to be the least number which is the Gödel number of an element $w \in \tilde{P}(V^*, \mathscr{R}^*)$ which represents p.

Exercises

4.9. Our definitions of Gödel numbers make it possible for an integer to be a Gödel number in the Turing machine sense and also in the propositional algebra sense—e.g., $11 = G(q_1) = G(c_0)$. Modify our definitions of Gödel numbers so that each $n \in \mathbb{N}$ is a Gödel number in at most one way.

4.10. Show that the function $f: \mathbb{N} \times \mathbb{N} \to \mathbb{N}$ defined by $f(m, n) = 1$ if m, n are Gödel numbers of elements of $\tilde{P}(V^*, \mathscr{R}^*)$ which represent the same element of $P(V^*, \mathscr{R}^*)$, $f(m, n) = 0$ if m, n are Gödel numbers of elements of $\tilde{P}(V^*, \mathscr{R}^*)$ which represent different elements of $P(V^*, \mathscr{R}^*)$, $f(m, n) = 2$ if either of m, n is not the Gödel number of an element of $\tilde{P}(V^*, \mathscr{R}^*)$, is a recursive function.

§5   Insoluble Problems in Mathematics

We consider various ways in which a mathematical problem can be insoluble, and we begin with two well-known examples—the classical problem of trisecting an angle, and the problem of solving quintic equations. The trisection problem is insoluble by Euclidean construction, but admits a simple solution if a quite minor extension of method is permitted (see [14]). Although there is no formula for the solution of quintic equations by radicals, there is one in terms of elliptic functions (see [5]). Clearly, insolubility of a particular problem depends on a precise statement as to what constitutes a solution.

Each of the above problems is in fact a family of problems. Since a right angle can be trisected, not every angle is impossible to trisect. There exist quintic equations whose solutions are expressible in terms of radicals. The trisection problem asks for a construction which works for every angle, and the non-existence of such a construction follows from the proof that an angle of $\pi/3$ cannot be trisected. Likewise, the existence of a single quintic equation that is insoluble by radicals suffices to demonstrate the non-existence of a general solution by radicals of quintic equations.

Our concern is with the problem of determining for a mathematical theory $\mathscr{T}$ whether or not elements $p_n \in \mathscr{L}(\mathscr{T})$ are theorems of $\mathscr{T}$. In the case of a single element $p \in \mathscr{L}(\mathscr{T})$, let us consider what would constitute a solution to our problem. If p actually is a theorem of $\mathscr{T}$, then we must show that there is a proof of p within $\mathscr{T}$. A proof of p would clearly suffice, provided we can check that it really is a proof. An alleged proof involves only finitely many symbols of our universal alphabet. We can test if a particular step is obtained from earlier steps by modus ponens, or if it is a logical axiom. (We can devise a Turing machine for the purpose.) We could also test the use of Generalisation if we could identify the mathematical axioms of $\mathscr{T}$. In short, proof checking can be performed by a Turing machine provided it can test for mathematical axioms.

Definition 5.1. Let $\mathscr{T}$ be a countable theory expressed in the universal alphabet. We say that $\mathscr{T}$ is *effectively axiomatised* if the characteristic function of the set of Gödel numbers of mathematical axioms of $\mathscr{T}$ is recursive.

This means that if $\mathcal{T}$ is effectively axiomatised, then there is a Turing machine which, when given the Gödel number of an element $q \in P(V^*, \mathcal{R}^*)$, tells us whether or not q is an axiom of $\mathcal{T}$. From the discussion above, it follows that for an effectively axiomatised theory $\mathcal{T} = (\mathcal{R}, A, C)$, there is a Turing machine which, when given the Gödel numbers of $p \in \mathcal{L}(\mathcal{T})$ and of the sequence $p_1, p_2, \ldots, p_n$ of elements of $P(V, \mathcal{R})$, tells us if $p_1, p_2, \ldots, p_n$ is a proof of p in $\mathcal{T}$. Furthermore, there is a Turing machine which, when given the Gödel number of a theorem p of $\mathcal{T}$, computes the smallest number which is the Gödel number of a proof of p.

Suppose now that p is not a theorem of $\mathcal{T}$. If it is the case that $\sim p$ is a theorem of $\mathcal{T}$, then finding a proof of $\sim p$ will not by itself solve our problem, because we would also have to show that $\mathcal{T}$ is consistent, i.e., that F is not a theorem of $\mathcal{T}$. However, if p is not a theorem of $\mathcal{T}$, then the theory $\mathcal{T}' = (\mathcal{R}, A \cup \{\sim p\}, C)$ is consistent, and hence has a model. The construction of a model of $\mathcal{T}'$ would clearly show that p is not a theorem of $\mathcal{T}$. Thus, for any effectively axiomatised theory $\mathcal{T}$ and any p, the problem of deciding whether or not p is a theorem $\mathcal{T}$ is soluble: find a proof of p if p is a theorem, or a model of $\mathcal{T}$ in which p is false if p is not a theorem. Of course, we have not given a general procedure for finding the solution— that is a different problem.

We now consider the case of a family $\{p_n | p \in \mathbb{N}\}$ of propositions of $\mathcal{T}$. The minimal requirement of a solution to the decision problem for the family is clearly that we should know for each n whether or not p_n is a theorem of $\mathcal{T}$. This requirement can be met by simply requiring the solution to be the determination of the function $f : \mathbb{N} \to \{0, 1\}$ such that $f(p_n) = 1$ if and only if p_n is a theorem. For the determination to be satisfactory, the function f must be capable of calculation in some routine manner, which means that f must be a recursive function. For this to be so, the family $\{p_n | n \in \mathbb{N}\}$ must be able to be systematically computed, i.e., there is a condition on the family in order that our decision problem be well posed. With these considerations in mind, we make the following definitions.

Definition 5.2. The family $\{p_n | n \in \mathbb{N}\}$ is called *recursively enumerable* if $\{G(p_n) | n \in \mathbb{N}\}$ is a recursively enumerable subset of $\mathbb{N}$, and it is *recursive* if $\{G(p_n) | n \in \mathbb{N}\}$ is a recursive subset of $\mathbb{N}$. If $G(p_n)$ is a recursive function of n, the family is called *recursively enumerated*.

Definition 5.3. Let $\mathcal{F} = \{p_n | n \in \mathbb{N}\}$ be a recursively enumerated family of propositions of the theory $\mathcal{T}$. We say the decision problem for $\mathcal{F}$ is *recursively soluble* if the characteristic function of $\{n \in \mathbb{N} | \mathcal{T} \vdash p_n\}$ is recursive.

If $\mathcal{T}$ is a countable theory (written in the universal alphabet), then the Gödel numbering of elements of $\mathcal{L}(\mathcal{T})$ orders $\mathcal{L}(\mathcal{T})$ and so provides a recursive enumeration of $\mathcal{L}(\mathcal{T})$. The theory $\mathcal{T}$ is then called *decidable* if the family $\mathcal{L}(\mathcal{T})$ has recursively soluble decision problem.

Our decidability criterion is based on the minimum answer we could expect—a "yes/no" answer. We do not require the Turing machine which provides this answer also to prove it by giving either a proof or a counter-model for each p_n. If the present formulation of a solution produces undecidable theories, then any more rigorous requirement must be expected to render even more problems insoluble.

Exercise 5.4. The consistent theory $\mathcal{T}$ is effectively axiomatised and complete. Show that $\mathcal{T}$ is decidable. Show further that there is a Turing machine which, given the Gödel number of an element $p \in \mathcal{L}(\mathcal{T})$, answers the question of whether or not p is a theorem, and also provides a proof of its answer.

The notion that a family of objects can have a recursively insoluble decision problem of some kind can be applied to situations in our informal mathematics, as the following example shows. Later, we shall find examples within formal mathematical structures.

Example 5.5. Let M_n denote the nth Turing machine. The problem is to determine for all integers n, r, whether or not there is a computation of M_n beginning with the state q_0 code(r). I.e., the problem is to determine whether or not an arbitrary Turing machine M_n, fed with an arbitrary integer r, stops. We show that this stopping problem is recursively insoluble. Put $f_n = \Psi_{M_n}^{(1,\,1)}$. The problem is to determine those (n, r) for which $f_n(r)$ exists, and we show that there is no Turing machine which computes for each n whether or not $f_n(n)$ exists. Suppose the function $f : \mathbb{N} \to \mathbb{N}$ defined by

$$f(n) = 1 \quad \text{if} \quad f_n(n) \text{ exists},$$
$$f(n) = 0 \text{ otherwise},$$

is recursive. Then the function $g : \mathbb{N} \to \mathbb{N}$ defined by

$$g(n) = f_n(n) + 1 \quad \text{if} \quad f_n(n) \text{ exists},$$
$$g(n) = 0 \text{ otherwise},$$

is also recursive, since $f_n(n)$ can be computed when it exists. We now have a contradiction, for since $\{f_n | n \in \mathbb{N}\}$ contains the set of recursive functions, $g = f_m$ for some integer m, and then

$$f_m(m) = g(m) = f_m(m) + 1.$$

Hence f is not recursive, and so there is no Turing machine which determines for all n, whether or not $f_n(n)$ exists. In fact, since $h(n) = f_n(n)$ is partial recursive, there is a Turing machine M which computes it, and we have proved that M has a recursively insoluble stopping problem.

Exercises

5.6. Use $h(n) = f_n(n)$ to construct a recursively enumerable set E which is not recursive.

5.7. Let $A = \{n \in \mathbb{N} | f_n \text{ is recursive}\}$. Prove that A is not recursively enumerable.

5.8. Let E be a recursively enumerable subset of $\mathbb{N}$. Show that there exists a Turing machine which, started in the state $q_0 s_1^n$, stops with blank tape if $n \in E$, and fails to stop if $n \notin E$. (Hint: if $f: \mathbb{N} \to \mathbb{N}$ is a recursive function with $f(\mathbb{N}) = E$, then, for given n, compute in turn $f(0), f(1), \ldots$ until the first r (if any) for which $f(r) = n$ is found.)

§6   Insoluble Problems in Arithmetic

In §5 we gave an example of a family of objects in informal mathematics for which a decision problem is recursively insoluble. We now wish to convert this example into an example within formal arithmetic. We do this in a way which will allow us to apply some of our ideas and results to other interesting systems. For that reason, we shall be concerned with theories which formalise some aspects of the theory of $\mathbb{N}$ (which is our underlying object of study) and it is convenient to set down first some notational conventions and some definitions. Throughout this section, θ, s, a, m respectively denote the property of being 0, the successor relation, the addition relation and the multiplication relation. Whenever $\mathbb{N}$ is given as a model of a theory $\mathscr{T}$, it is understood that any of θ, s, a, m which are relation symbols of $\mathscr{T}$ have their standard interpretations. Axioms which we will use in our constructions are

1)   the Peano axioms P_1, P_2, P_3, P_4, P_5 of the first-order theory $\mathscr{P}$ of §4 of Chapter VI. (Recall that the scope of the axiom scheme of induction (P_5) depends on the theory under consideration.)

2)   the addition axioms

$$\text{add}_1 = (\forall x)(\forall y)(\exists! z)a(x, y, z),$$
$$\text{add}_2 = (\forall x)(\forall y)(\theta(y) \Rightarrow a(x, y, x)),$$
$$\text{add}_3 = (\forall x)(\forall y)(\forall z)(\forall t)(\forall u)(s(z, y) \wedge a(x, z, t)$$
$$\wedge\, a(x, y, u) \Rightarrow s(t, u)).$$

3)   the multiplication axioms

$$\text{mult}_1 = (\forall x)(\forall y)(\exists! z)m(x, y, z),$$
$$\text{mult}_2 = (\forall x)(\forall y)(\theta(y) \Rightarrow m(x, y, y))$$
$$\text{mult}_3 = (\forall x)(\forall y)(\forall z)(\forall t)(\forall u)(s(z, y) \wedge m(x, z, t)$$
$$\wedge\, m(x, y, u) \Rightarrow a(u, x, t)).$$

4)   for theories with $\mathbb{N}$ contained in the set of constants, the identification axioms

$$e_n = (\exists x_0)(\exists x_1) \cdots (\exists x_{n-1})(\theta(x_0) \wedge s(x_1, x_0) \wedge \cdots \wedge$$
$$s(x_{n-1}, x_{n-2}) \wedge s(n, x_{n-1})).$$

We shall deal mainly with the theory $\mathscr{N}$ with relation symbols $\mathscr{R}(\mathscr{N}) = \{=, \theta, s, a, m\}$, constants $\mathbb{N}$, and axioms $A(\mathscr{N})$ being all those listed above.

We call $\mathcal{N}$ recursive arithmetic. To assist us, we shall also use the theory $\mathcal{N}_0$, which differs from $\mathcal{N}$ only in that the axiom scheme of induction is excluded from the axioms. Both $\mathcal{N}$ and $\mathcal{N}_0$ are effectively axiomatised. As we need to compare theories, we make the following definitions.

Definition 6.1. Let $\mathcal{T} = (\mathcal{R}, A, C)$ and $\mathcal{T}' = (\mathcal{R}', A', C')$ be first-order theories. We say $\mathcal{T}'$ *extends* $\mathcal{T}$, and write $\mathcal{T}' \supseteq \mathcal{T}$, if $\mathcal{R}' \supseteq \mathcal{R}$, $A' \supseteq A$ and $C' \supseteq C$.

Definition 6.2. Let $\mathcal{T}' \supseteq \mathcal{T}$, and let $M = (M, v, \psi)$ be a model of $\mathcal{T}$. We say that M *extends to a model of* $\mathcal{T}'$ if there exist $v' : C' \to M$ and $\psi' : \mathcal{R}' \to \mathrm{rel}(M)$, extending v, ψ respectively, such that (M, v', ψ') is a model of $\mathcal{T}'$.

Definition 6.3. Let $\mathcal{T} = (\mathcal{R}, A, C) \supseteq \mathcal{N}_0$ and have N as model. Let $f : U \to$ N be a function defined on some subset U of N. We say that f is *strongly definable* in $\mathcal{T}$ if there is an element $p(x, y) \in P(V, \mathcal{R})$ such that, for all $m, n \in$ N, $\mathcal{T} \vdash p(m, n)$ if and only if $m \in U$ and $f(m) = n$. The definition is extended in the obvious way for functions of several variables.

The key result we intend to prove is that if $\mathcal{T} \supseteq \mathcal{N}_0$ and has N as model, then every partial recursive function is strongly definable in $\mathcal{T}$. The proof is tedious, although the idea is simple—we build up descriptions of the state of a given Turing machine as a function of the input and the number of steps performed.

Definition 6.4. The *state function* corresponding to the state $\left[s_{\beta_\ell} \cdots s_{\beta_1} q_i s_{\alpha_1} \cdots s_{\alpha_k} \right]$ is the function $f : $ N $\to$ N given by

$$f(0) = G(q_i),$$
$$f(2i + 1) = G(s_{\alpha_{i+1}}), 0 \leqslant i \leqslant k - 1,$$
$$f(2i + 1) = G(s_0), i \geqslant k,$$
$$f(2i + 2) = G(s_{\beta_{i+1}}), 0 \leqslant i \leqslant \ell - 1,$$
$$f(2i + 2) = G(s_0), i \geqslant \ell.$$

State functions are always strongly definable, as they take the value $G(s_0)$ except on a finite set. For a given Turing machine, it is easy to construct a description of the state function f_1 produced from an initial state f after one step of the computation. Continuing, one can produce, for any n, a description of the state function f_n after n steps. The difficulty in this approach is that the complexity of the description so obtained increases with n, whereas we need a single description of $f_x(y)$ as a function of the two variables x, y. Fortunately, there is a trick which allows us to give bounded definitions of arbitrary finite sequences.

Lemma 6.5. (The Sequence Number Lemma). *There exists a strongly definable function* $\mathrm{seq} : $ N$^+ \times$ N $\to$ N *such that, for any n and $a_0, a_1, \ldots, a_n \in$ N, there exists $b \in$ N$^+$ with the property that* $\mathrm{seq}(b, r) = a_r$ *for* $r = 0, \ldots, n$.

Proof. Let $T(n)$ denote the nth triangular number:

$$T(n) = 1 + 2 + \cdots + n = \tfrac{1}{2}n(n + 1).$$

For each $z > 0$, there is a unique n such that

$$T(n) < z \leqslant T(n + 1) = T(n) + n + 1.$$

Thus z is uniquely expressible as $z = T(n) + y$ with $0 < y \leqslant n + 1$. (We choose this range for y because later we shall need $y \neq 0$.) Put $x = n + 2 - y$. Then x, y are uniquely determined functions of z, which we denote by $L(z)$, $R(z)$ respectively. Put $P(x, y) = T(x + y - 2)$. P, L, R are strongly definable functions, for we may regard $z = P(x, y)$ as an abbreviation for

$$(x > 0) \wedge (y > 0) \wedge (2z = (x + y - 2)(x + y - 1) + 2y),$$

$x = L(z)$ as one for

$$(x > 0) \wedge (z > 0) \wedge (\exists y)((y > 0) \wedge (2z = (x + y - 2)(x + y - 1) + 2y)),$$

and $y = R(z)$ as one for

$$(y > 0) \wedge (z > 0) \wedge (\exists x)((x > 0) \wedge (2z = (x + y - 2)(x + y - 1) + 2y)).$$

The function $\mathrm{seq}(b, r)$ is defined to be the remainder on division of $L(b)$ by $1 + (r + 1)R(b)$. This is strongly definable, the relation $z = \mathrm{seq}(x, y)$ being given by

$$(x > 0) \wedge (z < 1 + (y + 1)R(x)) \wedge (\exists t)(L(x) = t(1 + (y + 1)R(x)) + z).$$

Finally, given $a_0, a_1, \ldots, a_n \in N$, we have to find $b \in N^+$ such that $\mathrm{seq}(b, r) = a_r$ for $0 \leqslant r \leqslant n$. Pick $c \in N$ such that $c > a_r$ for $0 \leqslant r \leqslant n$ and such that c is divisible by each of $1, 2, \ldots, n$. Put $m_r = 1 + (r + 1)c$, $r = 0, \ldots, n$. m_r and m_s are relatively prime for every pair r, s such that $0 \leqslant r < s \leqslant n$, for if d is a common divisor of m_r and m_s, d also divides $(s + 1)m_r - (r + 1)m_s = s - r$. Hence d divides c, and the definition of m_r shows now that $d = 1$. We may therefore apply the Chinese Remainder Theorem (see $[10]$, p 135) to the system of congruences

$$x \equiv a_r \bmod m_r \, (r = 0, \ldots, n).$$

Let e be a positive solution to this system, and put $b = P(e, c)$. Then $e = L(b)$, $c = R(b)$, $L(b) \equiv a_r \bmod(1 + (r + 1)R(b))$, and $a_r < c < 1 + (r + 1)R(b)$, showing that $a_r = \mathrm{seq}(b, r)$ for $r = 0, \ldots, n$. $\square$

Exercises

6.6. Given $m, n, r \in N$ such that $m + n = r$, prove that $\mathcal{N}_0 \vdash a(m, n, r)$. Hence show that if $\mathcal{T} \supseteq \mathcal{N}_0$ and has N as a model, then $\mathcal{T} \vdash a(m, n, r)$ implies $\mathcal{N}_0 \vdash a(m, n, r)$ for $m, n, r \in N$. Do the same thing for multiplication.

6.7. For $m, n, r \in N$ and $\mathcal{T} \supseteq \mathcal{N}_0$ with N as model, show that $\mathcal{T} \vdash \mathrm{seq}(m, n) = r$ if and only if $\mathrm{seq}(m, n) = r$. (This shows that the formula given above as a definition in $\mathcal{T}$ of seq indeed strongly defines seq.)

The sequence number function defined in Lemma 6.5 enables us to give definitions in $\mathcal{T}$ of various functions describing a computation by a Turing machine M. We give the definitions and leave the reader to verify them. If M has a quadruple $(q_\alpha, s_\beta, a, b)$, we define $M_{\alpha, \beta}(x, y, z) \in P(V, \mathcal{R})$ as follows. We have $b = q_\gamma$ for some γ, $a = s_{\beta'}$ (for some β') or $a = L$ or $a = R$. Put

$$M_{\alpha, \beta}(x, y, z) = (\text{seq}(x, 0) = G(q_\alpha)) \wedge (\text{seq}(x, 1) = G(s_\beta))$$
$$\wedge (y = 0 \Rightarrow z = G(q_\gamma)) \wedge K(x, y, z),$$

where

$$K(x, y, z) = (y = 1 \Rightarrow z = G(s_{\beta'})) \wedge (y > 1 \Rightarrow z = \text{seq}(x, y)) \quad \text{if} \quad a = s_{\beta'},$$
$$K(x, y, z) = [((\exists k)(y = 2k + 1)) \Rightarrow z = \text{seq}(x, y + 2)] \wedge (y = 2 \Rightarrow z = \text{seq}(x, 1))$$
$$\wedge [((\exists k)(y = 2k + 4)) \Rightarrow z = \text{seq}(x, y - 2)] \quad \text{if} \quad a = R,$$
$$K(x, y, z) = (y = 1 \Rightarrow z = \text{seq}(x, 2) \wedge [((\exists k)(y = 2k + 3)) \Rightarrow z = \text{seq}(x, y - 2)]$$
$$\wedge [((\exists k)(y = 2k + 2)) \Rightarrow z = \text{seq}(x, i + 2)] \quad \text{if} \quad a = L.$$

Now put

$$M(x, y, z) = \bigvee_{\alpha, \beta} M_{\alpha, \beta}(x, y, z),$$

where the disjunction is taken over the finitely many pairs α, β for which there is a quadruple $(q_\alpha, s_\beta, a, b) \in M$. If there are no such quadruples, put $M(x, y, z) = F$.

Suppose that f, g are state functions such that $[f] \overset{M}{\to} [g]$. For $r \in \mathbb{N}$, let $u \in \mathbb{N}$ be such that $\text{seq}(u, i) = f(i)$ for $i = 0, \ldots, r + 2$. If $k \in \mathbb{N}$, we claim that $k = g(r)$ if and only if $\mathcal{T} \vdash M(u, r, k)$. We can now prove some results.

Lemma 6.8. *Let f be an initial state function (i.e., $f(0) = G(q_0)$) and let $g(n, r)$ be the value at r of the state function after n steps of the computation by M starting at $[f]$. Then g is strongly definable in $\mathcal{T}$.*

Proof. f is strongly definable, and so we give a definition of g in terms of the definition of f.
Put

$$\varphi(x, y, z) = (\exists u)[(\forall v)(v \leqslant y + 2x \Rightarrow \text{seq}(\text{seq}(u, 0), v) = f(v))$$
$$\wedge (\text{seq}(\text{seq}(u, x), y) = z) \wedge (\forall w)(\forall t)(((1 \leqslant w \leqslant x)$$
$$\wedge (t \leqslant y + 2(x - w))) \Rightarrow M(\text{seq}(u, w - 1), t, \text{seq}(\text{seq}(u, w), t)))].$$

Then $g(n, r) = k$ if and only if $\mathcal{T} \vdash \varphi(n, r, k)$, whence the result. ☐

Any initial state function f can be expressed in terms of two integers u, v, since we can always find u, v such that $f(x) = \text{seq}(u, x)$ if $x \leqslant v$, and $f(x) = G(s_0)$ if $x > v$. (If so desired, we can replace u, v by the single integer $w = P(u, v)$, using $u = L(w)$, $v = R(w)$.) If this definition of f is substituted into the element φ given above, we obtain a 5-variable formula, $\psi(u, v; x, y, z)$ say, which describes the behavior of M for any input. We can express the statement that M, started in the state given by (u, v), stops in the state of the function whose value at y is z, by

$$(\exists x)(\psi(u, v; x, y, z) \wedge (\forall x')(x' > x \Rightarrow (\forall t)(\sim \psi(u, v; x', y, t)))).$$

Theorem 6.9.   *Let* $\mathcal{T} \supseteq \mathcal{N}_0$ *be a theory with* N *as model. Then every partial recursive function is strongly definable in* $\mathcal{T}$.

Proof.   The formulae given above, together with a description of the input function in terms of the arguments of $\Psi_M^{(k,\,t)}$, can be adapted to give a definition of $\Psi_M^{(k,\,t)}$. The reader is asked to supply the details. $\square$

We are now able to provide an example of an insoluble decision problem within the formal theory $\mathcal{N}$. From Theorem 6.9, it follows that any relation on N whose characteristic function is recursive is also strongly definable in any theory $\mathcal{T}$ of the type considered above. In particular, there is a formula, $\text{comp}(x_1, x_2, x_3)$ say, defining the relation that the machine of Gödel number x_1, applied to the number x_2, computes x_3. Reference to Example 5.5 shows that the family $\{(\exists x)\text{comp}(n, n, x) | n \in \mathbf{N}\}$ has an insoluble decision problem.

Theorem 6.10.   *Let* $\mathcal{T} \supseteq \mathcal{N}_0$ *be a theory which has* N *as model. Then* $\mathcal{T}$ *is undecidable. In particular,* $\mathcal{N}$ *is undecidable.*

Proof.   A decision process for $\mathcal{T}$ would provide a decision process for the family $\{(\exists x)\text{comp}(n, n, x) | n \in \mathbf{N}\}$. $\square$

Theorem 6.11.   *Let* $\mathcal{T} \supseteq \mathcal{N}_0$ *be an effectively axiomatised theory with* N *as model. Then* $\mathcal{T}$ *is incomplete.*

Proof.   By Exercise 5.4 and Theorem 6.10. However, it is of interest to construct an element $q \in \mathcal{L}(\mathcal{T})$ such that neither q nor $\sim q$ is a theorem of $\mathcal{T}$. Let $\mathcal{T} = (\mathcal{R}, A, C)$, and write P for $P(V, \mathcal{R})$. Let $G:P \to \mathbf{N}$ denote the Gödel number function, and let $F:G(P) \to P$ denote its inverse (G is injective). Since proofs in $\mathcal{T}$ can be checked by Turing machine, the relation "x_2 is the Gödel number of a proof in $\mathcal{T}$ of $F(x_1)$" is recursive. Let $\text{proof}_{\mathcal{T}}(x_1, x_2)$ be a definition of this relation in $\mathcal{T}$, and put

$$\text{theorem}_{\mathcal{T}}(x_1) = (\exists x_2)\text{proof}_{\mathcal{T}}(x_1, x_2).$$

Then $\text{theorem}_{\mathcal{T}}(x_1)$ defines in $\mathcal{T}$ the property "x_1 is the Gödel number of a theorem of $\mathcal{T}$".

For any element $w \in P$, write $w(x_0)$ to denote its (possible) dependence on x_0. If $n \in \mathbf{N}$, then $n \in C$ and so $w(n) \in P$. We consider $w(n)$ as a function of both n and w, and denote it by $\text{sub}(n, w)$. Define $\varphi(m, n) = G(\text{sub}(m, F(n)))$, for $m \in \mathbf{N}$ and $n \in G(P)$. φ is then a partial recursive function, hence there is an element $p(x_1, x_2, x_3) \in P$ defining the relation $\varphi(x_1, x_2) = x_3$.

We now put

$$\pi(x_1, x_2) = (\exists x_3)(p(x_1, x_2, x_3) \wedge \text{theorem}_{\mathcal{T}}(x_3)),$$

and consider the meaning of $\pi(x_1, x_2)$ in certain cases. If w satisfies $\text{var}(w) \subseteq \{x_0\} \cup C$, then $w(m) \in \mathcal{L}(\mathcal{T})$ for all $m \in \mathbf{N}$. Choose such a w, and let $n = G(w)$. Then $\pi(m, n)$ is true in N if and only if, for some $a \in \mathbf{N}$, we have both $\varphi(m, n) = a$

and a is the Gödel number of a theorem of $\mathscr{T}$. Since $\varphi(m, n) = G(w(m))$, we see that a must be both the Gödel number of $w(m)$ and the Gödel number of a theorem. Hence $\pi(m, n)$ is true in N if and only if $\mathscr{T} \vdash w(m)$. We now choose $w(x_0) = \sim\pi(x_0, x_0)$, so that $n = G(\sim\pi(x_0, x_0))$, and put $q = w(n)$. Then $\pi(n, n)$ is true in N if and only if $\mathscr{T} \vdash q$. But $q = \sim\pi(n, n)$, hence $\mathscr{T} \vdash q$ if and only if q is false in N. Since N is a model of $\mathscr{T}$, $\mathscr{T} \vdash q$ implies q is true in N. Hence q cannot be a theorem of $\mathscr{T}$, which from the condition above implies q is true in N, which then implies that $\sim q$ cannot be a theorem of $\mathscr{T}$. Thus $q = \sim\pi(n, n)$ has the property required to demonstrate the incompleteness of $\mathscr{T}$. □

We note that this incompleteness cannot be cured by adding q as an axiom to form a new theory $\mathscr{T}' \supseteq \mathscr{T}$, because replacing theorem$_\mathscr{T}(x_3)$ by theorem$_{\mathscr{T}'}(x_3)$ in our construction provides another element q' with the requisite properties. The proof shows that no effective axiomatisation of N can lead to a complete theory.

The result of Theorem 6.11 is known as Gödel's Incompleteness Theorem.

Exercises

6.12. Show that $\{n \in \mathrm{N} | n = G(p) \text{ for some } p \in \mathscr{L}(\mathscr{N}) \text{ true in N}\}$ is not recursively enumerable.

6.13. Show that $\{n \in \mathrm{N} | n = G(p) \text{ for some } p \in \mathscr{L}(\mathscr{N}) \text{ such that } \mathscr{N} \vdash p\}$ is recursively enumerable but not recursive.

§7 Undecidability of the Predicate Calculus

We investigate the decidability of the predicate calculus by taking a known undecidable theory $\mathscr{T} = (\mathscr{R}, A, C)$, and trying to show that the theory $(\mathscr{R}, \varnothing, \varnothing)$ is also undecidable. The method is to suppose the existence of a decision process for $(\mathscr{R}, \varnothing, \varnothing)$ and to construct from it a decision process for $(\mathscr{R}, A, C)$. The following simple result will be used.

Lemma 7.1. Let $\mathscr{T}$, $\mathscr{T}'$ be theories, and let $\varphi : \mathscr{L}(\mathscr{T}) \to \mathscr{L}(\mathscr{T}')$ be a recursive function such that for all $p \in \mathscr{L}(\mathscr{T})$, we have $\mathscr{T} \vdash p$ if and only if $\mathscr{T}' \vdash \varphi(p)$. Suppose $\mathscr{T}'$ is decidable. Then $\mathscr{T}$ is decidable.

Proof: Clearly, to determine if p is a theorem, it suffices to calculate $\varphi(p)$ and to apply the decision process for $\mathscr{T}'$. □

Lemma 7.2. Let $\mathscr{N}_1$ be the theory formed from the theory $\mathscr{N}_0$ by omitting the constants and the axioms e_n which identify the constants. Then $\mathscr{N}_1$ is undecidable.

Proof: Put, for each $n \in \mathrm{N}$,

$$e_n(x) = (\exists x_0)(\exists x_1) \cdots (\exists x_{n-1})(\theta(x_0) \wedge s(x_1, x_0) \wedge \cdots \wedge s(x, x_{n-1})).$$

Now for any $p \in \mathscr{L}(\mathscr{N}_0)$, $\mathrm{var}(p) \subseteq \mathrm{N}$. Hence there is an element

$p(x_1, \ldots, x_r) \in P(V - \mathrm{N}, \mathscr{R})$, such that there exist integers $n_1, \ldots, n_r$ for which $p = p(n_1, \ldots, n_r)$. We define $\varphi : \mathscr{L}(\mathscr{N}_0) \to \mathscr{L}(\mathscr{N}_1)$ by

$$\varphi(p) = (\boldsymbol{\forall} x_1) \cdots (\boldsymbol{\forall} x_r)(e_{n_1}(x_1) \wedge \cdots \wedge e_{n_r}(x_r) \Rightarrow p(x_1, \ldots, x_r)).$$

In order to complete the proof by an appeal to the previous lemma, we have to show that $\mathscr{N}_0 \vdash p$ if and only if $\mathscr{N}_1 \vdash \varphi(p)$. Suppose that $\mathscr{N}_1 \vdash \varphi(p)$. Since $\mathscr{N}_0 \supseteq \mathscr{N}_1$, $\mathscr{N}_0 \vdash \varphi(p)$. Since $e_{n_i}(n_i)$ is an axiom of $\mathscr{N}_0$ for all i, it follows immediately that $\mathscr{N}_0 \vdash p(n_1, \ldots, n_r)$, i.e., that $\mathscr{N}_0 \vdash p$. Now suppose $\mathscr{N}_0 \vdash p$. Since $\mathscr{N}_1 \vdash (\boldsymbol{\exists}! x) e_n(x)$ for each $n \in \mathrm{N}$, then for each $n \in \mathrm{N}$ there is in any model M of $\mathscr{N}_1$ a unique element $m_n \in M$ such that $M \vDash e_n(m_n)$. By mapping n to m_n we make M a model of $\mathscr{N}_0$, so $p(m_{n_1}, \ldots, m_{n_r})$ is true in M. The uniqueness of the m_n now implies that $\varphi(p)$ is true in M. Thus $\varphi(p)$ is true in every model of $\mathscr{N}_1$, and so $\mathscr{N}_1 \vdash \varphi(p)$. $\square$

Lemma 7.3. *Let $V = \{x_0, x_1, \ldots\}$, and $\mathscr{R} = \{\rho\}$, where ρ is a 4-ary relation symbol. Then $\mathrm{Pred}(V, \mathscr{R})$ is undecidable.*

Proof: Since $\mathrm{Pred}(V, \mathscr{R})$ does not involve either the identity relation symbol or the axioms of identity, we first consider these axioms. The theory $\mathscr{N}_1$ has only finitely many relation symbols, hence the axiom scheme of substitution of identical elements is finite. Thus $\mathscr{N}_1$ has only finitely many axioms of identity. Denote the conjunction of all of these axioms by a.

The relation symbols of $\mathscr{N}_1$ will be replaced by ρ, which intuitively is regarded as follows: $\rho(x, y, z, t)$ means $xy + z = t$. Define a homomorphism $f : P(V, \mathscr{R}^{(1)}) \to P(V, \mathscr{R})$, where $\mathscr{R}^{(1)} = \{=, \theta, s, a, m\}$, by

$$f(\theta(x)) = \rho(x, x, x, x),$$
$$f(x = y) = (\boldsymbol{\forall} z)(\boldsymbol{\forall} t)(\rho(z, z, z, z) \Rightarrow \rho(z, t, x, y)),$$
$$f(s(x, y)) = (\boldsymbol{\forall} z)(\boldsymbol{\forall} t)((\rho(z, z, z, z) \wedge (\boldsymbol{\forall} u)\rho(t, u, z, u)) \Rightarrow \rho(t, y, t, x)),$$
$$f(a(x, y, z)) = (\boldsymbol{\forall} t)((\boldsymbol{\forall} u)(\boldsymbol{\forall} v)(\rho(u, u, u, u) \Rightarrow \rho(t, v, u, v)) \Rightarrow \rho(t, x, y, z)),$$
$$f(m(x, y, z)) = (\boldsymbol{\forall} t)(\rho(t, t, t, t) \Rightarrow \rho(x, y, t, z)),$$

for all $x, y, z \in V$. Then define $g : P(V, \mathscr{R}^{(1)}) \to P(V, \mathscr{R})$ by $g(p) = f(a) \Rightarrow f(p)$. We show that Lemma 7.1 applies, for then the undecidability of $\mathscr{N}_1$ suffices to complete the proof.

Suppose $\mathscr{N}_1 \vdash p$. A proof of p from a maps under f into a proof of $f(p)$ from $f(a)$. By the Deduction Theorem, $f(a) \Rightarrow f(p)$ is a theorem. Conversely, suppose $f(z) \Rightarrow f(p)$ is a theorem. If M is any model of $\mathscr{N}_1$, then interpreting $\rho(x, y, z, t)$ as $xy + z = t$ gives an interpretation of $P(V, \mathscr{R})$ in which $f(a)$ is true. Since $f(a) \Rightarrow f(p)$ is a theorem, we conclude that $f(p)$ is true. By the way the interpretation of ρ is defined, p is also true in M. Hence p is a theorem of $\mathscr{N}_1$. $\square$

Corollary 7.4. (Church's Theorem) *Let $\mathscr{R}^* = \{r_{ij} | i, j \in \mathrm{N}\}$, with r_{ij} an i-ary relation symbol, be the universal relation alphabet. Then $\mathrm{Pred}(V, \mathscr{R}^*)$ is undecidable.*

Proof: The inclusion $P(V, \mathscr{R}) \to P(V, \mathscr{R}^*)$ satisfies the conditions of Lemma 7.1. $\square$

We end the chapter with the proof of a stronger result due to Kalmar.

Theorem 7.5. *Let r be a binary predicate symbol. Then* $\mathrm{Pred}(V, \{r\})$ *is undecidable.*

Before giving the formal proof, we note that the result implies that if $\mathscr{R}$ contains at least one n-ary relation symbol with $n \geqslant 2$, then $\mathrm{Pred}(V, \mathscr{R})$ is undecidable. The theorem will be proved by constructing a function $f : P(V, \{\rho\}) \to P(V, \{r\})$, where ρ is a 4-ary relation symbol, such that $f(p)$ is a theorem if and only if p is a theorem, and in addition such that if $\mathrm{var}(p) = \varnothing$, then $\mathrm{var}(f(p)) = \varnothing$. The construction uses the following idea, which shows how to express a 4-ary relation ρ on a given set S in terms of a binary relation r on a related set S'. (For convenience we shall use ρ, r also to denote interpretations of the relation symbols ρ, r.)

Lemma 7.6. *Let ρ be a 4-ary relation on the non-empty set S. Put $S' = \{K\} \cup S^2 \cup S^4$. For $x \in S$, define $\Delta(x) = (x, x) \in S'$. Let r be the binary relation on S' consisting of those pairs (a, b) for which at least one of the following holds:*

(i) $a = (x, y), b = (z, t)$, *where* $x, y, z, t \in S$ *and* $x = y = z$ *or* $y = z = t$,
(ii) $a = (x, y), b = (x, y, z, t)$ *where* $x, y, z, t \in S$,
(iii) $a = (x, y, z, t), b = (z, t)$ *where* $x, y, z, t \in S$,
(iv) $a = K, b = (x, y, z, t)$ *where* $(x, y, z, t) \in \rho$.

Then the elements of $\Delta(S)$, and of ρ, can be characterised in terms of r.

Proof: An element $a \in S'$ is in $\Delta(S)$ if and only if $(a, a) \in r$. We claim that a quadruple (x, y, z, t) of elements of S is in ρ if and only if their images X, Y, Z, T under Δ satisfy the condition that there are elements $A, B, C, D \in S'$ such that all the pairs (X, A), (A, Y), (Z, B), (B, T), (A, C), (C, B) and (D, C) are in r, but (E, D) is not in r for any E. To show this, observe that (E, D) not in r for any E implies that $D = K$. Then (D, C) is in r if and only if C is a quadruple in ρ. (C, B) in r requires B to be the final pair of C. (A, C) in r shows that A is either K or the initial pair of C, and the former is excluded if (X, A) is in r. Hence if the condition is satisfied by X, Y, Z, T, then $C = (x, y, z, t)$ and is in ρ. $\square$

Proof of Theorem 7.5. Put

$$R(x, y, z, t) = (\exists a)(\exists b)(\exists c)(\exists d)(r(x, a) \wedge r(a, y)$$
$$\wedge r(z, b) \wedge r(b, t) \wedge r(a, c) \wedge r(c, b) \wedge r(d, c) \wedge (\forall e) \sim r(e, d)).$$

We define $f : P(V, \{\rho\}) \to P(V, \{r\})$ in terms of a prefix π and a kernel k. If $\mathrm{var}(p) \neq \varnothing$, we define π_p to be the conjunction of the $r(x, x)$ for which

$x \in \text{var}(p)$, while if $\text{var}(p) = \varnothing$, π_p is not defined. The kernel $k(p)$ is defined inductively by

$$k(F) = F,$$
$$k(\rho(x, y, z, t)) = R(x, y, z, t) \text{ for all } x, y, z, t \in V,$$
$$k(p \Rightarrow q) = k(p) \Rightarrow k(q),$$
$$k((\forall x)p) = (\forall x)(r(x, x) \Rightarrow k(p)).$$

We now put

$$f(p) = \pi_p \Rightarrow k(p) \quad \text{if} \quad \text{var}(p) \neq \varnothing,$$
$$f(p) = k(p) \quad \text{if} \quad \text{var}(p) = \varnothing,$$

and show f satisfies the conditions of Lemma 7.1. Suppose that $\text{var}(p) = \varnothing$ and $f(p)$ is a theorem. The truth or falsity of p in any interpretation depends only on the choice of the set S and of the 4-ary relation ρ on S. We construct S', and r on S', as in Lemma 7.6. Since the definition of f effectively limits consideration to elements of $\Delta(S)$, we find that p is true in S if and only if $f(p)$ is true in S'. Since $f(p)$ is a theorem, we conclude that p is true in every interpretation and so is also a theorem.

Conversely, suppose p is a theorem, and let $p_1, \ldots, p_n$ be a proof of p. We use induction over n to show that $f(p)$ is a theorem. (We do not assume $\text{var}(p) = \varnothing$, as this would upset the induction.) Suppose then that $f(p_1), \ldots, f(p_{n-1})$ are theorems. There are three possibilities for p_n: it is an axiom, it is obtained by modus ponens, or it is obtained by Generalisation.

If p is an axiom, then $f(p)$, although not an axiom, is easily seen to be provable. Similarly, if p follows from p_i and p_j by modus ponens, then (by use of truth functions or the Deduction Theorem) $f(p)$ is deducible from $f(p_i)$ and $f(p_j)$. Suppose finally that $p = (\forall x)q$ is obtained by Generalisation. Then $p_{n-1} = q$, and $f(q) = \pi_q \Rightarrow k(q)$ is a theorem (in the other case, $\text{var}(q) = \varnothing$ and $f(p)$ is trivially a theorem). Since π_q is either π_p or $\pi_p \wedge r(x, x)$, it follows that $\pi_p \Rightarrow (r(x, x) \Rightarrow k(q))$ is a theorem, and Generalisation yields $(\forall x)(\pi_p \Rightarrow (r(x, x) \Rightarrow k(q)))$. Since $x \notin \text{var}(p)$, this implies $\pi_p \Rightarrow (\forall x)(r(x, x) \Rightarrow k(q))$. But this is $f(p)$, and the proof is complete. $\quad \square$

Exercise 7.7. Suppose $\mathscr{R}$ contains only unary relation symbols. Prove that $\text{Pred}(V, \mathscr{R})$ is decidable. (If $p \in P(V, \mathscr{R})$ involves n distinct relation symbols, show that $\vDash p$ if and only if p is true in every interpretation in a set of at most 2^n elements. This can be done by taking any interpretation M, putting $m_1 \equiv m_2$ if $v(\rho(m_1)) = v(\rho(m_2))$ for all relevant ρ, and working with the equivalence classes.)

Chapter X

Hilbert's Tenth Problem, Word Problems

§1 Hilbert's Tenth Problem

A recursive function $f:\mathbb{N}^n \to \mathbb{N}$ has been defined as one for which there is a Turing machine, T_f say, which computes $f(x_1,\ldots,x_n)$ for all $(x_1,\ldots,x_n) \in \mathbb{N}^n$. Accordingly, in order to show that a particular function $g:\mathbb{N}^n \to \mathbb{N}$ is recursive, we must construct a Turing machine which computes g. This is a tiresome process, even for functions of relatively simple form, and consequently it is natural to seek an alternative characterisation of recursive functions that will facilitate their recognition.

We are accustomed to constructing or decomposing complicated functions in terms of simple functions in other branches of mathematics—for example, use of the chain rule in the differential calculus depends upon the possibility of expressing a function as a composition of simpler functions. We therefore ask if it is possible to build up the set of recursive functions by starting with a set of simple functions and applying certain permissible operations to them. The fact that this can be done is remarkable, for not only does it provide an algebraic characterisation of recursive functions, but it also offers strong support for a belief (known as "Church's Thesis") that all formulations of the concept of an "effectively computable" function must be equivalent (i.e., must produce the same set of functions). For a detailed proof of the characterisation given in Theorem 1.2 below, we refer the reader to [3]. Other accounts of the subject may be found, for example, in [6], [10] or [13].

As initial functions, we take the set I consisting of

(i) the zero function $c:\mathbb{N} \to \mathbb{N}$ given by $c(x) = 0$,

(ii) the successor function $s:\mathbb{N} \to \mathbb{N}$ given by $s(x) = x + 1$,

(iii) the projection functions $U_i^n:\mathbb{N}^n \to \mathbb{N}$ $(n \in \mathbb{N}^+, i = 1,\ldots,n)$ given by $U_i^n(x_1,\ldots,x_n) = x_i$.

Exercise 1.1. *Show that every initial function is a recursive function.*

The permitted operations are the following:

(i) composition: given $f_j:\mathbb{N}^m \to \mathbb{N}$ and $g:\mathbb{N}^n \to \mathbb{N}$ $(m, n \in \mathbb{N}^+, j = 1,\ldots,n)$, composition yields the function $h:\mathbb{N}^m \to \mathbb{N}$ defined by

$$h(x_1,\ldots,x_m) = g(f_1(x_1,\ldots,x_m),\ldots,f_n(x_1,\ldots,x_m)).$$

(ii) primitive recursion: given $f:\mathbb{N}^n \to \mathbb{N}$ and $g:\mathbb{N}^{n+2} \to \mathbb{N}$ $(n \in \mathbb{N})$, primitive recursion yields the function $h:\mathbb{N}^{n+1} \to \mathbb{N}$ given by

$$h(x_1,\ldots,x_n,0) = f(x_1,\ldots,x_n),$$
$$h(x_1,\ldots,x_n,t+1) = g(t,h(x_1,\ldots,x_n,t),x_1,\ldots,x_n) \ (t \in \mathbb{N}).$$

(iii) minimalisation: given $f, g : \mathbb{N}^{n+1} \to \mathbb{N}$ $(n \in \mathbb{N}^+)$ satisfying the condition that for each $(x_1, \ldots, x_n) \in \mathbb{N}^n$ there exists at least one y such that $f(x_1, \ldots, x_n, y) = g(x_1, \ldots, x_n, y)$, minimalisation yields the function $h : \mathbb{N}^n \to \mathbb{N}$ given by

$$h(x_1, \ldots, x_n) = \min_y(f(x_1, \ldots, x_n, y) = g(x_1, \ldots, x_n, y))$$
$$= \text{the least } y \in \mathbb{N} \text{ such that } f(x_1, \ldots, x_n, y)$$
$$= g(x_1, \ldots, x_n, y).$$

Theorem 1.2. *The set of recursive functions coincides with the set of functions obtainable from the set I of initial functions by finite interations of the above operations.*

Exercises

1.3. Prove that the functions $+(x, y) = x + y$, $\times(x, y) = xy$ and $c_k(x) = k$ $(k \in \mathbb{N})$ are recursive, by using Theorem 1.2. Deduce that every polynomial $P : \mathbb{N}^n \to \mathbb{N}$ with coefficients in $\mathbb{N}$ is a recursive function.

1.4. (Cf Lemma 6.5 of Chapter IX.) Define the pairing function $p : \mathbb{N}^2 \to \mathbb{N}$ by $p(x, y) = \sum_{r=0}^{x+y} r + y$. Prove that p is bijective, and hence show that the functions $\ell, r : \mathbb{N} \to \mathbb{N}$ given by $p(\ell(z), r(z)) = z$ are well-defined. Show that p, ℓ and r are recursive.

Write $z = p(x, y)$, and define the sequence number function $S : \mathbb{N}^2 \to \mathbb{N}$ by the rule that $S(z, i)$ is the least remainder on division of x by $1 + (i + 1)y$. Prove that S is recursive.

Hilbert's tenth problem seeks an algorithm which will determine whether or not an arbitrary polynomial equation with integral coefficients and in any number of variables has a solution in integers. In 1970, Matiyasevich provided the last step in an argument which proves that no such algorithm exists. We shall outline a method of proof given in full detail in a recent expository article [2] by Davis, which also contains a brief historical account and references.

By a polynomial $P = P(x_1, \ldots, x_n)$ we shall mean a polynomial with integral coefficients. By a solution to the diophantine equation $P(x_1, \ldots, x_n) = 0$ we mean a solution in integers $x_1, \ldots, x_n$. Since every $x \in \mathbb{N}$ is expressible as a sum of four squares of elements of $\mathbb{N}$, the existence of an algorithm to test for solutions implies the existence of an algorithm to test for non-negative solutions, for by testing $P(s_1^2 + t_1^2 + u_1^2 + v_1^2, \ldots, s_n^2 + t_n^2 + u_n^2 + v_n^2) = 0$ for solutions, we have tested $P(x_1, \ldots, x_n) = 0$ for non-negative solutions. Therefore we may restrict all variables to the set $\mathbb{N}$, and prove there does not exist an algorithm to test for solutions in $\mathbb{N}$. We interpret "algorithm" as meaning "Turing algorithm", i.e., a procedure that can be carried out by a suitably designed Turing machine. Since we have information about the set of Turing computable (i.e., recursive) functions, we shall try to relate this set to sets defined in terms of solubility criteria for polynomial equations.

Given a polynomial $P(x_1, \ldots, x_n)$, an obvious subset of N^n related to it is its solution set $S = \{(x_1, \ldots, x_n) | P(x_1, \ldots, x_n) = 0\}$. For $k = 1, \ldots, n - 1$, the projection S_k of S onto the first k coordinates is given by the set of $(x_1, \ldots, x_k)$ such that there exist $x_{k+1}, \ldots, x_n$ for which $P(x_1, \ldots, x_n) = 0$. Thus membership of the set S_k is related directly to the existence of a solution to P. The following definition generalises this relation.

Definition 1.5. (i) $S \subseteq N^n$ is *diophantine* if there is a polynomial $P(x_1, \ldots, x_n, y_1, \ldots, y_m)$ in $m + n \geqslant n$ variables such that $(x_1, \ldots, x_n) \in S$ if and only if there exist values $y_1, \ldots, y_m$ for which $P(x_1, \ldots, x_n, y_1, \ldots, y_m) = 0$.

(ii) A relation ρ on N^n is *diophantine* if the set $\{(x_1, \ldots, x_n) | \rho(x_1, \ldots, x_n)$ is true$\}$ is diophantine. In particular, a function $f : N^n \to N$ is diophantine if $\{(x_1, \ldots, x_n, y) | y = f(x_1, \ldots, x_n)\}$ is diophantine.

For brevity, we shall write the condition that S is diophantine informally as

$$(x_1, \ldots, x_n) \in S \text{ iff } (\exists y_1, \ldots, y_m)(P(x_1, \ldots, x_n, y_1, \ldots, y_m) = 0).$$

Example 1.6. The subset S of N, consisting of integers which are not powers of 2, is diophantine, because

$$x \in S \text{ iff } (\exists y, z)(x - y(2z + 1) = 0).$$

Exercises

1.7. Show that the composite elements of N form a diophantine set.

1.8. Prove that the ordering relations $\{(x, y) | x < y\}$ and $\{(x, y) | x \leqslant y\}$ are diophantine relations on N^2.

1.9. Prove that the divisibility relation $\{(x, y) | x$ divides $y\}$ is diophantine.

1.10. Show that the functions $c(x) = 0, s(x) = x + 1$, and $U_i^n(x_1, \ldots, x_n) = x_i$ $(i = 1, \ldots, n)$, are all diophantine.

1.11. $P, Q : N^n \to N$ are polynomials, with solution sets S, T respectively. Show that $S \cap T, S \cup T$ are the solution sets of $P^2 + Q^2 = 0$, $PQ = 0$ respectively. Deduce that diophantine sets are closed under finite unions and intersections.

1.12. Show that the functions p, ℓ, r, defined in Exercise 1.4, are diophantine, and then use Exercise 1.11 to show that the sequence number function $S(z, i)$ is also diophantine.

We have found (Exercise 1.3) that every polynomial function with coefficients in N is recursive. This result extends to diophantine functions.

Lemma 1.13. *Every diophantine function f is recursive.*

Proof. Write

$$y = f(x_1, \ldots, x_n) \text{ iff } (\exists t_1, \ldots, t_m)(P(x_1, \ldots, x_n, y, t_1, \ldots, t_m)$$
$$= Q(x_1, \ldots, x_n, y, t_1, \ldots, t_m)),$$

where P, Q are polynomials with coefficients in N. Denoting the sequence number function by $S(z, i)$, then Lemma 6.5 of Chapter IX shows that there exists, for every choice of $y, t_1, \ldots, t_m$, a value u such that $S(u, 0) = y, S(u, 1) = t_1, \ldots, S(u, m) = t_m$. Since f is a function, there is exactly one y for which $P = Q$, hence

$$f(x_1, \ldots, x_n) = y = S(\min_u(P(x_1, \ldots, x_n, S(u, 0), \ldots, S(u, m))$$
$$= Q(x_1, \ldots, x_n, S(u, 0), \ldots, S(u, m))), 0),$$

which, by Exercise 1.4 and Theorem 1.2, shows that f is recursive.   □

The essential difficulties arise in attempting to prove the converse to the above result. Using Theorem 1.2, it suffices to prove that every initial function is diophantine, and that the diophantine functions are closed with respect to the operations of composition, primitive recursion and minimalisation. Some of this is easy. Exercise 1.10 has dealt with the initial functions, while if $f_1, \ldots, f_n$ and g are diophantine, and if $h(x_1, \ldots, x_m) = g(f_1(x_1, \ldots, x_m), \ldots, f_n(x_1, \ldots, x_m))$, then so is h, because

$$y = h(x_1, \ldots, x_m) \text{ iff } (\exists t_1, \ldots, t_n)(t_1 = f_1(x_1, \ldots, x_m) \text{ and } \ldots \text{ and }$$
$$t_n = f_n(x_1, \ldots, x_m) \text{ and } y = f(t_1, \ldots, t_n)),$$

which, by Exercise 1.11, is sufficient to establish the result. So it remains to deal with the operations of primitive recursion and minimalisation, neither of which has yet been shown to be expressible in terms of operations which trivially preserve the property of being diophantine. Each of these operations is expressible in terms of the operation of bounded universal quantification, which is now known to preserve this property. A bounded universal quantifier is one which applies for those values of the quantified variable which are less than a given bound. We use the notation $(\forall y \leqslant x)(\ldots)$ to mean "for all $y \in N$, either $y > x$ or $(\ldots)$". The next theorem is proved in full in [2].

Theorem 1.14.   *Let* $P: N^{m+n+2} \to N$ *be a polynomial. Then*

$$S = \{(y, x_1, \ldots, x_n) | (\forall z \leqslant y)(\exists y_1, \ldots, y_m)$$
$$(P(y, z, x_1, \ldots x_n, y_1, \ldots, y_m) = 0))\}$$

is diophantine.

Corollary 1.15.   *The set of diophantine functions is closed under primitive recursion and minimalisation.*

Proof of the Corollary.   Suppose f, g are diophantine, and

$$h(x_1, \ldots, x_n, 0) = f(x_1, \ldots, x_n),$$
$$h(x_1, \ldots, x_n, t + 1) = g(t, h(x_1, \ldots, x_n, t), x_1, \ldots, x_n).$$

Using the sequence number function to represent the numbers $h(x_1, \ldots, x_n, 0)$, $\ldots, h(x_1, \ldots, x_n, z)$, we have $y = h(x_1, \ldots, x_n, z)$ if and only if

$$(\exists u)\big((\exists v)(v = S(u, 0) \wedge v = f(x_1, \ldots, x_n)) \wedge (\forall t \leqslant z)\big(t = z \vee$$
$$(\exists w)(w = S(u, t + 1) \wedge w = g(t, S(u, t), x_1, \ldots, x_n))\big) \wedge y = S(u, z)\big)$$

which, by Exercises 1.11 and 1.12, shows that h is diophantine.

Finally, if f, g are diophantine and

$$h(x_1, \ldots, x_n) = \min_y(f(x_1, \ldots, x_n, y) = g(x_1, \ldots, x_n, y)),$$

then $y = h(x_1, \ldots, x_n)$ if and only if

$$(\exists z)\big(z = f(x_1, \ldots, x_n, y) \wedge z = g(x_1, \ldots, x_n, y)\big) \wedge (\forall t \leqslant y)\big(t = y \vee$$
$$(\exists u)(\exists v)(u = f(x_1, \ldots, x_n, t) \wedge v = g(x_1, \ldots, x_n, t) \wedge (u < v \vee v < u))\big)$$

showing that h is diophantine. $\quad\square$

We may therefore state the following fundamental result.

Theorem 1.16. *A function is recursive if and only if it is diophantine.*

In chapter IX, we showed the existence of a subset E of $\mathbb{N}$ which is recursively enumerable but not recursive. That is, E is the range of some recursive function, but the characteristic function of E is not a recursive function. Theorem 1.16 implies that a subset of $\mathbb{N}$ is recursively enumerable if and only if it is diophantine. Hence E is diophantine, and so there is a polynomial P such that

$$x \in E \text{ iff } (\exists t_1, \ldots, t_m)(P(x, t_1, \ldots, t_m) = 0).$$

Suppose that there exists a Turing machine M which can test every polynomial equation for the existence of solutions. M, when applied to the sequence of polynomials $P(0, t_1, \ldots, t_m), P(1, t_1, \ldots, t_m), \ldots$, will then compute the characteristic function of E. Thus E has a recursive characteristic function and hence is a recursive set, which contradicts its definition. Therefore, no such Turing machine M can exist. This statement is to be considered as an explicit denial of the existence of any algorithm to test all polynomial diophantine equations for solutions, which therefore implies that Hilbert's tenth problem is insoluble.

Exercises

1.17. Prove that a subset of $\mathbb{N}$ is recursively enumerable if and only if it is diophantine.

1.18. Give an enumeration of the set of polynomials with integral coefficients and in an arbitrary finite number of variables chosen from $x, y_1, y_2, \ldots$. Hence obtain a sequence $\{D_n\}$ which contains all diophantine subsets of $\mathbb{N}$. Define a function $g : \mathbb{N}^2 \to \mathbb{N}$ by

$$g(x, n) = 0 \quad \text{if} \quad x \notin D_n,$$
$$g(x, n) = 1 \quad \text{if} \quad x \in D_n.$$

Use Theorem 1.16 to prove that g is not recursive. Obtain an alternative

proof that Hilbert's tenth problem is insoluble by showing that the existence
of a "Hilbert algorithm" would imply that g is recursive.

§2   Word Problems

A group G is often specified by giving a set X of generators of G together
with a set R of relations satisfied by these generators. The set R is required to
be such that every relation on the elements of X which holds in G is a consequence of those in R. Here, a relation is an equation $w_1(a_1, \ldots, a_n) =
w_2(a_1, \ldots, a_n)$ which holds in G, where $a_1, \ldots, a_n$ are particular elements of
X and w_1, w_2 are group theoretical words. We can express such an equation
in the form $w_1(a_1, \ldots, a_n)(w_2(a_1, \ldots, a_n))^{-1} = 1$, so we may always suppose
that each relation is given in the form $w(a_1, \ldots, a_n) = 1$, and identify the
relation with the word $w(a_1, \ldots, a_n)$.

Definition 2.1.   A *group presentation* is a set X together with a set R of
group theoretical words on the elements of X. The presentation (X, R) is
called *finite* if both X and R are finite.

Every group presentation (X, R) does determine a group: take the free
group F on X and the smallest normal subgroup K of F which contains R,
and then the group determined by (X, R) is the factor group F/K. We shall
write $G = \langle X | R \rangle$ to indicate that G is the group determined by the presen-
tation (X, R). (The group G has of course many different presentations.)
Henceforth, in order to avoid confusion between an element of G and a
particular construction of the element, a word w shall mean an element of the
free group F. The corresponding element of $G = F/K$ will be called the group
element represented by w. Two words w_1, w_2 will be called equivalent, written
$w_1 \sim w_2$, if they represent the same group element.

The properties of the group $G = \langle X | R \rangle$ may not be apparent from the
presentation. From the information in a given presentation of a group, we
may be able to obtain answers to various questions about the group, and we
are interested in finding procedures for doing this. M. Dehn in 1911 formulated three basic decision problems for a given presentation of a group
$G = \langle X | R \rangle$. These three problems are known as the Word Problem, the
Conjugacy Problem and the Isomorphism Problem.

Problem 2.2.   (The Word Problem) *Find an algorithm which, for each
word w in the elements of X, determines whether or not w represents the identity
element of G.*

Problem 2.3.   (The Conjugacy Problem) *Find an algorithm which, for
any two words w_1, w_2, determines whether or not w_1 and w_2 represent conjugate
elements of G.*

Problem 2.4.   (The Isomorphism Problem) *Find an algorithm which, for
any group presentation (X', R'), determines whether or not $\langle X' | R' \rangle$ is isomorphic
to G.*

These problems have been solved for certain suitably restricted classes of presentations. (The reader is referred to [9], Section 6.1, for details.) In general, however, these problems are insoluble, and we shall try to show in this section how the theory of Turing machines can be used to establish the insolubility. In order that the underlying ideas will not be obscured by details, we shall restrict ourselves to a demonstration that there is a finitely-presented semigroup S whose word problem is insoluble. The interested reader will find in Chapter 12 of [11] an account of the construction from S of a finitely presented group G with insoluble word problem. (This construction is purely algebraic, and makes no further use of the theory of Turing machines.)

Exercises

2.5. A presentation (X, R) is called *abelian* if, for every $x, y \in X$, we have $x^{-1}y^{-1}xy \in R$. Show that the word problem for a finite abelian presentation is soluble. Show also that the isomorphism problem is soluble for pairs of finite abelian presentations.

2.6. Given that the finitely presented group $G = \langle X|R \rangle$ is finite, prove that it has soluble word problem and soluble conjugacy problem.

We now show how to associate a finite semigroup presentation with a Turing machine M. The idea behind the construction is to regard instantaneous descriptions as words, and to introduce relations which will make an instantaneous description represent the same semigroup element as does the instantaneous description obtained from the former one by one operation of the machine.

We shall always work with the shortest description, thereby avoiding difficulties arising from different descriptions of the same state of M. Thus an instantaneous description shall neither begin nor end with s_0. However, this introduces some difficulty into the construction of the set of relations, which we resolve by use of an end symbol e. With the description $\sigma q_i \tau$, we shall associate the semigroup word $e\sigma q_i \tau e$. It is also convenient to introduce a new internal state symbol q_∞, meaning that the machine has stopped.

As we are dealing with semigroups and not groups, a relation necessarily involves two words, and has the form $w_1 \sim w_2$. For our purposes, it is convenient to regard this as an ordered pair of words, and so to treat $w_1 \sim w_2$ and $w_2 \sim w_1$ as different relations. Each relation then has a first word.

Let M be a Turing machine with alphabet $\mathfrak{S} = \{s_0, s_1, \ldots, s_m\}$ and set of internal states $\mathfrak{Q} = \{q_0, q_1, \ldots, q_n\}$. The *semigroup presentation associated with* M is the presentation with generator set $X = \mathfrak{S} \cup \mathfrak{Q} \cup \{e, q_\infty\}$ and with relation set R consisting of

(a) for each quadruple $q_i s_j s_k q_\ell \in M$, the relation
$$q_i s_j \sim q_\ell s_k,$$

(b) for each quadruple $q_i s_j L q_\ell \in M$ with $j \neq 0$, the relations
$$s_k q_i s_j \sim q_\ell s_k s_j \quad \text{(all } k\text{)},$$
$$e q_i s_j \sim e q_\ell s_0 s_j,$$

(b_0) for each quadruple $q_i s_0 L q_\ell \in M$, the relations

$$s_k q_i s_0 \sim q_\ell s_k s_0 \quad \text{(all } k\text{)},$$
$$e q_i s_0 \sim e q_\ell s_0 s_0,$$
$$s_k q_i e \sim q_\ell s_k e \quad \text{(all } k \neq 0\text{)}$$
$$s_0 q_i e \sim q_\ell e,$$
$$e q_i e \sim e q_\ell e,$$

(c)   for each quadruple $q_i s_j R q_\ell \in M$ with $j \neq 0$, the relation

$$q_i s_j \sim s_j q_\ell ,$$

(c_0) for each quadruple $q_i s_0 R q_\ell \in M$, the relations

$$s_k q_i s_0 \sim s_k s_0 q_\ell \quad \text{(all } k\text{)},$$
$$e q_i s_0 \sim e q_\ell ,$$
$$s_k q_i e \sim s_k s_0 q_\ell e \quad \text{(all } k\text{)},$$
$$e q_i e \sim e q_\ell e,$$

(d)   for each pair $q_i s_j$ for which there is no quadruple in M beginning with $q_i s_j$, the relation

$$q_i s_j \sim q_\infty s_j,$$

and, if $j = 0$, the relation

$$q_i e \sim q_\infty e.$$

Let $w_1 = \sigma a \tau$ be a word and let $a \sim b$ be a relation in the above list. Substitution of b for a in w_1 gives the equivalent word $w_2 = \sigma b \tau$. Such a substitution, where the second member of a relation is substituted for the first, will be called a *forward step*. We write $w_1 \to w_2$ to denote that w_2 is obtainable from w_1 by a forward step. The reverse substitution is called a *backward step*, and we write $w_2 \leftarrow w_1$ to denote that w_1 is obtainable from w_2 by a backward step. We write $w - w'$ to denote that w' is obtainable from w by a step which may be either forward or backward. A *path* from w to w' is a finite sequence of steps $w - w_1 - w_2 - \cdots - w_{n-1} - w'$ beginning with w and ending with w'. Clearly, two words w, w' are equivalent if and only if there exists a path from w to w'.

We now concentrate our attention on the words which correspond to an instantaneous description of the Turing machine.

Definition 2.7.   A *special word* on X is a word of the form $e \sigma q_i \tau e$, where σ, τ are words (possibly empty) on $\mathfrak{S}$, such that σ does not begin with s_0 and τ does not end with s_0. The special word $e \sigma q_i \tau e$ is called *terminal* if $i = \infty$.

Any word obtained from a special word by a step is again a special word. Forward steps on special words correspond to steps in the operation of the machine M.

Lemma 2.8. *Let w, w' be special words. Suppose w' is terminal. Then w, w' are equivalent if and only if there is a path from w to w' consisting only of forward steps.*

Proof. Trivially, if such a path exists, then $w \sim w'$. Suppose that $w \sim w'$. Then there exists a path

$$w = w_0 - w_1 - \cdots - w_n = w'$$

from w to w'. We may suppose the path is chosen so that the number n of steps is the least possible. (If $n = 0$, then the path consists only of forward steps.) If the path has any backward steps, then there is a last such, say $w_k \leftarrow w_{k+1}$. This cannot be the last step of the path, because there is no forward step away from a terminal word. Thus $k + 1 < n$ and $w_{k+1} \rightarrow w_{k+2}$ is a forward step. But there is at most one forward step away from any special word, since the machine operation is determined. This implies that $w_{k+2} = w_k$, and so

$$w = w_0 - w_1 - \cdots - w_k - w_{k+3} - \cdots - w_n = w'$$

is a shorter path from w to w', contrary to the original choice of path. Hence the shortest path consists only of forward steps. $\quad\square$

We are now able to produce a Turing machine whose associated semigroup presentation has insoluble word problem.

Theorem 2.9. *Let E be a recursively enumerable but non-recursive subset of N, and let M be a Turing machine which, when started in the state $q_0 s_1^n$, stops with blank tape if $n \in E$, and does not stop if $n \notin E$. Then the semigroup presentation associated with M has insoluble word problem.*

Remark. The existence of such a set E and Turing machine M was established in Exercises 5.6 and 5.8 of Chapter IX.

Proof. By Lemma 2.8, the special word $eq_0 s_1^n e$ is equivalent to $eq_\infty e$ if and only if there exists a forward path from $eq_0 s_1^n e$ to $eq_\infty e$. Such a path exists if and only if M, started in the state $q_0 s_1^n$, stops with blank tape—i.e., if and only if $n \in E$. Since E is non-recursive, the word problem (even for this restricted set of words) is recursively insoluble. $\quad\square$

References and Further Reading

[1] Bell, J. L., Slomson, A. B.: *Models and Ultraproducts*: *An Introduction*. First revised edition. Amsterdam: North-Holland 1971.

[2] Davis, Martin: Hilbert's Tenth Problem is Unsolvable. *Amer. Math. Monthly* **80**, 233–269 (1973).

[3] Davis, Martin: *Computability and Unsolvability*. New York: McGraw-Hill 1958.

[4] Halmos, P. R.: *Naive Set Theory*. Princeton: Van Nostrand 1960. New York–Heidelberg–Berlin: Springer 1974.

[5] Hermite, Charles: *Oeuvres, tII*, pp. 5–12 Paris: Gauthier-Villars 1908.

[6] Kleene, S. C.: *Introduction to Metamathematics*. Princeton: Van Nostrand 1952.

[7] Lyndon, Roger C: *Notes on Logic. Mathematical Studies*, no. 6. New York: Van Nostrand 1964.

[8] Machover, M., Hirschfeld, J.: *Lectures on Non-Standard Analysis. Lecture Notes in Mathematics* 94. Berlin–Heidelberg–New York: Springer 1969.

[9] Magnus, W., Karrass, A., Solitar, D.: *Combinatorial Group Theory*: *Presentations of Groups in Terms of Generators and Relations*. New York–London–Sydney: Wiley-Interscience 1966.

[10] Mendelson, E.: *Introduction to Mathematical Logic*. New York: Van Nostrand 1964.

[11] Rotman, Joseph J.: *The Theory of Groups*: *An Introduction*. Boston- Allyn and Bacon 1965.

[12] Stoll, Robert R.: *Set Theory and Logic*. San Francisco-London: Freeman 1963.

[13] Yasuhara, Ann: *Recursive Function Theory and Logic*. New York: Academic Press 1971.

[14] Yates, Robert C.: *The Trisection Problem*. Reston: The National Council of Teachers of Mathematics 1971.

Index of Notations

The following notations are used at points remote from their explanations, which are given on the pages indicated.

$A \models p$	p is a consequence of the assumptions A	13
$A \vdash p$	p is provable from the assumptions A	15
al(n)	formula expressing "at least n elements"	45
ar(t)	arity of t	2
Con(A)	set of consequences of A	13
$d(p)$	depth of quantification of p	28
$D\rho$	domain of the relation ρ	76
Ded(A)	set of all deductions from A	15
dist$(x_1, \ldots, x_n)$	formula expressing "$x_1, \ldots, x_n$ are distinct"	48
$G(p)$	Gödel number of p	91, 92
$\mathcal{I}$	identity relation symbol	38
I	(when unexplained) set of axioms of identity	38
$\mathcal{L}(\mathcal{T})$	language of the theory $\mathcal{T}$	40
$\mathcal{N}$	recursive arithmetic	96
$\mathcal{N}_0$	recursive arithmetic without induction axiom scheme	97
$\mathcal{O}^k(S)$	set of kth order objects on S	75
$\mathcal{P}$	Peano arithmetic	59
$P(\mathcal{T})$	algebra of the theory $\mathcal{T}$	40
$P(X)$	free proposition algebra on X	12
$\tilde{P}(V, \mathcal{R})$	full first-order algebra on $(V, \mathcal{R})$	27
$P(V, \mathcal{R})$	reduced first-order algebra on $(V, \mathcal{R})$	28
Pow(M)	power set of M	54
Pred$(V, \mathcal{R})$	first-order predicate calculus on $(V, \mathcal{R})$	30
Pred$_{\mathcal{I}}(V, \mathcal{R})$	first-order predicate calculus with identity, on $(V, \mathcal{R})$	39
Prop(X)	propositional calculus on X	18
$\Psi_M^{(k, \ell)}$	partial recursive function $\mathbf{N}^k \to \mathbf{N}^\ell$ defined by Turing machine M	89
$\mathcal{R}$	set of relation symbols	27
$\mathcal{R}_n$	subset of n-ary relation symbols	27
rel(M)	set of all relations of all arities on M	42
S^σ	enlargement of S with respect to σ	77
$*S$	full enlargement of S	77
$\mathcal{T}(S)$	theory of the system S	74
$v(p)$	truth value of p	13
var(p)	set of (free) variables of p	29

Subject Index

Addition, theory of 61
Adequacy 18
Adequacy Theorem 22, 36
Algebra, free 4
  full first-order 27
  Lindenbaum 22
  proposition 12
  reduced first-order 28
  relatively free 8
  universal 2
Algebraic theory 72
Alphabet of Turing machine 85
  universal 90
Arity 1
Axiom of Choice 55, 57
  of Extension 53, 56
  of Infinity 54, 57
  of Pairing 53, 57
  of Power Set 54, 57
  of Regularity 56
  of Union 54, 57
Axiom Schema of Replacement 55, 57
  of Restriction 55, 57
  of Subsets 53, 56
Axioms of $\mathrm{Pred}(V, \mathscr{R})$ 30
  of $\mathrm{Prop}(X)$ 15
  of a theory 40

Bounded universal quantifier 108

Cardinal of model 44
  of theory 45
Categorical 44, 47
  in cardinal χ 45, 47
Church's Theorem 102
Church's Thesis 105
Compactness Theorem 22, 37
Completeness 43, 44, 47, 51
Computation by Turing machine 88
Concurrent relation 76
Conjugacy Problem 110
Consequence 13, 30
  proper 39
Consistency 18
  of first-order theory 43
Consistency Theorem 19, 34
Consistent subset 40
  maximal 40

Constant 40
Continuity 83
  uniform 83

Decidability 19, 94
  of $\mathrm{Prop}(X)$ 24
  of Theory of Equality 51
Deduction 15, 31
Deduction Theorem 19, 34
Definable 43
  strongly 97
Dense Linear Order 47
Depth of quantification 28
Description, of relation 76
  instantaneous, of Turing machine 87
  shortest 87
Diophantine set 107
Domain of relation 76
Dual 42

Effectively axiomatised theory 93
Elementary 41
Elementary Group Theory 41
Elementary Theory of Fields 45
Enlargement of theory 77
  of standard object 77
  full 77
Extension of model 77
  of theory 77

Failure of Turing machine 88
Field, algebraic closure of 66
Fields, Elementary Theory of 45
Filter 63
  existence of 65
  Fréchet 63
Finite intersection property 65

Generalisation 31
Gödel number 91, 92
Gödel's Completeness Theorem 37
  see Adequacy Theorem 36
Gödel's Incompleteness Theorem 100

Hilbert's Tenth Problem 105
Homomorphism 13

119

Identical relation   7
Identity, axioms of   38
    Predicate Calculus with   39
Incompleteness of
        recursive arithmetic   100
Infinite real numbers   81
Infinitesimal real numbers   81
Initial state   88
Internal relation   80
Interpretation   30
    proper   39
Isomorphism of models   44
    of T-algebras   4
Isomorphism Problem   110

Kalmar's Theorem   103
kth-order objects   75

Language of theory   40
Law of an algebra   7
    of a variety   7
Logic   18
Limit, direct   70
        existence of   72
    of function   83
    of sequence   82
Löwenheim-Skolem Theorem   46

Minimalisation   106
Model of first-order theory   42
    of Peano arithmetic   61
    non-standard   75
    standard   75
Modus ponens   15
Monad   82

Natural numbers in ZF   54
Non-standard model   75
Normal form, conjunctive   24
    disjunctive   24
    prenex   37

Operation   1, 2

$\mathscr{P}$ (Peano arithmetic)   59
    completeness of   61
    models of   61
Pair   53
Pairing function   106
Path   112

Peano axioms   58, 59
Predicate   26
Presentation   110, 111
Primary proposition   49
Primitive recursion   105
Projective geometry   40
Problem, Conjugacy   110
    Decision   94
    Hilbert's Tenth   105
    Isomorphism   110
    Stopping   95
    Word   110
Proof   15, 31
Proper consequence   39
    interpretation   39

Quantifier, bounded universal   108
    elimination   48
    existential   26
    reduction   48
    universal   26

Recursive arithmetic   97
    function   89, 106
    subset   90
Recursively enumerable   90, 94
    but non-recursive   95
Recursively soluble   94
Relation, concurrent   76
    definable   43
    external   80
    identical   7
    internal   80
Russell Paradox   53

Satisfiability Theorem   21, 36, 39
Semantic implication   13, 30
Semigroup of Turing machine   111
Semi-homomorphism   32, 33
Sequence number function   97
Soundness   18
Soundness Theorem   19, 33
Special word   112
Standard model   75
Standard relation   78
Standard system   74
    higher-order   76
    language of   74
    theory of   74
State of Turing machine   87
    initial   88
    terminal   88

State function   97
Step   112
Stopping Problem   95
Strongly definable   97
Substitution Theorem   17, 33
Subsystem of standard system   77
Subultraproduct   69
Successor   54
Syntactic implication   15, 31

T-algebra   2
T-subalgebra   3
Tautology   14, 30
Terminal state   88
  word   112
Theorem   15, 31, 40, 100
Theory   40
Transitive set   55
Truth   13, 29
  function   23
Turing machine   85
  universal   92
Type   2

Ultrafilter   63
  α-complete   67

existence of   65
  principal   63
  restriction of   63
  uniform   65
Ultrapowers   67
Ultraproducts   64
Undecidability of
  recursive arithmetic   100
  of Predicate Calculus   101
Uniform convergence   84
Uniform ultrafilter   65
Universal algebra   1
  alphabet   90
  proposition   72
  Turing machine   92

Valid   14, 30
Valuation   13
Variable   6
Variables involved in element of $\tilde{P}$ 27
  of element of P 29
Variety of *T*-algebras   7

Word   6
  special   112
Word Problem   110

Graduate Texts in Mathematics

1 *Takeuti/Zaring*
   INTRODUCTION TO AXIOMATIC SET THEORY
   vii, 250 pages. 1971.

2 *Oxtoby*
   MEASURE AND CATEGORY
   viii, 95 pages. 1971.

3 *Schaefer*
   TOPOLOGICAL VECTOR SPACES
   xi, 294 pages. 1971.

4 *Hilton/Stammbach*
   A COURSE IN HOMOLOGICAL ALGEBRA
   ix, 338 pages. 1971.

5 *MacLane*
   CATEGORIES FOR THE WORKING MATHEMATICIAN
   ix, 262 pages. 1972.

6 *Hughes/Piper*
   PROJECTIVE PLANES
   xii, 291 pages. 1973.

7 *Serre*
   A COURSE IN ARITHMETIC
   x, 115 pages. 1973.

8 *Takeuti/Zaring*
   AXIOMATIC SET THEORY
   viii, 238 pages. 1973.

9 *Humphreys*
   INTRODUCTION TO LIE ALGEBRAS AND REPRESENTATION THEORY
   xiv, 169 pages. 1972.

10 *Cohen*
    A COURSE IN SIMPLE-HOMOTOPY THEORY
    xii, 114 pages. 1973.

11 *Conway*
    FUNCTIONS OF ONE COMPLEX VARIABLE
    xiii, 313 pages. 1973.

12 *Beals*
    ADVANCED MATHEMATICAL ANALYSIS
    xi, 230 pages. 1973.

13 *Anderson/Fuller*
    RINGS AND CATEGORIES OF MODULES
    ix, 339 pages. 1974.

14 *Golubitsky/Guillemin*
    STABLE MAPPINGS AND THEIR SINGULARITIES
    x, 211 pages. 1974.

15 *Berberian*
LECTURES IN FUNCTIONAL ANALYSIS AND OPERATOR THEORY
x, 356 pages. 1974.

16 *Winter*
THE STRUCTURE OF FIELDS
xiii, 205 pages. 1974.

17 *Rosenblatt*
RANDOM PROCESSES
2nd ed. x, 228 pages. 1974.

18 *Halmos*
MEASURE THEORY
xi, 304 pages. 1974.

19 *Halmos*
A HILBERT SPACE PROBLEM BOOK
xvii, 365 pages. 1974.

20 *Husemoller*
FIBRE BUNDLES
2nd ed. xvi, 344 pages. 1975.

21 *Humphreys*
LINEAR ALGEBRAIC GROUPS
xiv, 272 pages. 1975.

22 *Barnes/Mack*
AN ALGEBRAIC INTRODUCTION TO MATHEMATICAL LOGIC
x, 136 pages. 1975.

23 *Greub*
LINEAR ALGEBRA
4th ed. approx. 460 pages. Tentative publication date April 1975.

24 *Holmes*
GEOMETRIC FUNCTIONAL ANALYSIS AND ITS APPLICATIONS
approx. 350 pages. Tentative publication date November 1975.

25 *Hewitt/Stromberg*
REAL AND ABSTRACT ANALYSIS
3rd printing. viii, 476 pages. Tentative publication date July 1975.

26 *Manes*
ALGEBRAIC THEORIES
approx. 320 pages. Tentative publication date November 1975.

Soft and hard cover editions are available for each volume up to Vol. 14, hardcover only from Vol. 15.